Advances in Polymer Science

Advances in Polymer Science

Edited by
Jaden Baker

MURPHY & MOORE
www.murphy-moorepublishing.com

Published by Murphy & Moore Publishing,
1 Rockefeller Plaza,
New York City, NY 10020, USA

ISBN: 978-1-63987-022-6

Cataloging-in-Publication Data

Advances in polymer science / edited by Jaden Baker.
p. cm.
Includes bibliographical references and index.
ISBN 978-1-63987-022-6
1. Polymers. 2. Polymerization. 3. Macromolecules. I. Baker, Jaden.
QD381 .A38 2022
547.7--dc23

For information on all Murphy & Moore Publications
visit our website at www.murphy-moorepublishing.com

MURPHY & MOORE

Contents

Preface

Polymer is a chain of the basic building blocks of plastic. Polymer science, also known as macromolecular science, is a subfield of materials science that deals with polymers, especially synthetic polymers such as plastics and elastomers. The field of polymer science comprises three branches, namely, chemistry, physics, and engineering. Polymer chemistry or macromolecular chemistry is associated with the chemical synthesis and chemical properties of polymers. Polymer physics deals with the physical properties of polymer materials and engineering applications. Polymer characterization is concerned with the analysis of chemical structure, morphology, and the determination of physical properties in relation to compositional and structural parameters. This book elucidates new techniques and their applications in a multidisciplinary manner. It strives to provide a fair idea about this discipline and to help develop a better understanding of the latest advances within this field. This book will provide comprehensive knowledge to the readers.

After months of intensive research and writing, this book is the end result of all who devoted their time and efforts in the initiation and progress of this book. It will surely be a source of reference in enhancing the required knowledge of the new developments in the area. During the course of developing this book, certain measures such as accuracy, authenticity and research focused analytical studies were given preference in order to produce a comprehensive book in the area of study.

This book would not have been possible without the efforts of the authors and the publisher. I extend my sincere thanks to them. Secondly, I express my gratitude to my family and well-wishers. And most importantly, I thank my students for constantly expressing their willingness and curiosity in enhancing their knowledge in the field, which encourages me to take up further research projects for the advancement of the area.

Editor

1

Effect of Variation of Hard Segment Content and Graphene-Based Nanofiller Concentration on Morphological, Thermal, and Mechanical Properties of Polyurethane Nanocomposites

Michał Strankowski

Chemical Faculty, Polymer Technology Department, Gdansk University of Technology, Narutowicza 11/12, 80-233 Gdansk, Poland

Correspondence should be addressed to Michał Strankowski; micstran@pg.gda.pl

Academic Editor: Jeefferie Abd Razak

This study describes the development of a new class of high-performance polyurethane elastomer nanocomposites containing reduced graphene oxide (RGO) or graphene nanoplatelets (GNP). Two types of polyurethane elastomers with different contents of hard segments (HS) were used as a polymer matrix. The developed nanocomposites were characterized by thermal analysis (DSC, TG), dynamic mechanical testing (DMA), hardness testing, mechanical properties, rheology, FTIR spectroscopy, XRD, and microscopy investigation (TEM, SEM). Morphological investigation confirmed better compatibility of RGO with the polyurethane (PU) matrix compared to GNP. Both applied nanofillers influenced melting and crystallization of the PU matrix. The nonlinear viscoelastic behavior of the nanocomposites (Payne effect) was studied, and the results were compared with theoretical predictions.

1. Introduction

Many nanofillers are at present popular additives used to modify diverse properties of polymeric materials influencing their mechanical, physical, and chemical properties and durability parameters as well [1–3]. Nanofillers are often distinguished by the shape as nanoplatelets (1D), nanofibres (2D), or nanoparticles (3D) [3]. All of them should have at least one dimension down to 100 nm. Most commonly used nanofillers are silicates (includes the most popular montmorillonite), titanium dioxides, metallic nanoparticles, inorganic nanotubes, and all of the carbon nanofillers like carbon nanotubes, fullerenes, and finally graphene. When Geim and Novoselow in 2004 made a breakthrough in science with the discovery of graphene monolayers, the scientific community all over the world started focusing on these material properties and possible applications. Since then, improving methods of graphene synthesis and synthesis of different types of its derivatives were elaborated. Every single one of graphenes' derivatives, like graphene oxide (GO), chemically reduced graphene oxide (RGO), thermally reduced graphene (TRG), and chemically modified graphene (CMG), have different properties compared to pure graphene sheet. Every method of synthesis (from epitaxial methods, CVD, intercalation of graphite, exfoliation, etc.) provides a new material with specific characteristics.

Many polymers were already used to obtain graphene-based polymer nanocomposites. Most commonly used polymer matrices are polyethylene [4–6], polyamide 6 [7–9], poly(methyl methacrylate) [10–12], polystyrene [13–15], natural rubber [16–19], and polyurethanes [20–23]. Most of polymer nanocomposites showed better (comparing with nonmodified matrices) durability [24, 25] and good thermal properties [24] and gained some additional properties like electrical conductivity [25] and hydrophobic behavior [26]. Literature data reveal that using different types and contents of graphene-based nanofillers might improve a wide range of properties of polymers, especially polyurethanes [27–30].

Polyurethane elastomers are one of the most important polymeric materials, which can be applied in those areas, where use of classical elastomers is limited. Generally, polyurethane elastomer macromolecular chains are built up of soft (SS) and hard (HS) segments [31], and the proportion of hard to soft segments and the polarity of the segments determine especially elasticity, hardness [32, 33], mechanical [33, 34] or thermal [33, 34] properties, and the morphology of PU. Intra- and intermolecular hydrogen bonds within polyurethane materials influence additionally the strength and thermal stability of these elastomers [35]. The low miscibility between soft and hard segments leads to microphase separation of segmented polyurethanes. Increasing HS content above 30% often evidently changes long-range connectivity of HS, and this results in an increase in mechanical strength of polyurethanes. What is more, chemical components like isocyanates, polyols, and chain extenders (it is symmetric and presents reactive groups), used to obtain polyurethane materials, may introduce specific interactions between the segments, which determine the thermal and mechanical behavior [35]. Properties of polyurethane elastomers can be also modified during processing by creating polymer nanocomposites [21, 36, 37], where the polyurethane material acts as a matrix for dispersed nanoparticles.

The present paper is aimed at reporting the influence of two carbon nanofillers (reduced graphene oxide (RGO) and graphene nanoplatelets (GNP)) on two polyurethane elastomeric matrices (possessing different hard segment (HS) contents). This also focuses on nanocomposite systems while gaining changes in some properties, such as improved mechanical and thermal properties. The morphology and viscoelastic behavior with changing stiffness of the matrix was also studied. Nanofillers used in this work were thermally reduced graphene (RGO) and graphene nanoplatelets (GNP). This is a new approach to obtaining nanomaterials based on the wide use of polyurethane raw materials, which should be suitable for a commercial application. The main goal was to obtain high-performance nanocomposites by simple methods of synthesis, such as in situ polymerization method, without solvents. Also, some of the synthesized nanocomposites contained nanofiller RGO derived from graphite. Compared to GNP, provided by the commercial manufacturer, but still expensive [38], usage of low-cost laboratory-made RGO is an economical approach to estimating the way towards mass scale production of graphene derivative-based nanocomposites.

RGO is an economical approach to estimating the way towards mass scale production of graphene derivative-based nanocomposites. Incorporation of these types of nanofillers can extend polyurethane thermoplastic elastomers' property limits, improving the mechanical and tribological performance of common TPU applications like footwear, closure o-rings and seals, adhesives, automotive interiors, and sport and leisure items. There here is a high-potential, still needed investigation for using GNP or RGO containing polyurethane nanocomposites in these technical goods, where electrostatic dissipation ability is of special importance, like conveyer belts and technical fabrics [38–40].

It is the first performed detailed study of the influence of the used nanofillers on hard and soft segment phase formation and property changes in polyurethane elastomer nanocomposites.

2. Methods and Materials

2.1. Differential Scanning Calorimetry (DSC). Differential scanning calorimetry measurements have been carried out by a heat-flux Netzsch DSC 209F1 Proteus analyzer. The calibration process was performed for 6 points ($C_{10}H_{16}$, In, Sn, Bi, Zn, CsCl). The investigation temperature range was from −85°C to 250°C. 5–10 mg samples were used in alumina pans. Measurement was performed in the N_2 atmosphere. During the investigation, a heating/cooling rate equal to 10°C/min was used. A second heating scan was used for the all calculations.

2.2. Thermogravimetric Analysis (TG). Thermogravimetry of obtained materials was conducted by a Netzsch TG 209F3 Tarsus analyzer. The calibration process was performed for 6 points (In, Sn, Bi, Zn, Al, and Ag). Investigations were performed using the Al_2O_3 crucible. Mass loss was followed at a heating rate of 20°C/min for samples (5–10 mg) in the temperature range from 37°C to 600°C, under N_2 atmosphere.

2.3. Dynamic Mechanical Analysis (DMA). Dynamical mechanical analysis has been carried out using a TA Instruments DMA Q800 analyzer. Investigation has been performed using tension mode, in the temperature range − 100–100°C, frequency equal to 1 Hz, and heating rate at 4°C/min. Samples were cooled using liquid nitrogen. Strain sweep analysis, for nonlinear viscoelastic investigation, has been conducted at temperature 35°C, in the same tension mode and frequency of deformation but in the amplitude range from 0.5 to 6500 μm.

2.4. Fourier Transformation Infrared Spectroscopy (FTIR). Fourier transformation infrared spectroscopy of polyurethane materials has been carried out using a Thermo Electron Corporation Nicolet 8700 Spectrometer (equipped with ATR mode). Samples were scanned from 4000 to 500 cm^{-1}.

2.5. X-Ray Diffraction (XRD). A Bragg-Brentano X'Pert Philips diffractometer was used to characterize prepared materials (40 kV, 30 mA, λ Cu Kα = 0.1542 nm). All samples were scanned from the 5 to 35° 2θ range.

2.6. Scanning Electron Microscopy (SEM). Scanning electron microscopy was performed using FEI Quanta 250 FEG (20 kV, LFD detector) with EDAX Apollo X-SDD EDS module.

2.7. Transmission Electron Microscopy (TEM). Microscopy analysis was performed using STEM-EDX Transmission Electron Microscope FEI Europe, Tecnai F20 X-TWIN coupled with EDX spectrometer. All samples were cut in cryo-mode using an ultramicrotome (Leica EM UC7 Ultramicrotome). HRTEM (for nanocomposite systems) was from JEOL, model JEM-2100. The thickness of the samples

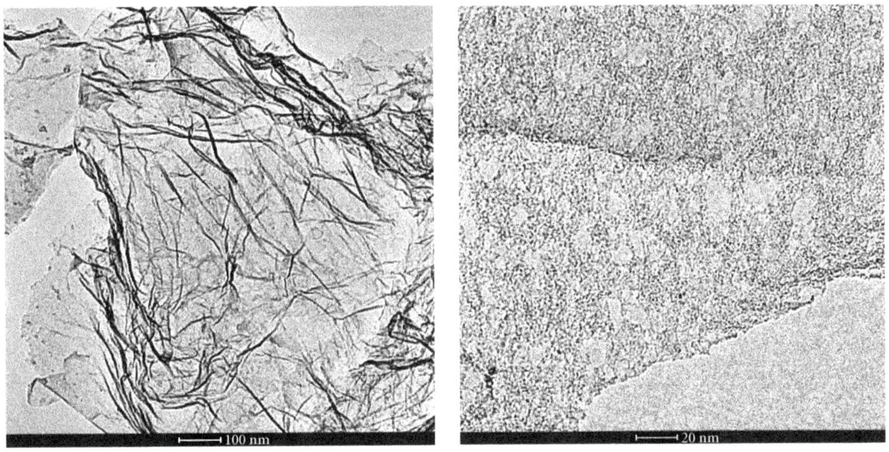

(a) (b)

FIGURE 1: TEM microphotographs of (a) thermally reduced graphene oxide (RGO) and (b) graphene nanoplatelets (GNP).

(a) (b)

FIGURE 2: SEM microphotographs of (a) thermally reduced graphene oxide (RGO) and (b) graphene nanoplatelets (GNP).

was 150 nm and was collected on a 200-mesh Cu grid before investigation.

2.8. Materials

2.8.1. Graphene Nanoplatelets. Graphene platelets (GNP) from ACS Materials (USA) [41] have 2–10 nm height (with 4–20 layers) and 20–40 m^2/g surface area (Figures 1(b) and 2(b)).

2.8.2. Reduced Graphene Oxide. Potassium permanganate ($KMnO_4$, >99%), hydrogen peroxide (H_2O_2, 30%), sulfuric acid (H_2SO_4, 35–38%), phosphoric acid (H_3PO_4, 85%), sulfuric acid (H_2SO_4, 96%), ethanol (C_2H_5OH, 96%), and diethyl ether ($(C_2H_5)_2O$,), all substrates from POCH SA., Poland, were used to obtain graphene oxide. Ar (>99% mol.) was used to perform thermal reduction of graphene oxide GO (Figures 1(a) and 2(a)).

2.8.3. Polyurethanes. The isocyanate used was 4,4′-methylene diphenyl diisocyanate (MDI), isocyanate content of 32 wt%, and an average functionality of 2.15, purchased from Zachem (Poland). The catalyst Dabco VR 33-LV-33 wt% solution of 1,4-diazobicyclo[2.2.2]octane in 2,2-dihydroxyisopropyl ether, short-chain hard-segment extender was 1,4-butanediol (1,4-BDO, from BASF, Germany), and polyol, poly(tetramethylene ether)glycol (PTMG, 1889 g/mol) was obtained from BASF, Hungary.

The method of the previous publication, Strankowski et al. [42], was followed.

3. Preparation

3.1. Reduced Graphene Oxide (RGO) Preparation. Two-step synthesis was used to obtain reduced graphene oxide (RGO). Based on modified Hummer's method, the GO nanofiller was prepared [43]. GO was thermally reduced to RGO at 200°C, using the Ar atmosphere.

3.2. Polyurethane (PU) Nanocomposite Preparation. Polyurethane materials were obtained using the two-step preparation

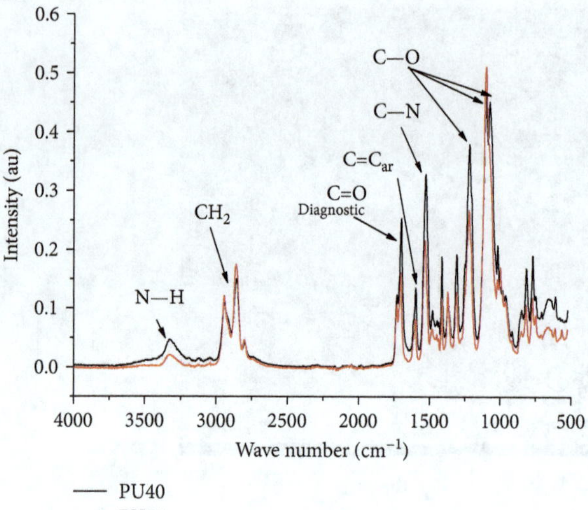

FIGURE 3: FTIR spectra for nonmodified PU with different hard segment ratios.

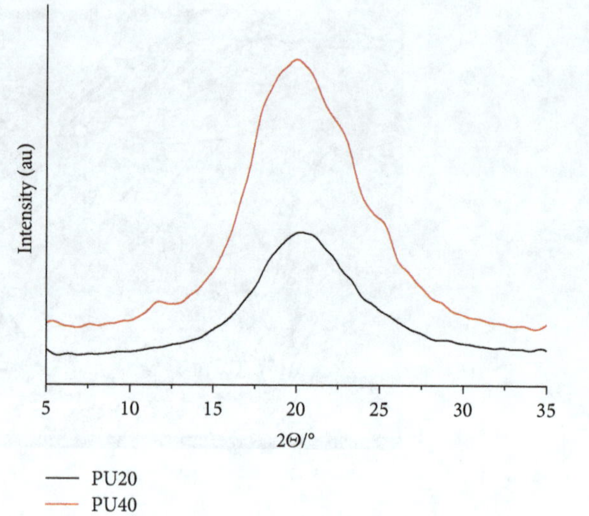

FIGURE 4: XRD diffractograms for nonmodified PU matrices with different hard segment ratios.

method with different hard segment (HS) contents of 20% (PU20) and 40% (PU40) by weight and at an isocyanate index equal to 1.05. The prepolymer was obtained at the first step by reacting PTMG polyol and MDI isocyanate at 70°C during 1 h under low pressure. The obtained prepolymer was used to prepare unmodified polyurethanes and nanocomposites. For unmodified systems at the second step, 1.2-butanediol (BD) as a chain extender and 1,4-diazabicyclo[2.2.2]octane (DABCO) as a catalyst were used. For preparing nanocomposite systems, a calculated amount of the nanofiller was mechanically (2500 rpm) dispersed in the prepolymer with ultrasound support for 30 min. At the last stage, to complete the reaction, BD and DABCO were added to the prepolymer containing the dispersed nanofiller. The reacting mixtures, when still fluid, were poured into the steel mold and annealed at 100°C for 24 h in a laboratory drier. After that, materials were compressed (2 tons) using a hydraulic press (Manual Hydraulic Press, Specac) to obtain thin 0.5 mm films, which were used for the investigations.

Samples designation: Notation PU20-1% GNP is the polyurethane matrix (PU) with proper amount hard segment content (20), containing 1 wt.% of the GNP nanofiller (1%GNP). Matrix|hard segment content|-|% weight of nanofiller|type of nanofiller|.

3.3. Unmodified Polyurethane Matrices (PU). Based on spectroscopy investigation, it was confirmed that general absorption bands are typical for the polyurethane matrices (Figure 3) [44, 45]. This study was performed to check that all reactions were completed and there were no visible bands for unreacted NCO groups from isocyanate substrates. What is more, X-ray diffraction analysis shows absence of diffraction maxima characteristic for crystalline regions (Figure 4) [46]. This study reveals that there are no long-range ordered structures in these polyurethane matrices. In Figure 4, a wide amorphous diffusion peak (at about $2\theta = 20°$) can be observed which is typical for amorphous phase in amorphous and partially crystalline polymers [47–49].

TABLE 1: Phase transitions parameters for nonmodified PUs.

PU matrix	T_g (°C)	T_{mSS} (°C)	ΔH_{mSS} (J/g)	T_{mHS} (°C)	ΔH_{mHS} (J/g)	T_{cHS} (°C)	ΔH_{cHS} (J/g)
PU20	−41	25	13.5	212	5.5	111	−7.5
PU40	−46	25	12.0	190	9.1	119	−10.3

T_m: melting temperature; T_c: crystallization temperature; ΔH_m: enthalpy of melting; ΔH_c: enthalpy of crystallization.

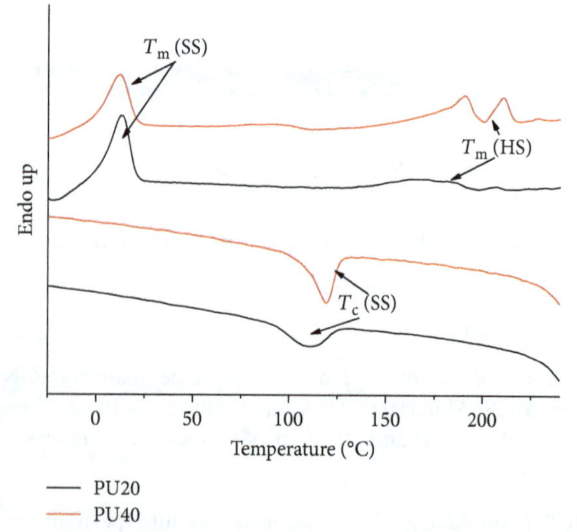

FIGURE 5: DSC thermograms for nonmodified PU matrices possess different hard segment contents.

For the polyurethane with higher hard segment content (PU40), higher temperature and enthalpy of melting point (T_m) were observed, in comparison to the softer matrix (PU20) (Table 1, Figure 5). The crystallization temperature for PU40 shifts to higher values (about 10°C) comparatively to the PU20 matrix, and enthalpy of crystallization has

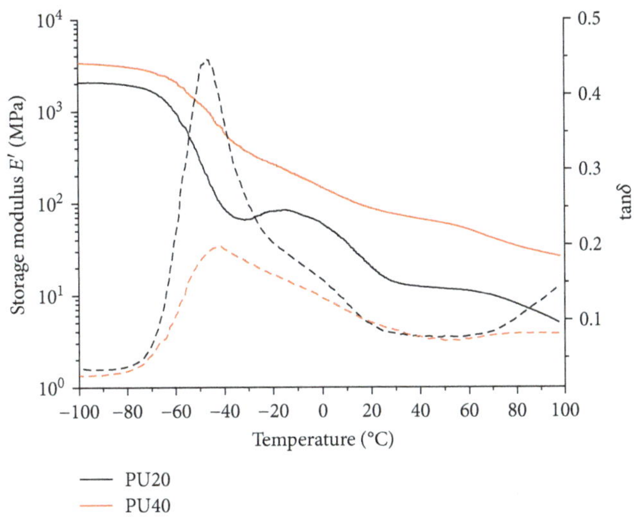

FIGURE 6: Storage modulus (E'/MPa—black line) and loss tangent (black dot line) versus temperature/°C for nonmodified PU matrices with different hard segment contents.

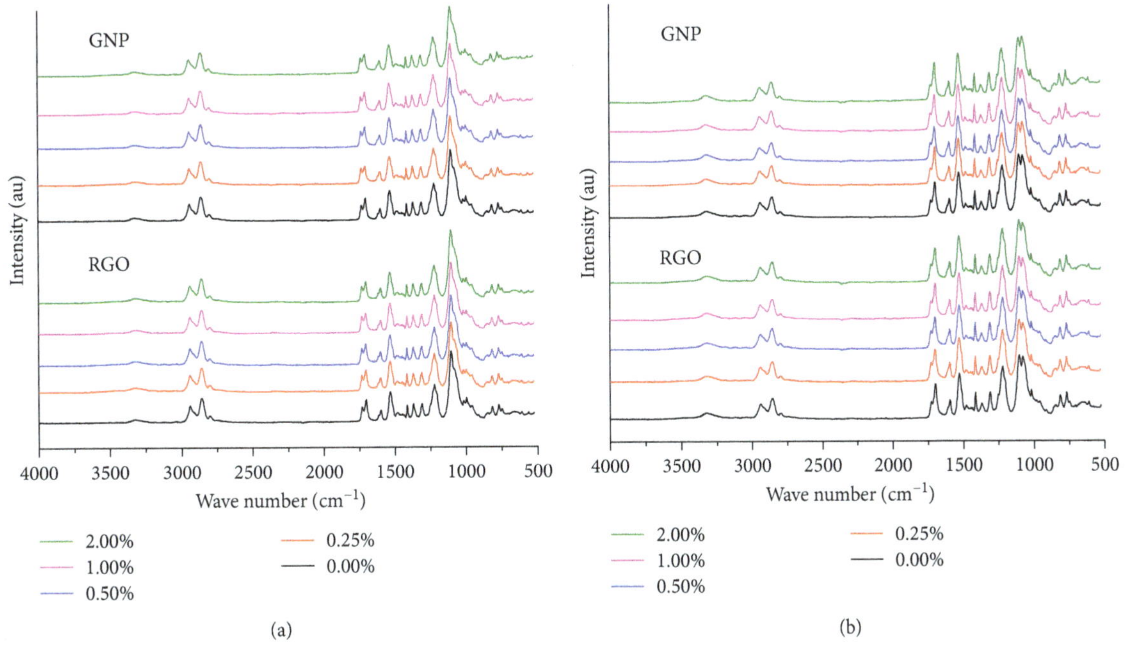

FIGURE 7: FTIR for (a) PU20 and (b) PU40 nanocomposites.

higher values for the harder matrix (PU40) (Figure 5). These results are in good agreement with the DMA study (Figure 6), where a higher storage modulus (E'/MPa) was observed for the PU40 matrix in the applied temperature range for this investigation.

4. Results

4.1. Polyurethane Nanocomposites

4.1.1. FTIR Spectroscopy. Infrared spectroscopy was performed for all nanocomposite materials containing RGO or GNP nanofiller. By comparing the FTIR spectra, it was confirmed that the addition of the nanofiller did not influence the chemical structure of the polyurethane matrix (Figure 7). However, there was a possible reaction of isocyanate groups with functional groups located on the surface of reduced graphene oxide. Nevertheless, the number of these reactive groups is negligible (max. 2%) and their influence of the nanocomposite creation is minor.

Based on the FTIR study where the absorption and diffraction of infrared radiation have low influence on waveforms (at such low content of the nanofiller) and, what is more, that the absorption coefficient of C=O and N–H bands

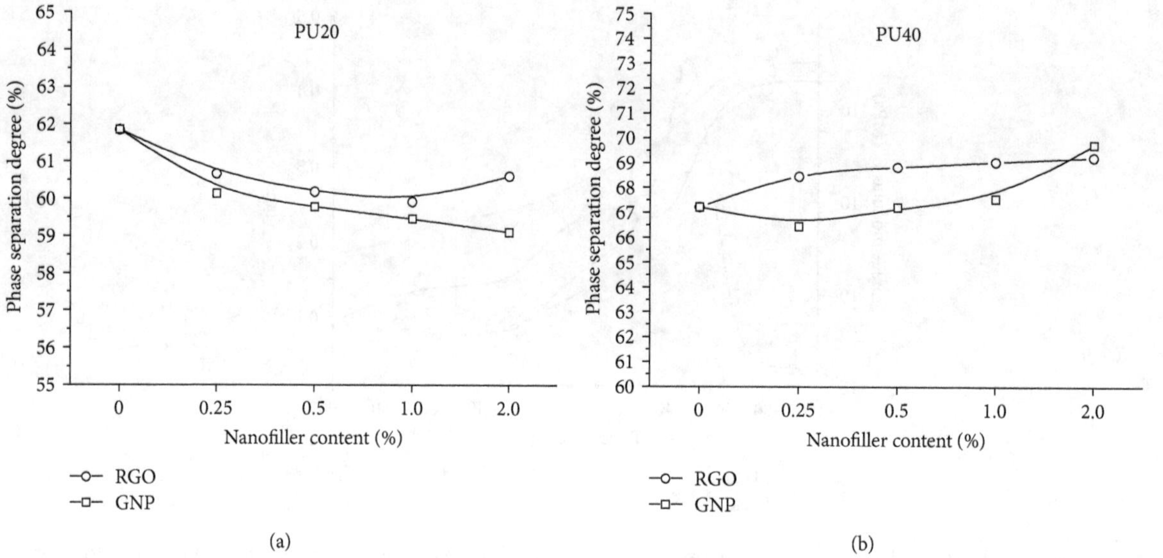

FIGURE 8: Phase separation degree (PSD) for (a) PU20 and (b) PU40 nanocomposites.

did not change, the phase separation behavior should be connected only with disorders of the domain structure caused by nanofiller addition. Phase separation was calculated according to [50–52] [for ~1730 cm^{-1} and ~1700 cm^{-1} bands; $PSD = R/(R+1)$, where $R = A_1/A_2 - NCO$ index, where A_1 and A_2 are surface areas of oscillating bands of the C=O group, hydrogen-bonded and nonhydrogen-bounded)], and the influence of nanofillers on this parameter is presented in Figure 8. It was confirmed that the phase separation degree (PSD) increased for PU40 nanocomposites and had lower values and decreased for PU20 materials (Figures 8(a) and 8(b)).

4.1.2. X-Ray Diffraction (XRD). The obtained diffractograms (Figure 9) for nanocomposites PU20-GNP and PU40-GNP show wide halo patterns originating from amorphous polyurethane materials. Moreover, for nanocomposites containing a GNP nanofiller, a diffraction maximum at about $2\theta = 27°$ was visible, originating from crystalline layers of GNP. The intensity of these maxima increases with increasing content of the GNP nanofiller in the polyurethane matrix. For polyurethane nanocomposites, no diffraction maxima were observed, which confirmed (with TEM investigation support) that all the systems possess an exfoliated structure with a well-dispersed RGO nanofiller.

4.1.3. Transmission Electron Microscopy (TEM). Based on transmission electron microscopy (TEM) analysis, good dispersion of the nanofiller (RGO or GNP) into polyurethane matrices with different hard segment (HS) contents was confirmed. This is a very important factor characterizing the morphology of the nanofiller (size, dispersion) in nanocomposite systems. Reduced graphene oxide (RGO) and graphene nanoplatelets (GNP) (Figure 10) are visible in the form of characteristic long, disordered dark lines into clear, gray, PU matrices. For PU/GNP systems (Figures 10(a) and 10(d)), long lines (about 500 nm) are observed, which are aggregated into PU matrices (PU20 or PU40). For PU/GNP systems (Figures 10(a) and 10(d)), larger agglomerates were observed in comparison to the RGO-filled systems. The incorporation of RGO into the PU matrix causes good dispersion, and a characteristic disordered structure (short, thin, black lines) was obtained, as clearly observed in Figures 10(b), 10(c), 10(e), and 10(f). The RGO nanofiller is affected by the hard and soft segment ratio in the PU matrix, and better exfoliation is archived for the PU matrix with lower content of hard segments (Figures 10(b) and 10(c)). The RGO nanofiller has better compatibility to the polyurethane matrix (what relates to the –O– or –OH group on the surface of the nanofiller), and therefore higher dispersion is achieved.

These results based on TEM analysis are in good correspondence to the XRD study where (no diffraction maxima; described in X-ray diffraction (XRD) paragraph) an exfoliated nanocomposite morphology was confirmed, especially for PU/RGO materials.

4.1.4. Differential Scanning Calorimetry (DSC). Crystalline and amorphous domain structure changes in the obtained polyurethane nanocomposites were investigated using DSC. Figures 11 and 12 show DSC and cooling and heating scans. Based on cooling thermograms, it is visible that nanofiller addition (RGO and GNP) influences on polyurethane hard and soft domains. The temperatures of crystallization and enthalpies of this process change when the nanofiller is incorporated into the polyurethane matrix. For PU20-GNP and PU40-RGO, the nanofiller affects the rate of crystallization, where maximum temperature rates are lowest and enthalpy of crystallization takes the lowest values. This behavior is connected with a smaller amount of crystalline phase creation during cooling. However, samples PU20 (0.25 wt% GNP) and PU40 (2 wt% RGO) exhibit a shift in crystallization temperature of HS to higher temperatures, in comparison to the other modified materials. This behavior can be

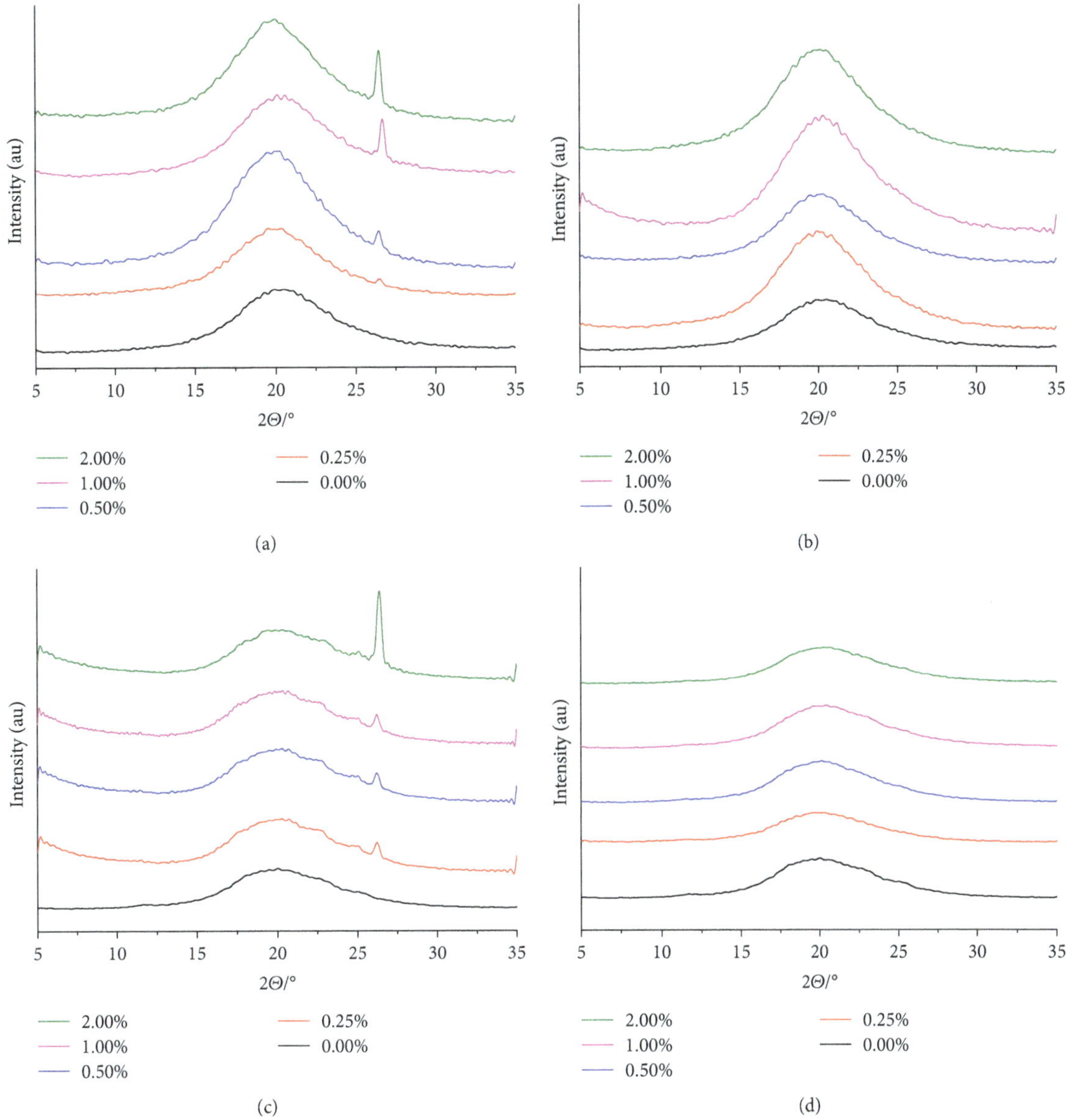

FIGURE 9: XRD for (a) PU20-GNP, (b) PU20-RGO, (c) PU40-GNP, and (d) PU40-RGO nanocomposites.

explained with possible presence of some impurities from the mold, which can infect the crystallization process of these materials. What is more, a reverse behavior exhibits PU20-RGO and PU40-GNP, where the crystallization temperatures are higher, which is related with the heterogenic nucleation process occurring at nanofiller surfaces. Additionally, for PU40-GNP nanocomposites, the observed crystallization peaks (shift to the higher temperatures) split with visible two maxima (Figure 11(c)), which is connected with creating crystalline areas of different size. Addition of the nanofiller also increases the crystallization rate of soft segments (SS), which is visible for all the PU40-GNP nanocomposites (Figure 11(c)), but not revealed for RGO-based nanocomposites, because this process could be extended in time after the controlled cooling step.

After the cooling process, all samples were heated at a controlled speed (second heating step, after clearing the thermal history of the samples) to verify the phase transition-melting behavior. For PU20-GNP, PU20-RGO, and PU40-RGO, the addition of the nanofiller causes a decrease in enthalpy of melting, which is connected with the lower amount of crystalline phase in the hard segment domains. Not very clear melting peaks for PU20 nanocomposites (at temperature range 125–210°C) were blurred in comparison to the nonmodified PU matrix. What is more, for PU40-RGO, there were revealed multiple peaks on the thermograms. The first maximum was visible at a temperature of about 90°C and the other two peaks at 190°C and 210°C (Figure 12(b)). The first maximum occurs before the melting point, and the next two maxima are connected with melting and recrystallization of the crystalline phase. The addition of the reduced graphene oxide (RGO) nanofiller, into the polyurethane matrix, causes an increase of the peak at 90°C, blurry peak at temperature 190°C, and extinction

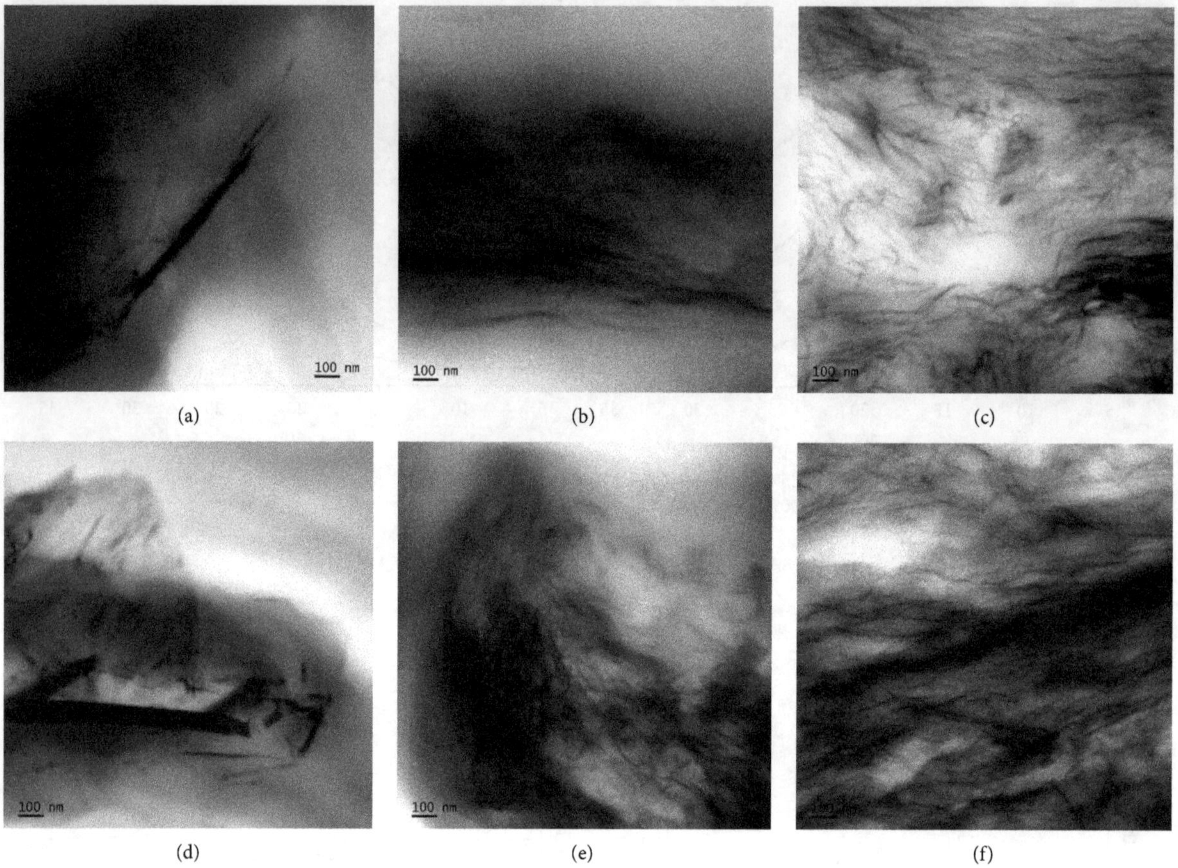

FIGURE 10: TEM microphotographs of (a) PU20/GNP1%, (b) PU20/RGO1%, (c) PU20/RGO2%, (d) PU40/GNP1%, (e) PU40/RGO1%, and (f) PU40/RGO2%.

of the maximum at 210°C (Figure 12(b)). Additionally, for the 2 wt% RGO nanocomposite, a wide region of melting at temperature 210–220°C is visible (Figure 12(b)). For PU40-GNP nanocomposites, enthalpy of melting shows no considerable difference in comparison to the PU40 matrix (Table 2). However, in the thermograms (Figure 12(b)), some configuration changes of melting peaks have been observed. At 190°C, the melting peak is lower and disappears with the addition of the 2 wt% GNP nanofiller. What is more, the second peak at 210°C is lower (up to 1 wt% GNP) and increases with the addition of 2 wt% in the nanofiller (GNP). Simultaneously for all nanocomposite systems, a third, well-defined, peak at 225°C is observed. For all heating runs, it is visible that the melting temperature of soft segment (SS) domains can be estimated at about 13°C (Figure 12(b)). Also, there was an observed decrease in the crystalline phase for nanocomposites PU20-GNP (above 0.5 wt% of the GNP nanofiller), PU40-RGO (for amount 0.25–1 wt% RGO), and PU20-RGO (for mount 0.5–1 wt% RGO), which is connected with lower values of melting enthalpy (Table 2). However, for PU40-GNP nanocomposites, this parameter (amount of crystalline phase within soft domains) remains at a similar level (Table 2).

4.1.5. Thermogravimetric Analysis (TG). Based on the performed TGA analysis, there was a confirmed slight improvement in the thermal stability of the nanocomposite system by the addition of the nanofiller (RGO or GNP). The onset degradation temperature of obtained materials is comparable for unmodified and nanocomposite systems (Figure 13, Table 3). For the modified system, this parameter shifts to higher temperatures at about $T_{5\%} = 5 - 7°C$ in comparison to the polyurethanes without the nanofiller. Visible improvement in thermal stability (11°C) is observed for the PU40-RGO system with the 2 wt% nanofiller (Figure 13(f)). Maximum degradation temperatures (T_{DTGmax}) of hard segments (HS) for PU20-RGO (Figure 13(d)) and PU40-RGO (Figure 13(h)) are shifted to higher values by 27°C and 14°C, respectively (Table 3). For all nanocomposite systems, the degradation temperature of soft segments possesses similar values. For filled systems, there is a visible higher mass content [5% for PU20–2.0% GNP/RGO (Figures 13(a) and 13(b)) and 9% for PU40–2% GNP (Figure 13(e))] after degradation (mass residue), which is connected with higher thermal stability of nanofillers in comparison to the pure polyurethane matrices (2% for PU20 and 5% for PU40) (Figure 13, Table 3).

4.1.6. Dynamic Mechanical Analysis (DMA)

(1) DMA (Temperature Investigation). Thermomechanical parameters were collected in Figures 14 and 15, and the

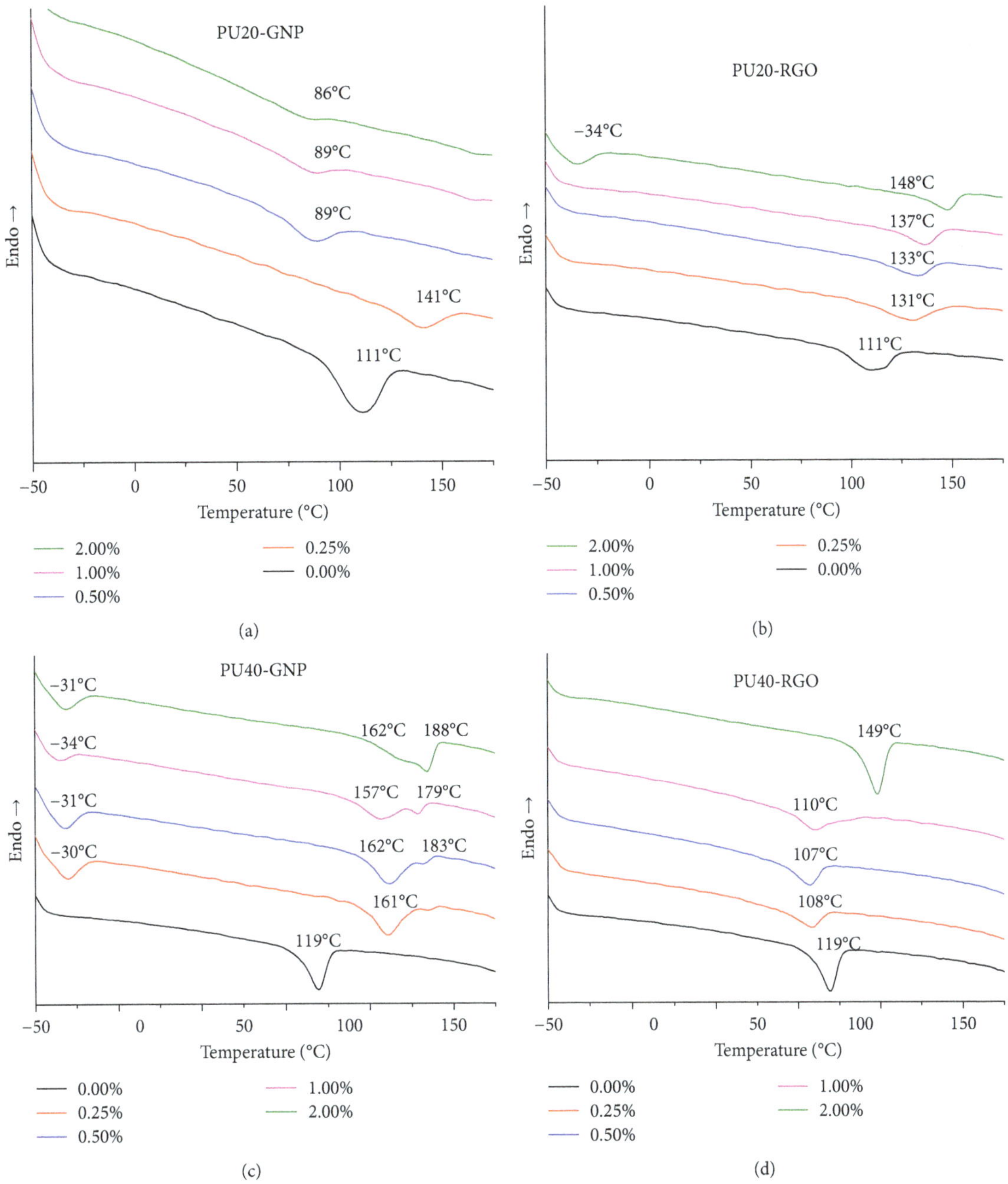

FIGURE 11: DSC thermograms for PU nanocomposites (cooling scans).

discussion of the results refers to specific regions which describe the polymer nanocomposite behavior (from glassy region to rubbery state, characteristic for polymeric materials).

(i) Glassy and viscoelastic area (temperature range: −100°C to −30°C)

Addition of the nanofiller does not cause significant changes in DMA characteristic parameters: storage modulus (E'), loss modulus (E''), and loss tangent (δ) in the glassy state temperature rage (Figures 14 and 15). Visible changes were observed above glass transition temperatures, where soft segments are at viscoelastic state and their movement is possible. It was confirmed that for polyurethane nanocomposites slightly lower values of the loss tangent were registered in comparison to the nonmodified PUs (Figures 14(e), 14(f) and 15(e), 15(f)). The glass transition temperature (calculated from the maximum of ($\tan\delta$)) changed with the addition of the nanofiller with a decrease of 6°C for PU20-2% RGO and an increase of about 4°C for PU20-2% GNP. For stiffer nanocomposite matrices with RGO and GNP nanofillers, a decrease in glass transition temperature in the range (1–9°C) was evident (Figures 14 and 15, Table 4). For PU40-0.5% GNP glass transition, the temperature shifts to the temperature of −51°C, with a value lower by 9°C compared to the nonmodified PU40 system (−42°C)

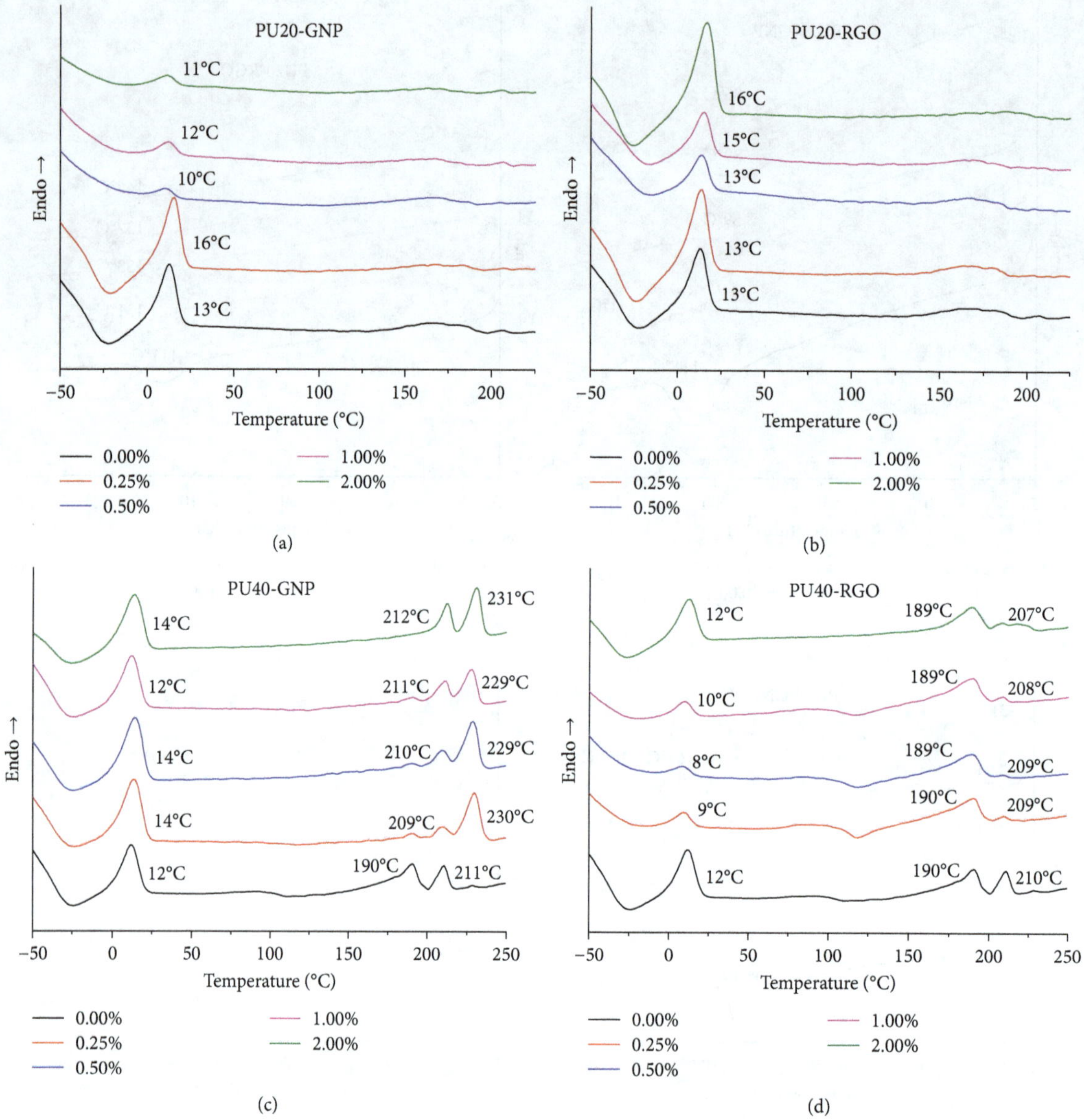

FIGURE 12: DSC thermograms for PU nanocomposites (heating scans).

TABLE 2: Enthalpy of melting for PU nanocomposites (heating scans).

% nanofiller	Heating scans							
	PU20-GNP		PU20-RGO		PU40-GNP		PU40-RGO	
	$\Delta H_{m,HD}$ (J/g)	$\Delta H_{m,SD}$ (J/g)	$\Delta H_{m,HD}$ (J/g)	$\Delta H_{m,SD}$ (J/g)	$\Delta H_{m,HD}$ (J/g)	$\Delta H_{m,SD}$ (J/g)	$\Delta H_{m,HD}$ (J/g)	$\Delta H_{m,SD}$ (J/g)
0.00	5.5	13.5	5.5	13.5	9.1	12.1	9.1	12.1
0.25	2.3	14.7	7.1	18.6	9.2	17.7	5.3	3.1
0.50	3.1	1.4	2.6	7.0	9.4	17.6	3.6	1.9
1.00	3.3	1.5	4.4	10.2	8.5	14.3	5.1	3.3
2.00	3.0	1.1	5.3	23.2	10.3	16.3	5.2	12.7

(Figure 14, Table 4). A higher decrease in glass transition temperature was observed for graphene nanoplatelet- (GNP-) based nanocomposites parallel to nanocomposites with reduced graphene oxide (RGO).

(ii) Soft domain melting area (temperature range: −30°C to 20°C)

At this specific range (−30°C to 20°C) within soft segment domains, crystalline forms are dispersed in the amorphous

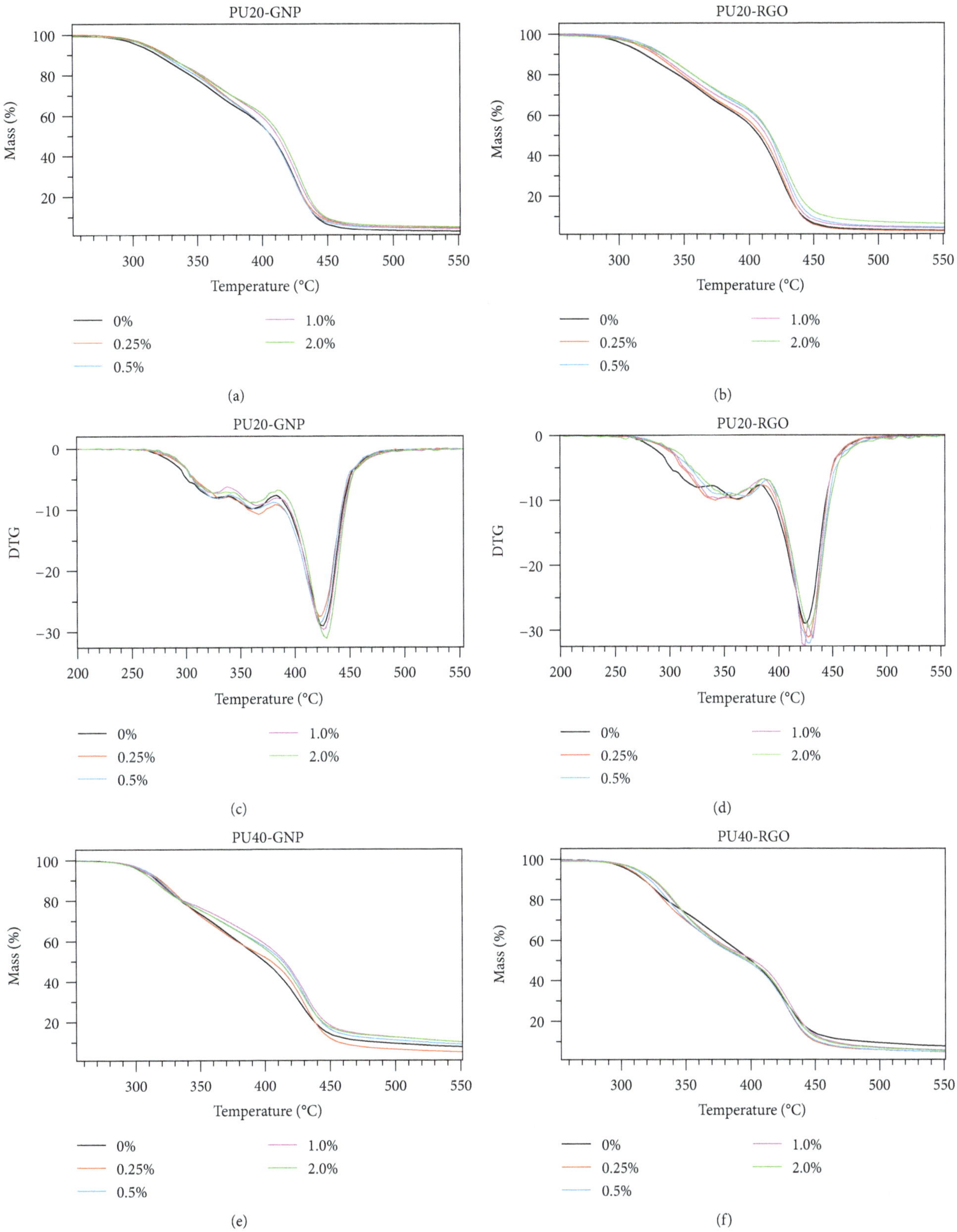

Figure 13: Continued.

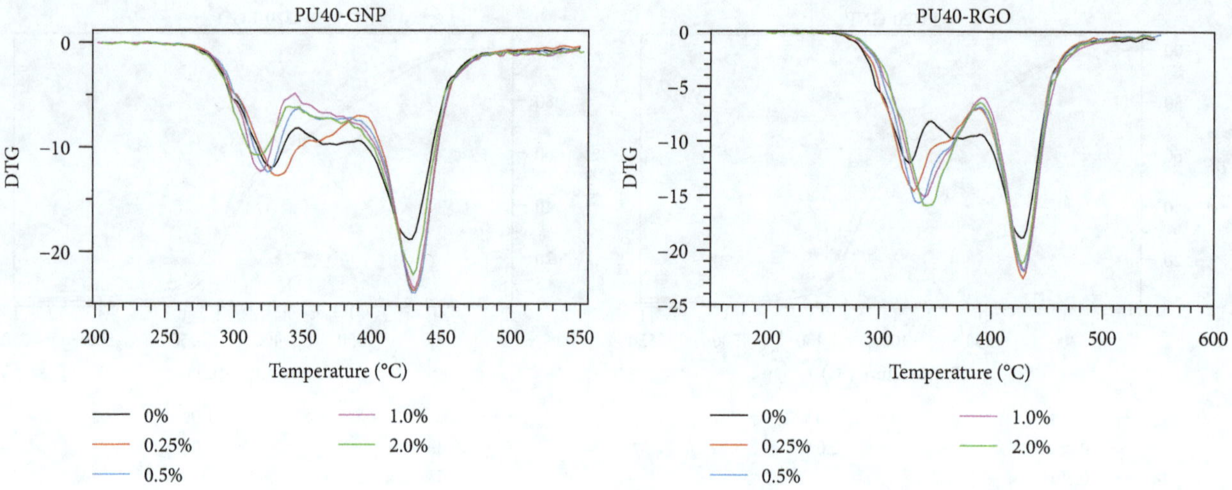

(g) (h)

Figure 13: TG curves [%mass versus temperature and derivative (DTG) versus temperature] for PU nanocomposites.

Table 3: TG data for PU nanocomposites.

% nanofiller	PU20-GNP $T_{5\%}$ (°C)	PU20-GNP T_{max} (°C)	PU20-RGO $T_{5\%}$ (°C)	PU20-RGO T_{max} (°C)	PU40-GNP $T_{5\%}$ (°C)	PU40-GNP T_{max} (°C)	PU40-RGO $T_{5\%}$ (°C)	PU40-RGO T_{max} (°C)
0.00	295	424	295	424	304	428	304	428
0.25	300	422	302	427	307	431	306	429
0.50	301	422	302	427	307	431	311	429
1.00	299	426	302	431	305	431	312	428
2.00	311	428	303	427	303	431	312	431

$T_{5\%}$: temperature 5% mass loss; T_{max}: temperature of maximum mass loss.

soft segment phase. What is more, at this region the mobility of SS is limited for both amorphous phase and crystalline phase of SS. However, with the temperature increase, the SS crystalline phase melts going completely into the amorphous phase at 20°C.

Addition of the nanofiller causes visible strain changes in comparison to stress behavior, defined as tanδ. For PU20-RGO nanocomposites (Figure 14(f)), values of tanδ are lower with increasing RGO nanofiller content, which is connected with a more elastic behavior of these materials at the investigated temperature range. This is similar for PU20-GNP series (Figure 14(e)), but with the difference that for a 2 wt% addition of GNP there is a rapid increase in tanδ values, which is higher in comparison to the other nanocomposite series and the nonmodified PU matrix. This distinct change probably is related with higher aggregation of the GNP nanofiller, and the increase in loss tangent parameter is the result of the sliding of nanoplatelets, resulting in a decrease in the elasticity of the system, which causes an increase in tanδ values. What is more, the decrease of this parameter for other systems (besides described above) probably is connected with a hydrodynamic effect, where nanofiller particles increase the viscosity of the nanocomposites causing the stiffening of the system.

In the case of PU20-RGO and PU20-GNP nanocomposites, there are noticeable storage modulus (E') changes (Figures 14(a) and 14(b)) in comparison to the PU20 matrix. These E' values are lower up to 1 wt% of the nanofiller content and increase for 2 wt% (RGO of GNP amount), reaching a similar level as for the PU20 matrix. Lower values of the storage modulus (E') (despite increased stiffness) are connected with the weaker adhesion of the nanofiller to the soft segment phase, with movement that is easier on the nanofiller surface.

For polyurethane nanocomposites (PU40-GNP and PU40-RGO), there was a visible (at all investigation temperature range) increase in storage modulus (E') values with the addition of the nanofiller (Figures 15(a) and 15(b)). On the other hand, tanδ changes are different, and for both series this parameter decreases. At up to 0°C for PU40-GNP nanocomposites, a high decrease in tanδ values is observed in comparison to the results for PU40-RGO. This behavior probably means that PU40-GNP possessed higher elasticity. These differences in tanδ values for the two types of the nanocomposite systems evidence a more effective interaction between bordering graphene nanoplatelets (GNP) than in the case of reduced graphene oxide (RGO). It can be expected that the GNP nanofiller creates a more effective network composed of nanofiller particles which are in contact, resulting in a drop of tanδ.

However, it should be highlighted that the stiffness of the nanocomposites, within the discussed temperature range, influences the crystalline phase of the soft domain, which contributes to the stiffening of the pure PU matrix, acting as the physical cross-link of the network. The addition of nanofillers can cause changes in the crystalline structure of these areas, which affect the thermomechanical properties. The fragmentation of the crystalline structure of the soft domain (SS) and the lowering of the crystallinity of these systems may cause an additional decrease in the stiffness of the nanocomposites.

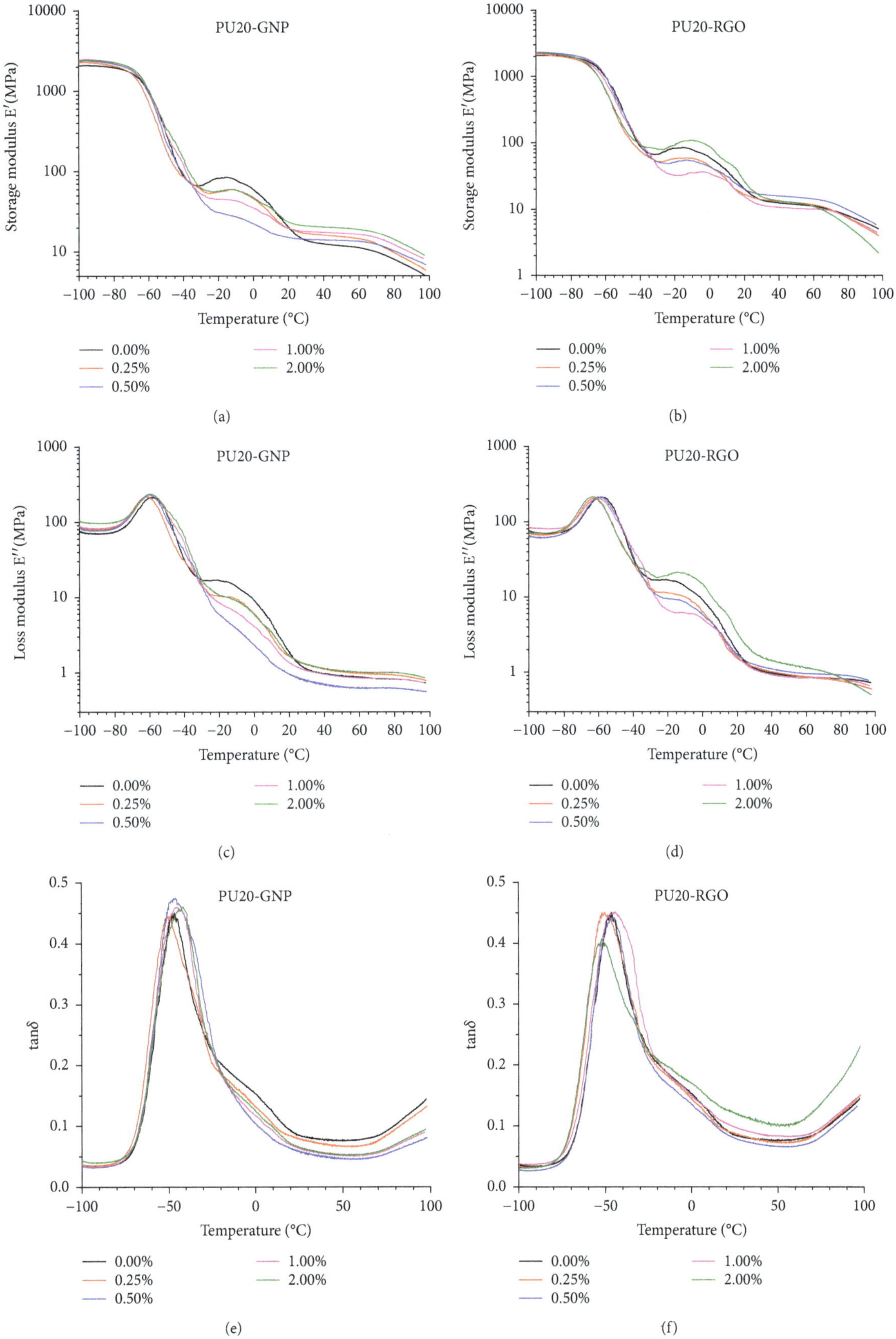

FIGURE 14: Thermomechanical behavior of PU nanocomposites PU20-GNP and PU20-RGO.

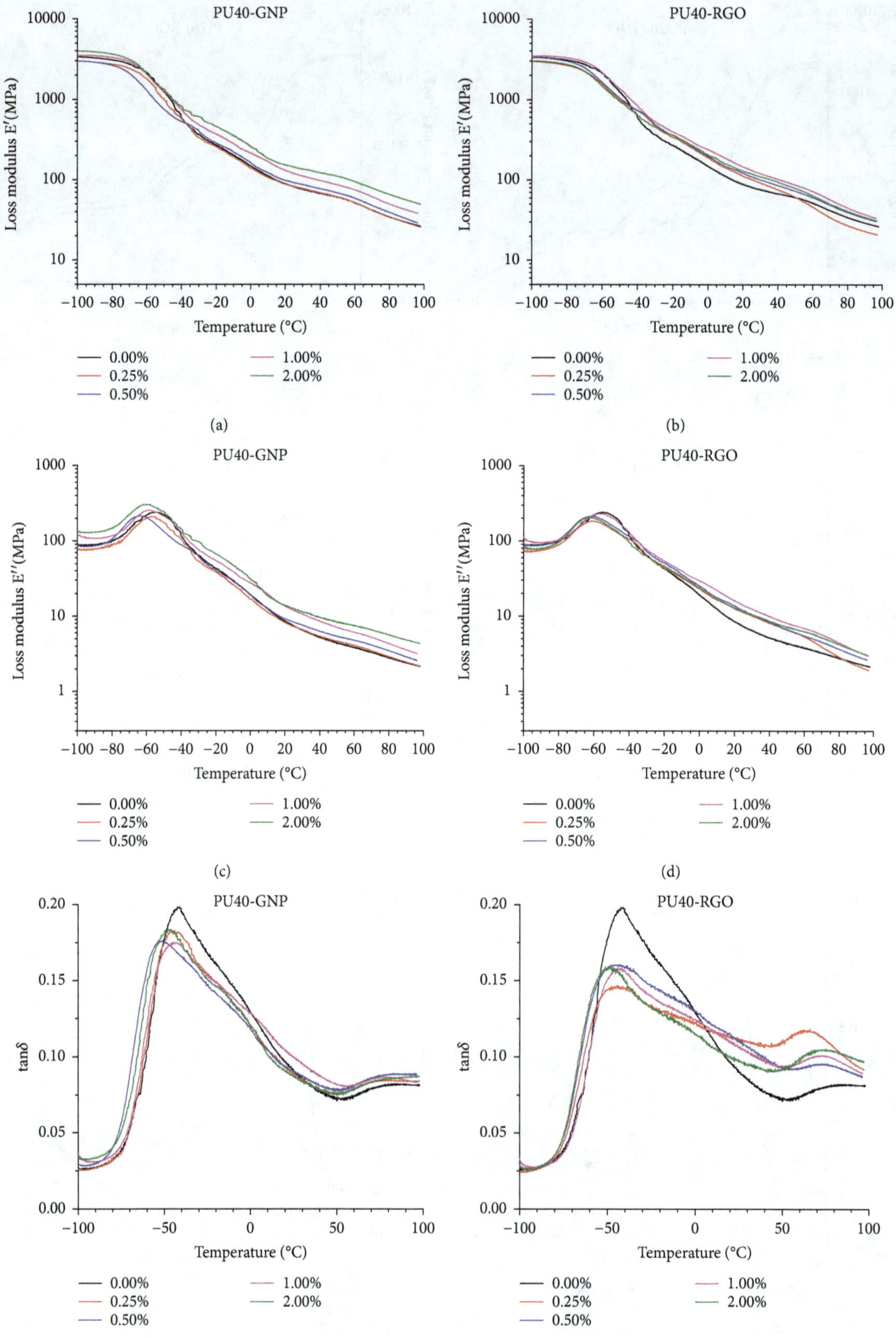

FIGURE 15: Thermomechanical behavior of PU nanocomposites PU40-GNP and PU40-RGO.

TABLE 4: Glass transition temperatures for PU and nanocomposite materials.

Materials	Glass transition temperature*/T_g (°C)				
Nanofiller (%)	0	0.25	0.5	1	2
PU20	−46	—	—	—	—
PU40	−42	—	—	—	—
PU20-GNP	—	−50	−46	−45	−42
PU40-GNP	—	−45	−51	−44	−46
PU20-RGO	—	−51	−46	−45	−52
PU40-RGO	—	−45	−45	−43	−48

*Calculated from maximum of loss tangent (tanδ).

(iii) Elastic area (temperature range: 20°C to 100°C)

At this elastic area, all crystallite structures represented by soft segments (SS) are in the amorphous state, above melting temperature. Soft segments have high a degree of free movement, which is limited only by the physical cross-linking network of the hard domain (HS). For all obtained nanocomposite systems, there was a confirmed increase in storage modulus (E') at the elastic plateau region (normal use of PU elastomers) with increasing concentration of the nanofiller. PU20-RGO nanocomposites exhibit a drop at tanδ for the nanofiller content up to 1 wt%, which is connected with a more stiffer behavior of these systems. Further addition of the nanofiller up to 2 wt% results in higher values of tanδ (Figure 14(f)) with a simultaneous increase in loss modulus (E'') (Figure 14(d)), which is the result of the sliding motion of the RGO resulting in the increase of the viscoelasticity. The decrease in tanδ values for the PU20-GNP series was also observed (Figure 14(e)). Likewise, in this case for 2 wt% GNP contents in the polyurethane matrix, there was a noticeable E'' increase, related with a sliding motion (Figure 14(c)). For PU40-RGO and PU40-GNP nanocomposites, the presence of the nanofiller causes an increase in storage modulus (E') values (Figures 15(a) and 15(b)) contributing to the mechanical strength of the nanocomposite. Some differences were also visible for tanδ curves (Figures 15(e) and 15(f)). For PU40-GNP nanocomposites, the shape of the tanδ curve for temperatures above 50°C is similar to the shape of the curve obtained for nonmodified PU (Figure 15(e)). These values also increase with a simultaneous increase in loss modulus (E''), which is correlated with a sliding motion of the GNP nanofiller (in contact with each other) and it causes an increase in E'' (Figure 15(c)). PU40-RGO nanocomposites exhibit an interesting behavior of the tanδ curves, where a maximum of loss tangent appears, connected with a relaxation transformation of melting behavior of the hard domains, which in this case melt at the lower temperature. This is correlated with a lower size of the crystallites, causing a shift of the melting temperature towards lower values. What is more, in comparison to the PU20 nanocomposites, for stiffer matrix (PU40), loss modulus (E'') values increase independently, regardless the amount of the nanofiller (Figure 15(c)).

(2) DMA (Nonlinear Viscoelastic Investigation). A nonlinear viscoelastic behavior of nanocomposite materials was investigated using dynamic mechanical analysis (DMA). All measured series possess three characteristic nonlinear regions (low-amplitude, transition, and high-amplitude). These regions were revealed for all pure polyurethane matrices PU20 and PU40 (Figures 16 and 17). Reinforcement of storage modulus (ΔE/MPa) was calculated as a difference between modulus values at $E'_{0.1\%}$ and $E'_{10\%}$. All curves (storage modulus versus strain for PU20 matrix) are presented in Figures 16(a) and 16(b).

For polyurethane nanocomposite PU20 containing a minimum 2 wt% GNP nanofiller, a reinforcement effect is observed (higher values of E'). In turn, for RGO nanocomposites, all E' values are similar in comparison to the pure polyurethane matrix (Figures 16(a) and 16(b)). In Figure 17, the difference between modulus change (ΔE) and the increasing amount of nanofiller content into the polyurethane matrix is presented.

A higher effect of the nanofiller content was observed for a harder polyurethane matrix (PU40) in comparison to the softer one (PU20) (Figures 18(a) and 18(b)). Both GNP and RGO nanofillers influence the PU40 matrix which is visible as higher storage modulus (E') values. For PU40 nanocomposites, storage modulus changes (ΔE) are most visible even at low contents of both nanofillers (Figure 19). A sudden drop of E' modulus at the transitional area was not observed, which is characteristic during breakdown of the nanofiller network. What is more, GNP shows a higher interaction with a harder polyurethane matrix (PU40), and a nonlinear viscoelastic behavior is visible for this type of nanofiller.

5. Discussion

The performed investigations of polyurethane nanocomposites, with different hard segment (HS) contents and modified with two types of graphene derivatives (RGO and GNP), revealed the explicit influence of applied nanofillers on morphological, thermal, and mechanical parameters on these systems.

The nanofillers (GNP and RGO) possess a nonpolar character; however, the PU matrix incorporates polar hard segments and the adhesion between these nanofillers and the polyurethane matrix is low. This restricted compatibility of both components (matrix and nanofiller) could be responsible for bigger particle aggregation or agglomeration (100–300 nm length) within the polyurethane matrix, especially for the GNP nanofiller (Figures 10(a) and 10(d)). FTIR investigation did not deliver any clues for the effective interaction between nanofiller (GNP) and matrix (PU) since not even a slight shifting in the position of characteristic bands was observed. There was some agglomeration of the nanofiller with its increasing amount. Based on XRD analysis, some diffraction maxima (representative for GNP nanoplatelets) are visible (Figures 9(a) and 9(c); $2\theta = 27°$), which are connected with multilayer structures, especially for higher content of the GNP (PU40-GNP and PU20-GNP samples). On the other hand, for nanocomposites PU40-RGO and PU20-RGO, there were not visible diffraction maxima (Figures 9(b) and 9(d)), which shows good dispersion and

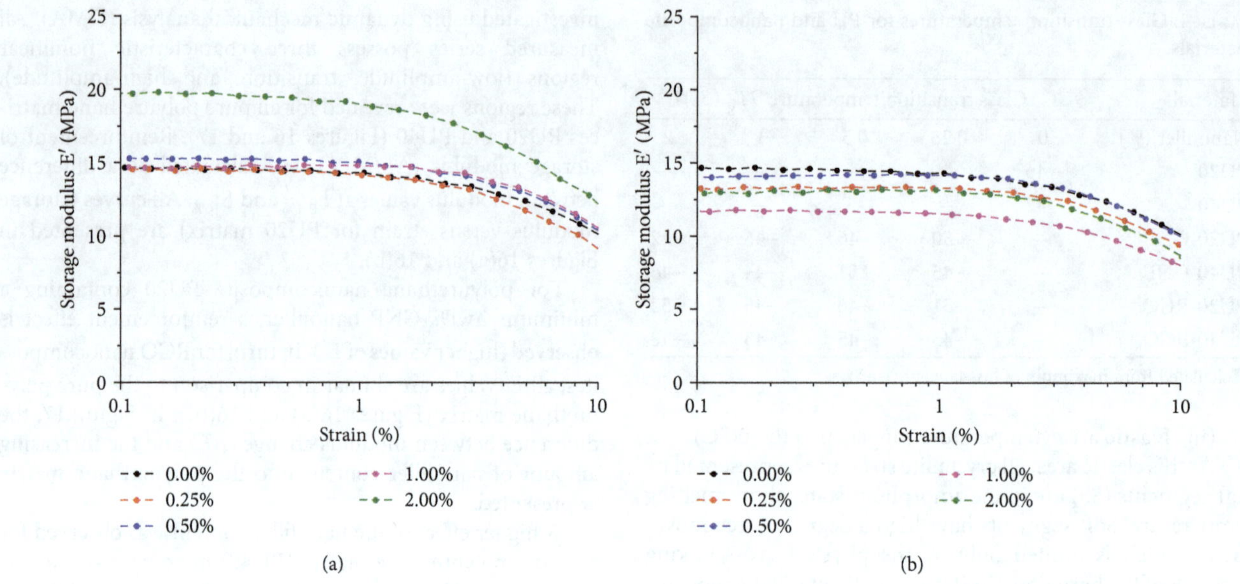

FIGURE 16: Storage modulus (E'/MPa) versus strain (%) for nanocomposites (a) PU20-GNP and (b) PU20-RGO.

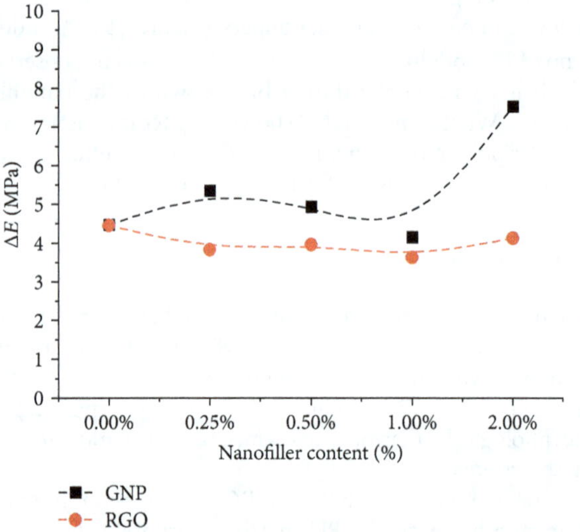

FIGURE 17: Payne effect improvement ΔE for PU20 nanocomposites.

better compatibility of this type of nanofiller. This result was confirmed using TEM analysis (Figures 10(b), 10(c), 10(e), and 10(f)), where good dispersion of the nanofiller in the polyurethane matrix was achieved. For all obtained nanocomposites (especially for GNP-based systems), it was observed that some part of the nanofiller is highly dispersed into the PU matrix and with higher concentrations of the nanofiller, more agglomerated structures were formed.

Dynamic mechanical analysis allows verifying changes of the basic viscoelastic parameters as a function of temperature for polyurethane nanocomposites with different amounts and different kinds of nanofiller (GNP, RGO). DMA analysis shows that GNP and RGO do not have the same compatibility with the polyurethane matrix. These nanofillers affect both areas of soft and hard segments, where dimensions are similar to the nanofiller surface, but in the glassy state segmental movements of the PU chains disappear and the addition of the nanofiller does not cause significant incensement of storage modulus (E').

Above the glass transition temperature (T_g), a relaxation of soft segments allowing free movement of the nanofiller within the liquid elastic phase was observed. Depending on the kind and amount of the nanofiller, a different effect on the polyurethane matrix was observed. At low nanofiller content (0.5–1.0 wt%), there was a visible increase in the stiffness, connected with the hydrodynamic effect, where stiff nanofiller particles cause higher viscosity of the "liquid" soft domain and that results in higher E' values (Figures 14(a), 14(b) and 15(a) and 15(b)). For higher content of the nanofiller (1.0–2.0 wt%), a further increase in storage modulus (E') and also higher values of the loss modulus (E'') were registered (Figures 14(c), 14(d) and 15(c) and 15(d)). This behavior is the result of a sliding motion of the nanoplatelets (at high nanofiller content) which can be submitted to aggregation and agglomeration. These nanoparticles can clash with one another, causing an increase in mechanical friction, which results in energy dissipation as a result of mechanical loss (increase of loss modulus, E'').

The Payne effect is not visible for all obtained nanocomposite systems. The big role of creating nanofiller network improvement plays as a type of the nanofiller and hard segment content of the polyurethane matrix. For PU20-RGO nanocomposites, the addition of the nanofiller did not change visibly the storage modulus (E'), which is probably connected with the absence of aggregation enlargement after exceeding the threshold of percolation, which allows for effective interactions between the platelets. Comparing PU20-RGO and PU20-GNP samples with 2 wt% nanofiller addition, it was observed that the percolation threshold for

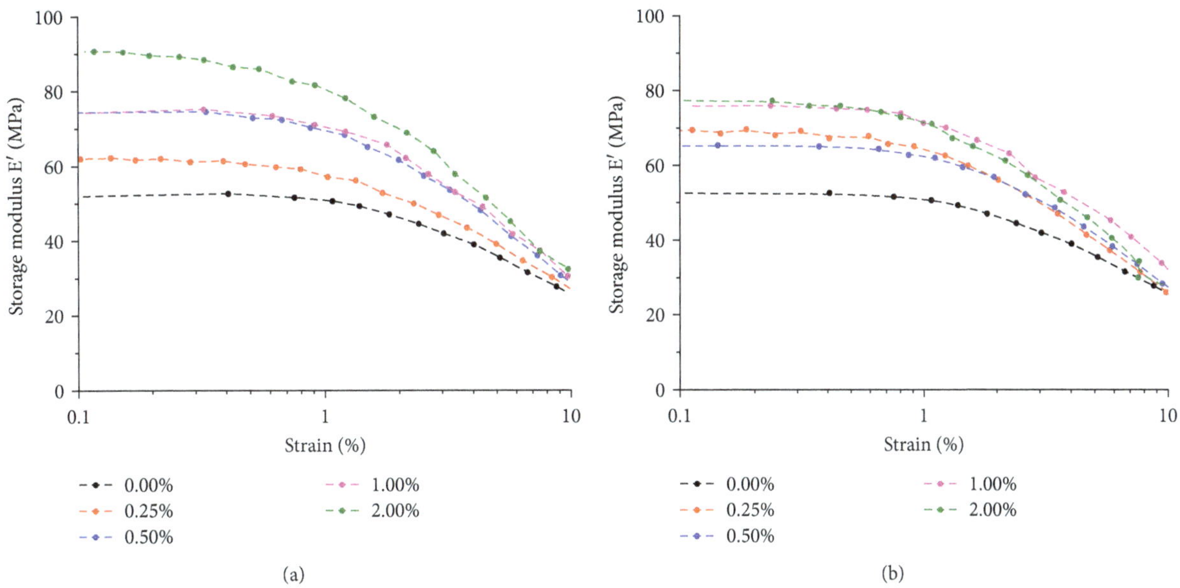

FIGURE 18: Storage modulus (E'/MPa) versus strain (%) for nanocomposites (a) PU40-GNP and (b) PU40-RGO.

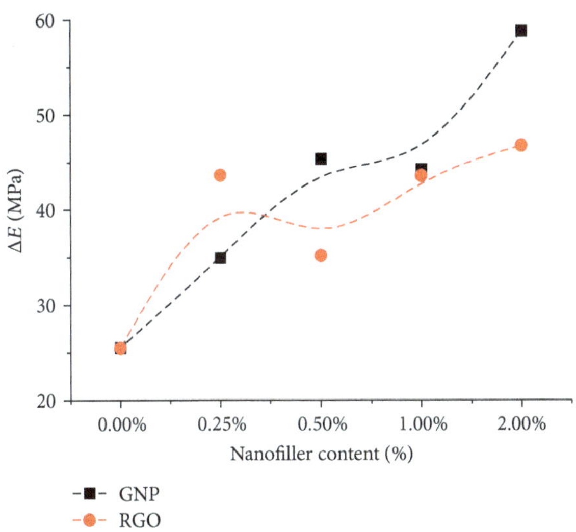

FIGURE 19: Payne effect improvement ΔE for PU40 nanocomposites.

RGO is higher than for the GNP nanofiller into the PU20 matrix. In PU40 samples, an increase in hard segment (HS) content causes a lower percolation threshold for both nanofillers, where an increase in E' is visible. Addition of these nanofiller leads to an instant increase in E'. The graphene nanoplatelet nanofiller was a more effective modificator, for the PU40 matrix, resulting in a higher E' than the RGO additive.

The hard domain behavior and crystallization process of polyurethane strongly depend on the type of introduced nanofiller and on hard segment (HS) content. Addition of the nanofillers impacts on the crystallization rate, degree of crystallization, and size of the crystallites formed. Table 5 shows the effect of the nanofiller additive on the following: crystallization rate—based on cooling thermograms, resulting from changes in the temperature at which they start to crystallize; effect on crystallinity—decrease in crystallinity was found by the lower values of the melting enthalpy; and influence on the size of forming crystallites—presented on the basis of shifting of melting temperatures (T_m).

Based on observed investigations for nanocomposite systems (PU20-RGO, PU20-GNP, and PU40-RGO), a decrease in crystallinity within the hard domain (Table 5) was conformed. This points to disruption of the hard domain crystallization process caused by the nanofiller, which acts as a physical barrier, reducing the tendency to crystallization. For PU40-GNP polyurethane nanocomposites, a significant influence on the crystallization degree within hard and soft domains was not conformed, but the rate of crystallization in both cases increased. For these samples, creation of hard domain crystallites, which can have larger sizes, was confirmed, which is connected with a shift to higher temperatures of melting temperature peaks with simultaneous disappearance of the preheating area. From the described changes in the crystalline structure of hard domain PU40 nanocomposites, it can be presumed that the crystallization process within these domains is different. Presented differences within HS can be explained by changes in the crystallization process, which depends on the type of nanofiller and also hard segment content in the polyurethane matrices.

For nanocomposites described above, a hard domain melting point is influenced by the fact that all segments (soft and hard) are mixed, while the nanofiller is dispersed between them. At cooling from the melt, the occurring phase separation becomes better with increasing concentration of hard segments. During crystallization of HS, nanoplatelets probably migrate into the soft domain. This behavior can be confirmed by no changes in HS crystallization parameters for PU40-GNP nanocomposites compared to the PU-40

Table 5: Enthalpy of crystallization for PU nanocomposites (cooling scans).

% nanofiller	Cooling scans							
	PU20-GNP		PU20-RGO		PU40-GNP		PU40-RGO	
	$\Delta H_{c,HD}$ (J/g)	$\Delta H_{c,SD}$ (J/g)	$\Delta H_{c,HD}$ (J/g)	$\Delta H_{c,SD}$ (J/g)	$\Delta H_{c,HD}$ (J/g)	$\Delta H_{c,SD}$ (J/g)	$\Delta H_{c,HD}$ (J/g)	$\Delta H_{c,SD}$ (J/g)
0.00	−7.5	—	−7.5	—	−10.3	0.0	−10.3	—
0.25	−5.9	—	−9.0	—	−8.1	−6.5	−8.0	—
0.50	−2.8	—	−4.6	—	−6.3	−6.3	−7.9	—
1.00	−1.6	—	−4.2	—	−5.2	−3.6	−5.2	—
2.00	−2.9	—	−3.3	−4.6	−11.7	−6.3	−12.5	—

Table 6: Comparison of the nanofiller's influence on PU hard and soft domains' behavior.

Samples	Hard domain			Soft domain	
	Crystallization rate	Degree of crystallinity	Crystallite size	Crystallization rate	Degree of crystallinity
PU20-RGO	↑	↓	↓	↑	↓
PU20-GNP	↓	↓	↓	—	↓
PU40-RGO	↓	↓	↓	<−>	↓
PU40-GNP	↑	<−>	↑	↑	<−>

Designation: (↑) increase parameter, (↓) decrease parameter, (<−>) parameter remains unchanged.

elastomer. Lower crystallization enthalpy values for PU40-RGO can be connected with a higher content of RGO, which remains in HS domains causing a decrease in crystallinity and higher crystallite fragmentation. RGO's less tendency to migrate to a soft segment domain (SS) can be due to its folding, smaller flake thickness, and higher polydispersity of its surface area. Also, polar groups on the RGO surface increase the affinity to the rigid segments of PU.

The nanoplatelets migrating into the hard segment domains can accumulate within these structures, and their impact can be enhanced by the hydrodynamic effect and by the creation of the Payne effect but only at a fitting amount of this nanofiller. In the case of nanocomposites PU20-GNP and PU20-RGO, a decrease in crystallinity was observed, indicating that nanoparticles are dispersed in HS domains. As explained above, the migration of RGO is more difficult due to its defected surface structure. The migration of nanofiller particles into hard segment domains for PU20-RGO and PU20-GNP can be less intense or may not occur, because of the small size of the crystalline domains forming from hard segments, which could be connected with a lower amount of HS in these matrices.

Based on the previous investigation, the influence of the nanofillers on phase separation was shown. It can be assumed that the used nanofillers do not have significant influence on the degree of phase separation, with parameter changes fluctuating around 3%. The increasing content of RGO and GNP in the PU20 matrix results in a decrease in the degree of phase separation. What is more, the addition of these nanofillers to the PU40 matrix results in the opposite effect and the increase in the degree of phase separation. The character of these changes correlates with the influence of the nanofillers on the crystallization behavior of HS. For PU20 nanocomposites, a decrease in crystallinity was observed, whereas for PU40 an increasing effect was found only for the GNP nanofiller, not for PU40-RGO samples. With a lower HS content in the PU20 matrix, nanoparticle segregation is smaller, causing more nanoparticles to be present in the hard segment domain thus reducing hydrogen bonding within these crystalline structures. A reverse effect is found for PU40 nanocomposites, where phase separation is higher causing more well-structured hard domains. However, this situation does not coincide for nanocomposites PU20-RGO and PU40-RGO, where phase separation is higher than for PU40-GNP and PU40-RGO. This could be the reason for the heterogeneity of the nanocomposites confirmed by FTIR investigation and of some agglomeration of the nanofiller on the surface of the samples.

6. Conclusions

A clear influence of the nanofillers on the structure, morphology, and thermal and mechanical behavior of the polyurethane nanocomposites was found for two polyurethane matrices possessing different hard segment contents (20% and 40% HS) modified with graphene derivatives (reduced graphene oxide (RGO) and graphene nanoplatelets (GNP)) (Table 6). Differences in the properties of the obtained nanocomposites result mainly from the change in the morphological structure of these nanocomposites as a result of the introduction of nanofillers. Obtaining nanocomposites with a different content of the nanofiller allowed following the changes in the structure and properties of nanocomposites. The addition of both types of graphene forms influenced the crystalline structure of hard and soft domains. The use of the polyurethane matrix with a different content of rigid segments revealed a different way in which crystallization and melting processes occur within both domains present. It has been found, depending on the filled matrix, that the graphene form may cause a decrease in the degree of crystallinity and delay the crystallization process or may exhibit a nucleating behavior.

Based on changes in crystallinity, the crystallization process occurring in the presence of nanofillers in polyurethane matrices was examined. A properly selected type of graphene nanofiller and its appropriate amount allows improving the elastic properties of polyurethane materials, as a result of hydrodynamic reinforcement connected with the creation of Payne's reinforcing aggregation network.

Conflicts of Interest

The author declares that there is no conflict of interest regarding the publication of this paper.

Acknowledgments

The author acknowledges Piotr Korzeniewski (Poland), who prepared the PU nanocomposite samples, and Anu A. S. (India) for performing TEM experiments. Special acknowledgments are due to Professor Sabu Thomas (India) for his support.

References

[1] M. Bhattacharya, "Polymer nanocomposites—a comparison between carbon nanotubes, graphene, and clay as nanofillers," *Materials*, vol. 9, no. 4, p. 262, 2016.

[2] M. Šupová, G. S. Martynková, and K. Barabaszová, "Effect of nanofillers dispersion in polymer matrices: a review," *Science of Advanced Materials*, vol. 3, no. 1, pp. 1–25, 2011.

[3] M. Damien, E. S. Guillaume, and C. Chivas-Joly, "Properties of nanofillers in polymer," in *Nanocomposites and Polymers with Analytical Methods*, 2011.

[4] H. Kim, S. Kobayashi, M. A. AbdurRahim et al., "Graphene/polyethylene nanocomposites: effect of polyethylene functionalization and blending methods," *Polymer*, vol. 52, no. 8, pp. 1837–1846, 2011.

[5] M. Z. Iqbal, A. A. Abdala, V. Mittal, S. Seifert, A. M. Herring, and M. W. Liberatore, "Processable conductive graphene/polyethylene nanocomposites: effects of graphene dispersion and polyethylene blending with oxidized polyethylene on rheology and microstructure," *Polymer*, vol. 98, pp. 143–155, 2016.

[6] S. Park, S. He, J. Wang, A. Stein, and C. W. Macosko, "Graphene-polyethylene nanocomposites: effect of graphene functionalization," *Polymer*, vol. 104, pp. 1–9, 2016.

[7] X. Fu, C. Yao, and G. Yang, "Recent advances in graphene/polyamide 6 composites: a review," *RSC Advances*, vol. 5, no. 76, pp. 61688–61702, 2015.

[8] X. Liu, X. Y. Shao, G. B. Fang, H. F. He, and Z. G. Wan, "Preparation and properties of chemically reduced graphene oxide/copolymer-polyamide nanocomposites," *E-Polymers*, vol. 17, no. 1, pp. 3–14, 2017.

[9] A. O'Neill, D. Bakirtzis, and D. Dixon, "Polyamide 6/graphene composites: the effect of in situ polymerisation on the structure and properties of graphene oxide and reduced graphene oxide," *European Polymer Journal*, vol. 59, pp. 353–362, 2014.

[10] T. Kuila, S. Bose, P. Khanra, N. H. Kim, K. Y. Rhee, and J. H. Lee, "Characterization and properties of in situ emulsion polymerized poly(methyl methacrylate)/graphene nanocomposites," *Composites Part A: Applied Science and Manufacturing*, vol. 42, no. 11, pp. 1856–1861, 2011.

[11] S. Morimune, T. Nishino, and T. Goto, "Ecological approach to graphene oxide reinforced poly (methyl methacrylate) nanocomposites," *ACS Applied Materials & Interfaces*, vol. 4, no. 7, pp. 3596–3601, 2012.

[12] R. Shah, A. Kausar, and B. Muhammad, "Characterization and properties of poly(methyl methacrylate)/graphene, poly(methyl methacrylate)/graphene oxide and poly(methyl methacrylate)/*p*-phenylenediamine-graphene oxide nanocomposites," *Polymer-Plastics Technology and Engineering*, vol. 54, no. 13, pp. 1334–1342, 2015.

[13] H. Hu, X. Wang, J. Wang et al., "Preparation and properties of graphene nanosheets-polystyrene nanocomposites via in situ emulsion polymerization," *Chemical Physics Letters*, vol. 484, no. 4-6, pp. 247–253, 2010.

[14] A. S. Patole, S. P. Patole, S. Y. Jung, J. B. Yoo, J. H. An, and T. H. Kim, "Self assembled graphene/carbon nanotube/polystyrene hybrid nanocomposite by in situ microemulsion polymerization," *European Polymer Journal*, vol. 48, no. 2, pp. 252–259, 2012.

[15] Z. He, B. Zhang, H. B. Zhang et al., "Improved rheological and electrical properties of graphene/polystyrene nanocomposites modified with styrene maleic anhydride copolymer," *Composites Science and Technology*, vol. 102, pp. 176–182, 2014.

[16] M. Hernández, M. . M. Bernal, R. Verdejo, T. A. Ezquerra, and M. A. López-Manchado, "Overall performance of natural rubber/graphene nanocomposites," *Composites Science and Technology*, vol. 73, pp. 40–46, 2012.

[17] A. Malas, "Rubber nanocomposites with graphene as the nanofiller," in *Progress in Rubber Nanocomposites*, pp. 179–229, Woodhead Publishing Series in Composites Science and Engineering, 2017.

[18] K. K. Sadasivuni, D. Ponnamma, J. Kim, and S. Thomas, *Graphene-Based Polymer Nanocomposites in Electronics*, Springer Series on Polymer and Composite Materials, Springer International Publishing, 2015.

[19] S. Thomas and R. Stephen, *Rubber Nanocomposites: Preparation, Properties, and Applications*, Wiley, 2010.

[20] H. Kim, Y. Miura, and C. W. MacOsko, "Graphene/polyurethane nanocomposites for improved gas barrier and electrical conductivity," *Chemistry of Materials*, vol. 22, no. 11, pp. 3441–3450, 2010.

[21] S. K. Yadav and J. W. Cho, "Functionalized graphene nanoplatelets for enhanced mechanical and thermal properties of polyurethane nanocomposites," *Applied Surface Science*, vol. 266, pp. 360–367, 2013.

[22] M. Kumar, J. S. Chung, B. S. Kong, E. J. Kim, and S. H. Hur, "Synthesis of graphene-polyurethane nanocomposite using highly functionalized graphene oxide as pseudo-crosslinker," *Materials Letters*, vol. 106, pp. 319–321, 2013.

[23] M. Verma, S. S. Chauhan, S. K. Dhawan, and V. Choudhary, "Graphene nanoplatelets/carbon nanotubes/polyurethane composites as efficient shield against electromagnetic polluting radiations," *Composites Part B: Engineering*, vol. 120, pp. 118–127, 2017.

[24] Y. R. Lee, A. V. Raghu, H. M. Jeong, and B. K. Kim, "Properties of waterborne polyurethane/functionalized graphene sheet nanocomposites prepared by an in situ method," *Macromolecular Chemistry and Physics*, vol. 210, no. 15, pp. 1247–1254, 2009.

[25] M. N. Muralidharan and S. Ansari, "Thermally reduced graphene oxide/thermoplastic polyurethane nanocomposites as photomechanical actuators," *Advanced Materials Letters*, vol. 4, no. 12, pp. 927–932, 2013.

[26] D. K. Pradhan, R. N. P. Choudhary, and B. K. Samantaray, "Studies of dielectric and electrical properties of plasticized polymer nanocomposite electrolytes," *Materials Chemistry and Physics*, vol. 115, no. 2-3, pp. 557–561, 2009.

[27] Y. Li, D. Pan, S. Chen, Q. Wang, G. Pan, and T. Wang, "In situ polymerization and mechanical, thermal properties of polyurethane/graphene oxide/epoxy nanocomposites," *Materials and Design*, vol. 47, pp. 850–856, 2013.

[28] P. Pokharel, S. Choi, and D. S. Lee, "The effect of hard segment length on the thermal and mechanical properties of polyurethane/graphene oxide nanocomposites," *Composites Part A: Applied Science and Manufacturing*, vol. 69, pp. 168–177, 2015.

[29] X. Qian, L. Song, Q. Tai, Y. Hu, and R. K. K. Yuen, "Graphite oxide/polyurea and graphene/polyurea nanocomposites: a comparative investigation on properties reinforcements and mechanism," *Composites Science and Technology*, vol. 74, pp. 228–234, 2013.

[30] I. S. Fahim, "A nanoscale investigation of mechanical, thermal stability and electrical conductivity properties of reinforced thermoplastic polyurethane/graphene nanocomposite," *American Journal of Nanoscience and Nanotechnology*, vol. 1, no. 1, pp. 31–40, 2013.

[31] M. A. Hood, B. Wang, J. M. Sands, J. J. la Scala, F. L. Beyer, and C. Y. Li, "Morphology control of segmented polyurethanes by crystallization of hard and soft segments," *Polymer*, vol. 51, no. 10, pp. 2191–2198, 2010.

[32] M. C. Lopes, H. Ribeiro, M. C. Gonçalves Santos et al., "High performance polyurethane composites with isocyanate-functionalized carbon nanotubes: improvements in tear strength and scratch hardness," *Journal of Applied Polymer Science*, vol. 134, no. 2, article 44394, 2017.

[33] S. Rana, J. W. Cho, and L. P. Tan, "Graphene-crosslinked polyurethane block copolymer nanocomposites with enhanced mechanical, electrical, and shape memory properties," *RSC Advances*, vol. 3, no. 33, p. 13796, 2013.

[34] D. A. Nguyen, Y. R. Lee, A. V. Raghu, H. M. Jeong, C. M. Shin, and B. K. Kim, "Morphological and physical properties of a thermoplastic polyurethane reinforced with functionalized graphene sheet," *Polymer International*, vol. 58, no. 4, pp. 412–417, 2009.

[35] D. B. Klinedinst, I. Yilgör, E. Yilgör, M. Zhang, and G. L. Wilkes, "The effect of varying soft and hard segment length on the structure–property relationships of segmented polyurethanes based on a linear symmetric diisocyanate, 1,4-butanediol and PTMO soft segments," *Polymer*, vol. 53, no. 23, pp. 5358–5366, 2012.

[36] I. V. Khudyakov, D. R. Zopf, and N. J. Turro, "Polyurethane nanocomposites," *Designed Monomers and Polymers*, vol. 12, no. 4, pp. 279–290, 2009.

[37] B. Xu, Y. Q. Fu, W. M. Huang et al., "Thermal-mechanical properties of polyurethane-clay shape memory polymer nanocomposites," *Polymers*, vol. 2, no. 2, pp. 31–39, 2010.

[38] A. V. Rane, K. Kanny, V. K. Abitha, S. Jadhav, S. Mulge, and S. Thomas, "Applications of polyurethane based composites and nanocomposites," in *Polyurethane Polymers*, S. Thomas, J. Datta, J. T. Haponiuk, and A. Reghunadhan, Eds., pp. 559–573, Elsevier, Amsterdam, 2017.

[39] P. K. Maji, "Graphene-based polymer nanocomposites: materials for future revolution," *MOJ Polymer Science*, vol. 1, no. 3, pp. 94–97, 2017.

[40] A. Kausar, "Emerging research trends in polyurethane/graphene nanocomposite: a review," *Polymer-Plastics Technology and Engineering*, vol. 56, no. 13, pp. 1468–1486, 2017.

[41] https://www.acsmaterial.com/graphene-nanoplatelets-2-10nm-1204.html.

[42] M. Strankowski, P. Korzeniewski, J. Strankowska, A. S. Anu, and S. Thomas, "Morphology, mechanical and thermal properties of thermoplastic polyurethane containing reduced graphene oxide and graphene nanoplatelets," *Materials*, vol. 11, no. 1, p. 82, 2018.

[43] D. C. Marcano, D. V. Kosynkin, J. M. Berlin et al., "Improved synthesis of graphene oxide," *ACS Nano*, vol. 4, no. 8, pp. 4806–4814, 2010.

[44] A. V. Raghu, G. S. Gadaginamath, N. Mathew, S. B. Halligudi, and T. M. Aminabhavi, "Synthesis, characterization, and acoustic properties of new soluble polyurethanes based on 2,2′-[1,4-phenylenebis(nitrilomethylylidene)]diphenol and 2,2′-[4,4′-methylene-di-2-methylphenylene-1,1′-bis(nitrilomethylylidene)]diphenol," *Journal of Applied Polymer Science*, vol. 106, no. 1, pp. 299–308, 2007.

[45] A. V. Raghu, H. M. Jeong, J. H. Kim, Y. R. Lee, Y. B. Cho, and K. Sirsalmath, "Synthesis and characterization of novel polyurethanes based on 4-{(4-hydroxyphenyl)iminomethyl}phenol," *Macromolecular Research*, vol. 16, no. 3, pp. 194–199, 2008.

[46] B. Janowski, K. Pielichowski, and R. Kwiatkowski, "Nanohybrid polyurethane (PUR)/functionalized silsesquioxane (PHIPOSS) systems. Part II. X-Ray structural investigations using WAXD and SAXS methods," *Polimery*, vol. 59, no. 2, pp. 147–159, 2014.

[47] P. K. Maji and A. K. Bhowmick, "Structure-property correlation of polyurethane nanocomposites: influence of loading and nature of nanosilica and microstructure of hyperbranched polyol," *Journal of Applied Polymer Science*, vol. 127, no. 6, pp. 4492–4504, 2013.

[48] D. Ponnamma, K. K. Sadasivuni, M. Strankowski, P. Moldenaers, S. Thomas, and Y. Grohens, "Interrelated shape memory and Payne effect in polyurethane/graphene oxide nanocomposites," *RSC Advances*, vol. 3, no. 36, p. 16068, 2013.

[49] K. Bagdi, K. Molnár, I. Sajó, and B. Pukánszky, "Specific interactions, structure and properties in segmented polyurethane elastomers," *Express Polymer Letters*, vol. 5, no. 5, pp. 417–427, 2011.

[50] J. A. Hiltz and J. P. Szabo, *FT-IR Study of Poly(ether)urethanes*, Def. Res. Est. Atl, 2001.

[51] H. S. Lee, Y. K. Wang, and S. L. Hsu, "Spectroscopic analysis of phase separation behavior of model polyurethanes," *Macromolecules*, vol. 20, no. 9, pp. 2089–2095, 1987.

[52] P. Zhao, Y. S. Wang, J. H. Zhu, X. Y. Hua, and Q. Z. Wen, "Characterization of graded polyurethane elastomer by FTIR," *Science in China Series B: Chemistry*, vol. 51, no. 1, pp. 58–61, 2008.

2

Composites from Thermoplastic Natural Rubber Reinforced Rubberwood Sawdust: Effects of Sawdust Size and Content on Thermal, Physical, and Mechanical Properties

Chatree Homkhiew,[1] **Surasit Rawangwong**,[1] **Worapong Boonchouytan**,[1] **Wiriya Thongruang**,[2] **and Thanate Ratanawilai**[3]

[1]*Department of Industrial Engineering, Materials Processing Technology Research Unit, Faculty of Engineering, Rajamangala University of Technology Srivijaya, Muang District, Songkhla 90000, Thailand*
[2]*Department of Mechanical Engineering, Faculty of Engineering, Prince of Songkla University, Hat Yai, Songkhla 90112, Thailand*
[3]*Department of Industrial Engineering, Faculty of Engineering, Prince of Songkla University, Hat Yai, Songkhla 90112, Thailand*

Correspondence should be addressed to Chatree Homkhiew; chatree.h@rmutsv.ac.th

Academic Editor: Cornelia Vasile

The aim of this work is to investigate the effects of rubberwood sawdust (RWS) size and content as well as the ratio of natural rubber (NR)/high-density polyethylene (HDPE) blend on properties of RWS reinforced thermoplastic natural rubber (TPNR) composites. The addition of RWS about 30–50 wt% improved the modulus of the rupture and tensile strength of TPNR composites blending with NR/HDPE ratios of 60/40 and 50/50. TPNR composites reinforced with RWS 80 mesh yielded better tensile strength and modulus of rupture than the composites with RWS 40 mesh. The TPNR/RWS composites with larger HDPE content gave higher tensile, flexural, and Shore hardness properties and thermal stability as well as lower water absorption. The TPNR/RWS composites with larger plastic content were therefore suggested for applications requiring high performance of thermal, physical, and mechanical properties.

1. Introduction

A thermoplastic natural rubber (TPNR) is a material blended from a natural rubber (NR) and a thermoplastic, such as polypropylene, polystyrene, and high-density polyethylene (HDPE) [1, 2]. Thus, it shows velvety surfaces and intermediate properties between the NR and the plastic, which provides flexibility in shaping, great recyclability of scrap, and low production cost [3]. Currently, the TPNR is still being developed for various applications such as in footwear, seals, hoses, automobiles, and marine engineering because the common thermoplastic machinery could be used without compounding or a vulcanization process [4]. However, due to the nature of the TPNR as a polymer, it has limitations in some physical and mechanical properties, particularly the stiffness and hardness. The addition of natural fiber (wood fiber or wood sawdust) as reinforcement can improve the performance of TPNR in the constructional and building applications. Jamil et al. [2] revealed that tensile modulus and Shore hardness increased with rice husk loadings in the TPNR matrix, and the addition of kenaf fiber in the TPNR matrix had significantly increased the tensile, flexural, and impact properties [5]. Further, the natural fibers also offer many advantages such as low cost, high toughness, lightweight, less abrasive on processing tools, and low energy consumption in manufacturing [6, 7].

Recently, the utilization of natural fiber such as bamboo, bagasse, flax, and rubberwood as a replacement to inorganic reinforcements in the plastic composites had been extensively performed [6]. For example, Srivabut et al. [8] applied the rubberwood flour as a reinforcement in recycled polypropylene composites, and the modulus and Shore hardness of recycled polypropylene composites increased linearly with rubberwood flour loadings [6]. Vu et al. [9] revealed that

Young's modulus, flexural strength, and flexural modulus of polypropylene were improved by blending polypropylene with cellulose fibers. Li et al. [10] found that the surface modification of sisal fiber had slightly increased the tensile strength of polylactide composites. Further, Luo et al. [11] showed that using the corn stem as the reinforcement in HDPE composites achieved the highest tensile break strength and flexural strength. Petchwattana and Covavisaruch [12] reported that the increasing additions of rubberwood flour in poly (lactic acid) composites increased the tensile modulus and strength. From this information, the use of rubberwood sawdust (RWS) as a reinforcement in the TPNR composites is of great interest. Because a huge amount of the RWS waste generated from sawmills and furniture industries was greatly found in Thailand, some of the wood waste is eliminated by burning and dumping in space areas [13].

Normally, the thermal, physical, and mechanical properties of wood-polymer composites are influenced by a few factors [14], that is, the mixing and molding parameters [14, 15], wood species [14, 16], and the ratio of the material component [14, 17, 18]. Jamil et al. [2] also concluded that the major factors affecting the critical performance properties of polymer composites were wood flour loading, wood particle size, interfacial adhesion, and wood structure. The chemical composition of wood is also a factor [19]. Therefore, the development of new composite materials to response the constructional and building purposes needs a better understanding for the process of creating wood-polymer composites from the material components [14].

Although extensive research in the area of the plastic composites reinforcing the natural fibers has been conducted, there are just a few studies in the use of the natural fibers to reinforce TPNR composites. No prior report on TPNR composite reinforced RWS was found to focus on the effects of content and particle size as well as the blend ratio of TPNR based on the composites. Hence, the effects of filler content and size on the properties of the composites are needed to be further studied. In the present work, the effects of RWS size and content as well as the ratio of NR/HDPE blend on the thermal, physical, and mechanical properties of TPNR/RWS composites were investigated. The overall results of the current work will facilitate the development of those composites for constructional and building applications, and they will potentially replace expensive building materials like concrete, bricks, and wood lumber in the future.

2. Materials and Methods

2.1. Materials. Natural rubber used in this study was STR 5L grade from the Rubber Estate Organization (Nakhon Si Thammarat, Thailand). High-density polyethylene granules with a melt flow index of 15 g/10 min at 190°C and a density of 0.957 g/cm^3 were purchased from IRPC Public Company Limited (Rayong, Thailand). Rubberwood sawdust was supplied by sawmill industry in the Songkhla Province of Thailand. Before compounding, the sawdust was sieved through a standard sieve and classified into two categories: passing 40 mesh (L) and passing 80 mesh (S), and then was dried in an oven at 120°C for 10 h.

2.2. Preparation of NR/HDPE Blends. TPNR was prepared via melt blending of NR/HDPE ratios of 60/40 (R60/P40), 50/50 (R50/P50), 40/60 (R40/P60), and 30/70 (R30/P70) in a corotating twin-screw extruder (Model CTE-D25L40 from Chareon Tut Co. Ltd., Samutprakarn, Thailand). The temperature of seven processing zones in the extruder was set in the range of 155–185°C from the feeding to the die zone, while the screw rotation speed was controlled at 60 rpm. Then, the extruded strand would pass through an air blower and was subsequently pelletized.

2.3. Preparation of TPNR/RWS Composites. TPNR/RWS composites were compounded using the same twin-screw extruder and parameters as with the preparation of the TPNR matrix. Prior to mixing, the TPNR was dry-blended with various loadings (30, 40, 50, and 60 wt%) and different sizes (L and S) of RWS (formulations in Table 1) in a mixer machine. Further, the TPNR/RWS composite pellets were molded in a compression molding machine with a temperature of 180°C and a pressure of 1000 psi for 10 min. The molding was then transferred to a cold compression set and pressed further under the pressure of 1000 psi for 8 min. Subsequently, the sample panels were processed as specimens for mechanical and physical tests, according to the American Society for Testing and Materials (ASTM) standard, such as ASTM D638-99, ASTM D790-92, ASTM D2240-91, and ASTM D570-88.

2.4. Thermal Analysis. Thermogravimetric (TG) tests were carried out using a Perkin Elmer (TGA-7, USA) thermal analyzer. The tests were conducted in the temperature range of 45–700°C under a nitrogen atmosphere at a heating rate of 10°C/min. Approximately 5-6 mg samples were heated in the sample pan. The onset temperature and thermal stability of the NR, HDPE, TPNR, and TPNR/RWS composites were determined from thermogravimetric analysis curves.

2.5. Morphological Analysis. Morphological analysis of the TPNR and the TPNR/RWS composites was carried out using a scanning electron microscope (Model FEI Quanta 400 from FEI Company, Oregon, USA). To avoid phase deformation and any damage, all specimens of the TPNR and the composites were broken in liquid nitrogen. The specimen surfaces were then coated with gold in order to eliminate electron charging during the imaging. They were imaged with magnifications of 500x at an accelerating voltage of 20 kV.

2.6. Characterizations

2.6.1. Mechanical Tests. Tensile and flexural properties were measured by using a Mechanical Universal Testing Machine (Model NRI-TS500-50 from Narin Instruments Co. Ltd., Samutprakarn, Thailand). The tensile testing of the type-IV specimens was conducted according to ASTM D638-99 with a crosshead speed of 5 mm/min. Three-point flexural testing of specimens with nominal dimensions of 4.8 mm × 13 mm × 100 mm was performed according to ASTM D790-92 with a span of 80 mm and a crosshead speed of 2 mm/min.

Table 1: Formulation of TPNR and TPNR/RWS composites in experiment.

Sample code	RWS size (mesh)	Sample code	RWS size (mesh)	Thermoplastic natural rubber (wt%) R60/P40	R50/P50	R40/P60	R30/P70	RWS (wt%)
R6P4	—	—	—	100	—	—	—	—
R6P4W30L	40	R6P4W30S	80	70	—	—	—	30
R6P4W40L	40	R6P4W40S	80	60	—	—	—	40
R6P4W50L	40	R6P4W50S	80	50	—	—	—	50
R6P4W60L	40	R6P4W60S	80	40	—	—	—	60
R5P5	—	—	—	—	100	—	—	—
R5P5W30L	40	R5P5W30S	80	—	70	—	—	30
R5P5W40L	40	R5P5W40S	80	—	60	—	—	40
R5P5W50L	40	R5P5W50S	80	—	50	—	—	50
R5P5W60L	40	R5P5W60S	80	—	40	—	—	60
R4P6	—	—	—	—	—	100	—	—
R4P6W30L	40	R4P6W30S	80	—	—	70	—	30
R4P6W40L	40	R4P6W40S	80	—	—	60	—	40
R4P6W50L	40	R4P6W50S	80	—	—	50	—	50
R4P6W60L	40	R4P6W60S	80	—	—	40	—	60
R3P7	—	—	—	—	—	—	100	—
R3P7W30L	40	R3P7W30S	80	—	—	—	70	30
R3P7W40L	40	R3P7W40S	80	—	—	—	60	40
R3P7W50L	40	R3P7W50S	80	—	—	—	50	50
R3P7W60L	40	R3P7W60S	80	—	—	—	40	60

Note: NR/HDPE ratios of 60/40 (R60/P40), 50/50 (R50/P50), 40/60 (R40/P60), and 30/70 (R30/P70). RWS: rubberwood sawdust; wt%: percent by weight.

The hardness of the composite samples was measured by using a mechanical Shore D Durometer (Model GS-702G from Teclock Corporation, Nagano, Japan), according to ASTM D2240-91. The dimensions of the specimens tested were approximately 30 mm × 30 mm × 6 mm. All mechanical tests were performed with five replications for each composite formulation at a temperature of 25°C.

2.6.2. Water Absorption Test. The measurement of water absorption was carried out according to ASTM D570-88. Prior to testing, five specimens of each formulation (10 mm × 20 mm × 6 mm) were dried in an oven at 50°C for 24 h. Their weights were then measured with a precision of 0.001 g before being immersed in water at ambient room temperature. After 24 h soaking, all specimens were removed from the water, thoroughly wiped off using tissue papers, and immediately, the weights were measured again. The results would indicate the percentages of water absorption in relation to the dry weight of the specimens.

3. Results and Discussion

3.1. Thermal Degradation Behavior of TPNR and TPNR/RWS Composites. Thermal degradation of the pure NR, pure HDPE, and TPNR was studied using thermogravimetric analysis, as shown in Figure 1 and Table 2. The pure NR displayed the lowest degradation temperature, while degradation of the pure HDPE occurred at the highest temperature. The temperature of the maximum weight loss for the NR occurred at approximately 378.1°C, while the HDPE degraded at approximately 491.0°C. Therefore, it was clearly exhibited that degradation temperature increased with an increase of plastic contents in the NR/HDPE blends. Table 2 also illustrates the temperature at different weight losses for TPNR. For 40% weight loss, the temperatures of R6P4, R5P5, R4P6, and R3P7 are 393.8°C, 401.4°C, 460.8°C, and 464.5°C, respectively.

Thermogravimetric analysis of TPNR/RWS composites is also needed to evaluate thermal stability for determining their application and end use [20]. Figures 2(a) and 2(b) illustrate the variation of thermogravimetric curves and derivative thermogravimetric (DTG) curves, respectively, of TPNR/RWS composites affected by RWS contents and sizes. The TG curves of TPNR/RWS composites show four mass loss steps. The first step is weight loss due to the evaporation of moisture in the range of around 90 to 190°C. The second and third steps are weight loss owing to the decomposition of wood sawdust and natural rubber in the range of around 290 to 430°C. Likewise, at the third peak, the maximum weight loss rates of the NR/HDPE blend of the R60/P40 ratio reinforced by 30 wt% (case R6P4W30L) and 60 wt% (case R6P4W60L) RWS contents with 40 mesh size occur at 382.3 and 368.8, respectively. Generally, the thermal decomposition of the natural fibers occurs in the temperature range of 194–386°C, due to the degradation of cellulose, hemicelluloses, and lignin compositions [21, 22]. The last weight loss step is the decomposition of HDPE in the range of around 450 to 510°C.

The TG curves (Figure 2(a)) representing the TPNR composites with 60 wt% RWS loading (lines (B) and (D))

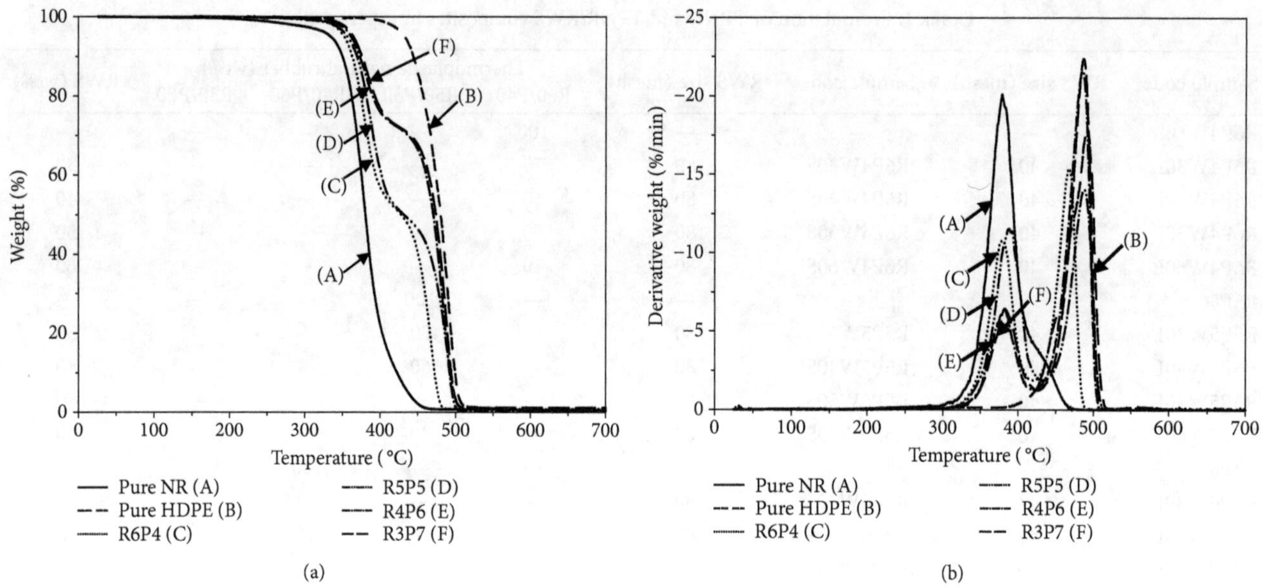

FIGURE 1: Curves of (a) TGA and (b) DTG for pure NR, pure HDPE, and NR/HDPE blends at various blend ratios.

TABLE 2: Thermal data obtained from TGA thermograms of TPNR and TPNR/RWS composites.

Formulation	Temperature at different weight loss (°C)				Temperature for maximum weight loss			
	T_{20}	T_{40}	T_{60}	T_{80}	1st peak	2nd peak	3rd peak	4th peak
Pure NR	361.8	374.9	385.6	402.7	378.1	—	—	—
Pure HDPE	456.9	473.5	485.2	493.9	491.0	—	—	—
R6P4	373.9	393.8	450.4	466.1	379.2	467.1	—	—
R5P5	381.1	401.4	466.0	485.6	387.8	487.3	—	—
R4P6	389.5	460.8	476.5	486.7	382.7	485.1	—	—
R3P7	392.4	464.5	479.7	489.1	382.1	486.3	—	—
R6P4W30L	357.7	383.1	426.8	487.9	159.2	311.3	382.3	491.3
R6P4W60L	335.1	370.6	408.2	494.8	156.5	305.2	368.8	491.9
R4P6W30L	365.5	394.4	475.2	494.5	162.1	300.8	383.7	493.5
R4P6W60L	334.1	371.0	456.3	498.6	162.8	298.4	367.9	492.6
R4P6W30S	362.8	393.1	476.4	495.0	164.4	302.3	376.4	493.9

Note: T_{20} (temperature for 20% weight loss).

show the weight loss 40% at 370.6–371.0°C, while the composites with 30 wt% RWS (lines (A), (C), and (E)) exhibit the weight loss 40% at 383.1–394.4°C. These results can indicate that an increase of RWS content in TPNR composites will decrease the thermal stability. Because the wood had more degradation than the HDPE, thus, the increasing addition of wood into the composites reduced the thermal stability. Likewise, stronger interfacial adhesion between the wood sawdust and the polymer matrix increased the thermal stability [13, 23]; an addition of sawdust content in the polymer composites resulted in poorer interfacial bonding between phases of the materials.

The TG curves also revealed that blend ratios of NR/HDPE affected the thermal stability of the composites. It can be observed that the TPNR composites with 30 wt% RWS loading showed the degradation at 394.4°C with 40% weight loss for the NR/HDPE blend of the R40/P60 ratio (line (C)) and 49% weight loss at 394.4°C for the NR/HDPE blend of the R60/P40 ratio (line (A)). Normally, the lower percentage of weight loss demonstrates a larger thermal stability of the composite material [23–25]; this is due to more degradation of natural rubber than plastic occurrence. In addition, the RWS sizes slightly affected the thermal stability of TPNR/RWS composites. It can be observed that the TPNR composites based on RWS 80 mesh (line (E)) degraded at 362.8°C with 20% weight loss, while the degradation of the composites based on RWS 40 mesh (line (C)) occurred with 20% weight loss at 365.5°C. This finding is in good agreement with Ratanawilai et al. [13], who found that the particle sizes insignificantly affected the thermal stability of wood-plastic composites. In fact, the chemical constituents (e.g., cellulose, lignin, and hemicelluloses) of wood as well as their contents mainly affect the thermal stability of wood sawdust [13, 23, 25], which the particle size influences lightly.

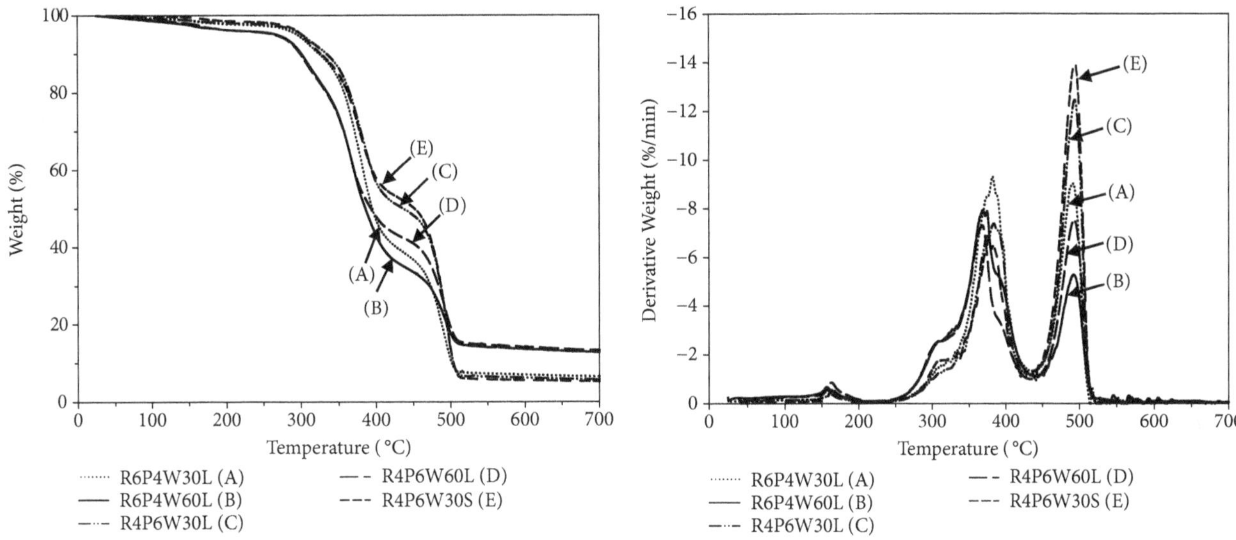

Figure 2: Curves of (a) TGA and (b) DTG for TPNR composites containing different RWS contents and sizes.

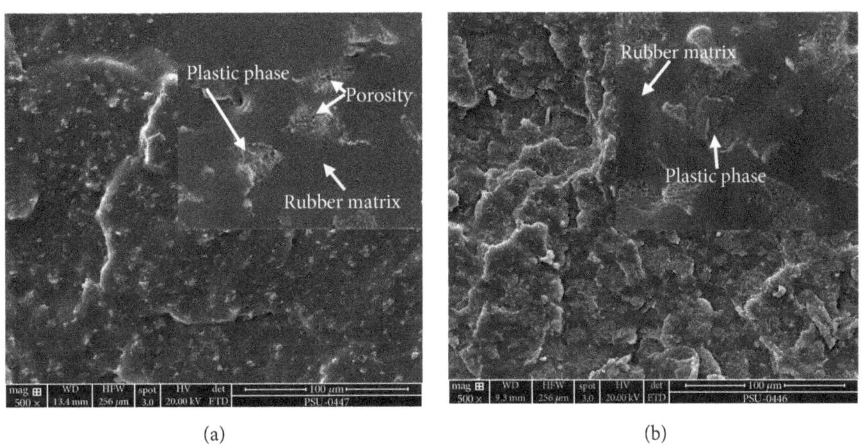

Figure 3: SEM images of TPNR blending with NR/HDPE ratios of (a) R60/P40 and (b) R30/P70.

3.2. Morphological Analysis of TPNR/RWS Composites. The morphology of the NR/HDPE blends reveals some interesting realities as shown in Figure 3. In NR/HDPE blends of R60/P40 and R30/P70 ratios, SEM images of both systems showed that plastic could not be melted as a continuous phase in the rubber matrix. Further, more some porosities were also found in the phase of the plastic (Figure 3(a)). In addition, the fractured surfaces of the composites from NR/HDPE blends of R60/P40 and R30/P70 ratios reinforced with RWS 40 mesh and 80 mesh are shown in Figures 4 and 5, respectively. It can be seen that the TPNR/RWS composites with blend ratios of higher rubber content (Figures 4(a), 4(b), 5(a), and 5(b)) show smoother surfaces and fewer voids. Besides, SEM micrographs were also exhibited that the RWS did not completely attach to the polymer matrix due to poor compatibility between the wood and the polymer matrix [26]. Ashori [26] said that plastic and wood did not merge well. Especially, for the TPNR/RWS composites with the NR/HDPE blend of the R30/P70 ratio, the morphology presented many voids and poor interfacial adhesion, indicating that the interaction between the polymer matrix and wood was weak [2]. Efficacy of interfacial adhesion significantly affected the mechanical and physical properties of wood-polymer composites.

The TPNR composites containing RWS sizes of 40 mesh (Figures 4(c) and 4(d)) showed more porosities and larger gaps between the wood sawdust and the TPNR matrix than the composites with RWS sizes of 80 mesh (Figures 5(c) and 5(d)). This is likely caused by agglomeration and entanglement of bigger wood sawdust [15]. Furthermore, the TPNR composites with 60 wt% RWS (Figures 4(b), 4(d), 5(b), and 5(d)) exhibited more voids than the composites with 30 wt% RWS (Figures 4(a), 4(c), 5(a), and 5(c)) because the addition of RWS content in the polymer composites increased the dispersing difficulty of the wood particles and their tendency to form strong agglomeration [23].

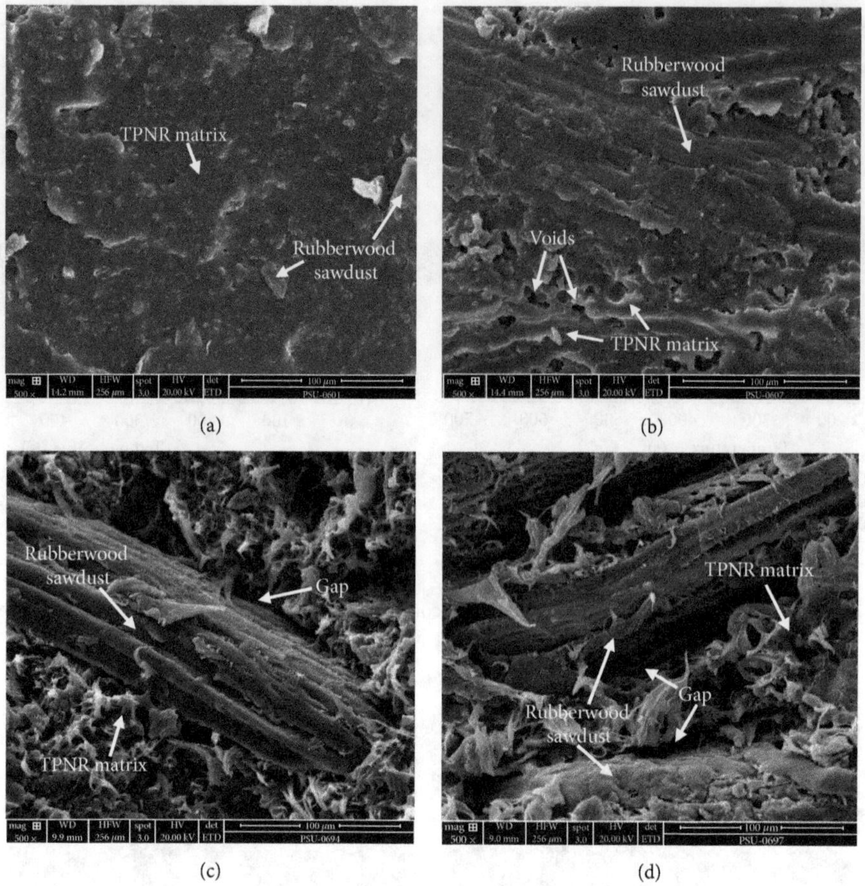

FIGURE 4: SEM images of R60/P40 composites with (a) 30 wt% RWS and (b) 60 wt% RWS and R30/P70 composites with (c) 30 wt% RWS and (d) 60 wt% RWS. All formulations reinforced with RWS 40 mesh.

Particularly for the TPNR/RWS composites with higher plastic content (R30/P70), it was observed that the wood sawdust displayed the shape of irregular short fibers in the composite structure. The composites containing higher wood sawdust content to TPNR blends (Figures 4(d) and 5(d)) seemed to exhibit a higher number of porosities and agglomerations as well as poorer interfacial bonding between the RWS and the polymer matrix.

3.3. Tensile Properties of TPNR/RWS Composites. The effect of RWS contents and sizes on tensile strength and modulus of the composites from NR/HDPE blends of R60/P40, R50/P50, R40/P60, and R30/P70 ratios is shown in Figures 6(a) and 6(b), respectively. The tensile strength of the TPNR/RWS composites with a blend of R30/P70 was clearly reduced with the increasing content of RWS, whereas the additions of RWS in the range of 30 to 50 wt% into the blends of R50/P50 and R60/P40 improved the tensile strength of the composites. These phenomena are caused by the domination of plastic in the blending which is strong; therefore, the incorporation of wood sawdust is less effective [4]. In contrast, for blending with rubber dominant which is softer than plastic, the addition of wood sawdust effectively improves the properties of TPNR composites [4]. These reasons can be confirmed by considering the morphology of TPNR/RWS composites in Figure 4 (Figures 4(a) and 4(b) for TPNR composites blending with R60/P40 and Figures 4(c) and 4(d) for TPNR composites blending with R30/P70). It is evident that the TPNR/RWS composites with R60/P40 clearly displayed smoother surfaces, lower voids, and stronger coupling between the TPNR and the RWS than the composites blending with R30/P70. Thus, the composites blending with R60/P40 possess better stress transfer from the polymer matrix to the wood sawdust. However, at the same RWS contents and sizes, the TPNR/RWS composites with blends of R30/P70 gave clearly better tensile strength and modulus than the composites with blends of R60/P40. It is rationalized because the HDPE plastic has more strength and stiffness than the natural rubber; thus, the composites with larger plastic concentration exhibited higher strength and modulus.

Figure 6(b) exhibits the increment of the tensile modulus with an increase of RWS contents in all systems of the composites. Because these composites have high stiffness wood sawdust and low stiffness polymer matrix, the increasing addition of wood filler volume into the composites increases the stiffness [22]. Jamil et al. [2] revealed that an increment of modulus causes a decrease in the flexibility or elasticity of the soft matrix. In addition, the TPNR composites based on RWS 80 mesh (solid lines) seem to give higher tensile strength than those based on RWF 40 mesh (dashed lines) for the same

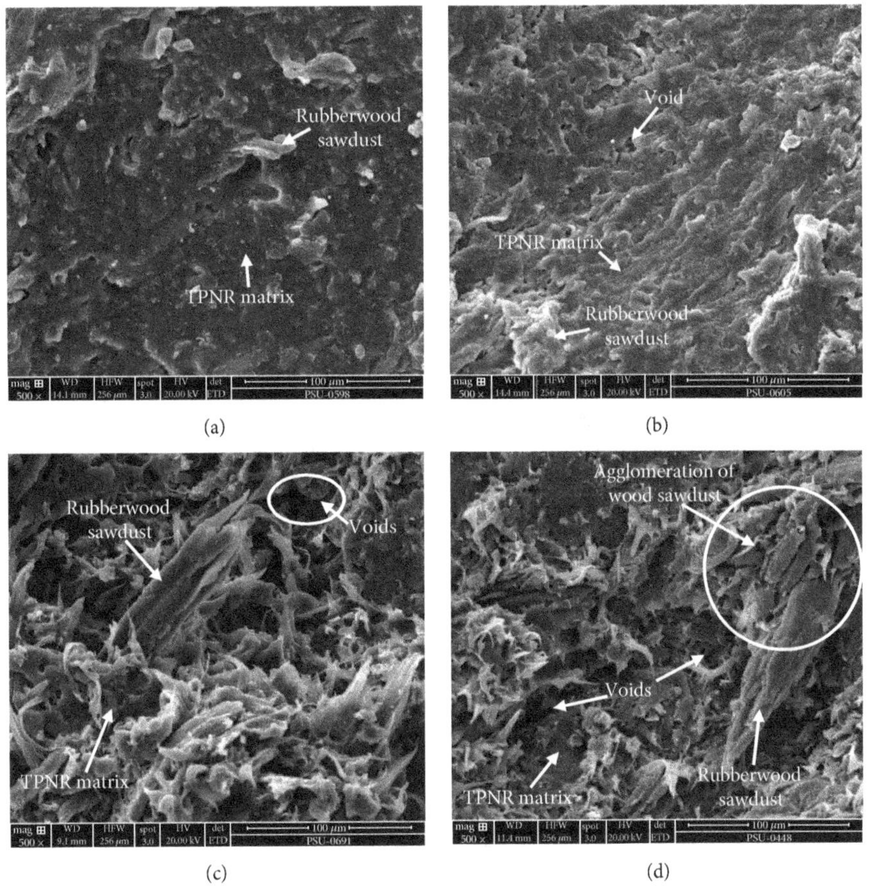

FIGURE 5: SEM images of R60/P40 composites with (a) 30 wt% RWS and (b) 60 wt% RWS and R30/P70 composites with (c) 30 wt% RWS and (d) 60 wt% RWS. All formulations reinforced with RWS 80 mesh.

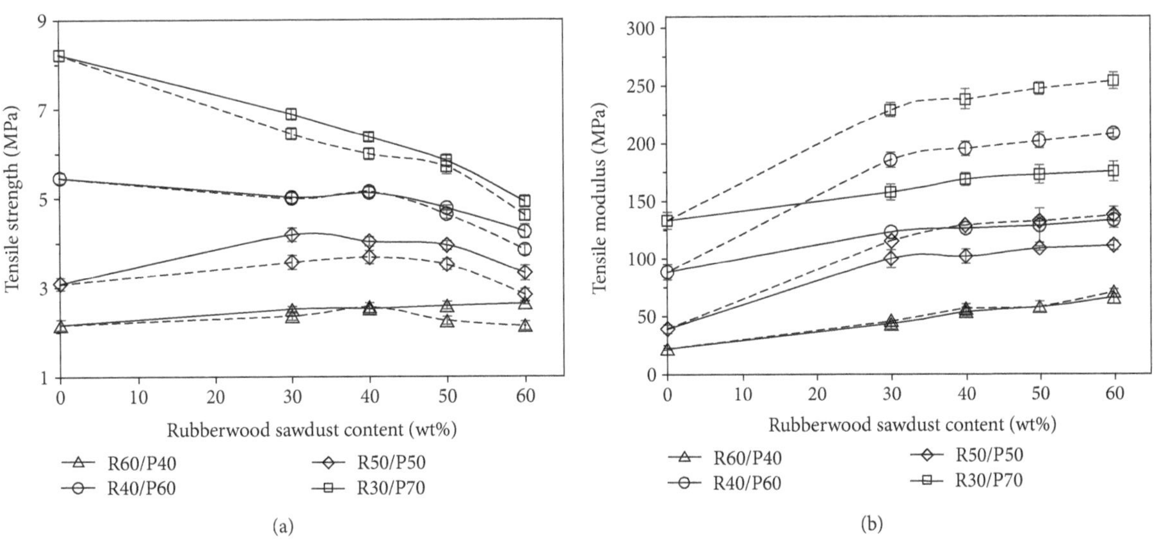

FIGURE 6: Effect of wood sawdust contents and sizes on (a) tensile strength and (b) tensile modulus for TPNR/RWS composites. Dashed lines show RWS 40 mesh, and solid lines represent RWS 80 mesh.

blend ratio of NR/HDPE. Probably, small sawdust size distributes more homogeneous than that of large sawdust size in the polymer matrix, resulting in a uniform stress transfer from the TPNR matrix to the dispersed wood sawdust [13, 27]. Thus, it could increase the load resistance performing from the tension. However, for the tensile modulus, the

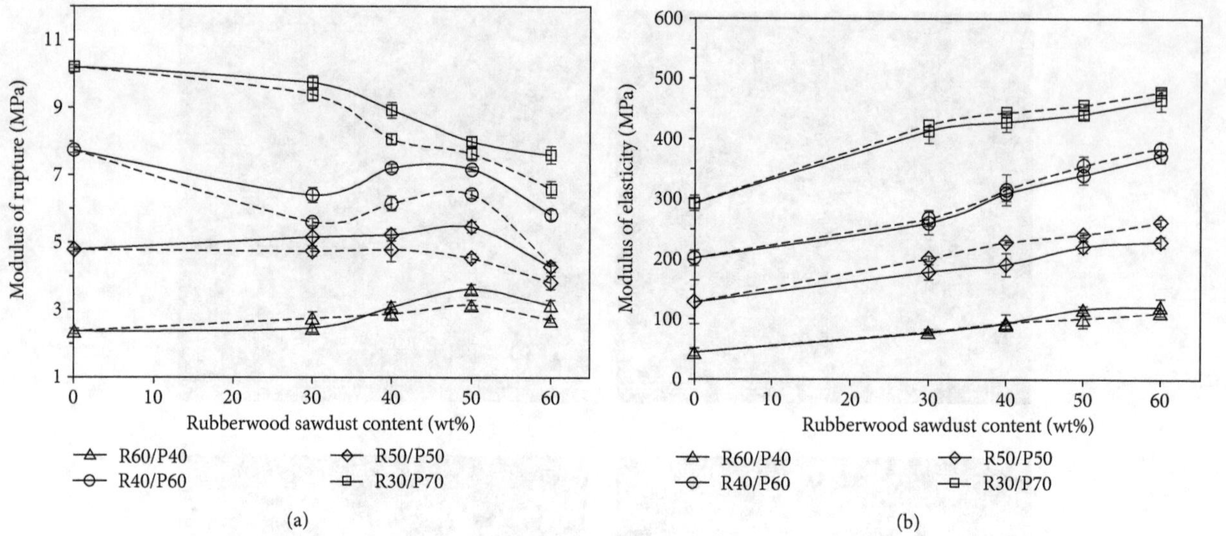

FIGURE 7: Effect of wood sawdust contents and sizes on (a) modulus of rupture and (b) modulus of elasticity for TPNR/RWS composites. Dashed lines show RWS 40 mesh, and solid lines represent RWS 80 mesh.

TPNR composites based on RWS 80 mesh (solid lines) exhibited lower values.

3.4. Flexural Properties of TPNR/RWS Composites. Figures 7(a) and 7(b) show the effect of adding RWS on the modulus of rupture (MOR) and modulus of elasticity (MOE), respectively, of the composites from NR/HDPE blends with R60/P40, R50/P50, R40/P60, and R30/P70 ratios. It can be seen that the MOR of TPNR composites blending with R30/P70 decreased with an increase of RWS content. The optimal RWS loading in the TPNR blending with R60/P40, R50/P50, and R40/P60 is about 50 wt%. The MOR reduced after 50 wt% RWS content indicates that there are small stress transfers from the matrix to the wood due to entanglement and agglomeration of RWS in the polymer matrix. The MOE of TPNR/RWS composites also showed a similar trend to the tensile modulus: the MOE increased greatly with wood sawdust content. Since wood sawdust has higher stiffness or modulus as compared to the polymer matrix, the TPNR composites with higher sawdust content demand higher stress for the same deformation [6, 28].

The MOR of TPNR composites containing RWS 40 mesh (dashed lines) is lower as compared to that of the composites containing RWS 80 mesh (solid lines). This is possible because poorer interfacial adhesion between the wood sawdust and the polymer matrix occurred in the composites with larger wood sawdust [15]. This can be observed in SEM micrographs which show that the composites with RWS 40 mesh (Figures 4(c) and 4(d)) had larger spaces between the wood sawdust and the TPNR matrix than those with RWS 80 mesh (Figures 5(c) and 5(d)). In contrast, the composites with RWS 40 mesh (dashed lines) presented higher MOE than those composites with RWS 80 mesh (solid lines) for the same TPNR to wood sawdust ratio. The composites reinforcing with larger wood sawdust gave a higher fiber aspect ratio. Furthermore, the fiber length-to-diameter ratio determined as the fiber aspect ratio is another factor influencing the mechanical properties of the composite material [29]. An increment of the fiber aspect ratio increases the stress transfer from the polymer matrix to the natural fibers, and then the modulus of composites material was improved [29]. In addition, the result obtained indicates that TPNR/RWS composites with a higher HDPE ratio gave higher values of MOR and MOE. It was observed that, for 60 wt% RWS with 40 mesh, the TPNR/RWS composites containing the NR/HDPE blend of the R30/P70 ratio exhibited higher MOR and MOE about 144% and 328%, respectively, as compared to composites containing blends of R60/P40.

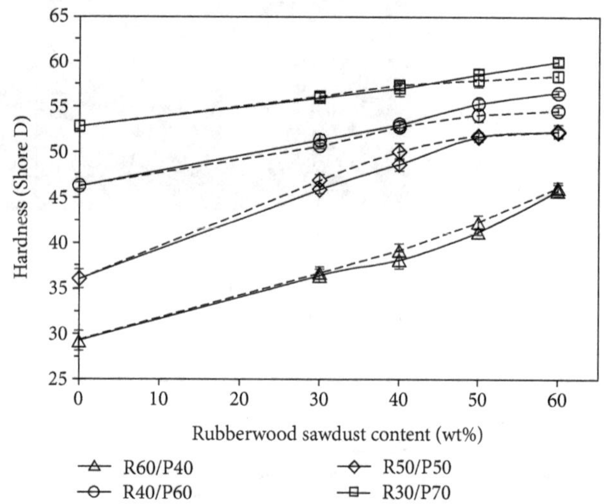

FIGURE 8: Effect of wood sawdust contents and sizes on Shore hardness for TPNR/RWS composites. Dashed lines show RWS 40 mesh, and solid lines represent RWS 80 mesh.

3.5. Hardness of TPNR/RWS Composites. The Shore hardness of the NR/HDPE blend and NR/HDPE composites with different RWS contents and sizes is shown in Figure 8. It was

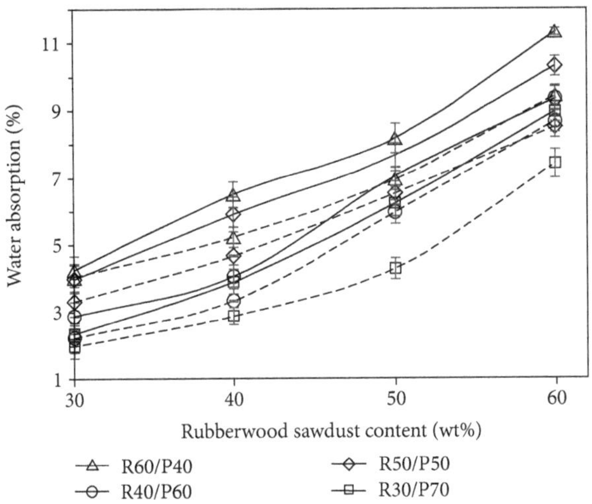

FIGURE 9: Effect of wood sawdust contents and sizes on water absorption for TPNR/RWS composites. Dashed lines show RWS 40 mesh, and solid lines represent RWS 80 mesh.

found that the Shore hardness of NR/HDPE blends increased with an increase of HDPE contents; the NR/HDPE blend of R30/P70 ratio has Shore hardness higher about 80% than the blend ratios of R60/P40. This is because the Shore hardness of natural rubber is low (5 Shore D) as compared to pure HDPE (65 Shore D), leading to higher flexibility TPNR. Moreover, it can be seen that the Shore hardness of the composites progressively increased as wood sawdust content increases in the TPNR/RWS composites. The reason is the wood sawdust has a significantly higher hardness than the thermoplastic elastomer matrix, and the addition of wood particles into TPNR phases decreases the flexibility or elasticity of polymer chains, resulting in more rigid composites [2]. Jamil et al. [2] reported that the presence of more natural rubber in the thermoplastic elastomer results in less Shore hardness.

For NR/HDPE blends of R60/P40 and R50/P50 ratios, the addition of RWS 40 mesh (dashed lines) into the TPNR composites seems to result in higher Shore hardness compared to the composites with RWS 80 mesh (solid lines); however, the TPNR composites reinforcing with RWS 40 mesh gave lower Shore hardness than those composites with RWS 80 mesh for the blend ratio of R40/P60 and R30/P70. This could be attributed to more homogeneous dispersion of the small particle size in the polymer matrix, which results in minimum voids [13]. Likewise, a stronger interfacial bonding between the wood filler and the polymer matrix can improve the Shore hardness of the composites.

3.6. Water Absorption of TPNR/RWS Composites. The water absorption of the TPNR/RWS composites was monitored by full water immersion for 24 hours as shown in Figure 9. Normally, the water absorption of the wood-polymer composites increased with wood sawdust content. It was found in the current work that as the amount of rubberwood sawdust increased, the water absorption of TPNR/RWS composites steadily increased due to the hydrophilic nature of wood sawdust, while the nature of the polymer is hydrophobic [30]. An increment of wood sawdust content increases the free OH groups in the composites, in which the formation of the free OH groups interacts with the H groups from water [30], resulting in the weight gain of the wood-polymer composites [31]. Additionally, an increase of wood sawdust content in the composites resulted more voids and poorer interfacial adhesion between the wood and the polymer matrix and consequently higher water absorption. Wan Busu et al. [30] claimed that the good interfacial bonding between wood particles and polymer can prevent the water molecules infiltrating into the composite system.

The water absorption of TPNR/RWS composites with particle sizes of 80 mesh (solid lines) is higher than that of 40 mesh (dashed lines), which is likely due to harder encapsulation of RWS 80 mesh into the TPNR matrix, which leads to pores in the composite structure. Likewise, the finer wood flour has a much larger surface area per weight [32], and therefore, it has more areas to contact with water molecules. Furthermore, at the same RWS contents and sizes, the TPNR composites with blend ratios of higher plastic content yielded lower water absorption. For example, with a RWS size of 40 mesh, the NR/HDPE blend of the R60/P40 ratio reinforced with 40 wt% RWS showed higher water absorption about 81% than that of the R30/P70 ratio with 40 wt% RWS. This higher water absorption is attributed to an increment of NR content in the plastic which results in larger holes and increases voids in the structure of the TPNR [33]. Therefore, the TPNR with higher rubber content allows easier penetration of water into the composite structure as well as an increase of water residence sites, resulting in larger water absorption.

4. Conclusions

The study has revealed that the RWS size and content as well as the ratio of NR/HDPE blends significantly affect the thermal, physical, and mechanical properties of the composites. The addition of RWS about 30–50 wt% improved the tensile strength and MOR of the composites from the NR/HDPE blend of R60/P40 and R50/P50 ratios, while the modulus and Shore hardness increased with an addition of RWS in the TPNR composites due to a stiffer wood sawdust than the polymer matrix. However, the incorporation of RWS into the TPNR composites negatively affected the thermal and physical properties. An increase of RWS content in TPNR composites decreased the thermal stability and increased the water absorption. In addition, TPNR composites with RWS 80 mesh gave better tensile strength and MOR than the composites with RWS 40 mesh due to less gaps between the wood sawdust and the TPNR matrix. The TPNR/RWS composites with larger plastic content yielded higher thermal stability and tensile, flexural, and Shore hardness properties but lower water absorption. SEM micrographs of the sample also revealed that the microgap formation was reduced by the addition of RWS 80 mesh in the TPNR composites. The TPNR/RWS composites from the NR/HDPE blend of R30/P70 ratio were therefore suggested for applications requiring high strength, modulus, hardness, and low water absorption.

The TPNR composites with a large RWS size is appropriate for applying in conditions of modulus and water resistances. The overall result of the composition will greatly facilitate the development of building materials from thermoplastic elastomer composites.

Conflicts of Interest

The authors declare that they have no conflicts of interest.

Acknowledgments

The authors would like to express their thanks to the Thailand Research Fund and National Research Council of Thailand (Grant no. RDG5950008) for the financial support throughout this work. Many thanks go to Rajamangala University of Technology Srivijaya (RMUTSV) for the research facilities and other supports. The authors also want to convey their thanks to Dr. Marwan Affandi from the Faculty of Engineering, RMUTSV, who has checked the paper.

References

[1] D. J. Elliot, "Developments with thermoplastic natural rubber blends," *NR technology*, vol. 12, no. 3, 59 pages, 1981.

[2] M. S. Jamil, I. Ahmad, and I. Abdullah, "Effects of rice husk filler on the mechanical and thermal properties of liquid natural rubber compatibilized high-density polyethylene/natural rubber blends," *Journal of Polymer Research*, vol. 13, no. 4, pp. 315–321, 2006.

[3] L. J. Yu, S. H. Ahmad, I. Kong, M. A. Tarawneh, and M. H. Flaifel, "Preparation and characterisation of NiZn ferrite/multiwalled nanotubes thermoplastic natural rubber composite," *International Journal of Materials Engineering Innovation*, vol. 4, no. 3-4, p. 214, 2013.

[4] I. Ahmad, R. Ismail, and I. Abdullah, "Effects of PBO fiber and clay on the mechanical, morphological, and dynamic mechanical properties of NR/HDPE blends," *Polymer Engineering and Science*, vol. 51, no. 3, pp. 419–425, 2011.

[5] H. Anuar and A. Zuraida, "Improvement in mechanical properties of reinforced thermoplastic elastomer composite with kenaf bast fibre," *Composites Part B, Engineering*, vol. 42, no. 3, pp. 462–465, 2011.

[6] C. Homkhiew, T. Ratanawilai, and W. Thongruang, "Composites from recycled polypropylene and rubberwood flour: effects of composition on mechanical properties," *Journal of Thermoplastic Composite Materials*, vol. 28, no. 2, pp. 179–194, 2015.

[7] S. A. Bahari, W. J. Grigsby, and A. Krause, "Flexural properties of PVC/bamboo composites under static and dynamic-thermal conditions: effects of composition and water absorption," *International Journal of Polymer Science*, vol. 2017, Article ID 2717848, 8 pages, 2017.

[8] C. Srivabut, T. Ratanawilai, and S. Hiziroglu, "Effect of nanoclay, talcum, and calcium carbonate as filler on properties of composites manufactured from recycled polypropylene and rubberwood fiber," *Construction and Building Materials*, vol. 162, pp. 450–458, 2018.

[9] N. D. Vu, H. T. Tran, and T. D. Nguyen, "Characterization of polypropylene green composites reinforced by cellulose fibers extracted from rice straw," *International Journal of Polymer Science*, vol. 2018, Article ID 1813847, 10 pages, 2018.

[10] Z. Li, X. Zhou, and C. Pei, "Effect of sisal fiber surface treatment on properties of sisal fiber reinforced polylactide composites," *International Journal of Polymer Science*, vol. 2011, Article ID 803428, 7 pages, 2011.

[11] Z. Luo, P. Li, D. Cai et al., "Comparison of performances of corn fiber plastic composites made from different parts of corn stalk," *Industrial Crops and Products*, vol. 95, pp. 521–527, 2017.

[12] N. Petchwattana and S. Covavisaruch, "Mechanical and morphological properties of wood plastic biocomposites prepared from toughened poly (lactic acid) and rubber wood sawdust (Hevea brasiliensis)," *Journal of Bionic Engineering*, vol. 11, no. 4, pp. 630–637, 2014.

[13] T. Ratanawilai, K. Nakawirot, A. Deachsrijan, and C. Homkhiew, "Influence of wood species and particle size on mechanical and thermal properties of wood polypropylene composites," *Fibers and Polymers*, vol. 15, no. 10, pp. 2160–2168, 2014.

[14] S. Y. Leu, T. H. Yang, S. F. Lo, and T. H. Yang, "Optimized material composition to improve the physical and mechanical properties of extruded wood-plastic composites (WPCs)," *Construction and Building Materials*, vol. 29, pp. 120–127, 2012.

[15] P. Y. Kuo, S. Y. Wang, J. H. Chen, H. C. Hsueh, and M. J. Tsai, "Effects of material compositions on the mechanical properties of wood-plastic composites manufactured by injection molding," *Materials and Design*, vol. 30, no. 9, pp. 3489–3496, 2009.

[16] J. W. Kim, D. P. Harper, and A. M. Taylor, "Effect of wood species on the mechanical and thermal properties of wood–plastic composites," *Journal of Applied Polymer Science*, vol. 112, no. 3, pp. 1378–1385, 2009.

[17] N. M. Stark and L. M. Matuana, "Characterization of weathered wood-plastic composite surfaces using FTIR spectroscopy, contact angle, and XPS," *Polymer Degradation and Stability*, vol. 92, no. 10, pp. 1883–1890, 2007.

[18] D. Harper and M. Wolcott, "Interaction between coupling agent and lubricants in wood-polypropylene composites," *Composites Part A: Applied Science and Manufacturing*, vol. 35, no. 3, pp. 385–394, 2004.

[19] B. B. Boonstra, "Reinfocerment by filler, rubber technology and manufacture," in *Butterworth*, C. M. Blow and C. Hepburn, Eds., Plastic and Rubber Institute, London, 1987.

[20] N. Iqbal, J. F. Mousumi, and M. A. Islam, "Effect of fiber length on properties of jute fiber reinforced polymer matrix composite," *International Journal of Industrial Engineering*, vol. 1, no. 6, pp. 201–207, 2017.

[21] S. Mohanty, S. K. Verma, and S. K. Nayak, "Dynamic mechanical and thermal properties of MAPE treated jute/HDPE composites," *Composites Science and Technology*, vol. 66, no. 3-4, pp. 538–547, 2006.

[22] Y. A. El-Shekeil, S. M. Sapuan, K. Abdan, and E. S. Zainudin, "Influence of fiber content on the mechanical and thermal properties of kenaf fiber reinforced thermoplastic polyurethane composites," *Materials and Design*, vol. 40, pp. 299–303, 2012.

[23] T. Ratanawilai, N. Thanawattanasirikul, and C. Homkhiew, "Mechanical and thermal properties of oil palm wood sawdust reinforced post-consumer polyethylene composites," *Science Asia*, vol. 38, no. 3, pp. 289–294, 2012.

[24] T. Jeevananda, "Thermal and morphological studies on ethylene-vinyl acetate copolymer–polyaniline blends," *Thermochimica Acta*, vol. 376, no. 1, pp. 51–61, 2001.

[25] M. A. Syed, S. Akhtar, Siddaramaiah, and A. A. Syed, "Studies on the physico-mechanical, thermal, and morphological behaviors of high density polyethylene/coleus spent green composites," *Journal of Applied Polymer Science*, vol. 119, no. 4, pp. 1889–1895, 2011.

[26] A. Ashori, "Wood–plastic composites as promising green-composites for automotive industries!," *Bioresource Technology*, vol. 99, no. 11, pp. 4661–4667, 2008.

[27] S. Mohanty and S. K. Nayak, "Interfacial, dynamic mechanical, and thermal fiber reinforced behavior of MAPE treated sisal fiber reinforced HDPE composites," *Journal of Applied Polymer Science*, vol. 102, no. 4, pp. 3306–3315, 2006.

[28] M. R. Rahman, M. M. Huque, M. N. Islam, and M. Hasan, "Improvement of physico-mechanical properties of jute fiber reinforced polypropylene composites by post-treatment," *Composites Part A: Applied Science and Manufacturing*, vol. 39, no. 11, pp. 1739–1747, 2008.

[29] M. Kociszewski, C. Gozdecki, A. Wilczynski, S. Zajchowski, and J. Mirowski, "Effect of industrial wood particle size on mechanical properties of wood-polyvinyl chloride composites," *European Journal of Wood and Wood Products*, vol. 70, no. 1–3, pp. 113–118, 2012.

[30] W. N. Wan Busu, H. Anuar, S. H. Ahmad, R. Rasid, and N. A. Jamal, "The mechanical and physical properties of thermoplastic natural rubber hybrid composites reinforced with *Hibiscus cannabinus, L* and short glass fiber," *Polymer-Plastics Technology and Engineering*, vol. 49, no. 13, pp. 1315–1322, 2010.

[31] K. B. Adhikary, S. Pang, and M. P. Staiger, "Long-term moisture absorption and thickness swelling behaviour of recycled thermoplastics reinforced with pinus radiata sawdust," *Chemical Engineering Journal*, vol. 142, no. 2, pp. 190–198, 2008.

[32] A. Nourbakhsh and M. Kouhpayehzadeh, "Mechanical properties and water absorption of fiber-reinforced polypropylene composites prepared by bagasse and beech fiber," *Journal of Applied Polymer Science*, vol. 114, no. 1, pp. 653–657, 2009.

[33] C. Nakason, S. Jamjinno, A. Kaesaman, and S. Kiatkamjornwong, "Thermoplastic elastomer based on high-density polyethylene/natural rubber blends: rheological, thermal, and morphological properties," *Polymers for Advanced Technologies*, vol. 19, no. 2, pp. 85–98, 2008.

3

Production and Modification of Cellulose Nanocrystals from *Agave tequilana* Weber Waste and its Effect on the Melt Rheology of PLA

Soledad-Cecilia Pech-Cohuo,[1] Gonzalo Canche-Escamilla,[1] Alex Valadez-González,[1] Victor Vladimir Amilcar Fernández-Escamilla,[2] and Jorge Uribe-Calderon[1]

[1]*Centro de Investigacion Cientifica de Yucatan, Calle 43 No. 130 entre 32 y 34. Col. Chuburna de Hidalgo, CP 97205 Mérida, YUC, Mexico*

[2]*Departamento de Ciencias Tecnológicas, Centro Universitario de la Ciénega, Universidad de Guadalajara, Av. Universidad No. 1115, 47819 Ocotlán, JAL, Mexico*

Correspondence should be addressed to Jorge Uribe-Calderon; jorge.uribe@cicy.mx

Academic Editor: Hossein Roghani-Mamaqani

The production and surface modification of cellulose nanocrystals (CNC) from *Agave tequilana* Weber waste (ATW) and their effect on the melt rheology of PLA were investigated. For that, CNC were obtained from ATW residues by acid hydrolysis and grafted with 2-ethyl hexyl acrylate (2-EHA) to promote compatibility with nonpolar polymers, such as PLA. The morphology of CNC occurred as rods, spheres, and the so-called porous network observed by scanning electron microscopy (SEM) and electron transmission microscopy (TEM), and the crystallinity of the cellulose fibers was ~73% and increased up to ~94% for CNC followed by X-ray diffraction (XRD). The acid hydrolysis and alkali extraction process produce changes on the native cellulose type I from ATW bagasse resulting in a mixture of cellulose type I and II and was preserved after CNC isolation and surface grafting. Polymer grafting was assessed using Fourier-transform infrared spectroscopy (FTIR) and X-ray electron spectroscopy (XPS). Upon surface modification, the grafted CNC presented better thermal stability than CNC pristine evaluated by thermogravimetric analysis (TGA), and the surface energy of the CNC was decreased which could promote CNC dispersion within polymers. Both pristine and surface modified CNC were melt mixed with PLA to produce nanocomposites at different concentrations of CNC. Dynamic time sweep tests showed that the introduction of CNC and CNC/2-EHA decreased the complex viscosity of PLA due to polymer degradation during compounding.

1. Introduction

Traditionally, CNC have been mainly obtained by acid treatments on cellulose obtaining crystalline nanoparticles virtually free of defects [1]. The conventional sources of cellulose for the production of CNC are wood and cotton fibers; however, crop wastes are alternative sources for obtaining CNC [2–5]. Thus, CNC has been produced from crop waste such as rice [6], banana [2], and coconut [5].

On the other, the blue agave (*Agave tequilana* Weber) is extensively cultivated for the production of the alcoholic drink denominated tequila. The agave bagasse is the result of agave heads cooked, crushed, and sugar extracted having an estimated cellulose content about 49 wt.% [7] with an estimated production of ~400,000 ton/year [8] which represents serious environmental problems [9]. There are very few studies concerning the CNC production from *Agave tequilana* Weber bagasse with promising results [10].

CNC can be used as a reinforcer for polymer matrices because of their excellent mechanical properties and unique morphology [11]. However, surface modification of CNC is required to improve its dispersion within hydrophobic matrices and to promote adhesion at the CNC-polymer interface [12, 13]. In this sense, CNC has been grafted with acetic anhydride [14], fatty acids [15, 16], poly(glycidyl methacrylate) [17], and other acrylic monomers to produce nanocomposites based on hydrophobic polymers such as low-density polyethylene (LDPE) and PLA [18]. Polymer nanocomposites containing modified CNC exhibited usually improved properties in comparison with those containing pristine CNC.

On the other hand, rheological characterization of polymer nanocomposites is a powerful technique to evaluate the dispersion of nanoparticles and to investigate the influence of the processing conditions of nanocomposites on the dispersion of the nanoparticles in the polymer matrices; since rheological properties depend strongly on the structure, surface characteristics, size, and shape of the nanoparticles [19].

Some studies have been conducted regarding the rheological behavior of PLA/pristine CNC nanocomposites reporting a significant increase in the complex viscosity, storage, and loss modulus for nanocomposites compared with pure PLA [20, 21]. Furthermore, it has been reported that the use of grafted CNC (i.e., hydrophobic grafts) improved the rheological properties of the nanocomposites due to the better dispersion of CNC within the polymer matrix [22].

In this study, CNC were obtained from ATW bagasse by acid hydrolysis and surface modified with poly(2-ethyl hexyl acrylate) to produce a biodegradable reinforcer for polymer matrices [23]. Poly(2-ethyl hexyl acrylate), a biocompatible polymer, was chosen due to its ability to hydrolysis degradation [23]; in addition, the monomers are able to react with the -OH groups from nanocellulose through free radical graft copolymerization [18]. The main physicochemical characteristics of pristine CNC and grafted CNC/2-EHA were determined. The dispersion of pristine and modified CNC within melt compounded PLA nanocomposites was assessed by SEM and TEM. In addition, the influence of CNC loading and surface modification of CNC on the complex viscosity as a function of time of PLA nanocomposites was evaluated.

2. Experimental

2.1. Materials. Raw fibers of ATW bagasse (Sol y Agave de Arandas, SA de CV, Jalisco, Mexico) were used to obtain cellulose using sulfuric acid (Fermont), sodium hydroxide (Fermont), and commercial sodium hypochlorite solution (NaClO). Inhibitor-free 2-EHA (Aldrich) and cerium and ammonium nitrate (CAN) (Aldrich) were used in the grafting reaction. PLA (NatureWorks 2002D) was used to prepare nanocomposite ($\rho = 1.25\,g/cm^3$, MFI = ~4–8 g/10 min at 190°C, according to the technical data sheet).

2.2. Procedures

2.2.1. Production of CNC Form ATW Bagasse. First, cellulose fibers were isolated from ATW bagasse fibers by acid hydrolysis and alkaline extraction [2, 24]: briefly, 1 g of ATW bagasse was mechanically dispersed in 19 mL of boiling H_2SO_4 solution (1% vol.) for 1 h and subsequently washed with distilled water; then 1 g of the above fibers was mechanically mixed with 10 mL of NaClO solution at room temperature (3.5 vol.%) until reaching a pH of ~9.2 and washed with distilled water until reaching a pH of ~7. Next, 1 g of the above fibers was mechanically mixed with 10 g of NaOH solution (20 wt.%) for 1 h and washed with distilled water; and finally, 1 g of the above fibers was bleached with 12 mL of NaClO solution at room temperature (0.5 vol.%) for 1 h and washed until reaching a pH of ~7. The resulting material was dried at room temperature for 24 h.

CNC were produced following the procedures reported by Bondeson et al. [1] and Hamad and Hu [25]: 10.2 g of cellulose was dispersed mechanically in 100 mL of deionized water (DI) in an ice bath, and then concentrated H_2SO_4 was slowly added until reaching a concentration of 63.5 wt.%. Subsequently, stirring was increased and the mixture was heated (44°C) for 130 min. The reaction was stopped by pouring 2000 mL of cold DI water. The suspension was allowed to precipitate for 24 h producing two phases. The clear phase was discharged and the turbid phase was washed with 2000 mL deionized water and centrifuged (15 min at 3500 rpm); this process was repeated three times. Supernatants from the second and third centrifugation cycles were collected and dialyzed against DI water using dialysis membranes (molecular weight 14,000 Da) until a neutral pH (1–4 days) was reached. Samples were sonicated using an ultrasonic probe (Sonics, Vibra-Cell, 5 min in an ice bath); the suspension was stored at 4°C and lyophilized to determine the concentration of the suspension.

2.2.2. 2-EHA Grafting on CNC. The 2-EHA grafting was performed using the procedure proposed by Canché-Escamilla et al. [26]: 250 mL of the CNC suspension (~0.3% *w/v*) was nitrogen bubbled for 1 h at 30°C. Subsequently, 0.75 g of 2-EHA (2-EHA : CNC, 1 : 1) and 0.8 g of CAN (dissolved in 4.6 mL of 1 N HNO_3) was added and stirred for 3 h at 30°C. The grafted CNC are washed/centrifuged with DI water three times. The suspension was frozen and lyophilized. Subsequently, the product was mixed with acetone and stirred for 24 h to dissolve the ungrafted homopolymer. The suspension was filtered and dried in a vacuum oven for 24 h at 50°C.

2.2.3. PLA/CNC Nanocomposites. PLA/CNC nanocomposites were obtained in two steps. First, a PLA/CNC (pristine or grafted) master batch (CNC 10 wt.%) was produced in a Brabender mixing chamber (model PLE-330) at 163°C and 30 rpm for 7 min. The masterbatch was then diluted with virgin PLA (0.5, 1, and 2 CNC wt.%) using a Brabender conical twin-screw extruder (model PLE-330). Processing temperatures were 150, 150, and 145°C in the extruder zones, the screws speed was 5 rpm, and a circular die (5 mm in diameter) was used at 140°C. Samples were compression molded (Carver laboratory press, at 180°C for 5 min, 2043 kPa clamping force for 5 min, water cooled) to produce 120 × 120 × 1 mm plates for further characterization.

2.2.4. Characterization. The CNC morphology was determined by SEM on dried and lyophilized samples using metal supports and copper tape. The samples were coated with a thin gold layer and observed in a SEM (JEOL JSM-6360, Japan, 10 kV). The CNC were also observed on a JEOL 1010 transmission electron microscope (TEM), at an

Table 1: Contact angles (°), dispersive (γ_s^d), polar (γ_s^p) components, and surface energy (γ_s) of CNC and CNC/2-EHA.

Sample	Contact angle θ (°) Diiodomethane	Ethylene glycol	Water	γ_s^d (mJ/m^{-2})	γ_s^p (mJ/m^{-2})	γ_s (mJ/m^{-2})
CNC	26.5 ± 1.6	27.2 ± 4.5	44.8 ± 5.4	30.7	22.9	53.6
CNC/2-EHA	25.2 ± 2.7	52.7 ± 9.0	73.2 ± 1.4	36.8	5.5	42.3

γ_s^d, γ_s^p, and γ_s correspond to the dispersive and polar components and the total surface tension, which are 49.5, 1.3, and 50.8 mJ/m^{-2} for diiodomethane; 29, 19, and 48 mJ/m^{-2} for ethylene glycol; and 21.8, 51, and 72.8 mJ/m^{-2} for water, respectively [14].

accelerating voltage of 100 kV, to examine the shape and size of dried CNC. A drop of the diluted solution (0.001% *w/v*) was deposited onto a carbon-coated copper grid and dried afterwards at room temperature for 12 h before the observation in the microscope. The morphology of PLA/CNC nanocomposites was observed thru the cryogenic fracture surface by SEM. The CNC embedded into the PLA matrix were also examined by TEM. Thin composite samples sliced with a microtome were deposited onto copper grids for observation in the microscope. The particle dimensions were determined using the Image-Pro Plus 6 software.

XRD patterns of the samples were determined using a Diffractometer Bruker D8 Advance (CuKα, $\lambda = 1.5418$ Å, 40 keV and 30 mA), and the crystallinity index was calculated using the Segal equation [27]:

$$X_c\% = \frac{I_{200} - I_{am}}{I_{200}} \times 100, \tag{1}$$

$$X_c\% = \frac{I_{1\bar{1}0} - I_{am}}{I_{1\bar{1}0}} \times 100, \tag{2}$$

where I_{200} is the maximum intensity of the (200) lattice diffraction ($2\theta = 22.6°$ in the case of cellulose I), $I_{1\bar{1}0}$ is the maximum intensity of the ($1\bar{1}0$) diffraction for cellulose II at $2\theta = 20.1°$, and I_{am} is the intensity of the diffraction peak at $2\theta = 18°$ (cellulose I) and $2\theta = 16°$ (cellulose II) [28, 29].

FTIR analysis was performed using a Thermo Scientific Model Nicolet 8700 spectrometer in the transmission mode, with a resolution of 4 cm^{-1} and 100 scans on 2 mg sample/135 mg KBr tablets. The XPS analysis was performed using Thermo Scientific K-Alpha Surface Analysis equipment, using a monochromatic Kα line, at 40 W and 12 kV.

The thermal stability of the CNC was determined on a Perkin-Elmer TGA 7 equipment. Samples were heated from 30°C to 700°C with a heating rate of 10°C/min under a nitrogen flow (100 mL/min).

The Owens-Wendt approximation was used to calculate the dispersive and polar contributions of the surface energy of the samples using the contact angles of sessile drops of diiodomethane, ethylene glycol, and water on tablets of pristine and grafted CNC. The liquid surface tension values were taken from literature [14] and are listed in the footnote for Table 1. 100 mg CNC tablets were obtained using a KBr die applying a compression force of 30 kN for 15 min at room temperature. Contact angle measurements were performed at room temperature using a goniometer (Ramé-Hart Instrument Co.).

Dynamic time-sweep tests were carried out using an AR-2000 rotational stress-controlled rheometer (TA Instruments, New Castle, DE, USA) employing the parallel plate geometry of 25 mm. The linear viscoelastic region (LVR) was initially confirmed by strain amplitude sweeps at constant frequency. Time-sweep experiments at strain amplitude of 0.05 and frequency of 0.5 rads^{-1} were conducted at 165°C for 3600 s using a gap between plates of 0.7 mm inside LVR. Specimens (25 mm disks) were obtained from compression molded plate.

3. Results and Discussion

The cellulose yield from ATW bagasse was 38% and for the CNC was 28% from obtained cellulose (10% from ATW bagasse); the above are similar to the study reported by Bondeson et al. [1] and Hamad and Hu [25], who used similar processing conditions to those used in this study but from Picea Norway microcrystalline cellulose and commercial softwood, both were CNC cellulose type I. The conversion of 2-EHA was 7.2%, the addition percentage was 5.4%, and the grafting efficiency was 81.5%. The reduced reactivity of 2-EHA [30] could contribute to the low 2-EHA conversion. The relative high grafting efficiency is due to the Ce(IV) ions which generate mainly free radicals on the backbone of cellulose [31, 32].

Figure 1(a) shows SEM photographs of a raw ATW fiber which is composed by several cellulose microfibers; after the acid hydrolysis and alkaline extraction, the cellulose fibers were obtained (Figure 1(b)). Figure 1(c) shows clusters of pristine CNC dried directly from a CNC solution; a close-up reveals that the CNC are stacked parallel among them (Figure 1(d)). Similarly, a sample of the CNC/2-EHA was observed after surface modification process, CNC/2-EHA were identified as aggregates covered by the grafted polymer (red dotted circles on Figures 1(e) and 1(f)). In addition, SEM was performed on lyophilized samples to determine possible changes in the morphology of CNC due to drying process; surprisingly, pristine CNC are seen as a lamellar structure (Figure 1(g)), which consists of a large number of thin CNC layers (~1.7 μm thick). No CNC agglomerates or microsize fibers were observed within the thin sheets suggesting that CNC were self-assembled into thin sheets during the lyophilization process; similar observations were reported by Li et al. [33] for cellulose I CNC. The self-assembly of CNC is produced due to the confinement of CNC at the boundaries of ice crystal during freezing process favoring the formation of a chiral nematic structures; the factors affecting

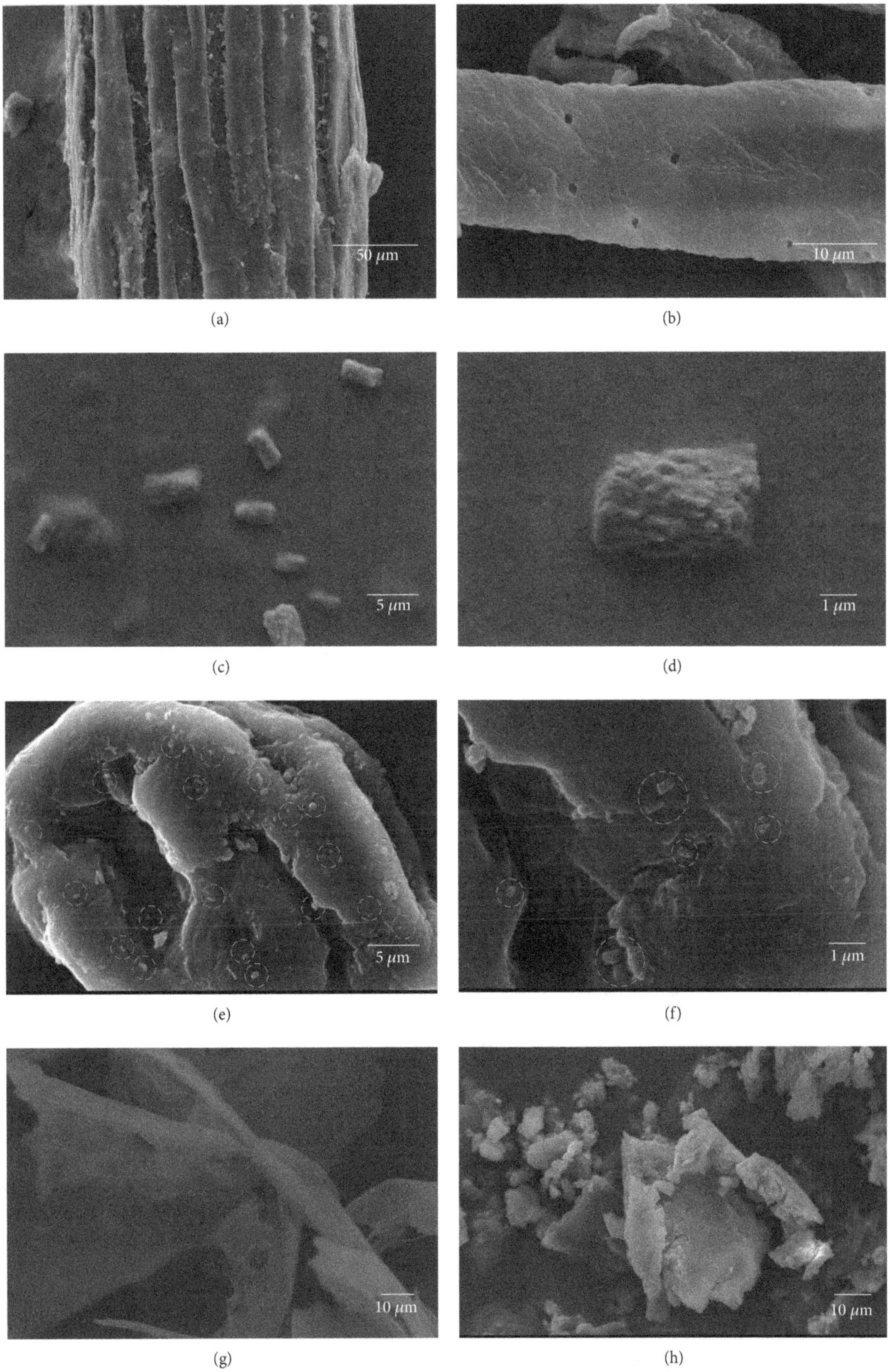

FIGURE 1: SEM micrographs of (a) raw ATW fiber, (b) extracted cellulose fiber, (c, d) CNC, (e, f) CNC/2-EHA, (g) lyophilized CNC, and (h) lyophilized CNC/2-EHA.

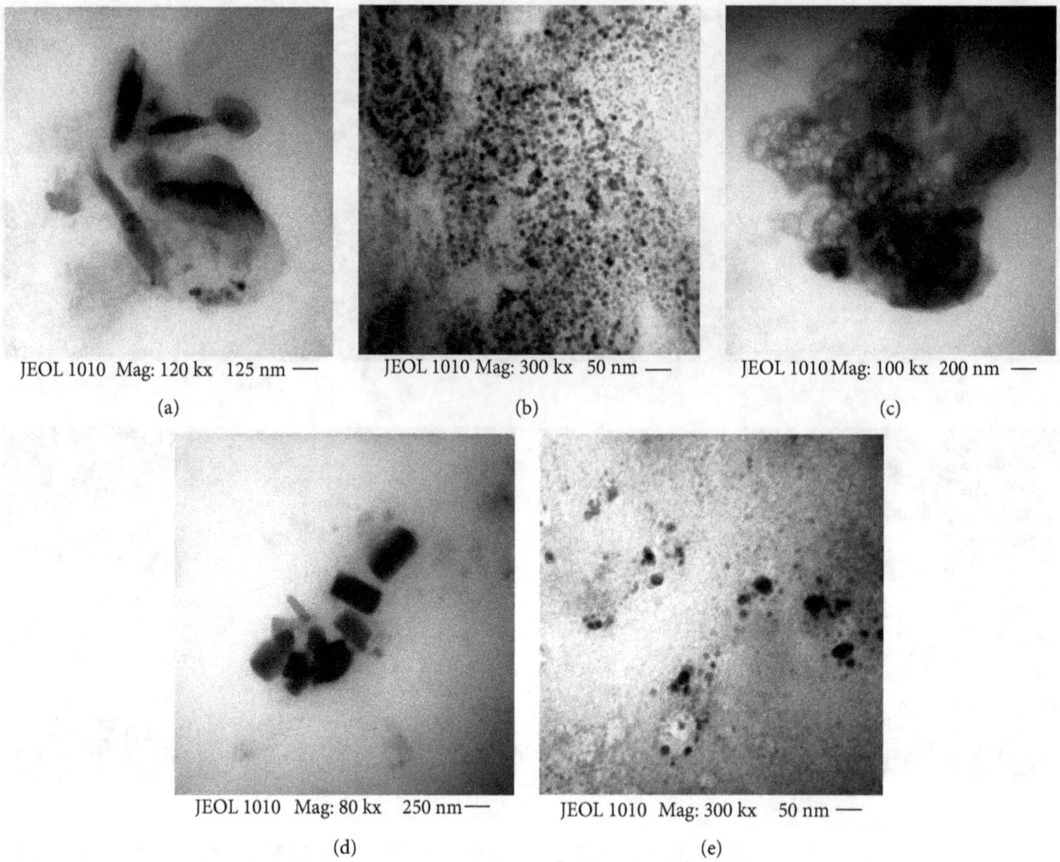

FIGURE 2: TEM micrographs of CNC (a–c), and CNC/2-EHA (d, e) exhibiting different morphologies.

the production of lamellar structure are the aspect ratio, concentration, hydrogen bonds, crystallinity, and surface charge of CNC [34].

A completely different morphology is observed for the CNC/2-EHA (Figure 1(h)) upon lyophilization, which is presented as aggregates of nanoparticles; no self-assembly was observed as pristine CNC due to the reduction of hydrogen bonds as a result of the polymer grafting.

Morphological details were obtained by TEM; Figure 2 shows TEM images of the pristine CNC and CNC/2-EHA. The results indicate that pristine CNC exhibited three different morphologies: rods (Figure 2(a)) with length ranging 590 nm and diameter ranging 100 nm, spheres (Figure 2(b)) with apparent diameters in the order of 20 nm, and the so-called porous network (Figure 2(c)) with apparent diameter ranging 80 nm. The former morphologies has been related to the self-assembly process of short cellulose rods through interfacial hydrogen bonds and depends on the preparation method of cellulose fiber [35], source of cellulose, and production process of CNC [4, 36]. On the other hand, the formation of the porous network morphology is the result of the combination of several factors such as the lyophilization process, degree of dispersion of the CNC in water, and drying in the preparation of TEM sample [36, 37]. A more detailed study is required to determine the proportion of each kind of morphology in the obtained CNC; it is worthy to mention that the isolation of CNC fractions exhibiting different morphology is not possible by traditional methods of separation such as filtration or centrifugation so far [36, 37].

Our results are quite different from the those reported by Espino et al. who produced CNC from ATW bagasse fiber in the form of rods with lengths of 335 ± 165 nm and diameters of 12 ± 4 and by means of acid hydrolysis [10], it is possible that the high shearing during preparation reduced the CNC size. On the other hand, two morphologies were found for CNC/2-EHA: rods (Figure 2(d)) with lengths and apparent diameters in the order of 500 nm and ~140 nm, respectively and spheres (Figure 2(e)) with apparent diameters of ~20 nm. Rods and spheres of CNC/2-EHA were observed like separated units; surface modification prevented agglomeration as reported by others [38].

Surface grafting was assessed by FTIR and XPS. Figure 3 shows the FTIR spectra of CNC and CNC/2-EHA which present a band in the region of 3500 to 3400 cm^{-1} corresponding to the –OH groups of the cellulose; the intensity of this band decreased upon 2-EHA grafting. An intense band is observed in ~2897 cm^{-1} attributable to stretching vibrations of the C-H bonds for both CNC and CNC/2-EHA. In addition, CNC/2-EHA presented a band at ~2930 cm^{-1} corresponding to the characteristic stretching of the C-H bonds of the pendant groups [18].

The vibrations of deformation, wagging, and flexion of anhydroglucopyranose units are observed from 1800 to

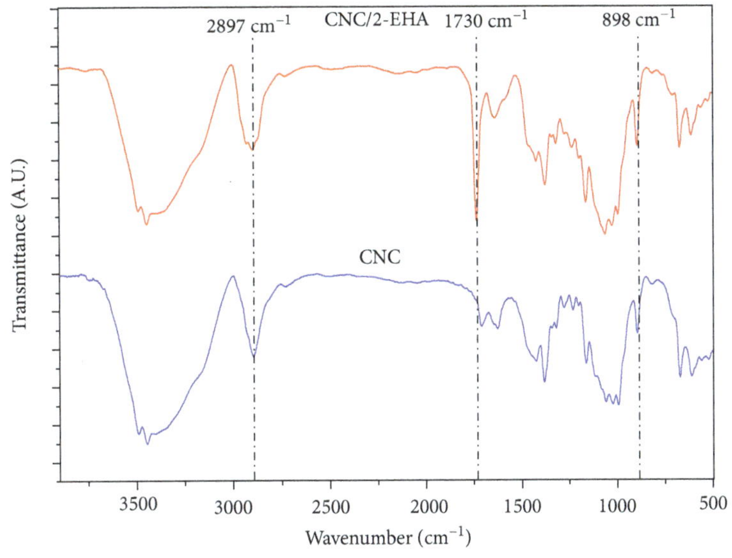

FIGURE 3: FTIR spectra of CNC and CNC/2-EHA.

800 cm^{-1} and are consistent with other reports [3, 5, 37, 39, 40]. The band at ~892–898 cm^{-1} is associated to β-glycosidic linkages of cellulose [3] and to cellulose type II [39]. The CNC/2-EHA spectrum exhibited a strong band at 1730 cm^{-1}, corresponding to the stretching of the carbonyl group of grafted 2-EHA [26, 41].

XPS was performed on the samples corroborate the 2-EHA grafting. Figure 4 shows the survey spectra (a) and the deconvolution of the C1s peak for the CNC (b) and CNC/2-EHA (c). Survey spectra present the peaks for O1s, C1s, and Sp2. The C1s peak deconvolution was carried out using the PeakFit v4.12 software and the Gaussian-Lorentzian method generating several groups: ***C1*** corresponding to C-C and C-H bonds, ***C2*** to C-O bonds of alcohols and ethers, ***C3*** to O-C-O and C=O bonds, and ***C4*** corresponding to O-C=O ester bond [15]. The O/C ratio of CNC was 0.54 which is similar to that reported by other authors [15, 42]; lower values of O/C could be attributable to impurities [43]. The O/C ratio for CNC/2-EHA was 0.45, which is lower than that CNC, due to the increase in the amount of carbon atoms from 2-EHA. On the other hand, the contribution of the ***C1*** group of CNC/2-EHA decreased slightly in comparison with CNC upon grafting, from 49.4% to 47.8%, possibly due to the impurities removing upon surface grafting [42].

The ***C2*** group exhibited a slight decrease in the CNC/2-EHA (30%) sample compared with CNC (29.4%) due to the loss of one glycol group of the anhydroglucopyranose at the position of carbons 2 or 3 as a consequence of grafting [44]. ***C3*** was observed to slightly increase after grafting from 13.3% for CNC to 14.3% for CNC/2-EHA, due to the formation of carbonyl groups in the cellulose unit and those from the grafted polymer chains [45]. The percentage of the bond type ***C4*** increased from 7.3% to 8.5% for CNC and CNC/2-EHA, respectively, due to the contribution of the ester groups of the poly(2-EHA). Sulfur is present in both samples as result from sulfate groups during acid hydrolysis; sulfate groups remained after the grafting reaction since grafting position of on C2 or C3 of anhydroglucose [44], while sulfate groups are joined to C6 position [37].

On the other hand, TGA was used to estimate the grafting content and thermal stability of samples. Figure 5 shows the TGA thermograms (a) and DTGA (b) for CNC and CNC/2-EHA. Figure 5(a) shows three major mass loss events attributable to the elimination of adsorbed water (~50–120°C), cellulose degradation (~269–384°C), and the generation of carbonized residues (~450°C). The thermal degradation of the cellulose occurs in two stages: (a) the reduction of the chain size due to the rupture of bonds by the generation of free radicals, carboxyl groups, and carbonyls and (b) the depolymerization of the cellulose by the rupture of the glycosidic units and the formation of levoglucosan [40]. Figure 5(b) shows the derivative TGA traces for CNC exhibiting two decomposition events associated to the decomposition of sulfonated CNC and CNC (~125–180°C and 398–448°C, resp.). CNC/2-EHA exhibited two decomposition events related to thermal decomposition of CNC and grafted polymer (270–395°C and 396–448°C, resp.), as reported by others [38]. The grafting of poly(2-EHA) on the CNC surface improved the thermal stability of CNC because the poly(alkyl acrylates) exhibit higher degradation temperature (405–415°C) [46]. The TGA results suggested that the 2-EHA grafted corresponded to 5 wt.%.

The change in the surface energy of CNC due to poly(2-EHA) grafting was evaluated through the Owens-Wendt approximation using the average of six measurements of the contact angles of sessile drops of various liquids on the surface of CNC samples at room temperature. The average values of the contact angles and the calculated surface energies are shown in Table 1. It was observed that the contact angle of diiodomethane was lower in the CNC/2-EHA compared to the CNC; in the case of ethylene glycol and water, an opposite behavior was observed. The droplet of water exhibited a larger contact angle when deposited on CNC/2-EHA which is consistent with the literature related

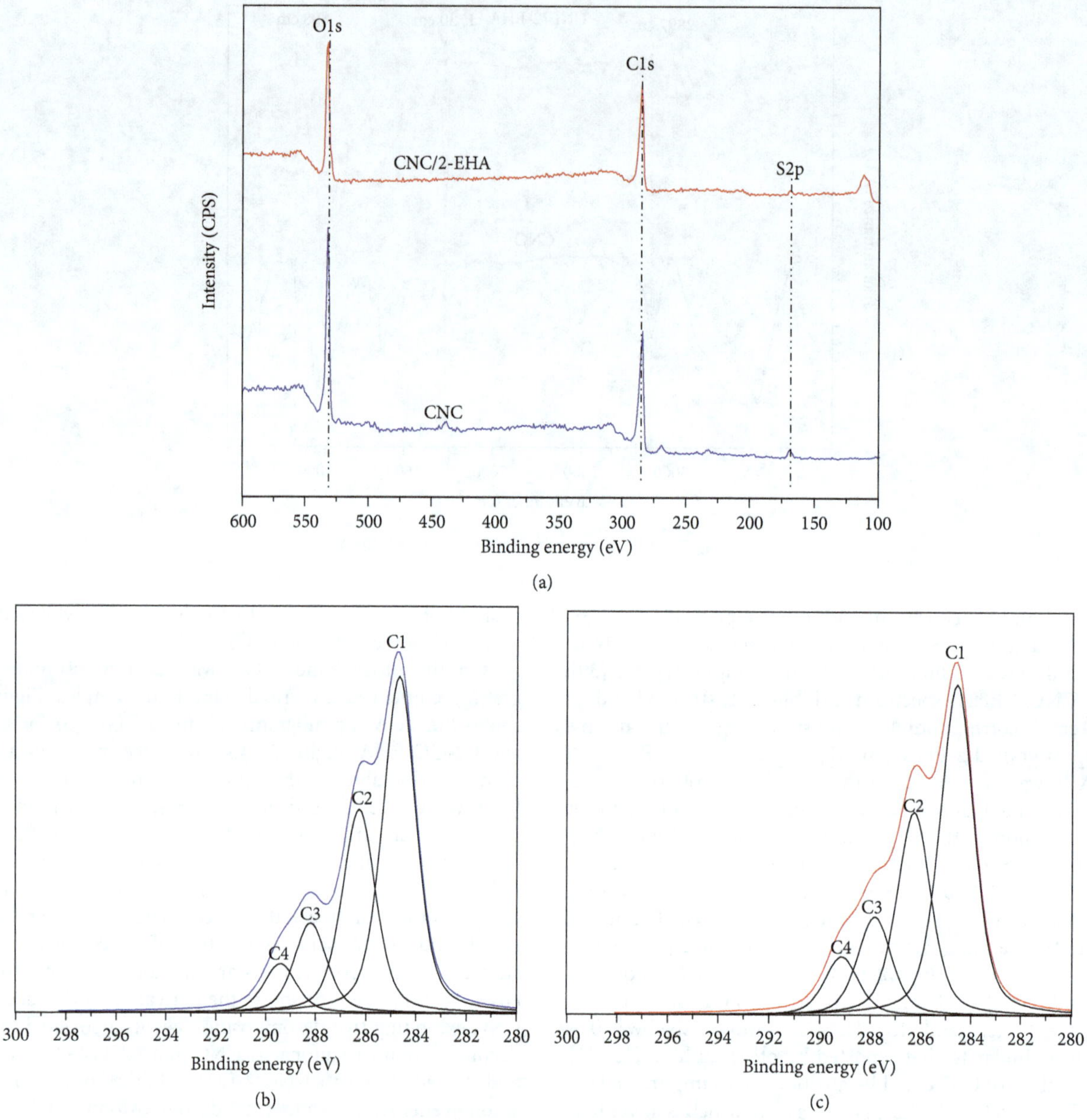

FIGURE 4: XPS survey spectra (a) and high resolution of C1s photoelectron spectra of CNC (b) and CNC/2-EHA (c).

to the surface-modified CNC [14, 15]. The R^2 values for the linear regression for Owens-Wendt model were 0.94 and 0.84 for CNC and CNC/2-EHA, respectively; low R^2 value for CNC/2-EHA could be attributed to the roughness effect of sample surface [47]. The roughness of samples was determined through AFM analysis (AFM Bruker model Innova SPM in tapping mode) to be 23.8 and 31.8 nm for CNC and CNC/2-EHA, respectively.

The calculated dispersive component increased from 30.7 to 36.8 mJ/m^{-2} upon grafting, whereas the polar component decreased from 22.9 mJ/m^{-2} to 5.5 mJ/m^{-2} as a result of the substitution of –OH by poly(2-EHA) and the formation of intramolecular hydrogen bonding of the carbonyl groups [14]. The decrease of the polar component and the increase of the dispersive component in the modified CNC suggest that CNC/2EHA could be better dispersed in hydrophobic polymer matrices such as PLA.

The changes in the crystallinity degree of the cellulose upon extraction and grafting processes were evaluated by XRD. Figure 6 shows XRD diffractograms for bagasse fiber *Agave tequilana* Weber (FATW), extracted cellulose, CNC, and CNC/2-EHA. FATW exhibited the characteristic peaks for cellulose type I ($2\theta = 15.2°$, $22.6°$), and the peak at $2\theta = 24.7°$ corresponds to silica [48] from the rich silica soils required for this type of crop [49]. The extracted cellulose shows 3 peaks at $2\theta \sim 12.4°$, $20.2°$, and $22.2°$ corresponding to the structure of cellulose type II [28, 50, 51]; however, it is possible to observe some overlapped peaks corresponding

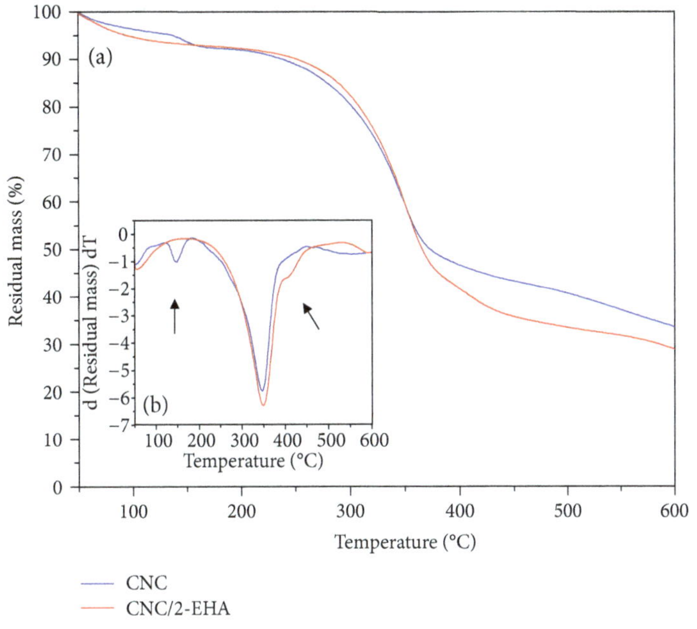

FIGURE 5: (a) TGA and (b) DTGA traces of CNC and CNC/2-EHA (nitrogen atmosphere, heating rate 10°C/min); arrows indicate the mass loss of water and grafted poly(2-EHA).

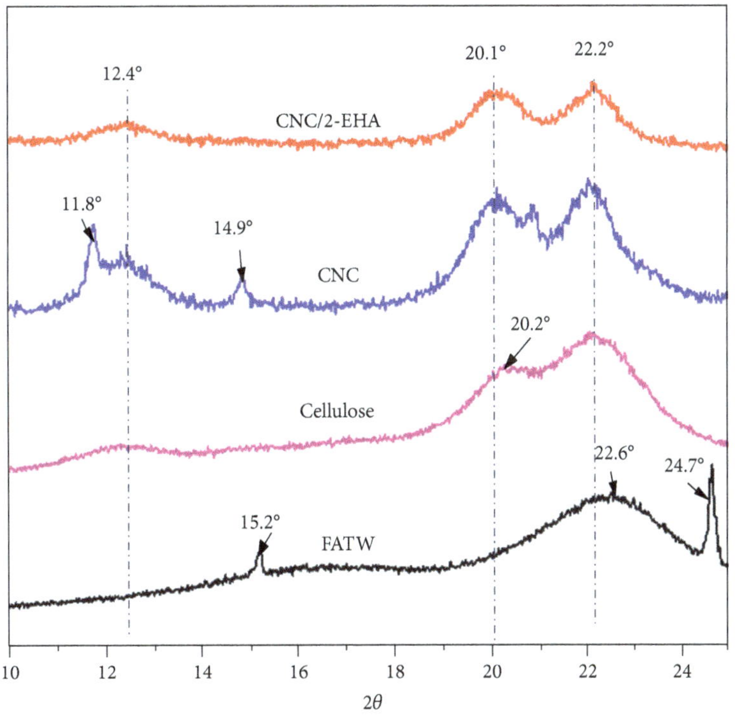

FIGURE 6: X-ray diffractograms of FATW, cellulose, CNC, and CNC/2-EHA.

to cellulose type I ($2\theta = 15.2°$, $22.6°$). The partial conversion from polymorphism I to II is due to alkali conditions (sodium hydroxide) and subsequent washing and drying processes employed to obtain cellulose [51]. CNC show both polymorphisms with defined peaks ($2\theta = 11.8°$, $12.4°$, $14.9°$, $20.1°$, $20.9°$, and $22.2°$); cellulose type I was preserved in part after the acid hydrolysis (isolation of CNC); other authors have obtained CNC exhibiting a mixture of cellulose type I and II by using different contents of H_2SO_4 during the isolation process [52]. CNC/2-EHA presents peaks at $2\theta = 12.2°$, $20.1°$, and $22.2°$ corresponding mainly to cellulose type II; however, it is possible that some peaks corresponding for cellulose I are overlapped within the amorphous zone. The intensity of the diffraction peaks for CNC/2-EHA is lower

TABLE 2: Crystallinity index for cellulose I and II and estimated percentage of the polymorphisms in the samples.

Sample	Crystallinity index (%)		Percentage of the polymorphisms (%)	
	Cellulose I	Cellulose II	Cellulose I	Cellulose II
FATW	64	—	100	—
Cellulose	74	73	57	43
CNC	89	94	48	52
CNC/2-EHA	69	69	49	51

due to the amorphous characteristic of poly(2-ethyl hexyl acrylate). Cellulose II is the most important crystalline form from a technical and commercial point of view, because it is thermodynamically more stable [51], and suspensions of cellulose type II CNC have been found to have better rheological properties than their type I counterparts [53].

Crystallinity index (Table 2) was calculated using corresponding peaks for cellulose I ($I_{am} = 18^\circ$, $I_{200}{\sim}22^\circ$) and II ($I_{am} = 16^\circ$, $I_{1\bar{1}0}{\sim}20^\circ$). FATW presents low values of crystallinity corresponding to cellulose type I due to the presence of lignin and other organic materials; extracted cellulose presents an increase in the crystallinity index in the cellulose type I and type II. It is worthy to notice that the crystallinity index of both polymorphism increased being the type II with higher values upon the isolation of CNC, and after polymer grafting, both decreased at the same value due the presence of the amorphous polymer [54].

An estimation of the percentage of the concentration of the polymorphism type I and II in the samples was carried using the area of the diffraction peaks of the corresponding cellulose type. The deconvolution of the diffraction peaks was performed using the PeakFit v4.12 software and the Gaussian-Lorentzian method; in consequence, FATW exhibited 100% cellulose I, extracted cellulose presented 57% type I and 43% type II, and after acid hydrolysis, the percentage of cellulose I diminished and that for cellulose II slightly increased (CNC) and remained at the same level after polymer grafting (CNC/2-EHA).

CNC was dispersed within PLA resin by melt blending in order to prepare PLA/CNC nanocomposites at different contents of CNC. Figure 7 presents SEM images of the fracture surface of PLA/CNC and PLA/(CNC/2-EHA) nanocomposites. Figures 7(a)–7(c) correspond to the surface of PLA/CNC at different loads of pristine CNC which was distributed as agglomerates indicating that the shear during processing was insufficient to disperse the CNC individually. On the other hand, Figures 7(d)–7(f) show the distribution of agglomerates CNC/2-EHA at different loadings. Figure 7(f) (at different magnifications) shows a CNC/2-EHA agglomerate at the highest concentration of modified CNC suggesting that the surface modification of CNC allowed its mixing until certain nanoparticle content.

Figures 8(a)–8(c) present TEM images of PLA/CNC nanocomposites at different lyophilized CNC contents, where the different morphologies of CNC discussed above coexisting in the nanocomposites. Indeed, Figures 8(c) and 8(e) show the coexistence of rods (red dotted circles) and spheres. TEM images showed CNC poorly dispersed in the PLA matrix consisted with the SEM observations.

Figures 8(d)–8(f) show the distribution of CNC/2-EHA within PLA exhibiting the three morphologies; in Figure 8(e), the porous network (red arrows) coexisting with the other morphologies can be observed. Figure 8(f) presents a better dispersion of (CNC/2-EHA), but it seems to be poorly distributed. PLA/(CNC/2-EHA) nanocomposites exhibited a better dispersion of modified CNC suggesting that the polymer grafted not only hindered the CNC aggregation but also allowed the dispersion of modified CNC during processing.

The effect of the degree of CNC dispersion on the rheological properties of PLA was evaluated [19]; it was reported that the rheological behavior of polymer nanocomposite is influenced by the type, concentration, surface area, aspect ratio, and chemical surface of nanoparticles [55]. In this sense, the surface modification of CNC is expected to produce an important impact on the rheological properties of polymer nanocomposites since the interactions among polymer-nanofiller are promoted favoring the formation of a nanofiller network within polymer matrix [55]. Thus, the potential use of CNC form ATW as nanorereinforcer for PLA was evaluated incorporating CNC (pristine and modified) at different proportions and determining its effect on the rheological properties of PLA/CNC nanocomposites. It is expected that grafted polymer promotes the polymer chain entanglement increasing the rheological properties.

Figure 9(a) presents the data for PLA and PLA/CNC nanocomposites showing that complex viscosity of PLA decreased continuously accounting for the ~65% of initial value at the end of the experiment. The complex viscosity dropped rapidly during the first 15 min; the molecular weight of PLA resin usually decreases during processing due to thermal degradation as reported by others [23, 56–59]. On the other hand, the complex viscosity of PLA/CNC nanocomposites showed lower values than PLA sample suggesting that CNC promoted polymer degradation during the processing due to the lower thermal stability of CNC; in fact, sulfate groups on the CNC promote the degradation of the PLA as reported by others [60, 61]. It was noticed that the CNC concentration has a slight effect on the viscosity decrement. The presence of CNC stabilized PLA/CNC nanocomposites since viscosity remained unchanged at early stages of experiment over 12 min.

Figure 9(b) shows the complex viscosity behavior of PLA/(CNC/2-EHA) nanocomposites with time at different CNC/2-EHA contents. The values of the complex viscosity of PLA decreased with CNC/2-EHA being slightly higher than those corresponding to PLA/CNC nanocomposites; it is possible that modified CNC promoted polymer degradation but in different manner than pristine CNC due to the ability of grafted poly(2-ethyl hexyl acrylate) to be hydrolyzed during processing [23]. It is interesting to notice that complex viscosity of PLA/(CNC-2EHA) nanocomposites increased with modified CNC content; this fact

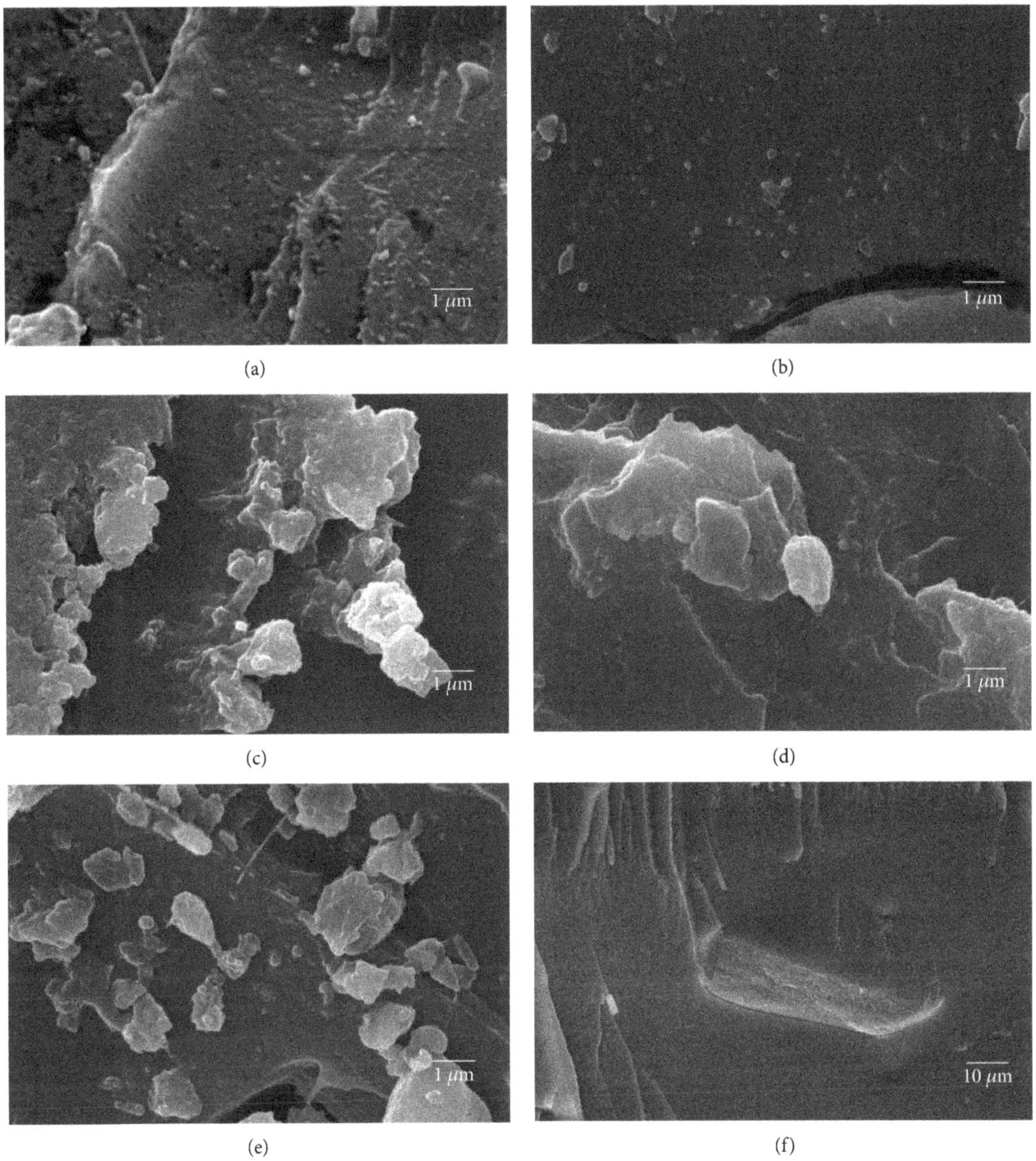

FIGURE 7: SEM micrographs of PLA/CNC nanocomposites (a) 0.5%, (b) 1%, and (c) 2% and PLA/(CNC/2-EHA) nanocomposites (d) 0.5%, (e) 1%, and (f) 2%.

suggests that polymer grafting promoted not only better CNC dispersion but also matrix-nanoparticle interactions [56]. Probably, higher aspect ratio of CNC could produce better rheological properties.

4. Conclusions

CNC were obtained from ATW bagasse presenting three morphologies: rods, spheres, and porous networks with a wide range in nanometric dimensions. 2-EHA was successfully grafted (~5 wt.%) on the surface of CNC as a result the surface energy and the crystallinity index decreased, whereas the thermal stability was improved. The cellulose type I from the ATW bagasse yielded to a mixture of cellulose I and II after acid hydrolysis and alkali extraction process; the resulting CNC are composed by a mixture of both polymorphism which remained after surface grafting. The decrease of the polar component of the modified CNC suggests that CNC/2-EHA could be dispersed and promotes adhesion in hydrophobic polymer matrices, such as PLA.

The addition of CNC or CNC/2-EHA to PLA caused a decrease in the complex viscosity of the nanocomposites depending on the concentration of the nanofiller. Complex viscosity reduction was attributed to PLA degradation during processing promoted by thermal degradation of pristine and modified CNC. CNC concentration had a slight effect on the viscosity decrement for the PLA/CNC nanocomposites;

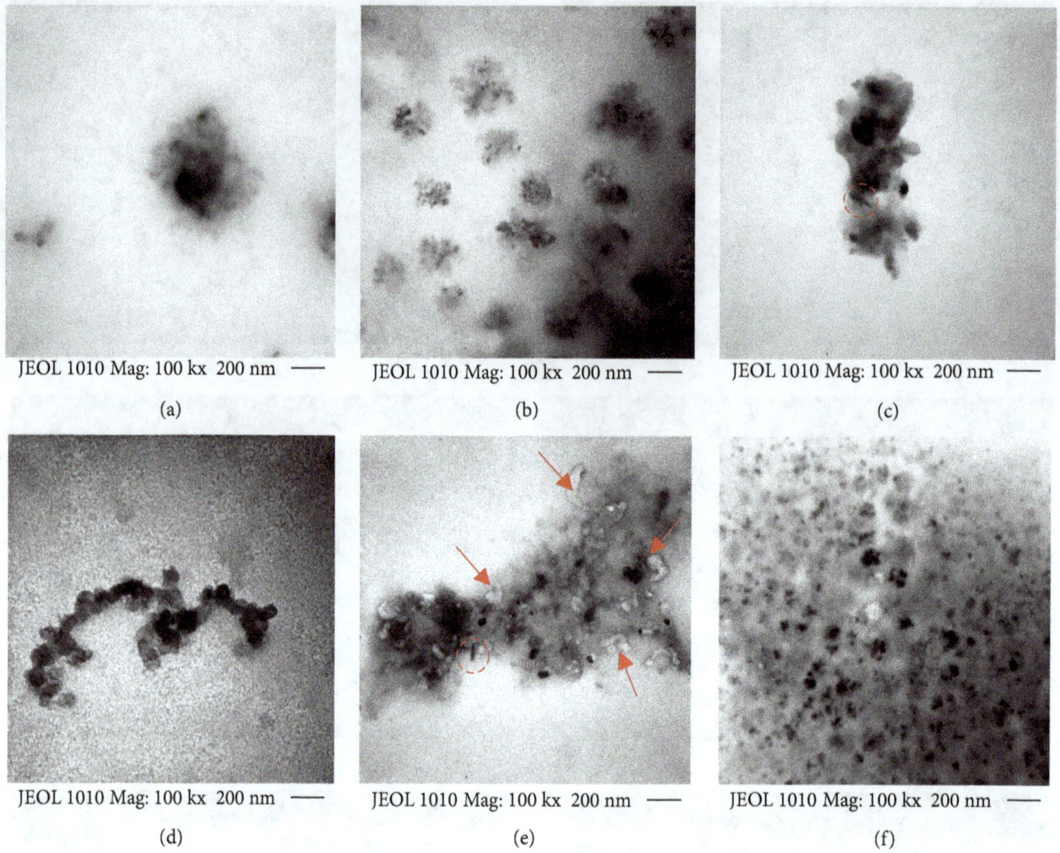

FIGURE 8: TEM micrographs of PLA/CNC nanocomposites (a) 0.5%, (b) 1%, and (c) 2% and PLA/(CNC/2-EHA) nanocomposites (d) 0.5%, (e) 1%, and (f) 2%.

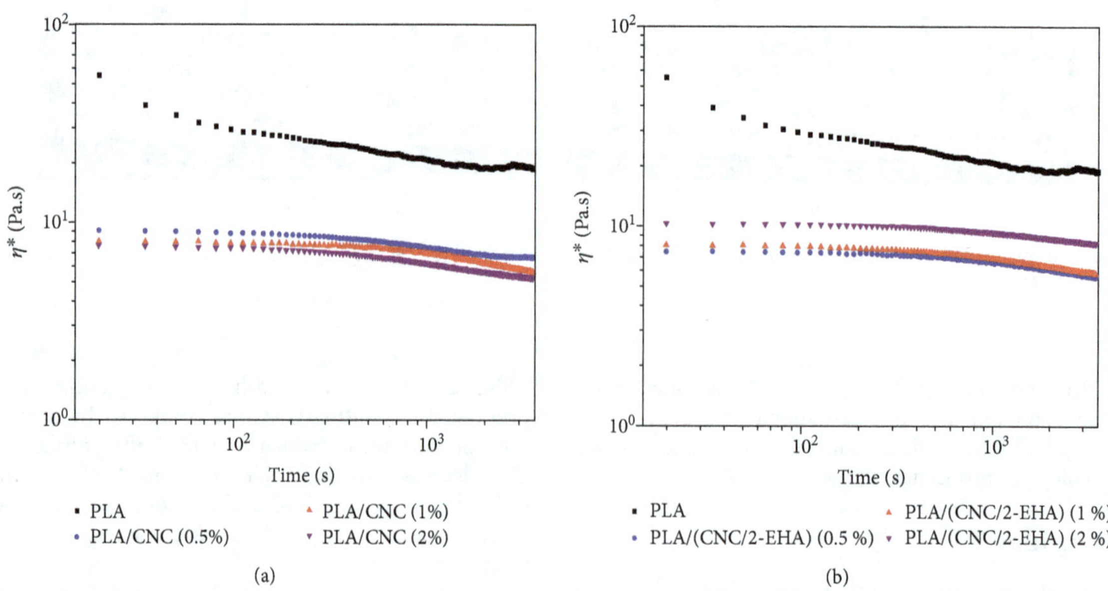

FIGURE 9: Complex viscosity as a function of time for PLA nanocomposites containing (a) CNC and (b) CNC/2-EHA at different contents.

however, the complex viscosity of PLA/(CNC-2EHA) nanocomposites seems to increase with modified CNC content which suggests that the polymer grafting improves the interactions of PLA-CNC. Finally, the results suggest that CNC/2-EHA could be used as an alternative nanomaterial to modify melt viscosity of polymers.

Conflicts of Interest

The authors declare that they have no conflicts of interest.

Acknowledgments

The technical support from Santiago Duarte-Aranda (SEM), Wilberth Herrera Kao (TGA), Alejandro May-Pat (AFM), Daniel Aguilar (XRD), and William Cauich (XPS) is highly appreciated. The XRD and XPS analysis was performed at the National Laboratory of Nano and Biomaterials (financed by Fomix-Yucatan and Conacyt-Mexico), CINVESTAV-IPN Merida Unit. The authors thank Dr. Patricia Quintana for the access to LANNBIO. This study was partially financed by the project FOMIX Yucatan CICY YUC-2014-C17-247046.

References

[1] D. Bondeson, A. Mathew, and K. Oksman, "Optimization of the isolation of nanocrystals from microcrystalline cellulose by acid hydrolysis," *Cellulose*, vol. 13, no. 2, pp. 171–180, 2006.

[2] G. Bolio-López, A. Valadez-González, L. Veleva, and A. Andreeva, "Whiskers de celulosa a partir de residuos agroindustriales de banano: Obtención y caracterización," *Revista mexicana de ingeniería química*, vol. 10, no. 2, pp. 291–299, 2011.

[3] A. Kumar, Y. S. Negi, V. Choudhary, and N. K. Bhardwaj, "Characterization of cellulose nanocrystals produced by acid-hydrolysis from sugarcane bagasse as agro-waste," *Journal of Materials Physics and Chemistry*, vol. 2, no. 1, pp. 1–8, 2014.

[4] H. Lu, Y. Gui, L. Zheng, and X. Liu, "Morphological, crystalline, thermal and physicochemical properties of cellulose nanocrystals obtained from sweet potato residue," *Food Research International*, vol. 50, no. 1, pp. 121–128, 2013.

[5] M. F. Rosa, E. S. Medeiros, J. A. Malmonge et al., "Cellulose nanowhiskers from coconut husk fibers: effect of preparation conditions on their thermal and morphological behavior," *Carbohydrate Polymers*, vol. 81, no. 1, pp. 83–92, 2010.

[6] P. Lu and Y.-L. Hsieh, "Preparation and characterization of cellulose nanocrystals from rice straw," *Carbohydrate Polymers*, vol. 87, no. 1, pp. 564–573, 2012.

[7] G. Iñiguez, N. Acosta, L. Martinez, J. Parra, and O. González, "Utilización de Subproductos de la Industria Tequilera. Parte 7. Compostaje de Bagazo de Agave y Vinazas Tequileras," *Revista Internacional de Contaminación Ambiental*, vol. 21, no. 1, pp. 37–50, 2010.

[8] P. Mignani, "Aprovechar los Desechos del Tequila," *La Gaceta de la Universidad de Guadalajara*, vol. 11, no. 713, p. 14, 2012.

[9] G. Iñiguez-Covarrubias, S. E. Lange, and R. M. Rowell, "Utilization of byproducts from the tequila industry: part 1: agave bagasse as a raw material for animal feeding and fiberboard production," *Bioresource Technology*, vol. 77, no. 1, pp. 25–32, 2001.

[10] E. Espino, M. Cakir, S. Domenek, A. D. Román-Gutiérrez, N. Belgacem, and J. Bras, "Isolation and characterization of cellulose nanocrystals from industrial by-products of Agave tequilana and barley," *Industrial Crops and Products*, vol. 62, pp. 552–559, 2014.

[11] Y. Habibi, L. A. Lucia, and O. J. Rojas, "Cellulose nanocrystals: chemistry, self-assembly, and applications," *Chemical Reviews*, vol. 110, no. 6, pp. 3479–3500, 2010.

[12] J.-M. Raquez, Y. Habibi, M. Murariu, and P. Dubois, "Polylactide (PLA)-based nanocomposites," *Progress in Polymer Science*, vol. 38, no. 10-11, pp. 1504–1542, 2013.

[13] D. Bondeson and K. Oksman, "Polylactic acid/cellulose whisker nanocomposites modified by polyvinyl alcohol," *Composites Part A: Applied Science and Manufacturing*, vol. 38, no. 12, pp. 2486–2492, 2007.

[14] N. Lin, J. Huang, P. R. Chang, J. Feng, and J. Yu, "Surface acetylation of cellulose nanocrystal and its reinforcing function in poly(lactic acid)," *Carbohydrate Polymers*, vol. 83, no. 4, pp. 1834–1842, 2011.

[15] A. Junior de Menezes, G. Siqueira, A. A. S. Curvelo, and A. Dufresne, "Extrusion and characterization of functionalized cellulose whiskers reinforced polyethylene nanocomposites," *Polymer*, vol. 50, no. 19, pp. 4552–4563, 2009.

[16] K.-Y. Lee, J. J. Blaker, and A. Bismarck, "Surface functionalisation of bacterial cellulose as the route to produce green polylactide nanocomposites with improved properties," *Composites Science and Technology*, vol. 69, no. 15-16, pp. 2724–2733, 2009.

[17] M. Martínez-Sanz, M. A. Abdelwahab, A. Lopez-Rubio et al., "Incorporation of poly(glycidylmethacrylate) grafted bacterial cellulose nanowhiskers in poly(lactic acid) nanocomposites: improved barrier and mechanical properties," *European Polymer Journal*, vol. 49, no. 8, pp. 2062–2072, 2013.

[18] K. Littunen, U. Hippi, L.-S. Johansson et al., "Free radical graft copolymerization of nanofibrillated cellulose with acrylic monomers," *Carbohydrate Polymers*, vol. 84, no. 3, pp. 1039–1047, 2011.

[19] P. Cassagnau, "Melt rheology of organoclay and fumed silica nanocomposites," *Polymer*, vol. 49, no. 9, pp. 2183–2196, 2008.

[20] M. R. Kamal and V. Khoshkava, "Effect of cellulose nanocrystals (CNC) on rheological and mechanical properties and crystallization behavior of PLA/CNC nanocomposites," *Carbohydrate Polymers*, vol. 123, pp. 105–114, 2015.

[21] D. Bagheriasl, P. J. Carreau, B. Riedl, C. Dubois, and W. Y. Hamad, "Shear rheology of polylactide (PLA)–cellulose nanocrystal (CNC) nanocomposites," *Cellulose*, vol. 23, no. 3, pp. 1885–1897, 2016.

[22] N. Bitinis, R. Verdejo, J. Bras et al., "Poly(lactic acid)/natural rubber/cellulose nanocrystal bionanocomposites Part I. Processing and morphology," *Carbohydrate Polymers*, vol. 96, no. 2, pp. 611–620, 2013.

[23] Y. Lumelsky and M. S. Silverstein, "Biodegradable porous polymers through emulsion templating," *Macromolecules*, vol. 42, no. 5, pp. 1627–1633, 2009.

[24] M. N. Cazaurang-Martinez, S. R. Peraza-Sánchez, and C. A. Cruz-Ramos, "Dissolving-grade pulps from henequen fiber," *Cellulose Chemistry and Technology*, vol. 24, pp. 629–638, 1990.

[25] W. Y. Hamad and T. Q. Hu, "Structure–process–yield interrelations in nanocrystalline cellulose extraction," *The Canadian Journal of Chemical Engineering*, vol. 88, no. 3, pp. 392–402, 2010.

[26] G. Canché-Escamilla, G. Rodríguez-Trujillo, P. J. Herrera-Franco, E. Mendizábal, and J. E. Puig, "Preparation and characterization of henequen cellulose grafted with methyl methacrylate and its application in composites," *Journal of Applied Polymer Science*, vol. 66, no. 2, pp. 339–346, 1997.

[27] A. D. French and M. Santiago Cintrón, "Cellulose polymorphy, crystallite size, and the Segal crystallinity index," *Cellulose*, vol. 20, no. 1, pp. 583–588, 2013.

[28] A. Kljun, T. A. S. Benians, F. Goubet, F. Meulewaeter, J. P. Knox, and R. S. Blackburn, "Comparative analysis of crystallinity changes in cellulose I polymers using ATR-FTIR, X-ray diffraction, and carbohydrate-binding module probes," *Biomacromolecules*, vol. 12, no. 11, pp. 4121–4126, 2011.

[29] L. Segal, J. J. Creely, A. E. Martin Jr., and C. M. Conrad, "An empirical method for estimating the degree of crystallinity of native cellulose using the X-ray diffractometer," *Textile Research Journal*, vol. 29, no. 10, pp. 786–794, 1959.

[30] A. Agirre, J. Nase, E. Degrandi, C. Creton, and J. M. Asua, "Miniemulsion polymerization of 2-ethylhexyl acrylate. Polymer architecture control and adhesion properties," *Macromolecules*, vol. 43, no. 21, pp. 8924–8932, 2010.

[31] A. Hebeish and J. T. Guthrie, "The chemistry and technology of cellulose copolymers," *Starch-Stärke*, vol. 33, no. 12, p. 432, 1981.

[32] M. Shah, M. Majid, and M. Bari, *Ceric-ion initiated graft copolymerization of methyl methacrylate on to cellulose [in Bangladesh]*, Chittagong University Studies, 1991.

[33] M.-C. Li, Q. Wu, K. Song, S. Lee, Y. Qing, and Y. Wu, "Cellulose nanoparticles: structure–morphology–rheology relationships," *ACS Sustainable Chemistry & Engineering*, vol. 3, no. 5, pp. 821–832, 2015.

[34] J. Han, C. Zhou, Y. Wu, F. Liu, and Q. Wu, "Self-assembling behavior of cellulose nanoparticles during freeze-drying: effect of suspension concentration, particle size, crystal structure, and surface charge," *Biomacromolecules*, vol. 14, no. 5, pp. 1529–1540, 2013.

[35] J. Araki, M. Wada, S. Kuga, and T. Okano, "Flow properties of microcrystalline cellulose suspension prepared by acid treatment of native cellulose," *Colloids and Surfaces A: Physicochemical and Engineering Aspects*, vol. 142, no. 1, pp. 75–82, 1998.

[36] Y.-L. Hsieh, "Cellulose nanocrystals and self-assembled nanostructures from cotton, rice straw and grape skin: a source perspective," *Journal of Materials Science*, vol. 48, no. 22, pp. 7837–7846, 2013.

[37] P. Lu and Y.-L. Hsieh, "Preparation and properties of cellulose nanocrystals: rods, spheres, and network," *Carbohydrate Polymers*, vol. 82, no. 2, pp. 329–336, 2010.

[38] M. Haqani, H. Roghani-Mamaqani, and M. Salami-Kalajahi, "Synthesis of dual-sensitive nanocrystalline cellulose-grafted block copolymers of N-isopropylacrylamide and acrylic acid by reversible addition-fragmentation chain transfer polymerization," *Cellulose*, vol. 24, no. 5, pp. 2241–2254, 2017.

[39] Q. Lu, W. Lin, L. Tang, S. Wang, X. Chen, and B. Huang, "A mechanochemical approach to manufacturing bamboo cellulose nanocrystals," *Journal of Materials Science*, vol. 50, no. 2, pp. 611–619, 2015.

[40] N. A. Rosli, I. Ahmad, and I. Abdullah, "Isolation and characterization of cellulose nanocrystals from Agave angustifolia fibre," *BioResources*, vol. 8, no. 2, 2013.

[41] L. Qin, J. Qiu, M. Liu et al., "Mechanical and thermal properties of poly (lactic acid) composites with rice straw fiber modified by poly (butyl acrylate)," *Chemical Engineering Journal*, vol. 166, no. 2, pp. 772–778, 2011.

[42] G. Siqueira, J. Bras, and A. Dufresne, "New process of chemical grafting of cellulose nanoparticles with a long chain isocyanate," *Langmuir*, vol. 26, no. 1, pp. 402–411, 2010.

[43] M. Jonoobi, J. Harun, M. Mishra, and K. Oksman, "Chemical composition, crystallinity and thermal degradation of bleached and unbleached kenaf bast (Hibiscus cannabinus) pulp and nanofiber," *BioResources*, vol. 4, no. 2, pp. 626–639, 2009.

[44] T. Doba, C. Rodehed, and B. Ranby, "Mechanism of graft co-polymerization onto polysaccharides initiated by metal ion oxidation reactions of model compounds for starch and cellulose," *Macromolecules*, vol. 17, no. 12, pp. 2512–2519, 1984.

[45] X. Shen, J. L. Shamshina, P. Berton, G. Gurau, and R. D. Rogers, "Hydrogels based on cellulose and chitin: fabrication, properties, and applications," *Green Chemistry*, vol. 18, no. 1, pp. 53–75, 2016.

[46] N. Grassie and J. G. Speakman, "Thermal degradation of poly(alkyl acrylates). I. Preliminary investigations," *Journal of Polymer Science Part A-1: Polymer Chemistry*, vol. 9, no. 4, pp. 919–929, 1971.

[47] K.-Y. Lee, F. Quero, J. J. Blaker, C. A. S. Hill, S. J. Eichhorn, and A. Bismarck, "Surface only modification of bacterial cellulose nanofibres with organic acids," *Cellulose*, vol. 18, no. 3, pp. 595–605, 2011.

[48] T. Zheng, X. Xu, J. Pan, S. Xiao, and J. Lv, "Studies of preparation and characteristics in borosilicate photosensitive glass-ceramics," *Materials Sciences and Applications*, vol. 3, no. 2, pp. 98–102, 2012.

[49] M. Bautista Justo, L. García Oropeza, J. Barboza-Corona, and L. Parra Negrete, "El Agave tequilana Weber y la producción de tequila," *Acta Universitaria*, vol. 11, no. 2, 2001.

[50] A. M. Hindeleh and D. J. Johnson, "Crystallinity and crystallite size measurement in cellulose fibres: 1. Ramie and Fortisan," *Polymer*, vol. 13, no. 9, pp. 423–430, 1972.

[51] F. Torres, O. Troncoso, C. Torres, and C. Grande, "Cellulose based blends, composites and nanocomposites," in *Advances in Natural Polymers*, S. Thomas, P. Visakh, and A. P. Mathew, Eds., pp. 21–54, Springer, 2013.

[52] G. Sèbe, F. Ham-Pichavant, E. Ibarboure, A. L. C. Koffi, and P. Tingaut, "Supramolecular structure characterization of cellulose II nanowhiskers produced by acid hydrolysis of cellulose I substrates," *Biomacromolecules*, vol. 13, no. 2, pp. 570–578, 2012.

[53] Y. Yue, C. Zhou, A. D. French et al., "Comparative properties of cellulose nano-crystals from native and mercerized cotton fibers," *Cellulose*, vol. 19, no. 4, pp. 1173–1187, 2012.

[54] M. Golshan, M. Salami-Kalajahi, H. Roghani-Mamaqani, and M. Mohammadi, "Poly (propylene imine) dendrimer-grafted nanocrystalline cellulose: doxorubicin loading and release behavior," *Polymer*, vol. 117, pp. 287–294, 2017.

[55] Y. C. Ching, M. Ershad Ali, L. C. Abdullah et al., "Rheological properties of cellulose nanocrystal-embedded polymer composites: a review," *Cellulose*, vol. 23, no. 2, pp. 1011–1030, 2016.

[56] Q. Meng, M.-C. Heuzey, and P. J. Carreau, "Control of thermal degradation of polylactide/clay nanocomposites during melt

processing by chain extension reaction," *Polymer Degradation and Stability*, vol. 97, no. 10, pp. 2010–2020, 2012.

[57] L. D'Urso, M. Acocella, G. Guerra, V. Iozzino, F. de Santis, and R. Pantani, "PLA melt stabilization by high-surface-area graphite and carbon black," *Polymer*, vol. 10, no. 2, p. 139, 2018.

[58] K. S. Seo and J. D. Cloyd, "Kinetics of hydrolysis and thermal degradation of polyester melts," *Journal of Applied Polymer Science*, vol. 42, no. 3, pp. 845–850, 1991.

[59] V. Speranza, A. De Meo, and R. Pantani, "Thermal and hydrolytic degradation kinetics of PLA in the molten state," *Polymer Degradation and Stability*, vol. 100, pp. 37–41, 2014.

[60] E. Fortunati, F. Luzi, D. Puglia, R. Petrucci, J. M. Kenny, and L. Torre, "Processing of PLA nanocomposites with cellulose nanocrystals extracted from *Posidonia oceanica* waste: innovative reuse of coastal plant," *Industrial Crops and Products*, vol. 67, pp. 439–447, 2015.

[61] P. Dhar, S. M. Bhasney, A. Kumar, and V. Katiyar, "Acid functionalized cellulose nanocrystals and its effect on mechanical, thermal, crystallization and surfaces properties of poly (lactic acid) bionanocomposites films: a comprehensive study," *Polymer*, vol. 101, pp. 75–92, 2016.

4

Effects of Zinc Oxide Nanoparticles on the Properties of Pectin/Alginate Edible Films

Thi Minh Phuong Ngo,[1] Thi Mong Quyen Dang,[2] Thi Xo Tran,[3] and Pornchai Rachtanapun[4,5]

[1]*The University of Danang - University of Technology and Education, Da Nang, Vietnam*
[2]*College of Food Industry, Da Nang, Vietnam*
[3]*The University of Danang - University of Science and Technology, Da Nang, Vietnam*
[4]*Division of Packaging Technology, Faculty of Agro-Industry, Chiang Mai University, Chiang Mai 50100, Thailand*
[5]*Center of Excellent in Materials Science and Technology, Chiang Mai University, Chiang Mai 50200, Thailand*

Correspondence should be addressed to Pornchai Rachtanapun; pornchai.r@cmu.ac.th

Guest Editor: Ing Kong

This study aimed at developing the functional packaging properties of pectin/alginate films by adding zinc oxide nanoparticles (ZnO-NPs). The ZnO-NPs were added to the pectin/alginate film at the concentrations of 0.5, 2.5, 5, and 25 g/100 g of blended polymer. The effects of the ZnO-NP incorporation on the mechanical properties, hydration properties, oxygen permeability, ultraviolet transmission, transparency, and antimicrobial activity of the films were investigated. The addition of ZnO-NPs from 0 to 5 g/100 g of blended polymer to the pectin/alginate film increased 191.4% of the tensile strength, 169.8% of elongation at break, and antimicrobial properties (against *Aspergillus niger*, *Colletotrichum gloeosporioides*, *Escherichia coli*, and *Saccharomyces cerevisiae*). In contrast, it reduced the solubility from 30.38% to 22.49%, water vapor permeability from 1.01×10^{-14} to 0.414×10^{-14} kg·m/m^2·Pa·s, moisture absorption, ultraviolet light transmission, and oxygen permeability from 270.86×10^{-19} to 110.79×10^{-19} kg·m/m^2·Pa·s. The effects are the highest when the concentration of the ZnO-NPs in the film was 5 g/100 g of blended polymer. Attenuated total reflectance-Fourier transform infrared analysis indicated that some interactions between the ZnO-NPs and pectin/alginate matrices were formed. These results suggest that pectin/alginate/ZnO-NPs films can be used as active packaging for food preservation.

1. Introduction

Food packaging materials with proper mechanical strength; barrier properties against water, water vapor, and other gases; thermal stability, recyclability, and biodegradability; and functional properties, such as antimicrobial activities, are highly desirable for food safety and extending the shelf-life of packaged foods [1]. However, biopolymer-based packaging materials have not been widely used in the packaging industry, mainly because of their poor mechanical, barrier, and processing properties. They also have a higher production cost compared to commodity plastic films [2]. Edible films are increasingly used in food preservation, and they are largely produced from nontoxic polysaccharide sources. Among various polysaccharides, pectin and alginate are particularly attractive due to their good film-forming property, moderate mechanical strength, and gas barrier properties with a unique colloidal nature. Recently, zinc oxide nanoparticles (ZnO-NPs) have become well known for their antimicrobial and UV-light barrier properties. The addition of ZnO-NPs to biopolymers has improved the film's properties and showed strong antimicrobial and UV barrier properties [3]. For these reasons, ZnO-NPs have been applied to food packaging. However, the antimicrobial activity and membrane properties will depend on the amount of ZnO-NPs used. Therefore, the main purpose of this study was to

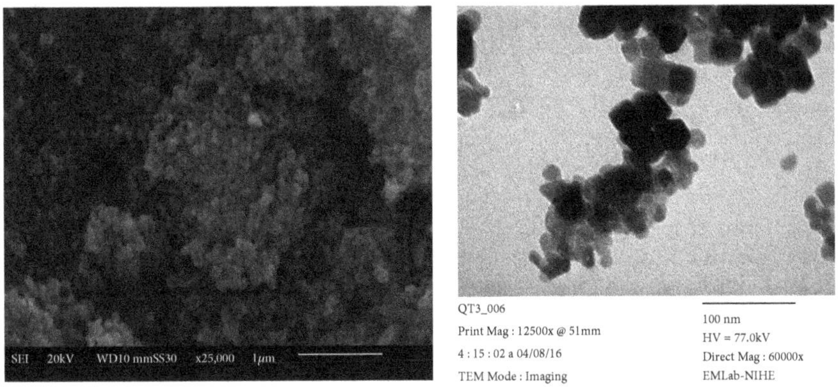

(a) (b)

FIGURE 1: (a) SEM photographs of ZnO-NPs at magnification of 25,000x and (b) TEM photographs of ZnO-NPs at magnification of 125,000x.

develop pectin/alginate films by mixing ZnO-NPs with different contents and evaluate the effect of them on the properties of the film.

2. Materials and Methods

2.1. Materials and Sample Preparation. Sodium alginate (AG), with a molecular weight (M_w) of 1.2×10^6 Da, was purchased from Lianyungang Zhongda Seaweed Industrial Co. Ltd. (China). Glycerol and calcium chloride were purchased from Xilong Co. Ltd., China. The BHI medium was purchased from Sigma-Aldrich (St. Louis, USA).

The pectin (P) powder and ZnO-NPs were obtained at the LAB of the University of Technology and Education, which belongs to the University of Danang, Vietnam.

The pectin powder was extracted from Yanang leaves using a heating method with an optimum temperature of 88°C and time of 75minutes and in a 6.5% citric acid solution. The extracted mixture was precipitated with 96% ethanol. Then, the mixture was centrifuged (4000 rpm for 10 min) and washed three times with 96% ethanol. The pectin precipitate was dried at 60°C, and the dried pectin was ground into a flour. The degree of methyl esterification (DE) measurement value was 48.36, and the M_w measurement value was 111,105 Da.

The ZnO-NPs were synthesized by the following sol-gel method: First, 1.1 g of $Zn(CH_3COO)_2 \cdot 2H_2O$ was dissolved with 50 ml diethylene glycol. The mixture was stirred continuously for one hour at 80°C. Then, 50 ml of a 0.1 M LiOH solution was added to the solution. After stirring for 5 minutes at 80°C, the white precipitate formed. The mixture was centrifuged (4000 rpm for 10 minutes) and washed three times with distilled water and then washed three times with 96% ethanol to remove the nonreactive and excess ions. The size of ZnO-NPs was from 20 to 40 nm (Figure 1).

The pectin/alginate film-forming aqueous solutions were prepared with 2% *w/v* at the proportion of pectin : alginate 1 : 1 *w/w*. Powdered calcium chloride was used at 0.01 g/g polymer and glycerol as a plasticizer at 50% (% *w/w* of polymer), and both were blended into the pectin-alginate film-solution. The ZnO-NPs were added in the pectin/alginate films with glycerol and fabricated with four levels of ZnO-NPs, namely, 0.5, 2.5, 5, and 25 g/100 g of blended polymer. The mixture was homogenized using a homogenizer at speed of 400 rpm in 15 minutes, and after that, it was stored at 4°C and 90% RH for degassing. To control the film thickness, the volume of each film-forming solution poured into a 15 cm × 15 cm mold was always 60 g. Films were dried at 25°C and at 52 ± 1% relative humidity (RH) for 24 h by using a fan at speed of 1500 rpm.

The dried films were removed from the mold and dipped in 5% $CaCl_2$ solution for 20 min to enable cross-linking [4]. Films were dried at 40°C for two hours in the oven (UFE 500, Memmert). Films were stored at 52 ± 1% RH and 25 ± 1°C for 5 days prior to testing.

2.2. Thickness. The thickness of the films was measured using a Mitutoyo Dial Thickness Gauge (Model PCM 137, No. 2046S, Japan) with an accuracy of 0.01 mm at 10 random positions for each film. The average standard deviation was about 5% of the average thickness. The tests were repeated three times.

2.3. Tensile Property Measurements and % Elongation at Break. The tensile properties and % elongation at break of the samples were measured using an Instron 1141 (Instron Ltd., Canton, USA) with the procedure according to ASTM D882-10 [5]. The samples were prepared as sheets using a gauge length and width of 100 mm and 25 mm, respectively. The initial grip separation was set at 125 mm/min. The elongation at break was calculated by dividing the elongation at the moment of rupture by the initial gauge length and multiplying by 100. The samples were conditioned at 52 ± 1% RH and 25°C for 48 h prior to testing. There were ten sample measurements.

2.4. Film Solubility. The films were cut into square 1 cm × 1 cm pieces and dried to a constant weight in a vacuum oven at 60°C for 24 hours to obtain the initial film dry weight (W_d). Each film was then placed in a bottle containing 20 ml distilled water for 24 hours under gentle agitation and controlled temperature at 25°C. The films were then dried under the same conditions to obtain the dry weight of water-leached film (W_{ws}). The tests were performed in triplicate. The film solubility was calculated using

$$\%\text{solubility} = \frac{(W_d - W_{ws})}{W_d} \times 100. \quad (1)$$

2.5. Water Contact Angle. Drop shape analysis (DSA30E, Krüss Co. Ltd., Germany) was used to observe the water contact angle. The samples were prepared as sheets with a dimension of 1 cm × 5 cm. Water was dropped onto the surface of the sheets, and images were taken automatically every 20 s for 2 min. Samples were kept at 25°C and 52 ± 1% RH at least 48 h prior to testing. The measurements were average from three samples for each formulation [6].

2.6. Water Vapor Permeability (WVP) and Water Vapor Transmission Rate (WVTR). The WVP of the films was measured gravimetrically at 25 ± 1°C according to ASTM E96-00 (ASTM, 2000). Film specimens with an approximately 8 cm diameter were mounted on the aluminum cups containing 10 g of silica gel. Paraffin was used to fix the film specimen on the wide rim of an aluminum cup [7]. The sealed cups were then weighed and kept in desiccators with a saturated solution of $Mg(NO_3)_2$ salt that provided 52 ± 1% RH at 25 ± 1°C throughout the experiment. The weight gain of the aluminum cups covered with the films was recorded every 10 hours for 5 days [8]. All tests were conducted in triplicate. The WVP and WVTR were calculated using

$$\text{WVP} = \frac{w \cdot x}{t \cdot A \cdot P_o \cdot (\text{RH1} - \text{RH2})} \left(\text{kg} \cdot \text{m/Pa} \cdot \text{m}^2 \cdot \text{s}\right), \quad (2)$$

$$\text{WVTR} = \frac{w}{t \cdot A} \left(\text{kg/m}^2 \cdot \text{s}\right), \quad (3)$$

where w/t is the permeation rate (g/day) that was calculated by a linear regression of the mass gain versus time ($R^2 > 0.99$), x is the film average thickness (mm), A is the permeation area (23.76 cm^2), P_o is the partial water vapor pressure at test temperature (3159 kPa at 25°C), and ($RH_1 - RH_2$) is the difference in relative humidity between the two sides of the film [9].

2.7. Moisture Sorption Isotherm. The films were cut into a 3 cm × 3 cm squares and dried in a hot air oven at 105°C for 3 h before being placed in desiccators containing dried silica gel for 2 days. The specimens were then placed in several different desiccators containing various kinds of saturated salt solutions to maintain the constant relative humidity at 11.2%, 33.5%, 53.1%, 64.8%, 75.1%, 85.5%, and 93.7%. A cotton wool soaked with 96% ethanol was used as a fungistatic agent in the desiccators containing the salt solutions with a high relative humidity. The specimens were weighed daily until equilibrium was obtained. Equilibrium was achieved when the change in weight did not exceed 0.1% after being weighed for three consecutive times. The equilibrium moisture content (EMC) (g water/100 g dry solid) of the films at each relative humidity was calculated using

$$\text{EMC} = \frac{W_e}{W_i}(M_i + 1) - 1, \quad (4)$$

where W_e is the equilibrium weight of the film specimen (g), W_i is the initial weight of the film specimen (g), and M_i is the initial moisture content of the film specimen (g water/100 g dry solid) [10, 11]. All measurements were performed in triplicate.

2.8. Transparency and Light Transmission of Films. Each film sample was cut into a rectangular piece and directly mounted between two spectrophotometer magnetic cells. The transparency and light transmission of the films were measured at selected wavelengths between 320 and 800 nm using a UV-Vis spectrophotometer (Model 8451A, Hewlett-Packard Co., Santa Clara, CA, USA). The film transparency was calculated with

$$\text{Transparency} = -\log \frac{T_{600}}{x} (1/\text{mm}), \quad (5)$$

where T_{600} is the fractional transmittance at 600 nm and x is the film thickness (mm) [12].

2.9. Oxygen Permeability and Oxygen Transmission Rate. The oxygen permeability (OP) and oxygen transmission rate (OTR) were measured using a Mocon Instruments tool (OX-TRAN® Model 2/21 Series, Mocon Inc., USA). The measurements were taken using 6 cm × 6 cm film samples at 23 ± 2°C, 75% RH, and 760 mmHg following the standard method [13]. The results are the average over at least three replicate measurements. The OP was calculated by

$$\text{OP} = \frac{\text{OTR} \cdot l}{\Delta P} \left(\text{kg} \cdot \text{m/Pa} \cdot \text{m}^2 \cdot \text{s}\right), \quad (6)$$

where OTR is the value of oxygen transmission rate, l is the thickness of the film, and ΔP is the difference between oxygen partial pressures between the two sides of the film (Pa).

2.10. Antimicrobial Action of Film-Forming Solutions. The antimicrobial activities of the film-forming solutions incorporating the ZnO-NPs were evaluated by the well diffusion method in agar against *E. coli*, *Saccharomyces cerevisiae*, *Colletotrichum gloeosporioides*, and *Aspergillus niger*. The microorganisms were prepared for the test as follows: *E. coli* was aseptically inoculated in BHI broth and subsequently incubated at 37°C for 16 h. *Saccharomyces cerevisiae*, *Aspergillus niger*, and *Colletotrichum gloeosporioides* were aseptically inoculated in Sabouraud broth and subsequently incubated at 30°C for 16 h. The evaluation process was as follows: The Sabouraud agar medium was used for *S. cerevisiae*, *Aspergillus niger*, and *Colletotrichum gloeosporioides*, and the BHI agar medium was used for *E. coli*. All were poured into petri dishes. After solidification, wells at 5 mm in diameter were cut from the agar and 0.1 ml of the film-forming

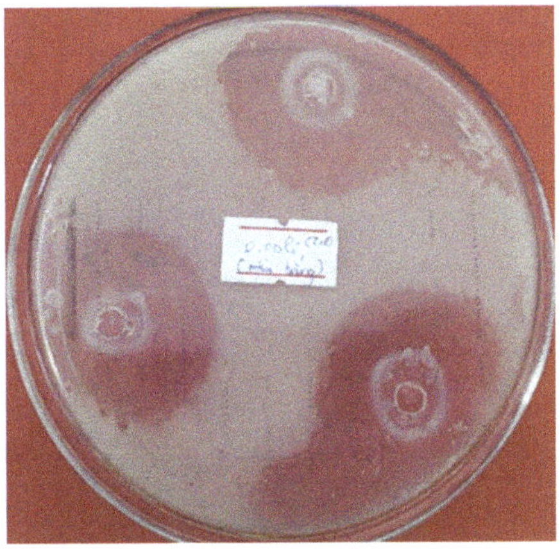

FIGURE 2: Evaluation image of antibacterial action by the "zone inhibition method."

solution was carefully poured inside. Next, a 1 ml suspension containing 10^6 CFU/ml of each investigated microorganism was spread in the specific agar medium. After incubating for 48 h at 30°C for *S. cerevisiae*, *Aspergillus niger*, and *Colletotrichum gloeosporioides* and 48 h at 37°C for *E. coli*, all plates were inspected. The diameters of the inhibition zones were measured (in cm) with a ruler on the underside of the petri dish. For example, the image that showed the antibacterial activity of film-forming solution against *E. coli* is shown in Figure 2. All measurements were performed in triplicate [14, 15].

2.11. Attenuated Total Reflectance-Fourier Transform Infrared (ATR-FTIR) Spectroscopy. The reaction between the pectin, alginate, and ZnO-NPs was monitored by ATR-FTIR (Thermo Nicolet 6700, USA). The samples were prepared as thin films (about 60 μm). The IR spectra were measured at wavelengths from 500 cm^{-1} to 4000 cm^{-1} and with a resolution of 4 cm^{-1}.

2.12. Statistical Analysis. Statistics were determined with a completely randomized design with an analysis of variance (ANOVA) using Minitab software (version 16.2.3.0; Minitab Inc., USA). Fisher's tests were used to compare the difference among the mean values of the film specimens' properties at the level of 0.05. All experiments were conducted in triplicate.

3. Results and Discussion

3.1. Thickness and Mechanical Properties. The thickness and mechanical properties of the P/AG films with various ZnO-NP contents are presented in Table 1. The thicknesses of the P/AG films that contain the ZnO-NPs is higher than for the P/AG only film. The increase in ZnO-NP content from 0 to 25 g/100 g of blended polymer led to an increase in the film thickness from 52.4 to 74.6 μm. The increase in the thickness was not only attributed to the higher solid content of nanocomposite films ZnO-NPs but also attributed to their structure. Similar results were reported by Shahmohammadi Jebel and Almasi [16].

The tensile strength (TS) and % elongation at break (E) for the films with various ZnO-NP contents equilibrated at 25 ± 1°C and 52 ± 1% RH are shown in Table 1. There was an increase in tensile strength from 21.71 to 41.56 MPa when increasing the ZnO-NP content from 0.5 to 5 g/100 g of blended polymer. On the other hand, the increase in the ZnO-NP concentration from 5 to 25 g/100 g of blended polymer reduced the tensile strength from 41.56 to 12.79 MPa. The tensile strength was the highest when the ZnO-NPs were added at a concentration of 5 g/100 g of blended polymer, where it reached 191.4% of the tensile strength for the P/AG film.

The change in elongation is similar to that of tensile strength. Changes in tensile strength and elongation values may be due to the formation of hydrogen links between the pectin, alginate's hydroxyl groups, and oxygen atoms in the ZnO-NPs [17, 18]. Moreover, a low concentration of ZnO-NPs had a negligible effect on the formation of hydrogen bonds between the OH and the COOH groups in the pectin/alginate matrix, as well as the formation of bonds between the Ca^{2+} and the COO^- groups of the pectin and alginate chains. The pectin and alginate chains thereby adopted a structure that resembles an "egg-box" model.

The decrease in TS and E of the film with 25 g ZnO-NPs/100 g polymer was probably due to agglomeration of the ZnO-NPs, the nonuniform distribution of the ZnO-NPs, and the recrystallization of the ZnO-NPs in the polymer matrix. These factors caused discontinuous cracks in the polymer matrix and resulted in a sudden decrease in the TS and E. A similar result was observed when adding CuO-NPs and ZnO-NPs into a carrageenan matrix [18].

3.2. Solubility and Water Contact Angle. The water contact angle (θ) is the most common parameter used to describe the hydrophilicity of film surfaces. Hydrophilic membranes possess a contact angle of $0° < \theta < 90°$, while hydrophobic membranes have a contact angle of $90° < \theta < 180°$ [19]. Table 2 shows the solubility and water contact angle for the film specimens.

Results showed that adding ZnO-NPs to the P/AG film at the amount of 0.5 to 5 g/100 g of blended polymer did not significantly change the water contact angle. However, the water contact angle for the film-incorporated ZnO-NPs with 25 g/100 g of blended polymer was lower than the other films. Reducing the water contact angle may be due to an increase in surface roughness at high ZnO-NP concentrations.

The solubility values for the specimens trended to the decrease with increasing ZnO-NP content, except for the ZnO-NP content of 25 g/100 g of blended polymer. These results indicate that the addition of the ZnO-NPs reduced the number of hydroxyl groups and charged groups ($-COO^-$). Moreover, the length of the polymer increased due to the formation of links between the ZnO-NPs, pectin, and alginate. Consequently, the solubility of the films was reduced. The ZnO-NP content in the 25 g/100 g of blended polymer was more than necessary for creating hydrogen bonds such that it probably had a negative effect on the cross-linking between the Ca^{2+} ion, the pectin, and the alginate carboxyl groups. This increased the solubility value.

Table 1: Thickness and mechanical properties (tensile strength = TS; % elongation = %E) of P/AG films with various ZnO-NP contents.

ZnO-NP content (g/100 g of blended polymer)	Thickness (μm)	TS (MPa)	%E (%)
0	43.8 ± 2.11^d	21.71 ± 1.23^d	0.86 ± 0.04^c
0.5	52.4 ± 2.5^c	32.59 ± 1.6^c	1.89 ± 0.05^b
2.5	56.8 ± 2.83^b	36.43 ± 0.98^b	1.93 ± 0.05^b
5	58.1 ± 2.08^b	41.56 ± 1.94^a	2.32 ± 0.06^a
25	74.6 ± 3.25^a	12.79 ± 0.58^e	0.85 ± 0.04^c

Alphabets in a column indicate a significant difference at $P < 0.05$ by Duncan's multiple.

Table 2: Effect of ZnO-NP contents on the solubility and water contact angle of P/AG films.

ZnO-NP contents (g/100 g of blended polymer)	Contact angle (°)	Solubility (%)
0	73.85 ± 6.4^a	30.38 ± 1.3^a
0.5	72.9 ± 4.5^a	27.34 ± 1.8^{ab}
2.5	72.5 ± 2.9^a	23.16 ± 2.1^b
5	70.73 ± 4.1^a	22.49 ± 3.1^b
25	54.07 ± 7.4^b	26.1 ± 1.5^{ab}

Alphabets in a column indicate a significant difference at $P < 0.05$ by Duncan's multiple.

Table 3: WVP and WVTR of P/AG films with various ZnO-NP contents.

ZnO-NP contents (g/100 g of blended polymer)	WVP $\times 10^{-14}$ (kg·m/m^2·Pa ·s)	WVRT $\times 10^{-7}$ (kg/m^2·s)
0	1.01 ± 0.029^a	4.226 ± 0.121^a
0.5	0.576 ± 0.018^b	2.148 ± 0.069^b
2.5	0.447 ± 0.008^c	1.949 ± 0.035^b
5	0.414 ± 0.02^{cd}	1.486 ± 0.069^c
25	0.381 ± 0.019^d	1.437 ± 0.069^c

Alphabets in a column indicate a significant difference at $P < 0.05$ by Duncan's multiple.

Thus, the film solubility trended to increase when the ZnO-NP content was 25 g/100 g of blended polymer.

3.3. *WVP and WVTR of P/AG Films.* Table 3 shows WVP and WVTR of film specimens. The WVP values of the specimens decreased from 1.01 to 0.414 kg·m/m^2·Pa·s (59%), and WVTR values decreased from 4.226 to 1.486 kg/m^2·s (64.8%), when increasing the ZnO-NP content from 0 to 5 g/100 g of blended polymer. This may be due to the formation of the hydrogen bonds between the pectin, alginate matrix, and oxygen atoms in the ZnO-NPs, which increases the adhesion of the biopolymer matrix and can reduce the diffusion of the water molecules in the material. A similar result was reported by Almasi et al. [20]. On the other hand, the even distribution of the ZnO-NPs with smaller sizes and

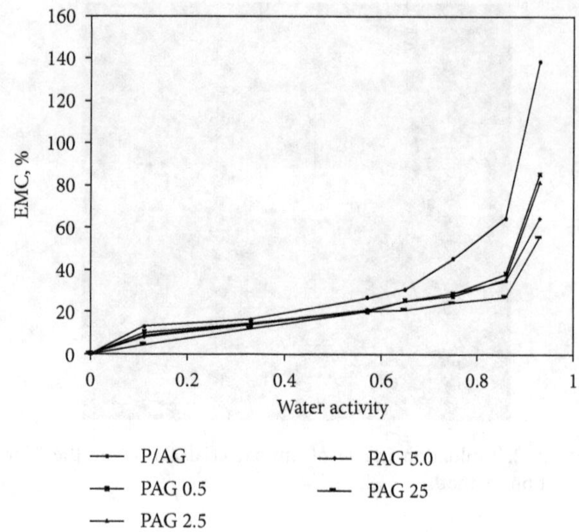

Figure 3: Moisture sorption isotherms of P/AG films incorporated with various ZnO-NP contents at 25 ± 1°C.

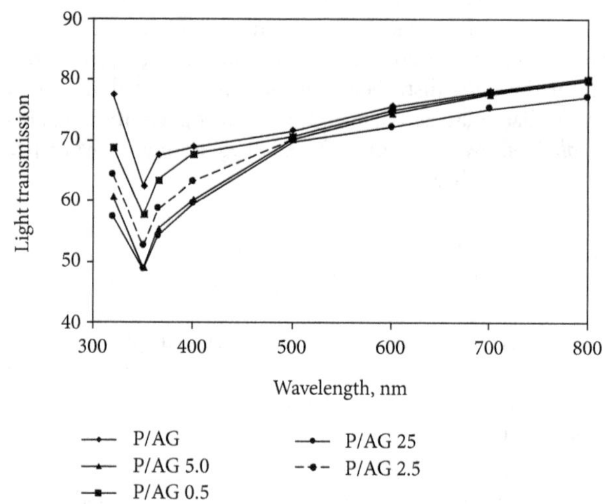

Figure 4: Light transmission of the films.

filling of the microvoids should reduce the length of a free path for water uptake. Similar results were reported for the effect of ZnO-NPs on the WVP of biopolymer-based nanocomposites [3, 21]. There was no significant change in the WVP and WVTR values of the sample when increasing the content of the ZnO-NPs from 5 to 25 g/100 g of blended polymer, which may be due to the maximum reduction of the hydrophilic groups, such as OH and COOH in the polymer.

3.4. *Moisture Sorption Isotherm.* The moisture absorption of the P/AG film was higher than that of P/AG films with the ZnO-NPs in the range of water activity between 0.86 and 0.93 (Figure 3). A similar result was reported by Galus and Lenart [22]. The film's moisture absorption decreased as the ZnO-NP content increased. However, there was no

Table 4: Transparency of P/AG films with various ZnO-NP contents.

ZnO-NP contents (g/100 g of blended polymer)	0	0.5	2.5	5	25
Transparency values (1/mm)	3.24 ± 0.02^{b}	3.26 ± 0.01^{b}	3.25 ± 0.03^{b}	3.17 ± 0.02^{ab}	3.01 ± 0.02^{a}

Alphabets in a column indicate a significant difference at $P < 0.05$ by Duncan's multiple.

Table 5: OP and OTR of P/AG films with various ZnO-NP contents.

ZnO-NP contents (g/100 g of blended polymer)	OP $\times 10^{-19}$ (kg·m/m^2·Pa·s)	OTR $\times 10^{-11}$ (kg/m^2·s)
0	270.86 ± 9.79^{a}	158.65 ± 16.17^{a}
0.5	253.85 ± 15.35^{a}	137.51 ± 6.95^{b}
2.5	191.87 ± 12.44^{b}	89.82 ± 3.68^{c}
5	110.79 ± 7.74^{d}	51.87 ± 3.98^{d}
25	132.92 ± 7.45^{c}	48.84 ± 3.55^{d}

significant change in the moisture absorption between the P/AG films and the P/AG films with the ZnO-NPs in the water activity range of 0 to 0.75. Based on the results obtained for the solubility, WVP, mechanical properties, and moisture sorption isotherm, it is recognized that these trends are due to the formation of hydrogen bonding between the OH groups of the pectin, alginate, and oxygen atoms of the ZnO-NPs. The reduction in moisture absorption may be due to the reduction of the free OH groups in the membrane. On the other hand, a uniform dispersion of smaller ZnO-NPs and filling of microvoids produces a decrease in the length of the free path for water uptake.

3.5. Transparency and Light Transmission of Films. Figure 4 shows the light transmission of film specimens. Light transmission decreases in the wavelength range of 320 to 350 nm (UV light) and reaches the lowest value at 350 nm. The light transmittance of films containing ZnO-NPs is always lower than that of films without the ZnO-NPs. The higher the ZnO concentration in the film, the lower the light transmission at the wavelength range of 320 to 350 nm. This result shows that ZnO-NPs in the film have absorbed ultraviolet light. This result is consistent with that of Khodaiyana et al. [23] who observed that the UV absorbance of the nanocomposite polymers increased as the ZnO-NP content increased. They also observed that a ZnO-Np content greater than 1% was necessary to induce UV filtration in a kefiran biopolymer. There is no significant difference in light transmittance between films containing ZnO-NPs and P/AG films in the wavelength range of 500 to 800 nm, except for P/AG films with a ZnO-NP content of 25 g/100 g of blended polymer.

Table 4 shows the transparency of the films with various ZnO-NP contents. The transparency for the films containing 0.5 to 5 g/100 g blended polymer of ZnO-NPs was not significantly different from the P/AG film. However, the addition of ZnO-NPs with 25 g/100 g of blended polymer into the films decreased the transparency value. The reduction of transparency of the films with a concentration of 25 g ZnO-NPs/100 g polymer may be due to the over-distribution of the ZnO-NPs in the pectin/alginate network.

3.6. Oxygen Permeability and Oxygen Transmission Rate. The OP and OTR of P/AG and P/AG films with various ZnO-NP contents are presented in Table 5.

The addition of ZnO-NP to the P/AG films led to a reduced oxygen permeability. The lowest reduction occurred when the ZnO-NP concentration reached 5 g/100 g of polymer. The OP values decreased by 59%, and OTR values decreased by 67.3% with the addition of 5 g ZnO-NPs/100 g of blended polymer. The permeability coefficient of a substance is determined by the formula $P = D \times S$, in which D is the diffusion coefficient and S is the solubility coefficient [24]. The reduction of OP and OTR can be attributed to the distribution of the ZnO nanoparticles in the film matrix, which led to a reduced oxygen diffusion. The increase in OP when the concentration of ZnO-NPs was 25 g/100 g of blended polymer was probably due to reducing the hydrogen bonds in the pectin/alginate matrix and reducing the interaction between P/AG and ZnO-NPs, due to structural changes in the polymer matrix. Therefore, the diffusion coefficient increased leading to an increase in the oxygen permeability of these films. This result shows the possibility of using P/AG membranes with ZnO-NPs in food preservation to inhibit oxidation.

3.7. Antimicrobial Action of Film-Forming Solutions. The antimicrobial activities of the P/AG ZnO-NPs films against *Escherichia coli, Saccharomyces cerevisiae, Colletotrichum gloeosporioides*, and *Aspergillus niger* are shown in Figure 5.

The inhibition zone increased with the increase of the ZnO-NP concentration in the range of 0.5 to 5 g/100 g of polymer. However, when increasing the concentration from 5 g to 25 g/100 g of polymer, the inhibitory zone did not increase. This may be related to the decrease of the diffusion ability of the ZnO-NPs in the agar at high concentrations. The antimicrobial effect was due to the ZnO-NPs because the pure P/AG film that did not contain ZnO-NPs was not antibacterial. The inhibitory effect of the ZnO-NPs against *Aspergillus niger* and *Colletotrichum gloeosporioides* was greater than that against *E. coli* and *Saccharomyces cerevisiae.* It is believed that the antimicrobial activity of ZnO-NPs is mainly dependent on the cell wall structure of the microorganisms. In almost mold, the hyphal cell wall consists of a single, 80 to 150 nm layer composed of chitin and β-glucan. In contrast, the 200 to 600 nm thick yeast cell wall has three layers [25]. The cell wall of *E. coli* (Gram-negative bacteria) is composed of a complex structure with thin peptidoglycan layer surrounded by an outer membrane [26, 27]. The antimicrobial activity of the P/AG films with ZnO-NPs was the highest when the ZnO-NP concentration was 5 g/100 g of blended polymer. The results of the study on the antimicrobial properties of P/AG films with ZnO-NPs support the use of this material to preserve food products, especially products that are susceptible to mold damage.

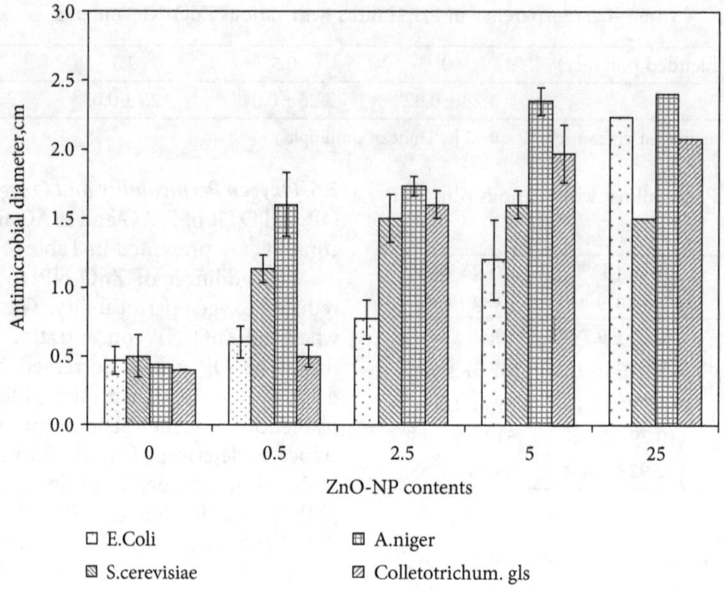

FIGURE 5: Antimicrobial activities of P/AG films with various ZnO-NP contents.

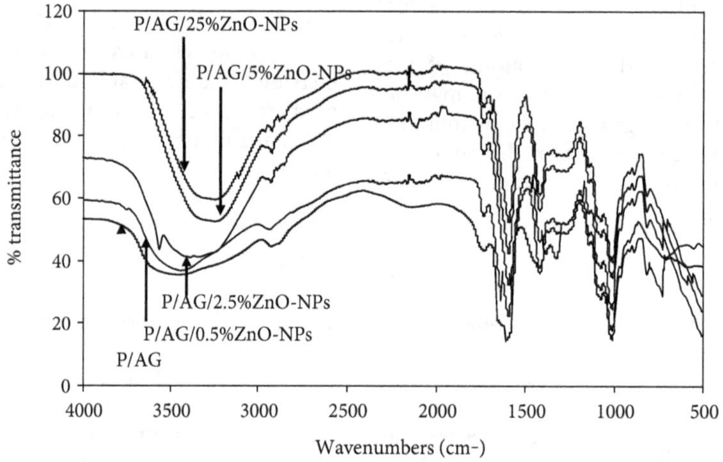

FIGURE 6: FTIR spectra of P/AG films incorporated with various ZnO-NP contents.

Moreover, ZnO-NPs have a negligible negative effect on human cells according to Reddy et al. [28].

3.8. FTIR Spectra of P/AG Films with ZnO-NPs. The ATR-FTIR analysis demonstrates the interaction between the ZnO-NPs, pectin, and alginate. The results are shown in Figure 6.

The results show that the spectra for all films have characteristic peaks at 4000–500 cm^{-1}. The P/AG films exhibited the major band at 1630 cm^{-1}, which is the characteristic peak of C=O stretching/hydrogen bonding coupled with a COO^- group of pectin and alginate [25]. The absorption peaks at 1400 and 1440 cm^{-1} are related to the stretching bands of the COO^- groups, and the peak at 3410 cm^{-1} was induced by the OH groups. The ATR-FTIR spectra of P/AG/ZnO-NPs films show that the peaks at 3410, 2890, 1630, and 1420 cm^{-1} were shifted to 3300–3200, 2920–2910, 1597, and 1410 cm^{-1}, respectively. Shifting of the adsorption peaks indicates that hydrogen bonds between the oxygen atoms of ZnO-NPs and the OH, the COOH, and the CH_3 (hexane) groups of pectin/alginate matrix occurred. These results confirm that the addition of ZnO-NPs can improve the water resistance, oxygen resistance, and mechanical properties of the P/AG films. The reason could be that the OH and the COOH groups of the pectin/alginate matrix are able to form hydrogen bonds with oxygen groups in the ZnO-NPs.

4. Conclusions

The addition of ZnO-NPs to pectin and alginate films changed some physicochemical properties of films compared to films without the ZnO-NPs. There was a decrease in the solubility, moisture absorption, UV light transmission, WVP, WVTR, OP, and the OTR of the P/AG films with ZnO-NPs compared to the films without them. In contrast, there was an increase in tensile strength, elongation at break, and antimicrobial properties of the P/AG films with ZnO-NPs compared to the films without them. In this work, the change was the highest when the ZnO-NP content in the film was 5 g/100 g of blended polymer. The ATR-FTIR analysis indicated that hydrogen bonds between the ZnO-NPs and pectin/alginate matrices were formed. These results suggest the possibility of using P/AG/ZnO-NPs film in food preservation and that they may be better than pectin/alginate films.

Conflicts of Interest

The authors declare that they have no conflicts of interest.

Acknowledgments

The authors gratefully acknowledge the financial support from the "911 Scholarship" for Ph.D. students from the Ministry of Education, Vietnam. We would also like to thank the Faculty of Chemical Engineering, University of Technology & Education (UTE) and University of Technology & Science (DUT), the University of Danang for the equipment and facility support. We wish to thank the Center of Excellence in Materials Science and Technology, Chiang Mai University, for the financial support under the administration of Materials Science Research Center, Faculty of Science, Chiang Mai University.

References

[1] J. W. Rhim, H. M. Park, and C. S. Ha, "Bio-nanocomposites for food packaging applications," *Progress in Polymer Science*, vol. 38, no. 10-11, pp. 1629–1652, 2013.

[2] S. Tunç and O. Duman, "Preparation of active antimicrobial methyl cellulose/carvacrol/montmorillonite nanocomposite films and investigation of carvacrol release," *LWT - Food Science and Technology*, vol. 44, no. 2, pp. 465–472, 2011.

[3] S. Shankar, X. Teng, G. Li, and J. W. Rhim, "Preparation, characterization, and antimicrobial activity of gelatin/ZnO nanocomposite films," *Food Hydrocolloids*, vol. 45, pp. 264–271, 2015.

[4] F. L. Seixasa, F. R. B. Turbianib, P. G. Salomãob, R. P. Souza, and M. L. Gimenes, "Biofilms composed of alginate and pectin: effect of concentration of crosslinker and plasticizer agents," *Chemical Engineering Transactions*, vol. 32, pp. 1693–1698, 2013.

[5] ASTM, "Standard test methods for tensile properties of thin plastic sheeting D882-10," in *Annual Book of ASTM*, American Society for Testing and Materials, Philadelphia, PA, USA, 2010.

[6] K. Jantanasakulwong, N. Leksawasdi, P. Seesuriyachan, S. Wongsuriyasak, C. Techapun, and T. Ougizawa, "Reactive blending of thermoplastic starch and polyethylene-graft-maleic anhydride with chitosan as compatibilizer," *Carbohydrate Polymers*, vol. 153, pp. 89–95, 2016.

[7] P. Rachtanapun and N. Rattanapanone, "Synthesis and characterization of carboxymethyl cellulose powder and films from Mimosa pigra," *Journal of Applied Polymer Science*, vol. 122, no. 5, pp. 3218–3226, 2011.

[8] K. N. Turhan and F. Sahbaz, "Water vapor permeability, tensile properties and solubility of methylcellulose-based edible films," *Journal of Food Engineering*, vol. 61, no. 3, pp. 459–466, 2004.

[9] C. M. Bitencourt, C. S. Fávaro-Trindade, P. J. A. Sobral, and R. A. Carvalho, "Gelatin-based films additivated with curcuma ethanol extract: antioxidant activity and physical properties of films," *Food Hydrocolloids*, vol. 40, pp. 145–152, 2014.

[10] P. Rachtanapun, "Shelf life study of salted crackers in pouch by using computer simulation model," *Chiang Mai Journal Science*, vol. 34, no. 2, pp. 209–218, 2007.

[11] P. Rachtanapun and R. Suriyatem, "Prediction models for moisture sorption isotherm of soy protein isolate/carboxymethyl chitosan blend films," *Journal of Agricultural Science and Technology*, vol. A2, pp. 50–57, 2012.

[12] S. Rawdkuen, P. Suthiluk, D. Kamhangwong, and S. Benjakul, "Mechanical, physico-chemical, and antimicrobial properties of gelatin-based film incorporated with catechin-lysozyme," *Chemistry Central Journal*, vol. 6, no. 1, pp. 1–10, 2012.

[13] A. S. T. M. D3985-10, "Standard test method for oxygen gas transmission rate through plastic film and sheeting using a coulometric sensor," in *Book of Standards*, ASTM International, 2010.

[14] S. Suganya, R. Bharathidasan, G. Senthilkumar, P. Madhanraj, and A. Panneerselvam, "Antibacterial activity of essential oil extracted from Coriandrumsativam (L.) and GC-MS analysis," *Journal of Chemical and Pharmaceutical Research*, vol. 4, no. 3, pp. 1846–1850, 2012.

[15] A. M. Cozmuta, A. Turila, R. Apjok et al., "Preparation and characterization of improved gelatin films incorporating hemp and sage oils," *Food Hydrocolloids*, vol. 49, pp. 144–155, 2015.

[16] F. Shahmohammadi Jebel and H. Almasi, "Morphological, physical, antimicrobial and release properties of ZnO nanoparticles-loaded bacterial cellulose films," *Carbohydrate Polymers*, vol. 149, pp. 8–19, 2016.

[17] L. Shi and S. Gunasekaran, "Preparation of pectin-ZnO nanocomposite," *Nanoscale Research Letters*, vol. 3, no. 12, pp. 491–495, 2008.

[18] A. A. Oun and J. W. Rhim, "Carrageenan-based hydrogels and films: effect of ZnO and CuO nanoparticles on the physical, mechanical, and antimicrobial properties," *Food Hydrocolloids*, vol. 67, pp. 45–53, 2017.

[19] H. Isawi, M. H. el-Sayed, X. Feng, H. Shawky, and M. S. Abdel Mottaleb, "Surface nanostructuring of thin film composite membranes via grafting polymerization and incorporation of ZnO nanoparticles," *Applied Surface Science*, vol. 385, pp. 268–281, 2016.

[20] H. Almasi, B. Ghanbarzadeh, and A. A. Entezami, "Physicochemical properties of starch-CMC-nanoclay biodegradable films," *International Journal of Biological Macromolecules*, vol. 46, no. 1, pp. 1–5, 2010.

[21] I. Shahabi-Ghahfarrokhi, F. Khodaiyan, M. Mousavi, and H. Yousefi, "Effect of γ-irradiation on the physical and mechanical properties of kefiran biopolymer film," *International Journal of Biological Macromolecules*, vol. 74, no. 3, pp. 343–350, 2015.

[22] S. Galus and A. Lenart, "Development and characterization of composite edible films based on sodium alginate and pectin," *Journal of Food Engineering*, vol. 115, no. 4, pp. 459–465, 2013.

[23] I. Shahabi-Ghahfarrokhi, F. Khodaiyan, M. Mousavi, and H. Yousefi, "Preparation of UV-protective kefiran/nano-ZnO nanocomposites:physical and mechanical properties," *International Journal of Biological Macromolecules*, vol. 72, pp. 41–46, 2015.

[24] R. J. Ashley, *Permeability and Plastics Packaging*, Elsevier Applied Science Publishers, London; New York, 1985.

[25] M. R. McGinnis and S. K. Tyring, *Medical Microbiology*, University of Texas Medical Branch at Galveston, Galveston, TX, USA, 1996.

[26] S. Anitha, B. Brabu, D. J. Thiruvadigal, C. Gopalakrishnan, and T. S. Natarajan, "Optical, bactericidal and water repellent properties of electrospun nanocomposite membranes of cellulose acetate and ZnO," *Carbohydrate Polymers*, vol. 87, no. 2, pp. 1065–1072, 2012.

[27] S. Paisoonsin, O. Pornsunthorntawee, and R. Rujiravanit, "Preparation and characterization of ZnO-deposited DBD plasma-treated PP packaging film with antibacterial activities," *Applied Surface Science*, vol. 273, pp. 824–835, 2013.

[28] K. M. Reddy, K. Feris, J. Bell, D. G. Wingett, C. Hanley, and A. Punnoose, "Selective toxicity of zinc oxide nanoparticles to prokaryotic and eukaryotic systems," *Applied Physics Letters*, vol. 90, no. 21, article 213902, 2007.

5

Proposition of Thin-Layer Repairing Methods Using Low-Elasticity Polymer Portland Cement Materials and Glue and Study on the Fatigue Resistance of Reinforced Concrete Slab

Tadashi Abe ,[1] Kiyoshi Ito,[2] Noriyuki Kozakai,[3] and Takayoshi Kodama[4]

[1]*College of Industrial Technology, Nihon University, Narashino City, Chiba 275-8575, Japan*
[2]*Nihon University Graduate School of Engineering, Nihon University, Narashino City, Chiba 275-8575, Japan*
[3]*Construction Materials Division, Sumitomo Osaka Cement Co., Ltd, Chiyoda-ku, Tokyo 102-8465, Japan*
[4]*Technical Sales Division, Kajima Road Co., Ltd, Bunkyo-ku, Tokyo 112-8566, Japan*

Correspondence should be addressed to Tadashi Abe; abe.tadashi@nihon-u.ac.jp

Guest Editor: Rishi Gupta

In this repair method, two types of adhesives, such as a penetrable adhesive and a high-durability adhesive, were applied to the adhesion interface of the materials to be repaired, and then cast. The repair method was then evaluated for fatigue resistance using wheel load running tests. It was developed to ensure the long-term integrity of the repair material and RC slabs. The results of the experiments confirmed that this repair method provides significantly better fatigue resistance than the conventional construction method. Moreover, this study proposes two wet repair cycles and one dry repair cycle with reinforcement measures to improve the load-bearing performance.

1. Introduction

Most of the Japanese road bridges were built between 1950 and 1970. By 2026, 46% of these bridges will be 50 years old or older, making the maintenance and management of aging bridges an important issue [1, 2]. Furthermore, implementation of waterproofing for the RC slabs was delayed. In fact, waterproofing in the design and during maintenance was required for expressways in 1988, but it was required for general national roads only after the road bridge specification was revised in 2002 [3]. For this reason, the upper surface of RC slabs covered with about 80 mm of asphalt pavement gets wet from rainwater and the damage from traffic load is accelerated. It was left untouched, and the damage due to cracks and potholes in the asphalt pavement has been left unrepaired. Japan is an elongated archipelago surrounded by water and has a varied climate with cold and temperate regions. As a result, bridges are affected by airborne salt near the coastline and by damage by salt attack. Bridges are also affected by freezing damage from the snow-melting and antifreezing agents that are used in snowy areas. Particularly in snowy cold regions, reinforcing bars become corroded, and the spraying of snow-melting agents and antifreezing agents causes daily freeze-thawing action. These factors accelerate damage to the concrete. As a result, the top surface concrete of RC slabs has been degraded into aggregate due to corrosion of the reinforcement caused by salt damage and due to freeze-thaw damage, and some of them have been removed after about 30 years of service. Moreover, a repair site can be redamaged and the damage can spread, resulting in the need to replace the RC slabs. The method for repairing a part of an RC slab when cracking has occurred or limestone has exuded from the asphalt is to remove the asphalt paving, and the part of the concrete that has degraded into aggregate is removed using a paving breaker. After removing the concrete, repair of the damaged part is carried out using a cementitious material such as an ultrarapid hardening nonshrink polymer cement mortar or ultrarapid hardening concrete. Finally, curing and waterproofing are carried out, and the asphalt paving is replaced. However, RC slabs are repaired with a thin layer of thickness of about 30 mm, so cracking is quickly caused by wheel loads

Table 1: Composition of repair materials.

				Unit content of water (kg/m^3)	Water bonding ratio (%)
U-M-36	Premix powder	Mixed material	Ultrarapid hardening cement	925	36
			Calcium hydrate	13	
		Other	Water reducer	2	
			Acrylic polymer powder	0	
			Silica sand	937	
	Fiber			—	
	Water			338	
U-FM-45	Premix powder	Mixed material	Ultrarapid hardening cement	581	48
			Calcium hydrate	37	
		Other	Water reducer	1	
			Acrylic polymer	6	
			Silica sand	1225	
	Fiber			5	
	Water			296	

due to the difference in elastic coefficients of the RC slab concrete and the repair material. Also, there have been many examples where peeling or cracking has occurred at the boundary between the existing RC slab and the repair layer, so repair was carried out again. Therefore, there is an urgent need to develop repair materials suitable for top surface damage of RC slabs and repair techniques capable of ensuring durability, because there is a concern that new cracks may occur due to the impact of a breaker used for cutting concrete in repair work. In this study, we examined the soundness of the adhesion interface of the existing RC slabs, the adhesion strength between repair materials and existing RC slabs, and the measures to be taken with attention to the repair materials. Wheel load running tests were conducted to evaluate the fatigue resistance of the repair material used in the conventional repair method and in the repair method proposed in this research. The repair technique using repair material developed exclusively for repairing RC slabs (fiber-reinforced, ultrahard, nonshrink, and polymer cement mortar) and the two types of adhesive (penetrable primer and highly durable epoxy resin adhesive) can reinforce the fine cracks in the RC slabs and integrate the repair material and the RC slabs. The repair technique ensures long-term durability and extends the repair cycle, thereby extending the service life of the bridge.

2. Materials and Methods

2.1. Material Properties of Repair Materials

2.1.1. Ultrarapid, Hardening, Nonshrink Polymer Cement Mortar. In Japan, road maintenance and repair work is restricted to about 8 hours. Rapid construction is required to demolish and remove road materials, repair the road, and release traffic from regulation within 8 hours. For this reason, an ultrarapid, hardening, nonshrink polymer cement mortar is used for the partial repair of conventional RC slabs. The materials of this cementitious mixture were mainly ultrarapid hardening cement and calcium hydrate, but in addition silica sand and water reducer were used. The curing characteristics of this repair material are as follows: initial setting time, 17 minutes; initial strength development, 45.2 N/mm^2 after 3 hours of aging; and static elastic modulus, 43.7 kN/mm^2 after 28 days of aging. Since this pot life is 30 minutes or less, prompt work is required. Note that this repair material is described as U-M-36 and is a conventional construction material. The compounding conditions are listed in Table 1. The expression intensity of U-M-36 is shown in Table 2.

2.1.2. Ultrahard, Fiber-Reinforced Polymer Cement Mortar. The ultrahard, fiber-reinforced polymer cement mortar is a new repair material. The materials of this cementitious mixture were mainly ultrarapid hardening cement and calcium hydrate, but in addition silica sand, water reducer, and acrylic polymer powder were used. This repair material is called U-FM-45. The compounding conditions are listed in Table 1. This repair material was premixed with high-strength vinylon fiber (fiber length: 12 mm). By setting the elastic coefficient to about the same as that of the existing RC slabs, resistance to cracking was improved. The material properties are as follows: curing, 45 minutes at setting completion time; compressive strength at 3 hours, 24.5 N/mm^2; and static elastic modulus at 28 days, 23.8 kN/mm^2. Therefore, compared with the ultrafast, hard, nonshrink mortar, adequate construction time is secured and cracking of thin-layer repairs is reduced. Table 2 shows the strength development of U-FM-45, and Figure 1 shows the relationship between compressive strength and static elastic modulus.

2.1.3. High-Durability Epoxy Resin Adhesive. A highly durable epoxy resin adhesive (hereinafter referred to as "bonding agent for placing") is applied to a penetrable adhesive that strengthens the adhering surface after damage. This firmly fixes the repairing material to the slab. This adhesive for adhesion was developed to help prevent fatigue in the steel deck [4, 5]. As reinforcement of the RC slab, it has been evaluated for its reinforcing effect by the bonding agent for

TABLE 2: Characteristic values of repair materials.

Test items		U-M-36	U-FM-45
Setting time	Initial setting	17 min	35 min
	Final setting	25 min	45 min
Compressive strength	2 hours	25.1 N/mm^2	19.1 N/mm^2
	3 hours	45.2 N/mm^2	24.5 N/mm^2
	4 hours	52.3 N/mm^2	27.1 N/mm^2
	28th	62.3 N/mm^2	49.9 N/mm^2
Young's modulus	28th	43.7 kN/mm^2	23.8 kN/mm^2

Concrete's age (Days)	Compressive strength (N/mm^2)	Static elastic modulus (kN/mm^2)
0.125	29	23.8
1	36.4	24.1
7	51.4	26
28	52.7	26.4

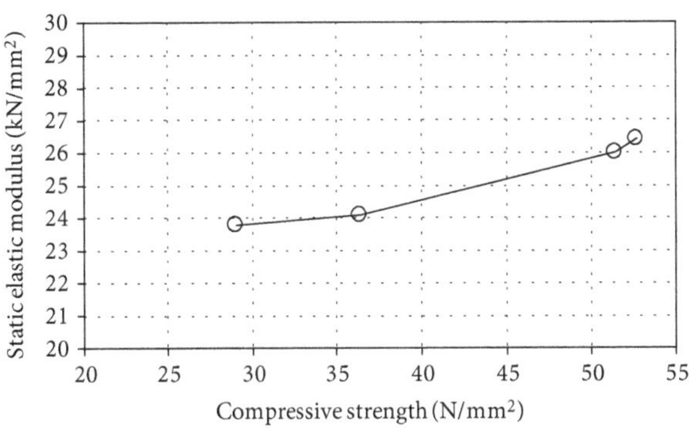

FIGURE 1: Relationship between compressive strength and static elastic modulus.

placing technique [6, 7]. An epoxy resin is a general term for a compound that contains two or more epoxides (oxiranes) in one molecule, and epoxy resin adhesives are epoxy resins in which denatured hardening agents and subsidiary materials are added and adjusted. The main features of the hardening agents used for epoxy resin adhesives and examples of typical subsidiary materials are summarized in Tables 3 and 4, respectively.

In the civil engineering field, bonding technologies using epoxy resin adhesives are applied in the precast concrete block method, in construction joints, in the steel plate bonding method, and in the CFRP bonding method. Also, the hardening agent used in these cases is almost always a polyamine type or a polyamine type combined with a polythiol (mercaptan) type. Table 5 lists the material property values of the adhesive for joining.

2.1.4. Penetrable Adhesive. When repairing the RC slab, a breaker is used to remove the damaged and associated fragile areas. The periphery becomes fragile due to fine cracks generated by the impact vibration of this breaker. A permeable adhesive has been developed and applied in order to reinforce the fine cracks generated by the shock vibration of this breaker [8]. This permeable adhesive was newly developed specially for the repair of RC slabs, with the viscosity adjusted so as to permeate fine cracks without using special equipment. The penetrative adhesive was compounded using as the main component a highly durable epoxy resin adhesive developed for fresh concrete construction joints, and adjusted so that the viscosity is lowered. The material properties of the permeable adhesive are listed in Figure 1. Figure 2 shows the status of the penetration in the test specimen U-M-36 that was coated with a permeable adhesive, after destruction.

2.2. Specification of Slab Specimen. A 3/5-scale model of the actual bridge has the following slab dimensions: length, 1600 mm; spacing, 1400 mm; and slab thickness, 150 mm. Based on the road bridge specifications for 2002 issued by the Japan Road Association, the reinforcing bars were made of SD 295A D13 as a double reinforcing bar arrangement. Then, D13 was arranged at 120 mm intervals in the direction perpendicular to the axis of the tensile side and in the axial direction, and 1/2 of the amount of tensile reinforcement was placed on the compression side. Figure 3 shows the dimensions of the slab, the rebar arrangement map, and the wheel load running fatigue testing machine. The concrete consisted of ordinary Portland cement, crushed sand (5 mm or less), and crushed stone (5 mm to 20 mm). The compressive strength of the concrete is 35 N/mm^2, the yield strength of the reinforcing bar is 377 N/mm^2, and the tensile strength is 511 N/mm^2.

2.3. Wheel Load Running Tests

2.3.1. Wheel Load Running Test Method. For the evaluation of fatigue resistance, the equivalent number of wheel loading cycles in a fatigue testing machine is used. In the wheel load running tests, a wheel load is repeatedly applied to the upper surface of the RC slabs and the repaired RC deck plate within a width of 300 mm for 1000 m in the axial direction. The wheel load starts traveling with an initial load of 100 kN. Loads are then added in increments of 20 kN every 20,000 runs. Deflection is measured for every 5000 runs when wheel loading runs are once, 10 times, 100 times, 1000 times, 5000 times, and more. A frame 600 mm in width and 1200 mm in length was placed around the wheel's traveling position to be filled with water. Wheel loading is performed under wet conditions after the first and second repairs. Holes 9 mm in diameter were drilled in several repair interfaces in order to observe fluctuations in the water level and check for

TABLE 3: Classification of hardening agents used for epoxy resin adhesives.

Classification	Features
Polyamine type hardening agents	Aliphatic polyamine: the reaction is fast and heat generation is remarkable. Durable to water and organic solvent. Good adhesion.
	Aliphatic polyamine: the reaction is fast and heat generation is remarkable. Durable to water and organic solvent. Good adhesion.
	Polyamide: long pot life and low cost
	Epoxy-adducted polyamine: high viscosity (useful to viscosity control).
	Mannich-modified polyamine: good to low-temperature hardening.
	Polythiol: good to low-temperature hardening polymer network structures are looser than those of polyamine. High water adsorption. Low durability to water.
Catalysts	Tertiary amine: catalyst of polyamine and as catalysts of polyamine and polyamide amine.
	Imidazole: reacts as hardening agents and as catalysts of anhydride hardening agents.

TABLE 4: Typical submaterials for epoxy resin adhesive.

	Classification
Additives to give functions	Aliphatic polyamine: the reaction is fast and heat generation is remarkable. Durable to water and organic solvent. Good adhesion.
Others	Pigments, dyestuff

TABLE 5: Material characteristic value of adhesive.

Test items		Penetrable adhesive	High-durability adhesive
Appearance	Base compound	Transparent liquid	White paste
	Curing agent	Transparent liquid	Blue liquid
Mixing ratio (main agent: curing agent)		10 : 3	5 : 1
Specific gravity of cured product		1.2	1.42
Compressive strength		$104.4\,N/mm^2$	$102.9\,N/mm^2$
Compressive modulus of elasticity		$3172\,N/mm^2$	$3976\,N/mm^2$
Flexural strength		$92.8\,N/mm^2$	$41.6\,N/mm^2$
Tensile shear strength		$58.2\,N/mm^2$	$14.9\,N/mm^2$
Concrete adhesion strength		$2.6\,N/mm^2$	$3.7\,N/mm^2$ or more of destruction in the base material

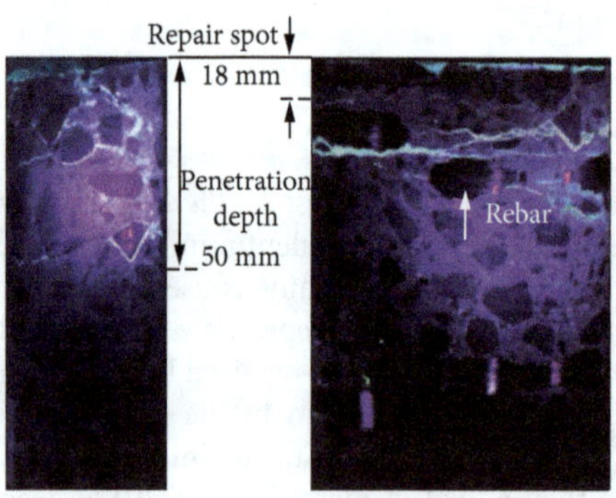

FIGURE 2: Result of permeation verification after specimen RC-U-36 destruction. (1) Diameter 50 mm. (2) Diameter 100 mm.

delamination. The specifications for the water filling is shown in Figure 4.

2.3.2. Repair Method and Repair Cycle. The specimen used in this experiment reproduced the damaged state in the actual RC slab. In the place where damage such as potholes repeatedly occurred on the asphalt pavement of the RC slab, when the asphalt pavement is removed, the concrete is separated into aggregate and sand (Figure 5(a)) and is in a fragile state. This weakened area was scraped off with a breaker accompanied by striking and vibration (Figure 5(b)) and was finished into a healthy and strong surface (Figure 5(c)). This area was then repaired with an ultrarapid hardening nonshrink polymer cement mortar or ultrarapid hardening concrete agent (Figure 5(d)), and water proofing was carried out. In order to reproduce such a state, the initial loading of the slab specimen was added to reproduce the condition requiring repair. From the previous research, it was determined that repair is required when a deflection corresponding to 1/400 of the span length occurs and two-way cracking is developed on the lower surface of the slab [9–11]. The repair method was verified using a slab surface test specimen subjected to this initial loading. The experimental situation is shown in Figure 6. Using the repair material U-M-36 of the conventional construction method, a breaker was used to break the

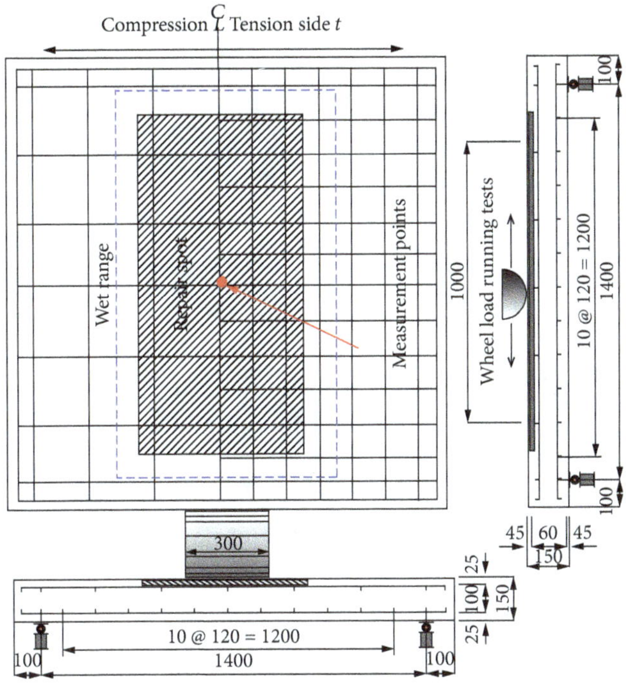

Figure 3: Dimensions of test specimen and plumbing diagram.

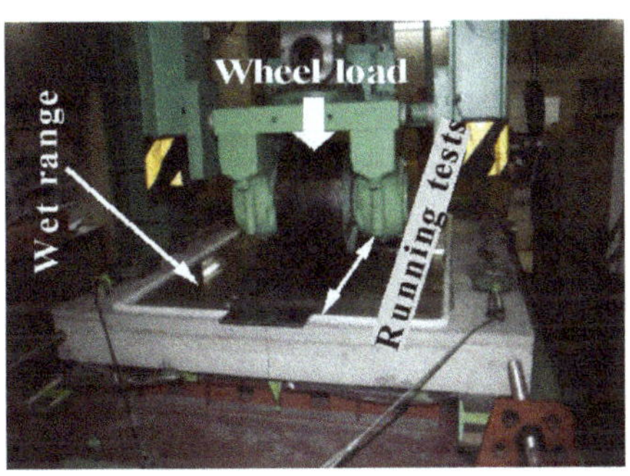

Figure 4: Appearance of running load fatigue testing machine and the specifications for the water filling.

repair area (Figures 6(a) and 6(b)), clean the surface, apply the repair material U-M-36, and then finish the surface. Next, in the work procedure of the repair method using the repair material U-FM-45 proposed in this research, a permeable adhesive is applied first to the adhesion interface which was scratched and cleaned, and the bonding agent for joining is repeatedly painted (Figures 6(c) and 6(d)). Then, the repair material U-FM-45 was poured onto the surface and the finish was the applied (Figure 6(e)). The cycle of repair and running wheel load tests was conducted twice under wet conditions and once under dry conditions. In the two wet-condition cycles, repair was performed when the wheel load caused unevenness or scaling or when the deflection reached 1/400 of the slab span L. The dry-condition cycle was performed until punching shear fracture occurred. Fatigue resistance is determined from the cumulative number of equivalent cycles obtained from these three tests.

2.3.3. Number of Equivalent Cycles in Running Fatigue Test. The wheel load running fatigue test uses a step load that increases every twenty thousand runs, and the equivalent running number is calculated to evaluate fatigue resistance. Assuming that the Miner's rule is applicable to the number of equivalent cycles, it is given by (1) [12]. Reference load P in (1) is set at 72 kN, which takes into consideration a safety factor of 1.2 based on 3/5 of the design active load, to calculate the equivalent traveling number. The value of 12.7, as proposed by Matsui, is applied to the absolute value of the reciprocal m of the slope of the SN curve.

$$N_{eq} = \sum_{i=1}^{n} (P_i/P)^m \times n_i. \quad (1)$$

The equation is expressed as follows: P_i is the loaded load (kN), P is the reference load (=72 kN), n_i is the number of experiment runs (times), and m is the reciprocal of the slope of the SN curve.

3. Results

3.1. Number of Equivalent Cycles. Table 6 lists the number of equivalent cycles and the reinforced equivalent running number used in this study.

3.1.1. Specimen RC-U-36 Using Conventional Repair Material U-M-36. The number of equivalent cycles of the initial loading is 7.865×10^6. With the repair material U-M-36 with wheel loading under wet conditions in the first cycle, the number of equivalent cycles in the running wheel load running tests is 1.685×10^6. Similarly, the number of equivalent cycles of the 2nd repair cycle is 1.621×10^6. The number

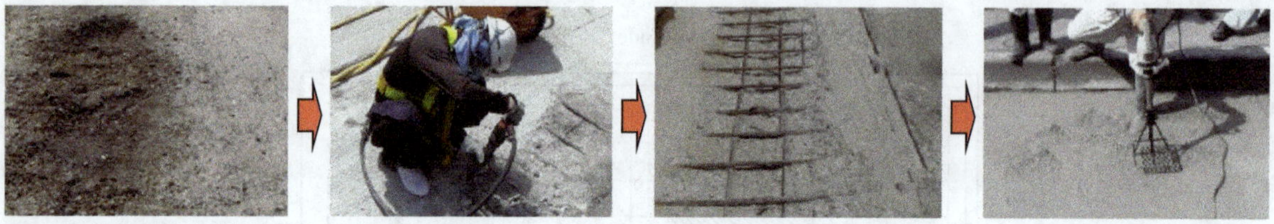

FIGURE 5: Example of thin-layer repair of upper layer of RC slab. (a) Degraded into aggregate. (b) Removal of damaged parts. (c) Removal of injured site. (d) Thin-layer repair.

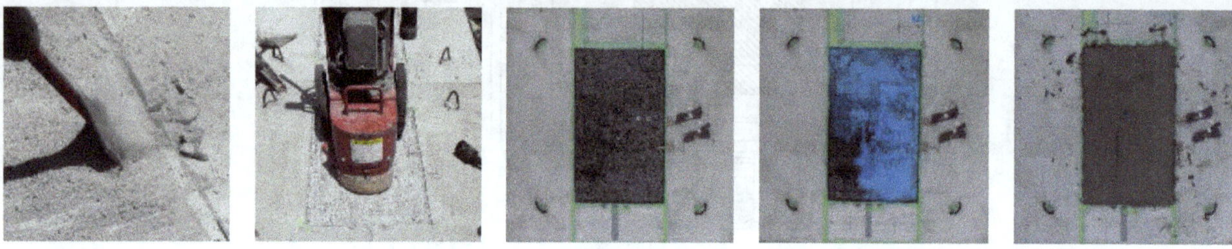

FIGURE 6: Procedure for manufacturing the RC slab specimen. (a) Removal of damaged parts. (b) Grinding of damaged parts. (c) Application of penetratable adhesive. (d) Application of bonding agent for placing. (e) Placing repair material and finish.

TABLE 6: The number of equivalent cycles and the reinforced equivalent running number used.

		RC-U-36	RC-UF-45
RC slab		7,865,598	7,865,598
Primary repair (wet state)	Equivalent runs, number of times	1,685,974	3,685,775
	Reinforcement effect	—	2.2
Secondary repair (wet state)	Equivalent runs, number of times	1,621,127	3,250,670
	Reinforcement effect	—	2.0
Tertiary repair (dry state)	Equivalent runs, number of times	10,493,077	42,350,384
	Reinforcement effect	—	4.0
Total until tertiary repair	Equivalent runs, number of times	21,665,776	57,152,427
	Reinforcement effect	—	2.64

of equivalent cycles for the third repair performed under dry conditions was 10.493×10^6, which resulted in punched shear failure. The cumulative number of the equivalent cycle count from 1 repair cycle to 3 repair cycles was 21.665×10^6, which resulted in a repair effect of 1.57 for the unrepaired slab plate. Wheel load data in Table 6 shows the number of equivalent cycles in the running wheel load running tests.

3.1.2. Specimen RC-UM-45 Using New Repair Material U-FM. The number of equivalent cycles of repairs using the proposed repair material U-FM and two kinds of adhesives was 3.685×10^6 in the first cycle of repair, 3.250×10^6 in the second cycle of repair, and 42.35×10^6 in the third cycle. This is 2.19 times more in the first cycle under dry conditions than with the conventional repairing method, 2.01 times in the second cycle, and 11.49 times in the third cycle. Also, the cumulative number of equivalent cycles until destruction in the 3rd cycle was 57.152×10^6. A reinforcement effect of 2.64 times that of the conventional construction method was obtained.

3.2. Relation between Deflection and Number of Equivalent Cycles

3.2.1. Specimen RC-U-36 Using Conventional Repair Material U-M-36. The residual deflection under initial loading was 1.42 mm. For repairs with repair material U-M-36, in the first cycle, the cumulative number of equivalent cycles is 9.551×10^6, the cumulative deflection is 3.44 mm, and the residual deflection is 1.96 mm. The cumulative number of equivalent cycles in repair cycle 2 is 11.172×10^6, the cumulative deflection is 4.10 mm, and the residual deflection is 2.74 mm. In the third repair performed under dry conditions, damage on the repair surface was suppressed. Therefore, the cumulative number of equivalent cycles is 21.655×10^6 and

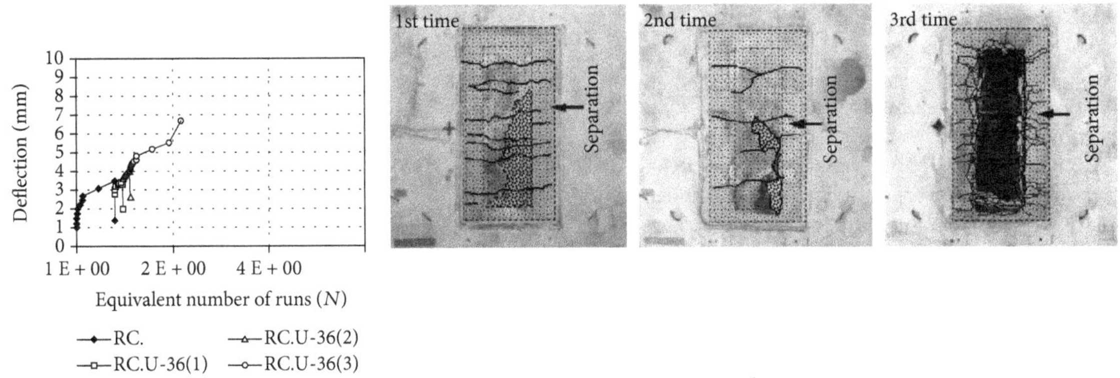

(a) (b)

FIGURE 7: RC-U-36. (a) Relationship between deflection and number of equivalent cycles. (b) Condition of damage caused on the upper surface.

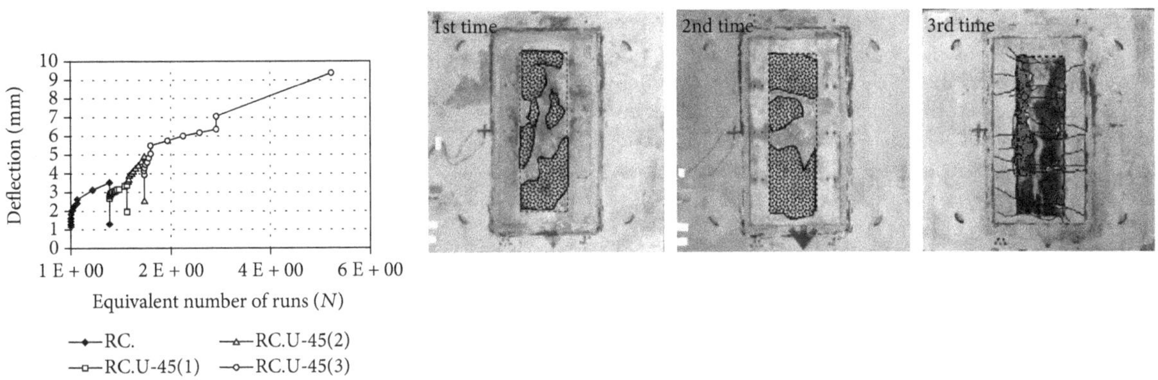

(a) (b)

FIGURE 8: RC-UF-45. (a) Relationship between deflection and number of equivalent cycles. (b) Condition of damage caused on the upper surface.

the cumulative deflection is 6.65 mm. Figure 7(a) shows the relationship between deflection and the number of equivalent cycles. Figure 7(b) shows the condition of damage caused on the upper surface.

3.2.2. About the Slab Using the Proposed Repair Material U-FM Punching Shear Strength. Residual deflection due to the initial loading of the proposed repair material U-FM-45 and the test specimen to which two types of adhesives are applied is 1.31 mm. The cumulative number of equivalent cycles in the first cycle of repair is 11.551×10^6, the cumulative deflection is 3.39 mm, and the residual deflection is 1.90 mm. Figure 8(a) shows the relationship between deflection and the number of equivalent cycles. As shown in Figure 8(a), the damaged upper surface of the specimen after the runs showed irregularities due to wear, but no cracks are observed. The cumulative number of equivalent cycles in repair cycle 2 is 14.802×10^6, the cumulative deflection is 4.85 mm, and the residual deflection is 2.53 mm. The only damage is the unevenness of the road surface. The cumulative number of the equivalent cycles of repair cycle 3 under dry conditions is 57.152×10^6, and the cumulative deflection is 9.33 mm. Since repair cycle 3 was conducted until the test specimen was destroyed by punching shear fracture, cracking occurred in the direction perpendicular to the axis on the upper surface of the test specimen. However, the number of equivalent cycles was four times that when using the conventional reinforcement method. These results confirm that the proposed repair method provides fatigue resistance and is practical for repairs using conventional reinforcement materials.

4. Discussion

4.1. Evaluation of Fatigue Resistance by the Number of Equivalent Cycles. A higher repair effect was obtained for RC slab test specimens repaired with the repair material U-FM-45 by applying 2 types of adhesive at the top surface repair boundary of the RC slab than the repair effect for RC slab test specimens repaired with the conventional repair material U-M-36, by a factor of 2.2 for the first repair, a factor of 2 for the second repair, a factor of 4 for the third repair under dry conditions, and a factor of 2.64 at failure.

4.2. Evaluation of Crack Resistance. In these tests, after the deflection of the RC slab reached 1/400 of the span L, 2 cycles of repair were carried out under wet conditions and one cycle under dry conditions. After the wheel load running tests, multiple cracks were observed extending in the direction normal to the axial direction in the top surface of the slab that was repaired using the conventional repair material U-M-36. However, on the top surface of the test specimens repaired with the repair material U-FM-45 after applying the two types of adhesive to the boundary, the number of cracks in the direction normal to the axial direction was greatly reduced compared with the repair material U-M-36, although damage due to wear could be seen on the running surface. As a result, the fatigue durability was increased.

5. Conclusions

In this study, we investigated countermeasure technologies presuming that redamage is caused by delamination of the adhesion interface and cracking of the repair material, further supposing that rainwater and others, intrude and induce damage. The first countermeasure technology is a permeable adhesive for strengthening the adhesion interface. The second countermeasure technology is an adhesive for fresh concrete that firmly bonds the existing repair material with the new repair material. Finally, the third is the development and application of a repair material that has an elastic modulus equivalent to that of existing slabs and further improves resistance to cracking by using resin fibers to add toughness. In this research, we proposed a repair method that combines these technologies and confirmed that the method provides much greater fatigue resistance than the conventional construction method. The repair method using the repair material U-M-45 after applying the two types of adhesive obtained the same repair effect as was verified in the tests when applied to a real bridge. Namely, the fragile concrete surface was strengthened by applying the penetrable adhesive, and the repair material was firmly bonded to the existing slab at the joint by applying the epoxy adhesive. Therefore, for repairing thin layers of RC slabs, a repair method that applies dedicated repair material for repairing the slab after application of two types of adhesive is practical.

Conflicts of Interest

The authors declare that there is no conflict of interest regarding the publication of this paper.

Acknowledgments

The authors are grateful for the advice from each of the engineers of Sumitomo Osaka Cement Co. Ltd and Kajima-road Co. Ltd.

References

[1] T. Tamakoshi, M. Okubo, and K. Ikuta, "Analysis of periodic inspection results of highway bridges in Japan," in *25th US-Japan Bridge Engineering Workshop*, pp. 61–68, Japan, 2009.

[2] H. Nakai, S. Matsui, T. Yoda, and A. Kurita, "Trends in steel-concrete composite bridges in Japan," *Structural Engineering International*, vol. 8, no. 1, pp. 30–34, 1998.

[3] Japan Road Association, *Specifications and cmmentary for road bridges I, II, III*, Yamato Planning Printing Co., Ltd., 2002.

[4] T. Kodama, O. Kamada, M. Nishi, Y. Zuzuki, and T. Fukute, "A study of bonded concrete overlays using epoxy adhesive in Japan," in *Sixth International Conference on Maintenance and Rehabilitation of Pavements and Technological Control (MAIREPAV6)*, p. 111, Turin, Italy, 2009.

[5] K. Ito, J. Takahagi, H. Chiba, and Y. Suzuki, "Study on a countermeasure for improving the fatigue durability of existing orthotropic steel deck plates using SFRC pavement," in *Proceedings of 5th China-Japan Workshop on Pavement Technologies (CJWPT 2009)*, pp. 221–226, China, 2009.

[6] T. Kodama, M. kagata, S. Higashi, K. Itoh, and Y. Ichinose, "Effect of reducing strains by SFRC pavement on Ohira Viaduct," in *Proceeding of the 25th US-Japan Bridge Engineering Workshop. Tsukuba: Public Works Research Institute 2009*, pp. 149–156, Tsukuba, Japan, 2009.

[7] T. Abe, H. Suzuki, Y. Kishi, and K. Nomoto, "The effect of adhesive on the fatigue resistance of RC slabs strengthened by SFRC upper surface thickness increasing method," *Proceedings of Structural Engineering*, vol. 59A, pp. 1084–1091, 2013.

[8] K. Ito, T. Abe, M. Fuchigami, N. Kozakai, and T. Kodama, "A study on fatigue durability of the top surface repair method using two types of adhesives for damaged RC slabs," in *Fourth Conference on Smart Monitoring, Assessment and Rehabilitation of Civil Structures (SMAR 2017)*, p. 144, Zürich, Switzerland, 2017.

[9] T. Abe, T. Kida, M. Takano, and Y. Kawai, "Evaluation of the punching shear strength and fatigue resistance of road bridge RC slabs," *Journal of Japan Society of Civil Engineers, Ser. A1 (Structural Engineering & Earthquake Engineering (SE/EE))*, vol. 67, no. 1, pp. 39–54, 2011.

[10] T. Abe, T. Kida, T. Sawano, A. Komori, and S. Hida, "Reinforcement effect and mechanical characteristics of CFS reinforced RC slabs with different number of layers," *Theoretical and Applied Mechanics Japan*, vol. 57, pp. 95–103, 2009.

[11] M. Takano, T. Abe, T. Kida, A. Kobayashi, and A. Komori, "Fatigue characteristics of CFS-reinforced RC slabs sustaining running vibration loads and running constant loads," *Theoretical and Applied Mechanics Japan*, vol. 58, pp. 31–39, 2010.

[12] S. Matui, "Design and construction of road bridge slabs and maintenance," *Mori Kita Publishing, Inc*, pp. 44–46, 2007.

6

Integration of Biomechanical and Biological Characterization in the Development of Porous Poly(caprolactone)-Based Membranes for Abdominal Wall Hernia Treatment

Federico Vozzi,[1] Tiziana Nardo,[2] Ilenia Guerrazzi,[1] Claudio Domenici,[1] Silvia Rocchiccioli,[3] Antonella Cecchettini,[3,4] Laura Comelli,[3] Giovanni Vozzi,[5] Carmelo De Maria,[5] Francesca Montemurro,[5] Gianluca Ciardelli,[2] and Valeria Chiono[2]

[1] *Laboratory of Biomimetic Materials and Tissue Engineering, Institute of Clinical Physiology IFC-CNR, Pisa, Italy*
[2] *Department of Mechanical and Aerospace Engineering, Polytechnic of Turin, Turin, Italy*
[3] *Laboratory of Proteomics, Institute of Clinical Physiology IFC-CNR, Pisa, Italy*
[4] *Department of Clinical and Experimental Medicine, University of Pisa, Pisa, Italy*
[5] *Research Center "E. Piaggio", University of Pisa, Pisa, Italy*

Correspondence should be addressed to Federico Vozzi; vozzi@ifc.cnr.it

Academic Editor: Vitor Sencadas

Aims. Synthetic meshes are the long-standing choice for the clinical treatment of abdominal wall hernias: the associated long-term complications have stimulated the development of a new generation of bioresorbable prostheses. In this work, polycaprolactone (PCL) porous membranes prepared by solvent casting/porogen leaching of PCL/poly(ethylene glycol) (PEG) blends with different compositions (different PCL/PEG weight ratios and PEG molecular weights) were investigated to be applied in the field. An optimal porous membrane structure was selected based on the evaluation of physicochemical, biomechanical, and *in vitro* biological properties, compared to a reference commercially available hernia mesh (CMC). *Findings.* Selected PCL7-2i membranes, derived from PCL/PEG 70/30 (PCL: M_w 70,000-90,000 Da; PEG: 35,000 Da), showed suitable pore size for the application, intermediate surface hydrophilicity, and biomimetic mechanical properties. *In vitro* cell tests performed on PCL7-2i membranes showed their cytocompatibility, high cell growth during 21 days, a reduced production of proinflammatory IL-6 with respect to CMC, and a significant secretion of collagen type I. *Conclusions.* PCL7-2i membranes showed biomimetic biomechanical properties and *in vitro* biological properties similar to or even better than - in the case of anti-inflammatory behavior and collagen production - CMC, a commercially available product, suggesting potentially improved integration in the host tissue.

1. Introduction

Ventral hernias are one of most frequent problems associated with major abdominal surgical procedures with an incidence between 2 and 20% of patients [1], with a recurrence rate of 20-25% [2]. The high incidence of this side effect generates an important cost for national healthcare systems, with a worldwide estimation of millions of hernia repairs every year (20 millions only for inguinal hernia treatment) [3].

Hernia treatments include a series of options as suture repair and prosthetic materials to reinforce the fascial defect [4]. This last option has shown its potentiality, limiting the recurrence rate with respect to suture repair [5, 6], in particular delaying this risk of 2-3 years [7]. Currently, implantation of prostheses, able to support damaged tissue in its mechanical and structural functions, represents the care standard treatment. Main advantages of meshes are related to their easy availability and resistance to patients' tissue stresses [8, 9].

Synthetic meshes, commonly based on nonresorbable polymers such as polypropylene (PP) and expanded polytetrafluoroethylene (ePTFE), are the long-standing choice in clinical practice since their introduction in the 50s [10]. These materials are effective in hernia treatment with respect to surgical suture but can cause long-term complications such as adhesion formation, graft infection/rejection, fistula formation, and hernia recurrence [11–13]. For all these reasons, research in soft tissue engineering is aimed at realizing a new generation of prostheses able to mimic mechanical properties of host tissue and recreate a physiological environment for its cellular components [14]. One of the possible improvements is development of scaffolds with resorbable properties, in terms of ability to integrate in the host, promotion of new tissue formation, and replacement after a well-defined period. Examples of these materials are VICRYL™® (Ethicon Inc., Somerville, NJ, USA) and DEXON™® (Covidien, Mansfield, MA, USA), based on copolymer formulations of poly(glycolic acid) and poly(lactic acid) (PLGA): although able to promote tissue integration, they were rapidly degraded, causing recurrence of hernia due to tissue strength loss [15–17].

Mechanical properties have been improved by Polylactide Mesh® (Ethicon, San Lorenzo, PR, USA), based on polylactide (95% lactide, 5% glycolide), which however has shown a more severe inflammatory response [14]. TIGR® (Novus Scientific, Uppsala, Sweden), a multifilament mesh based on a fast-degrading PLGA and slow-degrading poly(trimethylene carbonate) (PTMC), and GORE BIO-A® (Gore, Newark, DE, USA), an electrospun mesh based on polyglycolic acid (PGA)–PTMC copolymer, have shown limited mechanical load-bearing properties during application [14].

Phasix® (Bard, Warwick, RI, USA), a poly(4-hydroxybutyrate) (P4HB) mesh, has also been introduced having a long degradation time (>72 weeks in male Yucatan swine); however, its long-term *in vivo* response is unknown [14].

Additionally, bioresorbable polymers, such as polycaprolactone (PCL), polydioxanone (PDO), PLGA, poly(L-lactic acid) (PLLA), and PCL/collagen blends, have been investigated in scientific literature for the preparation of nanofibrous surgical meshes [18]. Although nanofiber texture mimics the extracellular matrix (ECM) structure and facilitates cell attachment, nanofibrous membranes suffer from poor mechanical properties and are unable to support the abdominal wall pressure, especially as soon as the polymer molecular weight decreases during degradation [18].

An alternative scaffold morphology to a typical filament or nanofibrous structure could offer superior mechanical properties allowing the use of bioresorbable polymers in the treatment of abdominal hernia.

PCL, a member of polyesters of the poly(α-hydroxyl acid) family, represents a promising candidate for bioresorbable hernia meshes. PCL is a US Food and Drug Administration-approved polymer, with biocompatible and biodegradable properties, able to favor attachment and growth of different cell types (fibroblasts [19, 20], myoblasts [21], chondrocytes [22], and smooth muscle cells [23]) and is commonly used in different biomedical applications [24]. In particular, PCL is widely used for the preparation of scaffolds applied in reparative surgery as treatment of wound healing [25].

In this work, polymeric blends based on PCL with poly (ethylene glycol) (PEG) of two different molecular weights (PEG1: M_w 10,000 Da; PEG2: M_w 35,000 Da) were developed for application in hernia treatment and tested for their structural, mechanical, and biological properties compared to a commercial product (CMC, Dipro Medical Devices, San Mauro Torinese, Italy), used as a reference system [26]. The CMC is a nonresorbable composite prosthesis used for ventral hernia repair and made up by a polypropylene lightweight macroporous mesh sewn on a smooth polypropylene film. PEG is a polymer frequently used in tissue engineering thanks to its nontoxicity properties, low protein adhesion, absence of immunogenicity, and solubility in water and organic solvents [27, 28]. In this work, PEG was used as a porogen blend component. Polymeric blends were analyzed for their physicochemical properties to evaluate morphology, thermal behavior, chemical characteristics, surface wettability, and tensile mechanical properties. *In vitro* cell tests using human fibroblasts, directly involved in the wound healing process [29, 30], were performed to evaluate cell biocompatibility of materials in terms of cytotoxicity and cell growth, inflammatory profile, and modulation of ECM by proteomic analysis.

The integration of biomechanical and biological characterization is fundamental in the development of a new generation of bioengineered scaffolds able to replicate and mimic the host tissue characteristics.

2. Materials and Methods

2.1. Materials. PCL (M_w: 70,000-90,000 Da) and PEG (M_w: 10,000 Da and 35,000 Da, coded as PEG1 and PEG2, respectively) were supplied from Sigma-Aldrich (Milan, Italy), in granular form. All solvents were of analytical grade and used as received (Sigma-Aldrich, St. Louis, MO, USA). CMC was gently furnished by Dipro Medical Devices (San Mauro Torinese, Italy).

2.2. Scaffold Fabrication. Porous PCL membranes were prepared using a solvent casting–leaching method. PCL/PEG1 and PCL/PEG2 100/0, 70/30, 60/40, and 0/100 (wt/wt) blends were prepared using solution mixing. The polymers were dissolved in chloroform at 10% (wt/v) concentration. Membranes were obtained by casting a volume of each solution (10 mL) on glass Petri dishes with 55 mm diameter. The solvent was removed via evaporation under a fume hood at room temperature for a period of 48 h, and the dried membranes were weighed (W_0). Each membrane was then immersed in deionized water for 48 h: water was replaced after this time, and membranes were immersed in deionized water for a further 24 h to assure PEG leaching out. The

Table 1: Samples nomenclature before and after the PEG leaching process.

Before leaching	After leaching
PCL	PCL
PEG1	—
PEG2	—
PCL/PEG1 70/30	PCL7-1i
PCL/PEG1 60/40	PCL6-1i
PCL/PEG2 70/30	PCL7-2i
PCL/PEG2 60/40	PCL6-2i

obtained porous membranes were dried at 37°C for a period of 24 h and weighed again (W_f). Samples codes are reported in Table 1. The code for leached samples is "PCL$x-yi$," where x indicates the PCL amount in the initial PCL/PEG blend ($x = 7$ or 6, indicating that the PCL content in the initial blend was 70% or 60%, respectively), while y indicates the PEG type used in the blend ($y = 1$ or 2, indicating PEG1 or PEG 2, respectively).

2.3. PEG Leached Out Percentage. The percentages of PEG leached out from the membranes were calculated according to the

$$\text{PEG leached out (\%)} = \frac{W_0 - W_f}{\text{PEG loading weight}} \bullet 100, \quad (1)$$

where W_0 and W_f are the weights of membranes before and after leaching.

2.4. Scanning Electron Microscopy. Scanning electron microscopy (SEM LEO 1430, Zeiss, Milan, Italy) was performed on the surface and fractured sections (in liquid nitrogen) of membranes before and after PEG leaching. Samples were sputter-coated with gold before SEM analysis. The accelerating voltage was 20 kV.

2.5. Differential Scanning Calorimetry. The thermal behavior of membranes, before and after leaching, was investigated by differential scanning calorimetry (DSC) using TA Instruments DSC Q20 (Milan, Italy). Membrane samples (5-10 mg) were weighed in aluminum pans, and nonisothermal scans were performed between −65°C and 100°C (first heating; cooling; second heating) at a heating rate of 10°C/min under nitrogen atmosphere. The melting temperature (T_ms) and enthalpy of fusion (ΔH_ms) of membranes were measured from the second DSC heating scan whereas crystallization temperature (T_Cs) and enthalpy (ΔH_Cs) were obtained from the DSC cooling scan. The glass transition temperature (T_gs) of PCL, PEG1, and PEG2 was not detected because it was lower than the minimum temperature reachable by the employed DSC apparatus.

Leached membranes were also analyzed by DSC by heating them from room temperature to 120°C at 10°C/min to test the efficiency of the drying procedure (24 h incubation in an oven at 37°C). The lack of an endothermic peak at 100°C due to water evaporation was considered the evidence of effective drying.

2.6. Fourier Transform Infrared Spectroscopy. Chemical analysis of samples before and after leaching has been carried out in a PerkinElmer Frontier Optical Spectrometer fitted with an attenuated total reflection sampler (FTIR-ATR, Milan, Italy). Spectra were recorded over the 400-4000 cm^{-1} range at room temperature, using diamond crystal, and 32 scans were performed for each sample.

2.7. Contact Angle. The static contact angles of the sample surfaces, before and after leaching, were determined through a sessile drop method, using a 5 μL double-distilled water droplet. Static contact angles of membranes were measured in air at room temperature in a CAM 200 KSV instrument (Milan, Italy) equipped with Attension Theta software for data acquisition. The static contact angle value was the average of 15 measures at random areas of each sample membrane.

2.8. Tensile Testing. Mechanical properties of membranes before and after leaching were evaluated by a tensile test on rectangular specimens using the uniaxial tensile testing machine ZwickRoell Z005 (Genoa, Italy). Samples presented a similar size both in the cross-sectional area (thickness 0.35 ± 0.05 mm, width 9.8 ± 0.2 mm) and in the length (13.5 ± 0.2 mm). The tensile force was applied along the length of the samples until fracture, using a preload of 0.05 N and deformation speed of 0.05%/s. At least 5 specimens for each type of membrane were tested. The elastic modulus (E), stress at failure, and deformation at failure were measured from the stress-strain curves, derived from the force-displacement curves using a specific sample size. E was evaluated as the slope of the first linear part of the stress-strain curve, and the failure stress and deformation were calculated as the maximum stress a material can withstand and the corresponding deformation, respectively.

CMC mesh tensile mechanical properties were also analyzed by testing dry rectangular specimens (50 × 300 mm) in a MT® QTest™/10 instrument (Turin, Italy). The traction force was applied along the length of the samples, at a constant cross-head displacement rate of 100 mm/min. At least 5 specimens were tested. Failure stress and failure deformation were measured from the stress-strain curves, while E was not evaluated due to the nonhomogeneous structure of samples, characterized by mesh holes.

2.9. Cell Culture. BJ human skin fibroblasts (ATCC-CRL-2522, Teddington, UK) were cultured in minimum essential medium (MEM) supplemented with 10% fetal bovine serum (FBS), 2 mM L-glutamine, 100 μU/mL penicillin, 10-1 μU/mL streptomycin, and 2.5 × 10-1 μU/mL amphotericin B. All these products were purchased at Sigma-Aldrich (St. Louis, MO, USA). Cells were amplified up to reaching confluence (34000 cells/cm^2), and, during this period, the medium was changed every 3 days.

2.10. Cytotoxicity. Cytotoxicity tests were conducted according to the ISO 10993-5 standard. In this study, membranes of each material after leaching (18 cm^2 surface) were sterilized with 70% ethanol and ultraviolet (UV) radiation treatment (20 minutes for each side) and then incubated at 37°C for 24 h in 3 mL of MEM complete medium. The medium was then extracted for further *in vitro* cell tests (extract medium). The positive control was a complete medium supplemented with 5% dimethyl sulfoxide (DMSO) (*v*/*v*). As negative control, cells cultured in standard medium were used.

Cells were seeded at confluence in a 96-well plate (~10.000 cells/well), and, once adhered, 200 μL of the medium was exchanged with the extract medium. Cells were placed in an incubator, and, after exposure to the extract medium for 72 h at 37°C, changes of cellular morphology were evaluated under an AX70 light-reverted microscope (Olympus, Milan, Italy) and cell viability tests were performed by using Promega CellTiter Blue® (Madison, WI, USA). In each well, 100 μL of fresh media was added together with 10 μL of reagent. After 150 minutes, the relative fluorescent unit (RFU) values were evaluated with a spectrophotometer plate reader (579ex/584em). Materials with a viability value of at least 70% were selected for cell culture growth profile evaluation.

2.11. Cell Culture Growth. Membranes were sterilized with 70% ethanol and UV radiation treatment (20 minutes for each side) and left in contact with the medium for 2 hours before cell seeding. Cells at confluence were washed with phosphate-buffered saline (PBS), detached by trypsin (0.25% (*w*/*v*)), and placed in contact with a fresh complete medium to deactivate enzyme action. Fibroblasts were counted with a hemocytometer and seeded on materials, placed in a 6-well plate at a cell density of 15% (equivalent to ~50.000 cells/well). At well-defined time (3, 7, 14, and 21 days) cell viability test was performed with CellTiter Blue®.

In order to count cells, a standard curve of fibroblasts on culture plates was obtained. Cells were seeded in 24-well plates at several concentrations (from 5000 to 70,000 cells/well), in duplicate. Once adhered, 50 μL of CellTiter Blue® was added to each well and, after 150 minutes, RFU emitted by dye was evaluated as above.

At the 21st day, samples were fixed with 4% paraformaldehyde solution for 30 minutes at room temperature and preserved in PBS at 4°C for immunohistochemistry protocol.

2.12. Cytokine Production. The medium at 7, 14, and 21 days was collected and stored at -80°C until cytokine analysis. IL-6 quantification was performed by use of Elecsys IL-6 assay (Roche Diagnostics, Milan, Italy), an electrochemiluminescent immunoassay (ECLIA).

Analysis of IL-10 was performed using ELISA kit assay (Boster Biological Technology, Fremont, CA, USA).

2.13. Immunohistochemistry. Paraformaldehyde-fixed samples were treated for evaluation of collagen type I and type III production. Collagen type I was recognized by anti-human COL1A1 goat primary antibody and relative FITC-conjugated secondary antibody (donkey anti-goat IgG-FITC); collagen type III was identified by anti-human COL3A1 goat primary antibody and relative rhodamine-conjugated secondary antibody (bovine anti-rabbit IgG-R). All the antibodies were purchased at Santa Cruz Biotechnology (Dallas, TX, USA). Samples were washed three times with PBS and treated with 0.1% Triton in PBS for 2 minutes at room temperature to permeabilize cell membranes. Samples were left in 5% BSA for 30 minutes and then treated with a mix of the two primary antibodies, 1 : 100 diluted in 1% BSA, for 1 hour at 37°C. Then, a mix of the secondary antibodies 1 : 200 diluted in 1% BSA was added to the samples for 1 hour at 37°C. Fifteen minutes before the end of incubation, DAPI (4′,6-diamidino-2-phenylindole) was added to each sample at 1 : 1000 dilution to detect cell nuclei. Finally, samples were analyzed by means of a fluorescence microscope (CX40, Olympus, Milan, Italy).

2.14. ECM Extraction. Cells were seeded on PCL7-2i membranes and CMC mesh and cultured for 7 days. Cells were removed and ECM recovered as described in [26].

2.15. Sample Processing and Liquid Chromatography-Tandem Mass Spectrometry with SWATH™-Based Analysis (LC-MS/MS SWATH) for Proteomics. For sample treatment and LC-MS/MS analysis, the same protocol described in Vozzi et al. [26] was used. Briefly, protein concentration was determined by bicinchoninic acid assay (Pierce, Thermo Scientific, Waltham, MA, USA). About 40 mg of protein was dissolved in ammonium bicarbonate 25 mM, reduced with dithiothreitol 5 mM at 80°C for 30 min, and alkylated using iodoacetamide 10 mM at 37°C for 20 min. Digestion was obtained incubating overnight with 1 : 100 Trypsin (Roche, Basel, Switzerland): substrate at 37°C. Peptide mixtures were acidified and then loaded on a C18 cartridge in order to eliminate debris and additionally purified with 0.22 μm filters. Peptides were diluted to 0.1 μg/μL by 2% acetonitrile (ACN, Romil, UK)/0.1% formic acid (FA); 5 μL was injected for library searching and 2 μL in duplicate for SWATH™ method analysis.

Chromatographic separation of peptides was performed using a nano-HPLC system (Eksigent, AB Sciex, Washington, DC, USA). Peptides eluted from chromatography were directly processed using a 5600 TripleTOF™ mass spectrometer (AB Sciex) equipped with a DuoSpray™ ion source (AB Sciex). Data were acquired using the SWATH™ method for shotgun data-independent MRM quantification. For library, MS/MS data were processed with ProteinPilot™ Software (AB Sciex). The false discovery rate (FDR) analysis was set to a confidence level of 95%.

2.16. Statistical Analysis. All data were expressed as mean ± standard deviation obtained by at least five different experiments. Differences between groups were evaluated by one-way ANOVA and Student *t*-test; Prism 7 software was used for statistical analysis. Values were considered statistically significant for p value <0.05. For proteomics data, to evaluate differentially expressed proteins between CMC and PCL7-2i at 7 days, MarkerView 1.2 software was used. Three technical replicates for biological sample were used. The two groups were compared by *t*-test statistics: paired *t*-test was used as a statistical parameter between the means of continuous

Table 2: Percentage of weight loss in blends due to PEG dissolution and percentage of undissolved PEG with respect to the initial PEG amount.

Samples	Weight loss after leaching (%)	Undissolved PEG (%)
PCL7-1i	29.4 ± 0.6	2.0 ± 0.2
PCL7-2i	28.8 ± 0.6	4.0 ± 0.2
PCL6-1i	38.9 ± 0.3	2.8 ± 0.8
PCL6-2i	38.8 ± 0.0	3.0 ± 0.0

variables to determine significant differences between the two categories of mass spectrometric data. A p value <0.05 and a fold change > 2 were considered significant to validate differences between categories.

3. Results

3.1. PEG Leached Out Percentage. Incubation of PCL/PEG1 and PCL/PEG2 blend films in a water bath almost caused an approximately complete dissolution of the PEG component leaving porous membranes (Table 2). The percentage of leached out PEG corresponds to the porosity degree of membranes. A small percentage of PEG with respect to the initial amount was still present in blends (Table 2).

3.2. Morphological Characterization by Scanning Electron Microscopy (SEM). Figure 1 shows the SEM images of the external surface and fractured section of PCL/PEG1 and PCL/PEG2 blends.

Blend films showed a compact section. No evident phase separation between blend components was observed in any of the blend compositions.

Morphological evaluation was also conducted on PCL-based membranes after PEG dissolution (Figure 1). Surfaces were scarcely porous. On the contrary, fractured sections were highly porous with homogeneous distribution of pores with spherical shape.

3.3. Thermal Properties by Differential Scanning Calorimetry (DSC). The first heating scan of leached samples confirmed the complete lack of water as the endothermic peak associated with water evaporation was not present (data not shown). Hence, the efficiency of the employed drying protocol (sample incubation for 24 h at 37°C) was demonstrated.

Figure 2 collects DSC second heating thermograms for PCL, PCL/PEG1, and PCL/PEG2 film samples before and after PEG removal. The analysis of the second heating scan allows a comparison of material properties after deleting the thermal history.

PEGs and PCL are semicrystalline polymers characterized by a melting temperature (T_m). PEG and PCL glass transition temperatures could not be detected by DSC analysis, as they are both at -67°C [31], corresponding to the limit cooling temperature of the DSC instrument used in this work. For each blend, a double melting peak (Table 3) was detected, due to the different melting temperatures of the blend components. In blends, T_m of PCL phase did not vary, while a slight decrease was observed for the melting peaks of PEG1 and PEG2 phases with respect to pure PEGs (Figure 2, B1 and B2).

The cooling scans of PCL, PEG1, and PEG2 samples showed a crystallization event (Figure 2, A1 and A2). On the contrary, PCL/PEG1 blend films were characterized by three T_c (Table 3): the lower T_c was related to the crystallization by homogeneous nucleation of PEG1 (1.6 ± 0.1°C for PCL/PEG1 70/30 and 2.4 ± 0.2°C for PCL/PEG1 60/40), the intermediate T_c was due to PCL crystallization (at around 23°C), and the higher T_c was associated to PEG1 heterogeneous nucleation (35.5 ± 0.4°C for PCL/PEG1 70/30 and 39.1 ± 1.4°C for PCL/PEG1 60/40) [32]. In blends, T_c values of PEG1 and PCL (considering for PEG1 the heterogeneous crystallization only) decreased compared to the pure polymers. The presence of the homogeneous PEG1 nucleation mechanisms was ascribed to a fine dispersion of PEG1 in the PCL matrix.

For blends containing PEG2, the crystallization of PEG2 through two different nucleation mechanisms, homogeneous and heterogeneous, was observed only for PCL/PEG2 60/40 blend films, respectively, at 1.4 ± 0.4°C and 40.0 ± 0.7°C. Similarly to that observed for PCL/PEG1 blends, T_c values of PCL and PEG2 (considering crystallization by heterogeneous nucleation mechanism for PEG2 in PCL/PEG2 60/40 blends) slightly decreased with respect to pure polymers.

On the other hand, the PCL/PEG2 70/30 blend only showed PEG2 crystallization by homogeneous nucleation. Hence, PCL/PEG2 70/30 blend samples only showed two distinct T_c at about 1.5°C (crystallization of PEG2 by homogeneous nucleation) and 25°C (crystallization of PCL). The presence of a unique crystallization temperature for PEG2 phase could be due to a finer dispersion of PEG2 in the PCL matrix (Table 3).

After PEG removal, all the membranes showed only a characteristic thermal event, both in the cooling scan and in the second heating scan, attributed to PCL crystallization and melting, respectively. In porous membranes, the cooling and melting enthalpies of PCL were slightly higher compared to the values of control PCL samples, suggesting a slightly higher crystallization degree of porous membranes.

After PEG dissolution (Table 3), the characteristic thermal transitions associated with PEG phase could not be detected. This could be due to a complete removal of PEG or to the permanence of a very small amount of PEG which thermal transitions cannot be detected by DSC analysis.

3.4. Chemical Analysis by FTIR-ATR. Characteristic FTIR-ATR absorption bands (Supplementary Material Figure S1) of PCL and PEG could be detected in the FTIR-ATR spectra of blend films. In particular, a sharp band at around 1726 cm^{-1} was observed, corresponding to the carbonyl-stretching of PCL. When the content of PEG increased, no appreciable shift of this band was observed suggesting poor interaction between the phases.

After PEG removal, absorption bands attributed to the PEG2 component were no more detected in the FTIR-ATR spectra of porous films, such as the asymmetric stretching vibrations of C-O-C bonds at around 1280 cm^{-1}. This result

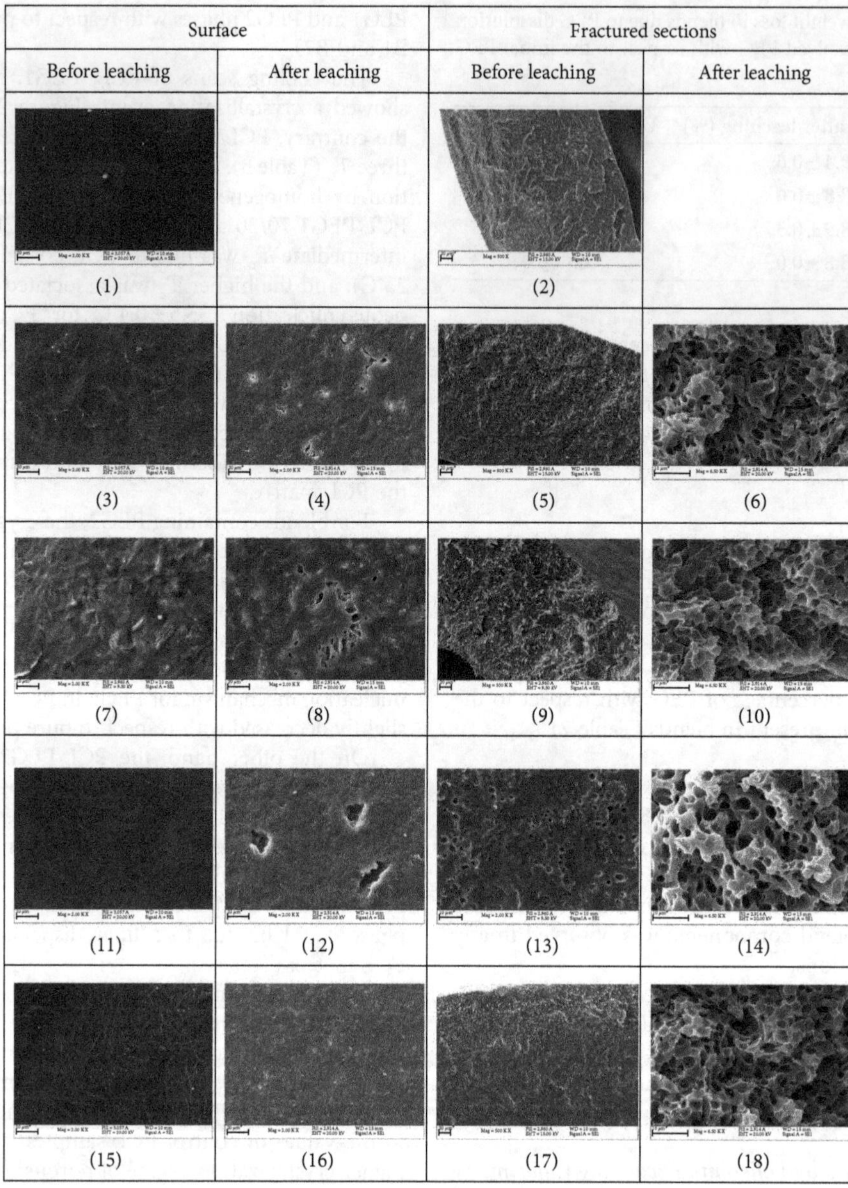

FIGURE 1: SEM images of the surface and fractured sections of materials before (odd numbers) and after (even numbers) leaching: PCL (1, 2), PCL/PEG1 70/30 (3, 5), PCL7-1i (4, 6), PCL/PEG1 60/40 (7, 9), PCL6-1i (8, 10), PCL/PEG2 70/30 (11, 13), PCL7-2i (12, 14), PCL/PEG2 60/40 (15, 17), and PCL6-2i (16, 18). Scale bar = 20 μm for surfaces and 10 μm for section images.

confirmed the complete removal of porogen from PCL/PEG blend films (Supplementary Material Figure S1).

3.5. Surface Wettability Properties. Table 4 collects the static contact angle for all samples, before and after leaching. PCL samples (73.1 ± 2.8°) were markedly less hydrophilic than PEG1 (41.5 ± 2.9°) and PEG2 (44.0 ± 1.9°) samples (for both comparison, $p < 0.05$). The surface static contact angle of PCL/PEG1 cast films decreased as the amount of PEG1 increased: it was 71.5 ± 2.4° for blends containing 30%wt. PEG1 and 44.7 ± 2.6° for blend containing 40%wt. PEG1. On the contrary, the static contact angle of PCL/PEG2 blend samples did not change significantly with increasing PEG2 content (49.1 ± 3.0° for PCL/PEG2 70/30 and 51.2 ± 2.5° for PCL/PEG2 60/40), although it was lower than for PCL.

After leaching, surface wettability decreased for all the samples. Porous films with PEG1 as porogen showed the highest contact angle values (110.0 ± 6.8° for PCL7-1i and 82.9 ± 7.4° for PCL6-1i) while PCL7-2i and PCL6-2i retained surface hydrophilicity with contact angles of 67.8 ± 3.9° and 66.8 ± 6.1°.

3.6. Mechanical Characterization. The mechanical properties of the tested materials including the elastic modulus, the maximum load, and the deformation at maximum load are reported in Figure 3.

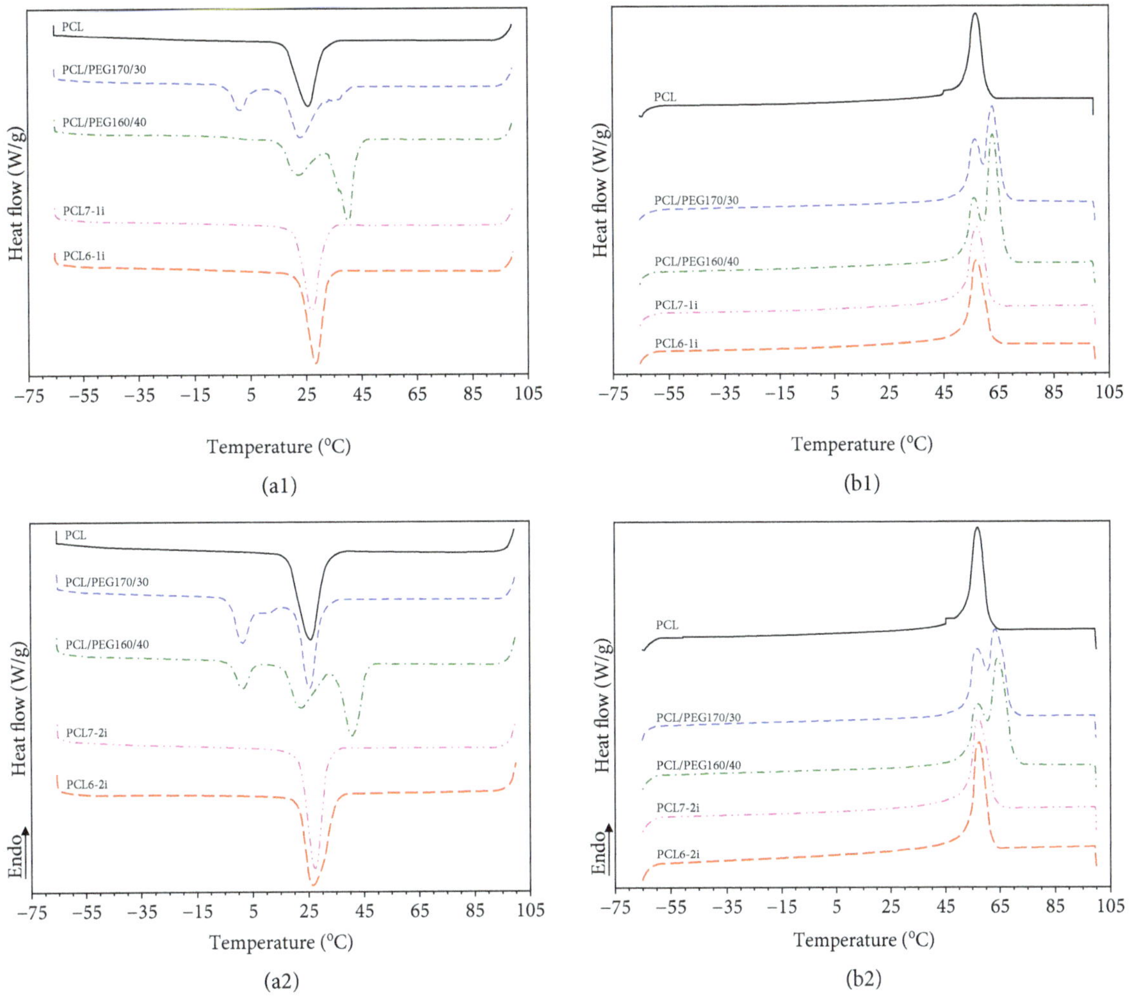

FIGURE 2: Comparison of DSC cooling thermograms (a1 and a2) and second heating thermograms (b1 and b2) for PCL/PEG1 (numbers 1) and PCL/PEG2 (numbers 2) samples, before and after leaching.

The elastic modulus increased with increasing PCL content (Figure 3(a)). The elastic modulus of PCL/PEG2 blends was higher than that of PCL/PEG1 blends with the same composition. The same trend is confirmed after the leaching process: PEG2-derived samples have a higher elastic modulus than the corresponding PEG1-derived ones. In both cases, the leaching process leads to less rigid structures with respect to nonleached samples.

Similarly, failure stress increased with increasing content of PCL (Figure 3(b)), with higher values for PCL/PEG2 blends, for both leached and nonleached samples. The leaching process decreases the failure stress. The stress at failure of CMC was 2.4 ± 0.3 MPa.

Before leaching, the deformation at failure (Figure 3(c)) was similar for all blend samples and significantly lower than that of pure PCL. After leaching, samples showed higher deformability: particularly, PCL/PEG2-based blends had higher failure deformation. CMC deformation at failure was 49.1 ± 3.5%, significantly higher than for PCL samples.

3.7. In Vitro Cytotoxicity. Cytotoxicity tests showed that all the porous and nonporous materials presented after 3 days the same viability (>95%) with respect to negative control (untreated cells, 100%) and to CMC (>95%), with values above the threshold of 70% indicated by ISO 10993-5 (data not shown).

3.8. In Vitro Cell Growth Tests. The tests were only performed on porous samples, as for the PCL/PEG blends: PEG would be released in the culture medium during the test, and their elastic modulus was too high with respect to reference meshes for hernia prostheses [33]. Results showed that only PCL7-2i presented a good cell growth trend, with a continuous increase in cell viability (Figure 4) superimposable to control and CMC.

3.9. Inflammatory Profile. Based on the results of cytocompatibility and cell growth tests, PCL7-2i membranes were selected to perform the analysis on inflammatory profile by anti- (IL-10) and pro- (IL-6) inflammatory markers. As shown in Figure 5, the initial anti-inflammatory profile of PCL7-2i (Figure 5(a)) was substituted by a proinflammatory trend (Figure 5(b)), with IL-6 final concentration 4 times

Table 3: Crystallization temperature and enthalpy of PCL and PEG phases as a function of blend composition measured from the DSC cooling scan (performed at 10°C/min); melting temperature and enthalpy of PCL and PEG phases as a function of blend composition measured from the second heating DSC scan (performed at 10°C/min).

Samples	Cooling cycle						Second heating cycle				
	T_c^{PCL} (°C)	P value	ΔH_c^{PCL} (J/g)	P value	T_c^{PEG} (°C)	ΔH_c^{PEG} (J/g)	T_m^{PCL} (°C)	P value	T_m^{PEG} (°C)	ΔH_m^{TOT} (J/g)	P value
PCL	26.1 ± 0.3	—	54.5 ± 2.9	—	—	—	57.4 ± 0.3	—	—	54.2 ± 1.9	—
PCL/PEG1 70/30	22.8 ± 0.7	<0.0001	49.9 ± 1.8	<0.0001	1.6 ± 0.1 35.5 ± 0.4	8.9 ± 0.3 Overlapped	56.9 ± 0.1	<0.0001	62.7 ± 0.1	81.1 ± 0.1	<0.0001
PCL7-1i	26.4 ± 0.7		57.9 ± 0.8		—	—	57.5 ± 0.1		—	59.8 ± 0.3	
PCL/PEG1 60/40	23.0 ± 0.4	<0.0001	22.1 ± 3.3	<0.0001	2.4 ± 0.2 39.1 ± 1.4	0.9 ± 0.5 36.3 ± 4.4	56.2 ± 0.2	0.0003	63.3 ± 0.2	90.9 ± 3.2	<0.0001
PCL6-1i	28.6 ± 0.2		57.6 ± 0.7		—	—	57.4 ± 0.4		—	59.0 ± 2.0	
PEG1	—	—	—	—	41.6 ± 0.5	160.9 ± 3.6	—	—	65.6 ± 0.4	164.7 ± 3.3	—
PCL/PEG2 70/30	25.6 ± 0.1	<0.0001	34.9 ± 1.9	<0.0001	1.2 ± 0.3 Not detected	10.6 ± 3.4 Not detected	56.9 ± 0.4	0.0266	63.8 ± 0.3	75.1 ± 1.2	<0.0001
PCL7-2i	27.5 ± 0.1		57.8 ± 0.8		—	—	57.4 ± 0.1		—	58.2 ± 2.5	
PCL/PEG2 60/40	22.8 ± 0.8	<0.0001	23.8 ± 0.8	<0.0001	1.4 ± 0.4 40.0 ± 0.7	9.5 ± 0.3 30.1 ± 2.5	56.6 ± 0.3	0.0029	64.5 ± 0.2	83.7 ± 1.9	<0.0001
PCL6-2i	26.8 ± 0.1		58.2 ± 0.6		—	—	57.4 ± 0.3		—	55.9 ± 1.1	
PEG2	—	—	—	—	41.9 ± 0.7	148.1 ± 3.2	—	—	66.5 ± 0.1	148.7 ± 3.3	—

TABLE 4: Static contact angle values of samples before and after leaching.

Before leaching	Contact angle (°)	After leaching	Contact angle (°)	P value
PCL	73.1 ± 2.8	—	—	—
PEG1	41.5 ± 2.9	—	—	—
PEG2	44.0 ± 1.9	—	—	—
PCL/PEG1 70/30	71.5 ± 2.4	PCL7-1i	110.0 ± 6.8	<0.0001
PCL/PEG1 60/40	44.7 ± 2.6	PCL6-1i	82.9 ± 7.4	<0.0001
PCL/PEG2 70/30	49.1 ± 3.0	PCL7-2i	67.8 ± 3.9	<0.0001
PCL/PEG2 60/40	51.2 ± 2.5	PCL6-2i	66.8 ± 6.1	0.0007

higher with respect to day 7, but with reduced levels with respect to CMC.

3.10. Immunohistochemistry. Finally, also type I and type III collagen production was semiquantitatively evaluated with immunohistochemistry and the type I collagen/type III collagen ratio quantified. In control conditions, BJ fibroblast cells produced type III collagen (neo-formed type) and, in small quantity, type I (stable) collagen (Figure 6(a)). PCL7-2i reduced the production of type III collagen favoring a strong and significant increase (±3 times, $p < 0.0001$) of type I collagen. This trend was also confirmed by the type I collagen/type III collagen ratio (Figure 6(b)), which showed a value near 0.8 ($p < 0.0001$).

3.11. ECM Protein Characterization. ECM proteins were extracted from CMC and PCL7-2i membranes and analyzed by mass spectrometry that allowed the identification of 243 proteins (Supplementary Material Table S1). Proteins were classified according to their localization (Supplementary Material Figure S2), and up to 125 were ECM, ECM-associated, or membrane proteins. Of these, four resulted to be differentially expressed when samples from CMC and PCL7-2i were compared (Table 5).

4. Discussion

This work was aimed at the preparation and characterization of porous PCL membranes to be used as a new generation of resorbable hernia meshes. Particularly, porous PCL membranes were characterized for their physicochemical, mechanical, and biological characteristics, comparing results with the performance of a commercially available CMC mesh: this lightweight polypropylene mesh is able to favor cell colonization, generating a series of biochemical modifications in terms of typical inflammatory reaction, collagen production, and ECM expression [26]. However, the lightweight mesh is not suitable to prevent adhesion to internal organs [34]. On the contrary, the polypropylene film does not induce the formation of adhesion to viscera due to its plane surface and its antiadhesive features shown in Wistar rats [33] and in clinical studies [35, 36].

In our work, porous membranes with an asymmetric structure characterized by a poorly porous skin layer and high internal porosity were obtained. As the amount of dissolved PEG was approximately equal to the amount of incorporated PEG, the porosity degree was 29.4 ± 0.7% and 28.8 ± 0.6% for PCL7-1i and PCL7-2i samples and 38.9 ± 0.3% and 38.8 ± 0.0% for PCL6-1i and PCL6-2i, respectively. PCL and PEG have a different polarity (PCL is poorly hydrophilic while PEG is highly hydrophilic as suggested by contact angle measurements) and consequent weak reciprocal interactions: for this reason, the PEG phase formed spherical dispersed domains in the PCL matrix, leading to spherical pores of approximately 5 μm size after PEG removal. Additionally, the pore surface was smooth. These results suggested that PEG and PCL were immiscible, which was in agreement with previous findings in the scientific literature [37]. Indeed, FTIR-ATR analysis showed that the carbonyl stretching band of PCL did not shift as a function of blend composition, confirming poor interaction between blend components. This result suggested that any interaction between PCL and PEG2 was weak, and this conclusion was consistent with DSC data. However, this result was opposite to the findings by Ozeki et al. [38] who found interactions between PEG and a carboxyvinyl polymer, having the same carboxyl functional group of PCL. Probably, the different polarities between PEG and PCL (PEG is highly hydrophilic while PCL is poorly hydrophilic as suggested by static contact angle measurements) hindered the formation of a sufficiently high number of hydrogen bonding to detect shifts of the PCL carbonyl FTIR-ATR band.

Thermal characterization by DSC analysis showed that the presence of partially crystallized PEG phase in blends decreased the PCL phase crystallization rate, causing a slight decrease in PCL T_c. This result was not observed for PCL/PEG2 70/30 blends where PEG2 was still in a molten state during PCL crystallization, while it underwent crystallization at lower temperature than PCL by a homogeneous nucleation mechanism. On the other hand, the PCL crystallization degree slightly increased in porous membranes with respect to compact PCL films, suggesting some nucleating effect of PEG component. However, the T_m of the PCL phase remained unchanged in blends before and after PEG removal, suggesting that the perfection degree of PCL crystals was similar in all the samples.

Importantly, DSC analysis suggested a different interaction between blend components, depending on the molecular weight of PEG: a higher amount of PEG tended to crystallize by a homogeneous crystallization mechanism in PCL/PEG2 blends vs. PCL/PEG1 blends with the same composition, as suggested by the corresponding crystallization enthalpy values. The homogeneous crystallization behavior of PEG2 in PCL/PEG2 blends was indicative of a distribution of PEG2 in domains with smaller size compared to the PEG1 ones in PCL/PEG1 blends. This difference was attributed to the higher molecular weight of PEG2 with respect to PEG1: during solvent evaporation for membrane formation, coalescence phenomena of PEG2 phase were partially hindered by its higher molecular weight. Hence, PEG2 domains retained a smaller size compared to PEG1 domains in PCL/PEG1 blends.

Porous films obtained using PEG1 as porogen showed surface hydrophobicity (static contact angle of 110.0 ± 6.8°

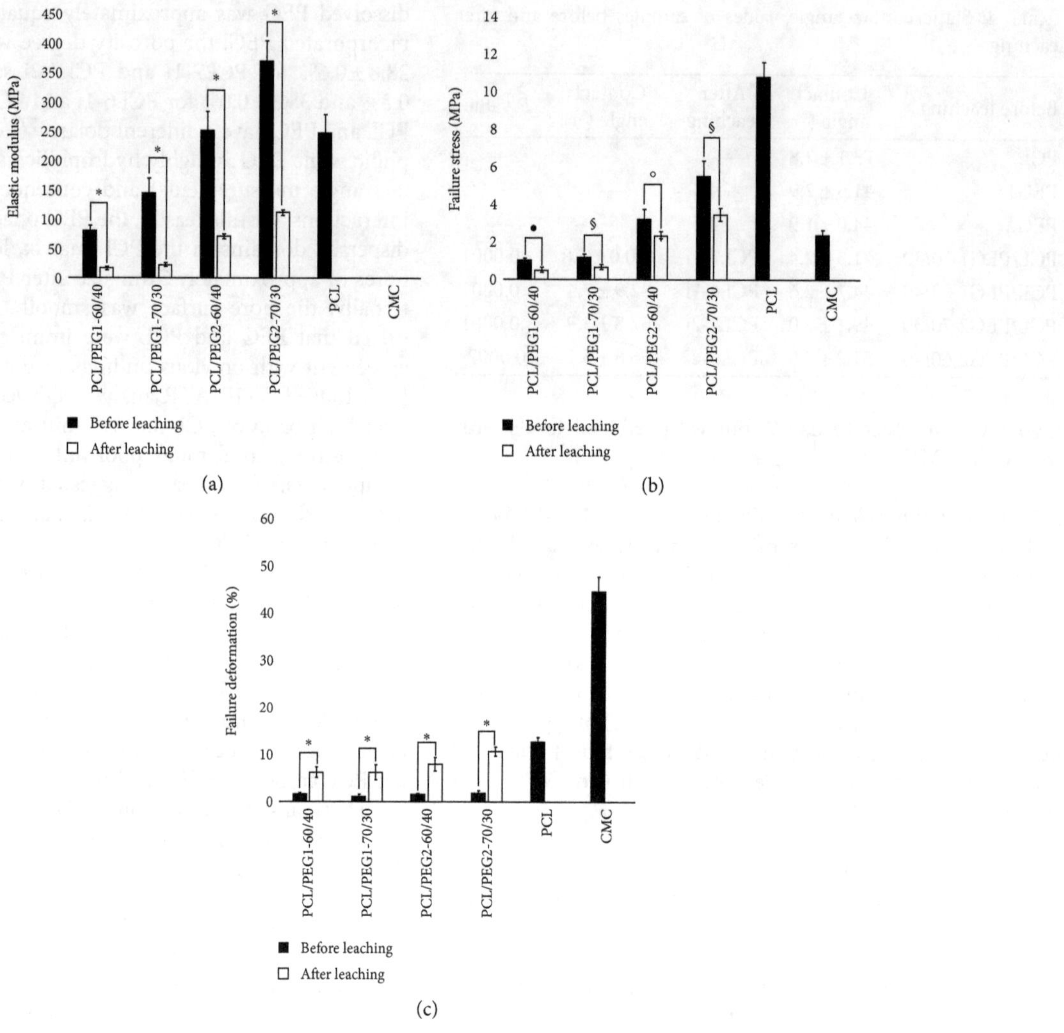

FIGURE 3: Comparative analysis of elastic modulus (a), failure stress (b), and failure deformation (c) for PCL/PEG blends before and after leaching. $^{*}p < 0.0001$, $^{\bullet}p = 0.0002$, $^{\$}p = 0.0008$, and $^{\circ}p = 0.0086$.

for PCL7-1i) or poor hydrophilicity (static contact angle of 82.9 ± 7.4° for PCL6-1i). On the other hand, PCL7-2i and PCL6-2i were more hydrophilic with contact angles of 67.8 ± 3.9° and 66.8 ± 6.1°, respectively. The different types of dispersion for PEG1 and PEG2 in PCL/PEG blends were probably the cause for a different organization of PCL chains on the surface: in the case of porous samples from PCL/PEG2 blends, hydrophilic carboxylic moieties were probably preferentially distributed on the sample surface. Additionally, differences in surface morphology also affected surface wettability.

Mechanical analysis was performed on PCL/PEG membranes and porous PCL films (Figure 3), in order to characterize and select biomechanical features similar to meshes for hernia repair currently used in clinics. Meshes should be strong enough to resist the state of stress present in the adnominal wall and at the same time not so stiff to cause discomfort and pain to patients. Literature reference values for the elastic modulus of hernia meshes are 20-120 MPa [33].

The elastic modulus increased with increasing PCL content, both before and after the leaching process. The molecular weight of PEG had influence on mechanical properties: the elastic modulus of PEG2-derived blends (PEG2: 35000 Da) was higher than that of PEG1-derived ones (PEG1: 10000 Da).

Interestingly, the elastic modulus of the PCL/PEG2 70/30 blend was higher with respect to that of PCL, while the elastic modulus of the PCL/PEG1 70/30 blend was significantly lower with respect to that of PCL. This behavior was probably due to the different interaction degrees between blend components and the different blend morphologies when using PEGs with different molecular weights. Indeed, DSC analysis suggested that the PEG2 component was finely distributed in the PCL/PEG2 70/30 blend; this was probably the reason for its superior mechanical performance respect to the PCL/PEG1 70/30 blend. After the leaching process, the elastic modulus of porous samples obtained from PCL/PEG2 blends was higher than for samples prepared from PCL/PEG1 blends, again suggesting a finer and more homogeneous pore

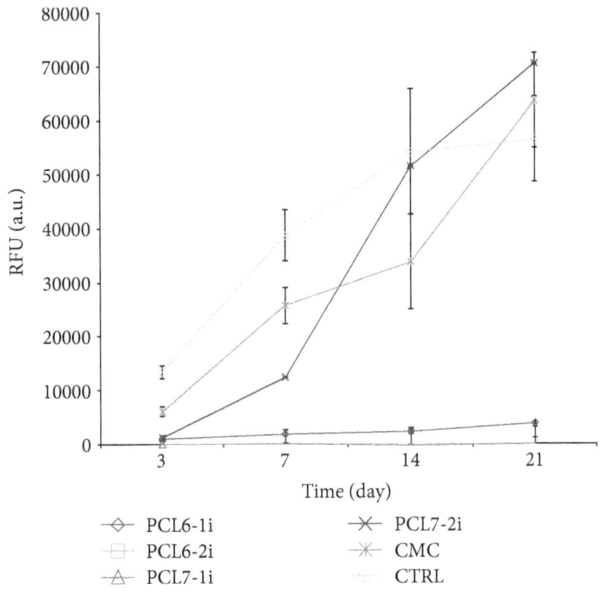

Figure 4: Cell growth curve (time = 21 days).

morphology. In both cases, after leaching, the resulting porous structures present an elastic modulus significantly lower than that of the corresponding nonleached samples.

Similarly, the failure stress increased with increasing PCL content (Figure 3(b)) due to the superior mechanical performance of PCL with respect to PEG in nonleached blends and due to the lower porosity in the case of leached samples. As for the elastic modulus, the failure stress of PCL/PEG2 was higher with respect to PCL/PEG1 blends, both before and after leaching. This behavior, again, can be addressed to a different pore morphology, depending on the molecular weight of PEG. The leaching process increases the deformation at failure (Figure 3(c)), which presented marked lower values in nonleached samples with respect to pure PCL samples, again suggesting a more homogeneous and finer pore structure.

As a conclusion, higher concentrations of PEG resulted into more porous and less stiff membranes, but interestingly PEG with the highest molecular weight (PEG2) resulted to be more finely dispersed and led to stiffer structures. After leaching, the membranes showed an elastic modulus in the same range of meshes for hernia prosthesis and thus suitable for applications in hernia repair. This is the first time that the solvent-casting/porogen leaching technique of PCL/PEG blends is exploited for the preparation of scaffolds with suitable properties for the treatment of abdominal wall hernia and as an alternative to electrospun scaffolds or filament-based meshes [14].

For biological characterization, BJ human skin fibroblasts were selected for tests on materials under investigation, according to their role in the wound healing process. In fact, they represent a cell barrier respect to prosthesis, regulate reparative process by secretion of growth factors, and modulate the surrounding cell environment [39].

All the samples showed an optimal cell biocompatibility compared to the negative control. Cell growth analysis, performed on porous samples, allowed a further selection of materials under investigation. In fact, PCL6-1i, PCL6-2i, and PCL7-1i showed nonadhesive and nonproliferative cell properties, while PCL7-2i was characterized by progressively increasing cell colonization on its surface. Hydrophobicity together with mechanical and structural properties of PCL7-2i seems to produce a positive effect on its capabilities to support cell colonization.

Based on these results, PCL7-2i membranes were selected for further analyses. The inflammatory profile showed an initial increase of the anti-inflammatory IL-10 cytokine [40] followed by a strong increase of the proinflammatory one, IL-6 [41], with reduced the magnitude with respect to CMC. This trend was highlighted in our previous work [26] and in literature [42]: inflammation is a typical body response to surgery implant and can support mesh colonization and integration [43]. This type of response probably could be related to their surface and mechanical properties, which, interacting with cells, modulate their chemomechanical signal integration and stiffness sensing and stimulate the secretion of inflammatory mediators [44, 45].

The biological characterization of PCL7-2i foresaw the immunohistochemical analysis of two of the main components of ECM produced by fibroblasts: collagen I and collagen III. PCL7-2i stimulated the production, at 21 days, of mature collagen type I with respect to type III in comparison to control and CMC mesh, confirmed also by the collagen type I/III ratio (value near 0.8). These two proteins are heavily involved in mesh integration during prosthesis implant where a remodeling process produces an initial increase in the secretion of immature type III, followed by production of mature type I [46–48]. Literature showed how a decreased ratio could be of significant importance in the hernia pathophysiology probably leading to structural and mechanical stability alteration of mesh in the host tissue [49]. The increased production of collagen type I observed in our experimentation and results of the inflammatory profile suggest a potential stable in vivo mesh implant favoring tissue integration of and a contemporary reduction of host rejection processes.

Interestingly, proteomics analyses were validated and, at the same time, confirmed the histological data: among the differentially expressed proteins (Table 5), collagen type III resulted to be significantly overexpressed at day 7 in ECM from PCL7-2i compared to CMC mesh. In our previous paper [26], it was highlighted that major changes in ECM deposition and modulation occurred during the first 14 days and, for these reasons, we focused on day 7 after cell seeding for mass spectrometry analyses. Besides collagen type III, also myosin regulatory light chain 12B and 78 kDa glucose-regulated protein resulted to be overexpressed on PCL7-2i, while 60 kDa heat shock protein was downregulated.

Myosin regulatory light chains are important for the maintenance of cell morphology and dynamic [50]. 78 kDa glucose-regulated protein is a endoplasmatic reticulum (ER) chaperone involved in assembly and quality control of proteins [51]. It is expressed on the cell membrane and released in the cell culture medium [52], and its synthesis is induced by the accumulation of unfolded polypeptides

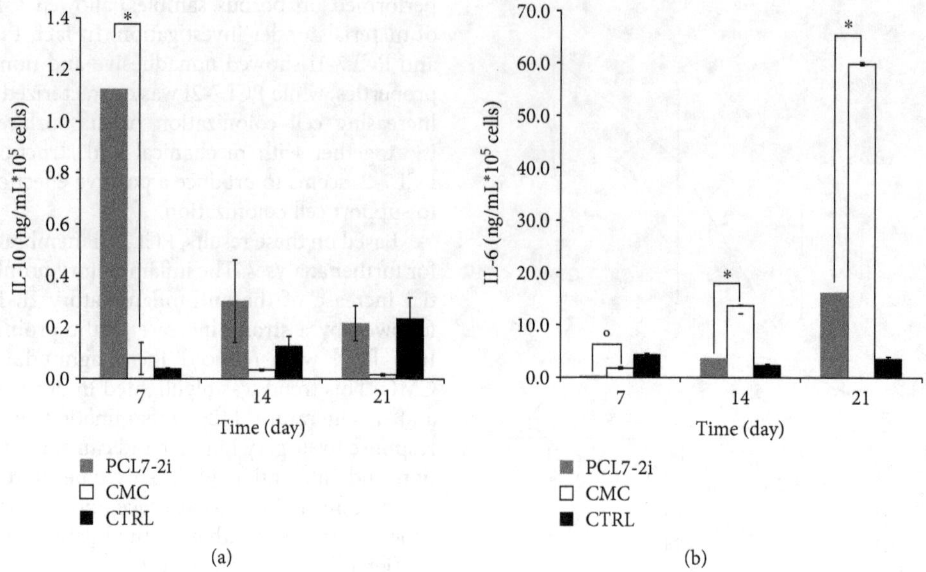

Figure 5: Interleukin-10 (a) and 6 (b) trend in PCL7-2i and control samples. $^{*}p < 0.0001$ and $^{\circ}p = 0.0002$.

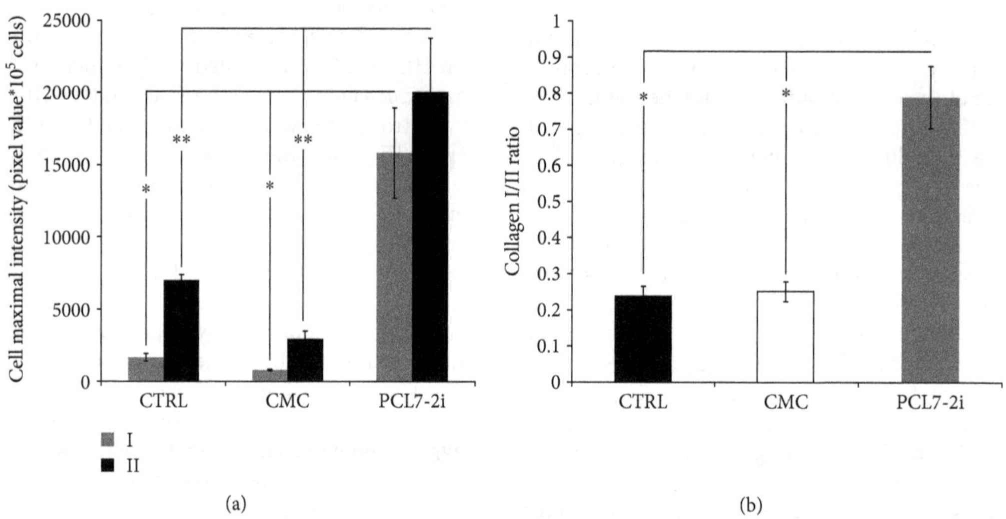

Figure 6: Collagen type I and type III production ((a); * and ** $p < 0.0001$) and in collagen type I/type III ratio ((b); $^{*}p < 0.0001$) in PCL7-2i and control samples.

Table 5: Differentially expressed proteins in ECM protein analysis.

Gene name	Protein name	t-test	P value	Fold change	
GRP78_HUMAN	78 kDa glucose-regulated protein	-35.557	0.00079	0.468	Upregulation in PCL7-2i
CO3A1_HUMAN	Collagen alpha-1 (III) chain	-12.545	0.00629	0.447	Upregulation in PCL7-2i
ML12B_HUMAN	Myosin regulatory light chain 12B	-10.756	0.00853	0.417	Upregulation in PCL7-2i
CH60_HUMAN	60 kDa heat shock protein, mitochondrial	4.495	0.04608	3.388	Downregulation in PCL7-2i

in the ER. It is interesting that this protein is overexpressed in cancer and stressed cells, in which it can be actively secreted and assume roles in controlling cell signaling, proliferation, invasion, inflammation, and apoptosis [53]. Its induction and secretion on PCL7-2i could represent a useful answer to a stress state, leading to cell proliferation and invasiveness of the membrane. Also, the 60 kDa heat shock protein is a chaperonin mainly involved in mitochondrial protein folding and translocation, but it is also suggested as a danger signal of stressed or damaged cells [54]. This protein resulted to be slightly underrepresented on PCL7-2i.

This work represents the first step for the development of a new generation of resorbable hernia scaffolds. In particular, a porous PCL membrane (PCL7-2i) was selected among four different PCL porous membranes, based on its physicochemical and biomechanical properties, cytocompatibility, and cell growth properties, and then further characterized for its detailed biological behavior. Results showed that the PCL7-2i membrane could be potentially applied in abdominal wall hernia treatment. In fact, the analysis of biological response of PCL7-2i membranes showed that this type of scaffold is able to mechanically and structurally replicate the reference tissue and to support cell colonization. This was highlighted by the high cell growth rate on the scaffold, its minimal inflammatory response, potentially avoiding undesirable side effects, and its modulation of ECM synthesis with the support of stable forms of collagen and of proteins related to cell proliferation and colonization. All these aspects could contribute to scaffold integration.

Further tests will be aimed at the *in vivo* validation of PCL7-2i in order to evaluate its effectiveness in surgical treatment.

Conflicts of Interest

The authors declare that they have no conflicts of interest.

Acknowledgments

This work was funded by MANUNET 2011 ERA-Net Project (MES-STAR-Morphologically Engineered Scaffold for Soft Tissue Application and Regeneration).

Supplementary Materials

Figure S1: a comparison between FTIR-ATR spectra of PCL, PEG2, and PCL/PEG2 60/40 films before and after PEG2 removal. The same results have been observed for samples containing 30% (wt) of PEG2 or PEG1 as porogen. Figure S2: the classification of proteins extracted from CMC and PCL7-2i membranes with respect to their cellular localization. Table S1: the 243 proteins extracted from CMC and PCL7-2i membranes and identified by mass spectrometry analysis. *(Supplementary Materials)*

References

[1] J. W. A. Burger, J. F. Lange, J. A. Halm, G.-J. Kleinrensink, and H. Jeekel, "Incisional hernia: early complication of abdominal surgery," *World Journal of Surgery*, vol. 29, no. 12, pp. 1608–1613, 2005.

[2] V. Singhal, P. Szeto, T. J. VanderMeer, and B. Cagir, "Ventral hernia repair: outcomes change with long-term follow-up," *Journal of the Society of Laparoendoscopic Surgeons*, vol. 16, no. 3, pp. 373–379, 2012.

[3] R. J. Fitzgibbons, A. T. Richards, and T. H. Quinn, "Open hernia repair," in *ACS Surgery: Principles and Practice*, W. S. Souba, P. Mitchell, M. P. Fink, G. J. Jurkovich, L. R. Kaiser, W. H. Pearce, J. H. Pemberton, and N. J. Soper, Eds., Decker Publishing Inc., Philadelphia, USA, 6th edition, 2002.

[4] F. Helgstrand, J. Rosenberg, H. Kehlet, L. N. Jorgensen, and T. Bisgaard, "Nationwide prospective study of outcomes after elective incisional hernia repair," *Journal of the American College of Surgeons*, vol. 216, no. 2, pp. 217–228, 2013.

[5] J. Dalenbäck, C. Andersson, D. Ribokas, and G. Rimbäck, "Long-term follow-up after elective adult umbilical hernia repair: low recurrence rates also after non-mesh repairs," *Surgery*, vol. 17, no. 4, pp. 493–497, 2013.

[6] M. W. Christoffersen, F. Helgstrand, J. Rosenberg, H. Kehlet, P. Strandfelt, and T. Bisgaard, "Long-term recurrence and chronic pain after repair for small umbilical or epigastric hernias: a regional cohort study," *American Journal of Surgery*, vol. 209, no. 4, pp. 725–732, 2015.

[7] D. R. Flum, K. Horvath, and T. Koepsell, "Have outcomes of incisional hernia repair improved with time?," *Annals of Surgery*, vol. 237, no. 1, pp. 129–135, 2003.

[8] H. W. Harris, "Clinical outcomes of biologic mesh: where do we stand?," *Surgical Clinics of North America*, vol. 93, no. 5, pp. 1217–1225, 2013.

[9] A. Lukasiewicz, J. Skopinska-Wisniewska, A. Marszalek, S. Molski, and T. Drewa, "Collagen/polypropylene composite mesh biocompatibility in abdominal wall reconstruction," *Plastic and Reconstructive Surgery*, vol. 131, no. 5, pp. 731e–740e, 2013.

[10] F. C. Usher, J. Ochsner, and L. L. Tuttle Jr., "Use of marlex mesh in the repair of incisional hernias," *The American Surgeon*, vol. 24, no. 12, pp. 969–974, 1958.

[11] B. P. Jacob, N. J. Hogle, E. Durak, T. Kim, and D. L. Fowler, "Tissue ingrowth and bowel adhesion formation in an animal comparative study: polypropylene versus Proceed versus Parietex Composite," *Surgical Endoscopy*, vol. 21, no. 4, pp. 629–633, 2007.

[12] J. E. Losanoff, B. W. Richman, and J. W. Jones, "Entero-colocutaneous fistula: a late consequence of polypropylene mesh abdominal wall repair: case report and review of the literature," *Surgery*, vol. 6, no. 3, pp. 144–147, 2002.

[13] A. Jezupovs, A. Jezupors, and M. Mihelsons, "The analysis of infection after polypropylene mesh repair of abdominal wall hernia," *World Journal of Surgery*, vol. 30, no. 12, pp. 2270–2278, 2006.

[14] S. Kalaba, E. Gerhard, J. S. Winder, E. M. Pauli, R. S. Haluck, and J. Yang, "Design strategies and applications of biomaterials and devices for hernia repair," *Bioactive Materials*, vol. 1, no. 1, pp. 2–17, 2016.

[15] R. D. Rice, F. S. Ayubi, Z. J. Shaub, D. M. Parker, P. J. Armstrong, and J. W. Tsai, "Comparison of Surgisis, AlloDerm, and Vicryl Woven Mesh grafts for abdominal wall defect repair in an animal model," *Aesthetic Plastic Surgery*, vol. 34, no. 3, pp. 290–296, 2010.

[16] M. W. Laschke, J. M. Häufel, C. Scheuer, and M. D. Menger, "Angiogenic and inflammatory host response to surgical meshes of different mesh architecture and polymer composition," *Journal of Biomedical Materials Research Part B: Applied Biomaterials*, vol. 91, no. 2, pp. 497–507, 2009.

[17] U. Klinge, "Functional assessment and tissue response of short- and long-term absorbable surgical meshes," *Biomaterials*, vol. 22, no. 11, pp. 1415–1424, 2001.

[18] K. Baylón, P. Rodríguez-Camarillo, A. Elías-Zúñiga, J. A. Díaz-Elizondo, R. Gilkerson, and K. Lozano, "Past, present and future of surgical meshes: a review," *Membranes*, vol. 7, no. 3, 2017.

[19] M. Chen, P. K. Patra, S. B. Warner, and S. Bhowmick, "Role of fiber diameter in adhesion and proliferation of NIH 3T3 fibroblast on electrospun polycaprolactone scaffolds," *Tissue Engineering.*, vol. 13, no. 3, pp. 579–587, 2007.

[20] M. Plencner, B. East, Z. Tonar et al., "Abdominal closure reinforcement by using polypropylene mesh functionalized with poly-ε-caprolactone nanofibers and growth factors for prevention of incisional hernia formation," *International Journal of Nanomedicine*, vol. 9, pp. 3263–3277, 2014.

[21] M. R. Williamson, E. F. Adams, and A. G. A. Coombes, "Gravity spun polycaprolactone fibres for soft tissue engineering: interaction with fibroblasts and myoblasts in cell culture," *Biomaterials*, vol. 27, no. 7, pp. 1019–1026, 2006.

[22] R. Jakubova, A. Mickova, M. Buzgo et al., "Immobilization of thrombocytes on PCL nanofibres enhances chondrocyte proliferation in vitro," *Cell Proliferation*, vol. 44, no. 2, pp. 183–191, 2011.

[23] A. Thapa, T. J. Webster, and K. M. Haberstroh, "Polymers with nano-dimensional surface features enhance bladder smooth muscle cell adhesion," *Journal of Biomedical Materials Research*, vol. 67, no. 4, pp. 1374–1383, 2003.

[24] M. D. Bhavsar and M. M. Amiji, "Development of novel biodegradable polymeric nanoparticles-in-microsphere formulation for local plasmid DNA delivery in the gastrointestinal tract," *AAPS PharmSciTech*, vol. 9, no. 1, pp. 288–294, 2008.

[25] T. K. Dash and V. B. Konkimalla, "Poly-є-caprolactone based formulations for drug delivery and tissue engineering: a review," *Journal of controlled release: official journal of the Controlled Release Society.*, vol. 158, no. 1, pp. 15–33, 2012.

[26] F. Vozzi, I. Guerrazzi, J. Campolo et al., "Biological and proteomic characterization of a composite mesh for abdominal wall hernia treatment: reference study," *Journal of Biomedical Materials Research Part B: Applied Biomaterials*, vol. 105, no. 7, pp. 2045–2052, 2017.

[27] A. N. Buxton, J. Zhu, R. Marchant, J. L. West, J. U. Yoo, and B. Johnstone, "Design and characterization of poly(ethylene glycol) photopolymerizable semi-interpenetrating networks for chondrogenesis of human mesenchymal stem cells," *Tissue Engineering*, vol. 13, no. 10, pp. 2549–2560, 2007.

[28] J. A. Beamish, J. Zhu, K. Kottke-Marchant, and R. E. Marchant, "The effects of monoacrylated poly(ethylene glycol) on the properties of poly(ethylene glycol) diacrylate hydrogels used for tissue engineering," *Journal of Biomedical Materials Research Part A*, vol. 92, no. 2, pp. 441–450, 2010.

[29] H. E. desJardins-Park, D. S. Foster, and M. T. Longaker, "Fibroblasts and wound healing: an update," *Regenerative Medicine*, vol. 13, no. 5, pp. 491–495, 2018.

[30] I. A. Darby, B. Laverdet, F. Bonté, and A. Desmoulière, "Fibroblasts and myofibroblasts in wound healing," *Clinical, Cosmetic and Investigational Dermatology*, vol. 7, pp. 301–311, 2014.

[31] "Polymer Properties Database," https://polymerdatabase.com/polymer%20classes/Intro.html.

[32] B. Wunderlich, *Macromolecular Physics, Crystal Nucleation, Growth, Annealing*, Academic, New York, NY, USA, 1976.

[33] C. De Maria, S. Burchielli, C. Salvadori et al., "The influence of mesh topology in the abdominal wall repair process," *Journal of Biomedical Materials Research Part B: Applied Biomaterials*, vol. 104, no. 6, pp. 1220–1228, 2016, 2nd ed.

[34] I. Felemovicius, M. E. Bonsack, G. Hagerman, and J. P. Delaney, "Prevention of adhesions to polypropylene mesh," *Journal of the American College of Surgeons*, vol. 198, no. 4, pp. 543–548, 2004.

[35] F. Agresta, A. Marzetti, S. Vigna, D. Prando, R. Porfidia, and S. Di Saverio, "Repair of primary and incisional hernias using composite mesh fixed with absorbable tackers: preliminary experience of a laparoscopic approach with a newly designed mesh in 29 cases," *Updates in Surgery*, vol. 69, no. 4, pp. 493–497, 2017.

[36] G. Munegato, L. Fei, M. Schiano di Visconte, D. Da Ros, L. Moras, and G. Bellio, "A new technique for tension-free reconstruction in large incisional hernia," *Updates in Surgery*, vol. 69, no. 4, pp. 485–491, 2017.

[37] W.-J. Lin and C.-H. Lu, "Characterization and permeation of microporous poly(ε-caprolactone) films," *Journal of Membrane Science*, vol. 198, no. 1, pp. 109–118, 2002.

[38] T. Ozeki, H. Yuasa, and Y. Kanaya, "Control of medicine release from solid dispersion composed of the poly(ethylene oxide)-carboxyvinylpolymer interpolymer complex by varying molecular weight of poly(ethylene oxide)," *Journal of controlled release*, vol. 58, no. 1, pp. 87–95, 1999.

[39] T. Velnar, T. Bailey, and V. Smrkolj, "The wound healing process: an overview of the cellular and molecular mechanisms," *The Journal of International Medical Research*, vol. 37, no. 5, pp. 1528–1542, 2009.

[40] R. Sabat, G. Grütz, K. Warszawska et al., "Biology of interleukin-10," *Cytokine and Growth Factor Reviews*, vol. 21, no. 5, pp. 331–344, 2010.

[41] J. Scheller, C. Garbers, and S. Rose-John, "Interleukin-6: from basic biology to selective blockade of pro-inflammatory activities," *Seminars in Immunology*, vol. 26, no. 1, pp. 2–12, 2014.

[42] R. A. Canuto, S. Saracino, M. Oraldi et al., "Colonization by human fibroblasts of polypropylene prosthesis in a composite form for hernia repair," *Surgery*, vol. 17, pp. 241–248, 2012.

[43] C. G. Pereira-Lucena, R. Artigiani-Neto, G. J. Lopes-Filho et al., "Experimental study comparing meshes made of polypropylene, polypropylene + polyglactin and polypropylene + titanium: inflammatory cytokines, histological changes and morphometric analysis of collagen," *Hernia*, vol. 14, no. 3, pp. 299–304, 2010.

[44] R. T. Hannan, S. M. Peirce, and T. H. Barker, "Fibroblasts: diverse cells critical to biomaterials integration," *ACS Biomaterials Science & Engineering*, vol. 4, no. 4, pp. 1223–1232, 2018.

[45] J. M. Anderson, "Exploiting the inflammatory response on biomaterials research and development," *Journal of Materials Science: Materials in Medicine*, vol. 26, no. 3, p. 121, 2015.

[46] A. Fachinelli and M. R. Maciel Trindade, "Qualitative and quantitative evaluation of total and types I and III collagens in patients with ventral hernias," *Langenbeck's Archives of Surgery*, vol. 392, no. 4, pp. 459–464, 2007, 3rd ed.

[47] M. Vaz, R. K. Krebs, E. N. Trindade, and M. R. M. Trindade, "Fibroplasia after polypropylene mesh implantation for

abdominal wall hernia repair in rats," *Acta Cirúrgica Brasileira*, vol. 24, no. 1, pp. 19–25, 2009.

[48] G. Pascual, B. Hernández-Gascón, M. Rodríguez et al., "The long-term behavior of lightweight and heavyweight meshes used to repair abdominal wall defects is determined by the host tissue repair process provoked by the mesh," *Surgery*, vol. 152, no. 5, pp. 886–895, 2012.

[49] E. Peeters, G. De Hertogh, K. Junge, U. Klinge, and M. Miserez, "Skin as marker for collagen type I/III ratio in abdominal wall fascia," *Hernia*, vol. 18, no. 4, pp. 519–525, 2014.

[50] I. Park, C. Han, S. Jin et al., "Myosin regulatory light chains are required to maintain the stability of myosin II and cellular integrity," *The Biochemical Journal*, vol. 434, no. 1, pp. 171–180, 2011.

[51] J. Yang, M. Nune, Y. Zong, L. Zhou, and Q. Liu, "Close and allosteric opening of the polypeptide-binding site in a human Hsp70 chaperone BiP," *Structure*, vol. 23, no. 12, pp. 2191–2203, 2015.

[52] A. Delpino and M. Castelli, "The 78 kDa glucose-regulated protein (GRP78/BIP) is expressed on the cell membrane, is released into cell culture medium and is also present in human peripheral circulation," *Bioscience Reports*, vol. 22, no. 3-4, pp. 407–420, 2002.

[53] A. S. Lee, "Glucose-regulated proteins in cancer: molecular mechanisms and therapeutic potential," *Nature Reviews Cancer*, vol. 14, no. 4, pp. 263–276, 2014.

[54] K. Ohashi, V. Burkart, and S. Flohé, "Cutting edge: heat shock protein 60 is a putative endogenous ligand of the toll-like receptor-4 complex," *The Journal of Immunology*, vol. 164, no. 2, pp. 558–561, 2000.

7

Preparation and Properties of DMF-Based Polyurethanes for Wet-Type Polyurethane Artificial Leather

Suk-Hun Sur,[1] Pil-Jun Choi,[1] Jae-Wang Ko,[1] Jae-Yeon Lee,[1] Young-Hee Lee,[2] and Han-Do Kim[2]

[1]*Industrial Materials Fusion Technology Center, Korea Institute of Footwear and Leather Technology, Busan 47154, Republic of Korea*
[2]*Department of Organic Material Science and Engineering, Pusan National University, Busan 46241, Republic of Korea*

Correspondence should be addressed to Han-Do Kim; kimhd@pusan.ac.kr

Academic Editor: Marta Fernández-García

DMF-based polyurethanes for wet-type polyurethane artificial leather were prepared using 4,4′-diphenylmethane diisocyanate (MDI) as diisocyanate, 1,4-butanediol (BD)/1,6-hexanediol (HD) as chain extender, and four different macroglycols (MW: 2000): poly(butylene/3-methyl-pentylene adipate) glycol based on 1,4-butanediol/3-methyl-1,5-pentanediol (MPD), poly(butylene adipate) glycol based on BD, polycarbonate glycol based on HD/MPD, and polycarbonate glycol based on HD. The properties of the as-polymerized polyurethane solutions and their films and the size/uniformity of cells of the foam layer of the resulting artificial leather were investigated. The viscosities of as-polymerized polyurethane solutions were significantly high in the range of 7500–15,000 cps, indicating that the molecular weight of the obtained polyurethanes is high. The tensile strengths of polyurethane films based on carbonate-polyol and BD chain extender were found to be higher than those of polyurethane films based on ester-polyol and HD chain extender. The elongation at the break of polyurethane films with the MPD component were higher than that of polyurethane films without the MPD component. The artificial leather containing MPD had smaller cells than the artificial leather without MPD and had higher uniformity. From these results, DMF-based polyurethanes prepared by using MPD components containing polyols are found to be most suitable for wet-type artificial leather.

1. Introduction

Since polyurethane is excellent in abrasion resistance, flexibility, durability, and toughness, interest in various fields such as textile, leather, plastic, wood, concrete, glass- and metal-coating materials, adhesives, and binders is increasing [1, 2]. Generally, polyurethanes are made from long chin polyols with an average molecular weight (MW) of 200–10,000, chain extenders with a MW of 60–4000, and polyisocyanates. The long flexible soft segment chiefly controls the low temperature properties, the solvent resistance, and the weather resistant properties of polyurethanes. Meanwhile, the choice of chain extender and diisocyanate determines mainly the mechanical properties of polyurethanes [3, 4].

There are mainly two types of soft segments such as the hydroxyl-terminated polyesters and the hydroxyl-terminated polyethers. The typical polyester diol is made from adipic acid and an excess of glycol such as ethylene glycol, 1,4-butanediol (BD), 1,6-hexanediol (HD), neopentyl glycol, or mixtures of these diols. Generally, polycarbonate diols are generally made from 1,6-hexanediol and phosgene or by transesterification with low MW carbonates like diethyl or diphenyl carbonate. In general, a homopolyester diol with a molecular weight of 2000 or more is a crystalline material with a melting temperature of about 50–60°C, whereas a copolyester diol prepared from an acid and two or more diols has no crystallinity. The branched chain structure such as 3-methyl-1,5-pentanediol (MPD) instead of the linear chain diol such as BD and HD in the polyol component interferes with the packing of the molecule, so the polyol cannot have crystallinity. The resulting polyurethane is amorphous and can have properties such as excellent softness, excellent durability, broad solvent compatibility, good transparency, excellent stain resistance, and improved mold resistance. The characteristics of the soft segment are determined according to the kind and content of the constituent, which is an important factor determining the properties of the resulting polyurethane. In particular, polycarbonate-based polyurethanes are superior in hydrolytic stability, mechanical properties, durability, and transparency to conventional polyol-based polyurethanes. Therefore, polycarbonate-based polyurethanes are widely applied to automotive coatings, dry coatings, and UV-curing resins [5–10].

Generally, artificial leather is prepared by impregnating a nonwoven fabric/knitted goods with a polyurethane resin dissolved in an organic solvent. In this case, artificial leather is made by wet process with dimethylformamide (DMF) as the main organic solvent. In particular, wet process polyurethane synthetic leather has the advantage of being lightweight, durable, sewable, and washable by using synthetic fibers (nonwoven fabric, knitted fabric). Synthetic leather is a substitute material of natural leather [11–13]. In the 1970s, Toray of Japan developed suede artificial leather called Ecsaine (trade name) using ultrafine fiber-nonwoven fabric and solvent-based polyurethane. Kuraray has developed Clarino (trade name) using microfiber fibers, and it was spotlighted not only for shoes but also for automotive interiors and interior materials [14, 15]. Generally, artificial leathers based on nonwoven support materials coated with poly(vinyl chlorides) or polyurethanes are used for apparels and shoe upper materials [16].

Research on polyurethanes containing MPD component has been done at many company laboratories, but it is hardly found in the open literature. Studies on artificial leather using such polyurethane could not be found in the open literature. There is little systematic academic study of the relationship between the structure and properties of polyurethanes with these various molecular structures. Therefore, in this study, the DMF-based polyurethanes were synthesized using four different types of polyols with/without MPD component, 1,4-butanediol and 1,6-hexanediol as chain extenders, and MDI as diisocyanate. The wet process artificial leathers based on nonwoven fabrics coated with DMF-based polyurethanes were prepared. The thermal properties, dynamic mechanical thermal properties, and mechanical properties of the eight polyurethane film samples were compared. The cell-forming properties of wet process artificial leather were also investigated.

2. Materials and Methods

2.1. Materials. Four different polyols used in this study are shown in Table 1: polycarbonate diol (MW: 2000, C2090, Kuraray, Japan) based on 1,6-hexanediol/3-methyl-1 and 5-pentane diol (MPD) and polyester diol (polyadipate diol, MW: 2000, P2010, Kuraray, Japan) based on 1,4-butanediol/MPD were used as polyols containing 3-methyl-1,5-pentanediol (MPD). And polycarbonate diol (MW: 2000, T6002, Asahi Kasei) based on 1,6-hexanediol and polyester diol (MW 2000, HP1020, Heung Il Polychem) based on 1,4-butanediol were used as polyols. All polyols used in this study were dried at 100°C under 65 mmHg for 3 hours until no bubbling was observed. Methylene diphenyl diisocyanate (MDI, BASF), dibutyltin dilaurate (DBTDL, Aldrich Chemical, Milwaukee, WI), 1,4-butanediol (Aldrich Chemical, Milwaukee, WI), 1,6-hexanediol (Aldrich Chemical, Milwaukee, WI), dimethylformamide (DMF, Junsei, Japan), and antioxidant (Irganox 1010, SA) were used without further purification. The characteristic of nonwoven fabric (HDF 3310, HACO, Korea) used in this study is as follows: method: needle punching, composition: PET/nylon (30/70 wt%), weight: 330 g/m^2, thickness: 1.30 mm, and width: 1410 mm.

2.2. Preparation of DMF-Based Polyurethane Solutions. Polyurethanes were synthesized by solution polymerization using DMF as a solvent (Scheme 1). The polyols (140 g), DBTDL (catalyst, 0.03 wt% based on solid content), and DMF (solvent, 30 g) were placed in a four-necked flask that was equipped with a thermometer, a stirrer, a condenser, an inlet and outlet for dry nitrogen, and a heat jacket, and then the system was degassed under vacuum at 80°C for 1 h to obtain stable solution. The mixture was allowed to cool to 45°C with moderate stirring (175–200 rpm). MDI was then slowly dropped into the flask, and the reaction mixture was allowed to react at 85°C under stirring (125–150 rpm) until the theoretical NCO content was reached. The change in the NCO value during the reaction was determined using standard dibutylamine back-titration method (ASTM D 1638). The prepolymer was chain-extended by dropping 1,4-butanediol/1,6-hexanediol at room temperature for 2 h, and the reaction continued until the NCO peak (2170 cm^{-1}) in the IR spectrum had completely disappeared. The solid content of the polyurethane solution was fixed at 30 wt%. The molar ratio of MDI/polyol/diol was fixed at 0.2/0.07/0.13. The sample designation and composition are shown in Table 2.

2.3. Preparation of Polyurethane Film and Wet Process Artificial Leather. The release paper (RP film) was coated with DMF-based polyurethane and dried in a dry oven at 80°C for 12 h in order to volatilize the solvent DMF. In this study, wet-type artificial leather was manufactured as follows: after coating DMF-based polyurethane on the nonwoven fabric, DMF was extracted by the osmotic principle using the coagulation of polyurethane and the solubility of DMF in water, so that continuous cells were formed in the polyurethane layer. The DMF-based polyurethane prepared in this study was coated to nonwoven fabrics and then coagulated for 8 minutes in a water bath containing 10% DMF solution at 25°C and washed in a 50°C water bath for one hour to remove residual DMF. Finally, it was dried in a dryer at 60°C for 3 h to produce artificial leather.

2.4. Characterization

2.4.1. Viscosity Measurement. A Brookfield digital viscometer (Brookfield LVDVII+, USA) was used to measure the

Table 1: Four different macroglycols used in this study.

Macroglycols		
PE/B/M	P2010	Polyadipate diol based on 1,4-butanediol/3-methyl-1,5-pentandiol (MPD)
PE/B	HP1020	Polyadipate diol based on 1,4-butanediol
PC/H/M	C2090	Polycarbonate diol based on 1,6-hexanediol/3-methyl-1,5-pentandiol (MPD)
PC/H	T6002	Polycarbonate diol based on 1,6-hexanediol

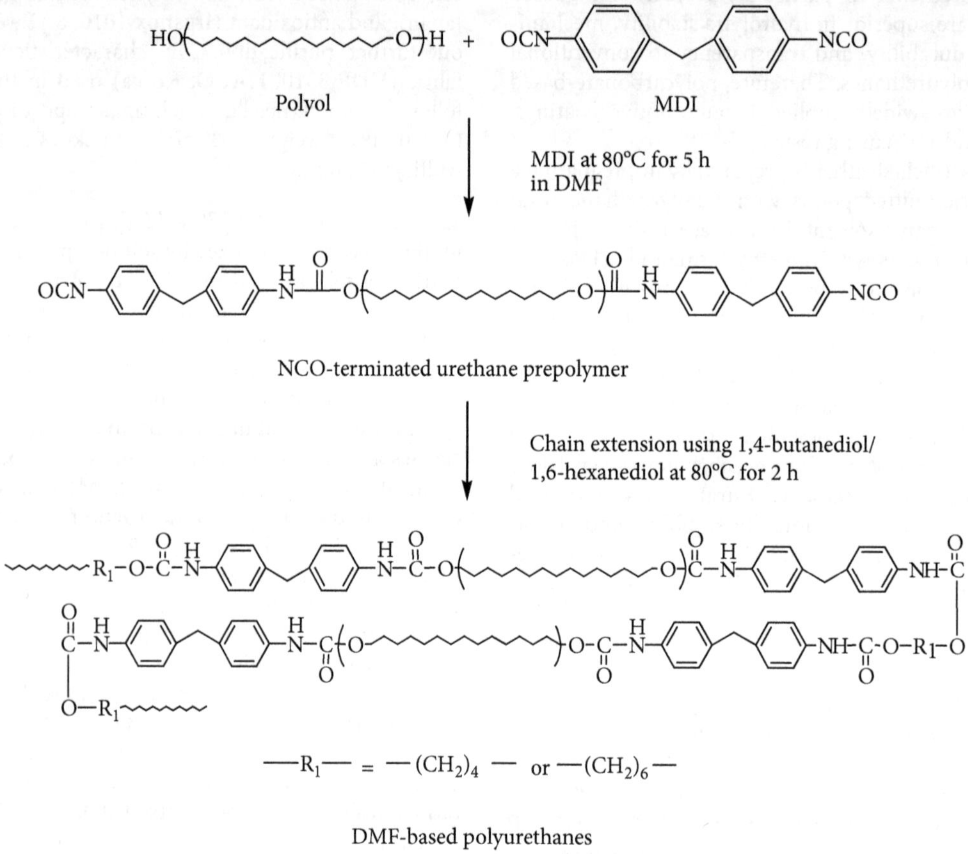

Scheme 1: Preparation process of DMF-based polyurethanes.

Table 2: Sample designation and composition of wet-type PU.

Sample designation	Isocyanate	Composition (molar ratio) Polyol				Chain extender	
	MDI	PE/B/M	PE/B	PC/H/M	PC/H	BD diol	HD diol
B-PE/B/M	0.2	0.07	—	—	—	0.13	—
B-PE/B	0.2	—	0.07	—	—	0.13	—
B-PC/H/M	0.2	—	—	0.07	—	0.13	—
B-/PC/H	0.2	—		—	0.07	0.13	—
H-PE/B/M	0.2	0.07	—		—	—	0.13
H-PE/B	0.2	—	0.07	—	—	—	0.13
H-PC/H/M	0.2	—	—	0.07	—	—	0.13
H-PC/H	0.2	—	—	—	0.07	—	0.13

viscosity of DMF-based polyurethane solutions (solid content: 30 wt%) at 25°C. The measurements were performed at 10 rpm using a LV-3 spindle.

2.4.2. FTIR Analysis. FTIR spectroscopy was used to confirm the chemical components of the CWPU/P samples by Fourier transform infrared spectrometer (FT-IR 3D analysis system, MB 104, BOMEM). For each sample, 8 scans at 4 cm^{-1} resolution were collected in the transmittance mode and recorded in the range of 4000–650 cm^{-1}.

2.4.3. Mechanical Test. The tensile properties were measured at room temperature using a universal testing machine (UTM, Model M-130, Instron Co., USA) with a 5582 system according to ASTM D412. A cross-head speed of 100 mm/min was used throughout these investigations to determine the ultimate tensile strength and elongation at break for all samples. The values quoted are the average of three measurements.

2.4.4. Differential Scanning Calorimetry Analysis. Differential scanning calorimetry (DSC, Model Q-25, TA Instruments) was used for thermal characterization. The sample (10.0 ± 1.5 mg) was placed in an aluminum pan and quenched to −80°C, and then the change in calories was measured while raising the temperature from −80°C to 100°C at a rate of 10°C/min under nitrogen atmosphere at 50 ml/min of N_2 flow rate.

2.4.5. Dynamic Mechanical Thermal Analyzer Analysis. The dynamic mechanical properties of film samples were measured at 1 Hz using dynamic mechanical thermal analyzer (DMA, Q800, TA Instruments, USA) with a heating rate of 5°C/min in the temperature range from −80 to 150°C.

2.4.6. Scanning Electron Microscope (SEM) Observation. In order to confirm the cell shape of the polyurethane film layer of the artificial leather produced by the wet process, the cross section of the film was cut and analyzed by a field-emission scanning electron microscope (SEM, Hitachi TM-1000, Japan).

3. Results and Discussion

3.1. Analysis of Chemical Structure of Polyurethane Films. The results of the qualitative analysis of the synthesized polyurethane structure using FT-IR are shown in Figure 1, since the content of polyols acting as a soft segment in the polyurethane structure is relatively larger than that of the other components. The peaks of the polyol component were found to be very large in the spectroscopic analysis. As shown in Figure 1, B-PE/B/M, B-PE/B, H-PE/B/M, and H-PE/B samples are polyurethanes synthesized using ester polyols as polyols. It can be seen that a characteristic peak appeared at 2800–3000 cm^{-1} corresponding to CH stretching vibrations of asymmetric and symmetric stretching modes of methylene groups and a characteristic peak due to stretching vibration of the ester carbonyl group at 1733 cm^{-1} is observed. Typical peaks of polyadipate were also observed at 1200 and 1300 cm^{-1}. It was found that the characteristic absorption peaks of the carbonate group (-O-COO-C) were observed at around 1250 cm^{-1} in the polyurethane B-PC/H/M, B-PC/H, H-PC/H/M, and H-PC/H samples synthesized by using polycarbonate diol as the polyol. Based on the results of the IR analysis, it was confirmed that polyester polyol and polycarbonate polyol components were well incorporated in the polyurethane synthesized in this study. In addition, it was confirmed that the NCO absorption peak at 2270 cm^{-1} disappears as the reaction progresses. The urethane group formed was identified from NH peaks at 3200 and 1500 cm^{-1}.

3.2. Viscosity of As-Polymerized Polyurethane Solutions. Figure 2 shows the viscosity measured using a Brookfield viscometer of DMF solutions of polyurethanes synthesized with the same solid content (30 wt%). Generally, the viscosity of a polymer solution is proportional to the molecular weight and polarity of the polymer. The viscosities of as-polymerized polyurethane solutions were in the range of 7500–15,000 cps at 25°C. This very high viscosity value means that the molecular weight of the obtained polyurethane is considerably high. The viscosity of the polyurethane solution containing MPD was slightly lower than that without MPD. This can be attributed to the fact that the methyl groups present in the MPD structure interfere with intermolecular forces. The viscosity of the as-polymerized polyurethane solution was found to be higher in the case of BD as a chain extender than in HD. Also, the viscosity of the carbonate-polyol-based polyurethane solution was higher than that of the ester-polyol-based polyurethane solution. The higher viscosity of BD/carbonate-polyol-based polyurethanes might be due to the higher molecular weight or higher polarity.

3.3. The Glass Transition Temperature/Storage Modulus of Prepared Polyurethane Films. Figure 3 shows the DSC curves of the synthesized polyurethane films. The dynamic mechanical thermal analysis curves are shown in Figures 4 and 5. The soft segment glass transition temperature (Tgs) of the polyurethane films determined by DSC and DMA is summarized in Table 3. The DSC analysis showed that the melting temperature (Tm) was not observed. The Tgs change tendency of polyurethanes according to the composition measured by DSC and DMA was the same. The DSC-Tgs of B-PE/B/M, B-PE/B, B-PC/H/M, and B-PC/H samples was observed respectively at −40.1, −42.4, −10.9, and −14.0°C, and the DSC-Tgs of H-PE/B/M, H-PE/B H-PC/H/M, and H-PC/H samples appeared respectively at −35.2, −31.9, −13.11, and −7.6°C, while the Tgs of B-series samples measured by DMA was observed at −37.4, −40.2, −11.5, and −15.1°C and the Tgs of H series samples appeared at −37.6, −36.9, −15.8, and −12.8°C. In the case of chain extender BD-based polyurethane films, the Tgs of polyurethane films containing the MPD component was a little higher than that of polyurethane films without MPD. This is presumed to be attributable to the limited segment motion of MPD-contained polyurethanes due to the side chain methyl group of MPD. However, in the case of chain extender HD, polyurethane films with the MPD component showed lower Tgs than polyurethane films

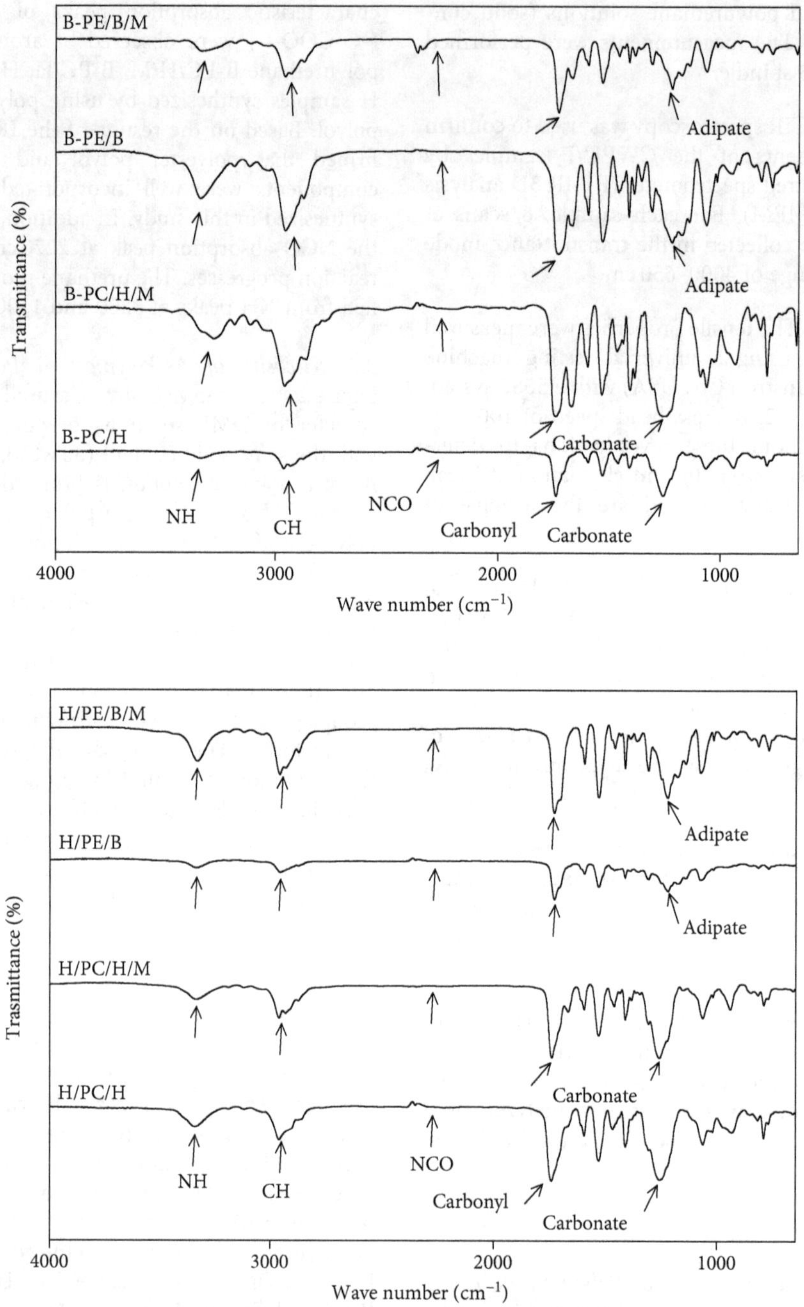

Figure 1: FT-IR spectra of polyurethane films.

without MPD. This might be due to the number of carbons in the HD of 6 more than the 4 carbons in BD. The carbonate-polyol-based polyurethane films showed the higher Tgs than ester-polyol-based polyurethane films. It was found that the polyurethane film containing the MPD component had a higher storage modulus value than the corresponding polyurethane film without the MPD component. The glassy plateau region of carbonate-polyol-based polyurethane films was found to be wider than that of ester-polyol-based polyurethane films. This should be due to the higher polarity of carbonate-glycol than ester-glycol.

3.4. Mechanical Properties of Prepared Polyurethane Films. Figure 6 shows the stress-strain curves of polyurethane films. The tensile strength (kgf/cm^2) and elongation (%) at the break of samples are summarized in Table 3. The tensile strength of the polyurethane film containing no MPD component was higher than that of the film containing the MPD component. In the same series, the tensile strength of carbonate-polyol-based polyurethane was higher than that of ester-polyol-based polyurethane. It was found that the tendency of the change in tensile strength of the polyurethane film was almost consistent with the change in

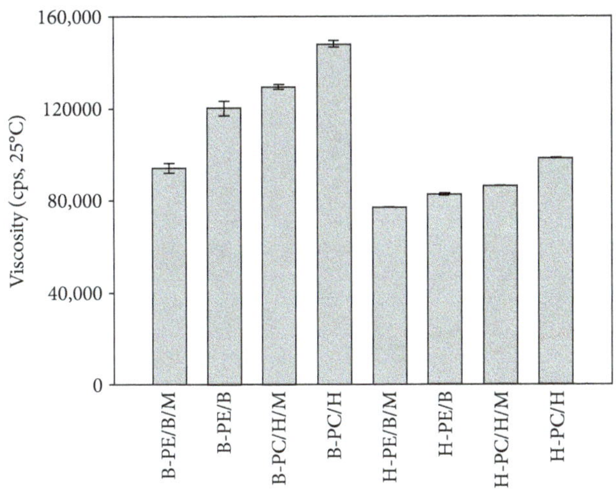

FIGURE 2: Viscosity of DMF-based polyurethane solutions (average value of five tests).

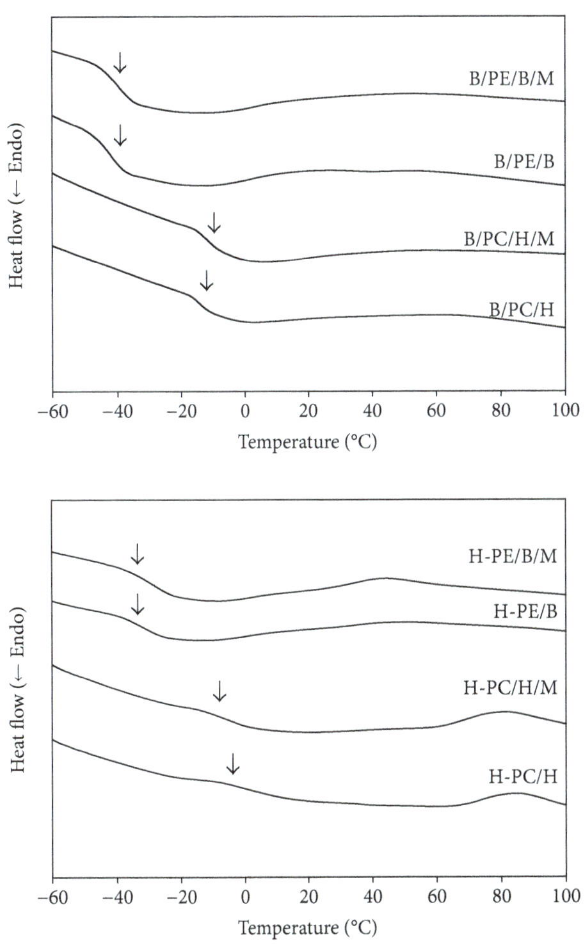

FIGURE 3: DSC thermograms of polyurethane films.

viscosity of the DMF-polyurethane solution. These results are in good agreement with the fact that strength depends on molecular weight. The elongation at the break of the polyurethane film with the MPD component was larger than that of the film without the MPD component. And the ester-polyol-based polyurethane film showed higher elongation at the break than the carbonate-polyol-based film. This is thought to be due to both the branched structure in the MPD component and the ester group with lower polarity than the carbonate group.

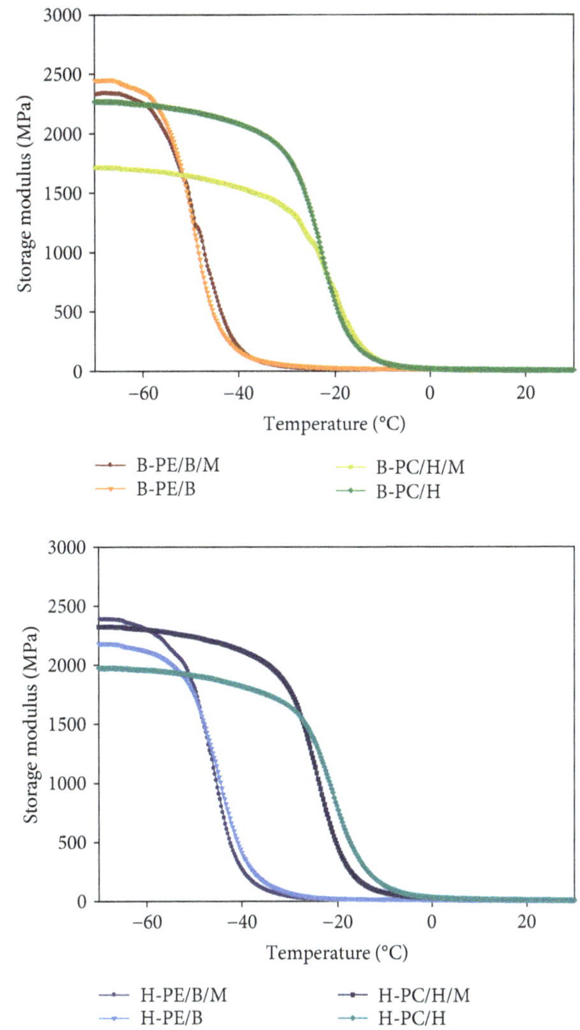

FIGURE 4: Storage modulus of polyurethane films.

3.5. Cross Sections of Wet Process Artificial Leathers Based on Nonwoven Fabrics Coated with DMF-Based Polyurethanes. Figure 7 shows the cross-sectional SEM micrographs of the wet process artificial leathers based on nonwoven fabrics coated with DMF-based polyurethanes prepared in this study. Generally, artificial leather having a large number of cells per unit volume, a small cell size, and a uniform distribution is preferable. This is because such a characteristic can impart an excellent tactile touch to the synthetic leather. In this study, wet process artificial leathers based on nonwoven fabrics coated with DMF-based polyurethanes are prepared for shoe upper materials. The polyurethane-based artificial leather containing the MPD component prepared

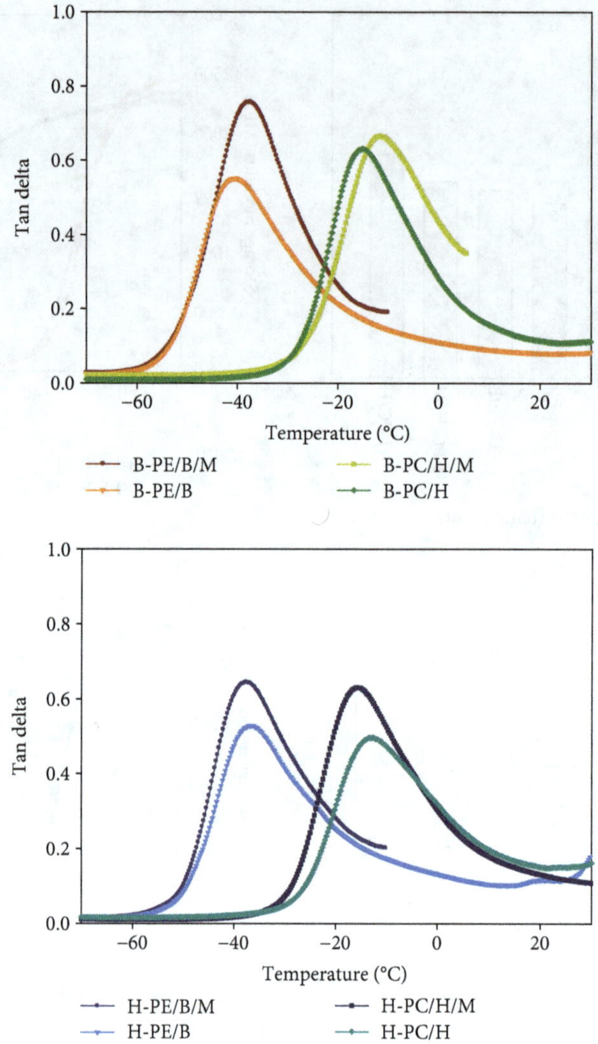

Figure 5: Tan delta of polyurethane films.

Table 3: Thermal and mechanical properties of wet-type PU films.

Sample designation	DSC result Tgs (°C)	DMA result Tgs (°C)	Mechanical properties Tensile strength (kgf/cm^2)	Elongation at break (%)
B-PE/B/M	−40.1	−37.4	41	483
B-PE/B	−42.4	−40.2	62	391
B-PC/H/M	−10.9	−11.5	68	342
B-PC/H	−14.0	−15.1	101	162
H-PE/B/M	−35.2	−37.6	39	669
H-PE/B	−31.9	−36.9	46	346
H-PC/H/M	−13.1	−15.8	61	429
H-PC/H	−7.6	−12.8	96	218

under the same coating conditions was found to have a large coating layer and a large number of cells and a uniform distribution. In general, the ester-polyol-based polyurethane has a slightly higher cell layer than the carbonate-polyol. In the case of using BD as a chain extender, the coating layer was thicker than that of using HD, and the number of cells per unit volume was larger than HD. From these results, it was found that DMF-

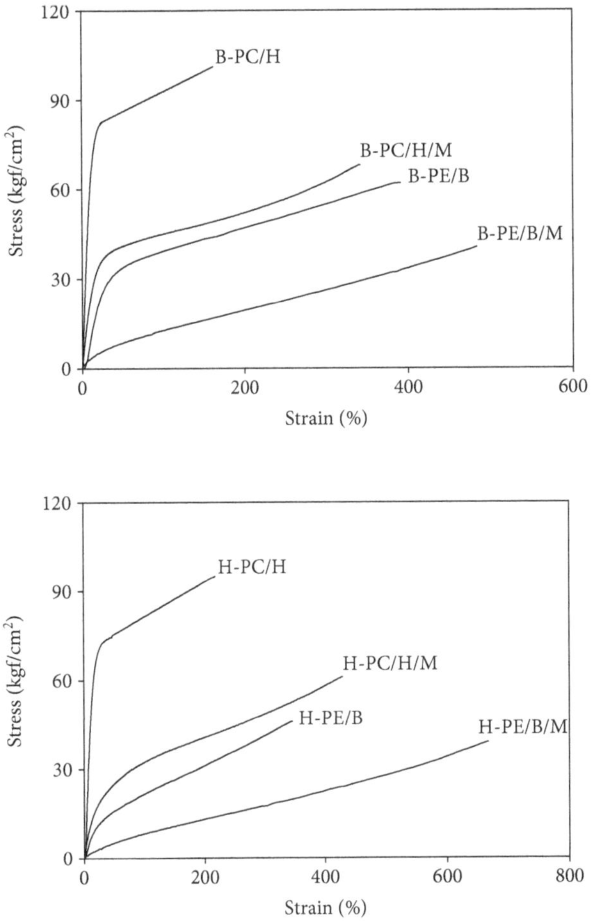

FIGURE 6: Stress-strain curves of polyurethane films.

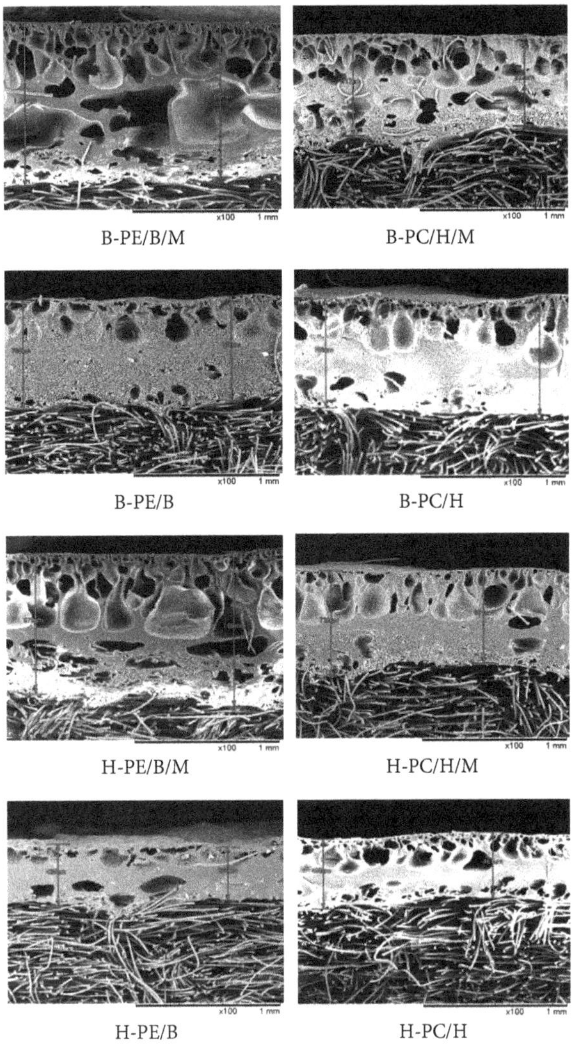

FIGURE 7: SEM micrographs of wet-type artificial leathers based on nonwoven fabrics coated with various DMF-based polyurethanes.

based polyurethanes prepared by using MPD component-contained polyols are most suitable for the production of wet-type artificial leather.

4. Conclusions

In this study, DMF-based polyurethanes for wet process artificial leather were prepared by using 1,6-hexanediol/1,4-butanediol as chain extender and polycarbonate diol and polyester diol containing 3-methyl-1,5-pentanediol (MPD) component, which is known to give excellent adhesion, durability, low temperature flexibility, and transparency. The effects of these compositions on the physical properties of polyurethane films and the cell-forming properties of wet-type artificial leather were investigated. The viscosities of as-polymerized polyurethane solutions were significantly high in the range of 7500–15,000 cps, indicating that the molecular weight of the obtained polyurethanes is high. It was confirmed that the Tgs obtained from the peak of tan delta and the Tgs obtained by DSC were almost the same. The tensile strength of the polyurethane film containing no MPD component was higher than that of the film containing the MPD component. In the same series, the tensile strength of carbonate-polyol-based polyurethane was higher than that of ester-polyol-based polyurethane. The MPD component-contained synthetic leather prepared under the same coating conditions was found to have a large coating layer and a large number of cells and a uniform distribution. These results indicate that MPD component-contained polyurethane prepared in this study is most suitable for the production of wet-type artificial leather.

Conflicts of Interest

The authors declare that they have no conflicts of interest.

References

[1] D. K. Chattopadhyay and K. V. S. N. Raju, "Structural engineering of polyurethane coatings for high performance applications," *Progress in Polymer Science*, vol. 32, no. 3, pp. 352–418, 2007.

[2] M. M. Rahman, A. Hasneen, W. K. Lee, and K. T. Lim, "Preparation and properties of sol-gel waterborne polyurethane adhesive," *Journal of Sol-Gel Science and Technology*, vol. 67, no. 3, pp. 473–479, 2013.

[3] K. C. Frisch and D. Klempner, *Advances in Urethane Science and Technology, Volume 14*, Technomic Publishing Company, USA Lancaster, PA, USA, 1998.

[4] G. Oertel, *Polyurethane Handbook*, Carl Hanser Verlag, Munich, 1985.

[5] Z. Hrdlicka, A. Kuta, R. Poreba, and M. Spirkova, "Polycarbonate-based polyurethane elastomers: temperature-dependence of tensile properties," *Chemical Papers*, vol. 68, no. 2, pp. 233–238, 2014.

[6] R. Poręba, M. Špírková, and Z. Hrdlička, "Mechanical and thermomechanical properties of polycarbonate-based polyurethane-silica nanocomposites," *Processing and Application of Ceramics*, vol. 5, no. 3, pp. 155–159, 2011.

[7] P. A. Gunatillake, G. F. Meijs, S. J. Mccarthy, R. Adhikari, and N. Sherriff, "Synthesis and characterization of a series of poly(alkylene carbonate) macrodiols and the effect of their structure on the properties of polyurethanes," *Journal of Applied Polymer Science*, vol. 69, no. 8, pp. 1621–1633, 1998.

[8] H. Tanaka and M. Kunimura, "Mechanical properties of thermoplastic polyurethanes containing aliphatic polycarbonate soft segments with different chemical structures," *Polymer Engineering and Science*, vol. 42, no. 6, pp. 1333–1349, 2002.

[9] M. Špírková, J. Pavličević, A. Strachota et al., "Novel polycarbonate-based polyurethane elastomers: composition–property relationship," *European Polymer Journal*, vol. 47, no. 5, pp. 959–972, 2011.

[10] A. Eceiza, M. Larrañaga, K. de la Caba et al., "Structure–property relationships of thermoplastic polyurethane elastomers based on polycarbonate diols," *Journal of Applied Polymer Science*, vol. 108, no. 5, pp. 3092–3103, 2008.

[11] J. Hemmrich, J. Fikkert, and M. van den Berg, "Porous structural forms resulting from aggregate modification in polyurethane dispersions by means of isothermic foam coagulation," *Journal of Coated Fabrics*, vol. 22, no. 4, pp. 268–278, 1993.

[12] O. Fukushima, H. Hayanami, and K. Nagoshi, "Method for manufacturing synthetic leather," U. S. Patent No. 3, 424, 604, 1969.

[13] J. Kesselmeier and M. Staudt, "Biogenic volatile organic compounds (VOC): an overview on emission, physiology and ecology," *Journal of Atmospheric Chemistry*, vol. 33, no. 1, pp. 23–88, 1999.

[14] http://www.kuraray.us.com/products/fibers/clarino.

[15] http://www.ecsaine.com/en/.

[16] M. Chen, D. L. Zhou, Y. Chen, and P. X. Zhu, "Analyses of structures for a synthetic leather made of polyurethane and microfiber," *Journal of Applied Polymer Science*, vol. 103, no. 2, pp. 903–908, 2007.

8

"*Pillaring Effects*" in Cross-Linked Cellulose Biopolymers: A Study of Structure and Properties

Inimfon A. Udoetok,[1] Lee D. Wilson,[1] and John V. Headley[2]

[1]*Department of Chemistry, 110 Science Place, University of Saskatchewan, Saskatoon, SK, Canada S7N 5C9*
[2]*Water Science and Technology Directorate, 11 Innovation Boulevard, Environment and Climate Change Canada, Saskatoon, SK, Canada S7N 3H5*

Correspondence should be addressed to Lee D. Wilson; lee.wilson@usask.ca

Academic Editor: Yiqi Yang

Modified cellulose materials (CLE-4, CLE-1, and CLE-0.5) were prepared by cross-linking with epichlorohydrin (EP), where the products display variable structure, morphology, and thermal stability. Adsorptive probes such as nitrogen gas and phenolic dyes in aqueous solution reveal that cross-linked cellulose has greater accessible surface area (SA) than native cellulose. The results also reveal that the SA of cross-linked cellulose increased with greater EP content, except for CLE-0.5. The attenuation of SA for CLE-0.5 may relate to surface grafting onto cellulose beyond the stoichiometric cellulose and EP ratio since ca. 30% of the hydroxyl groups of cellulose are accessible for cross-linking reaction due to its tertiary fibril nature. Scanning electron microscopy (SEM) results reveal the variable surface roughness and fibre domains of cellulose due to cross-linking. X-ray diffraction (XRD) and ^{13}C NMR spectroscopy indicate that cellulose adopts a one-chain triclinic unit cell structure (P1 space group) with *gauche-trans* (*gt*) and *trans-gauche* (*tg*) conformations of the glucosyl linkages and hydroxymethyl groups. The structural characterization results reveal that cross-linking of cellulose occurs at the amorphous domains. By contrast, the crystalline domains are preserved according to similar features in the XRD, FTIR, and ^{13}C NMR spectra of cellulose and its cross-linked forms. This study contributes to an improved understanding of the role of cross-linking of native cellulose in its structure and functional properties. Cross-linked cellulose has variable surface functionality, structure, and textural properties that contribute significantly to their unique physicochemical properties over its native form.

1. Introduction

Cellulose is one of the most abundant and renewable biopolymers on earth [1, 2]. It is environmentally benign, biodegradable, recyclable, and renewable. Notwithstanding the structural complexity and recalcitrant nature of cellulose, its relative abundance and low cost have drawn interest to its alternative and sustainable use as a raw material over petrochemical-based feedstocks [1–3]. While there are various strategies for the synthetic modification of cellulose, cross-linking is known to alter the physicochemical properties due to *pillaring effects* [4, 5]. Previous studies have shown that the introduction of cross-links within a biopolymer network alters the textural, hydration, and mechanical properties, along with the chemical stability toward biodegradation [6–10]. The structure and physicochemical properties of cross-linked polymers related to adsorption (e.g., morphology, thermal stability, solvent swelling, crystallinity, and surface chemistry) differ compared with unmodified (native) polymers. The enhancement of such properties has led to diverse industrial and biomedical applications [11], aerospace and electronics [12], sorbents for wastewater treatment (decolorization and chelation of pollutants), or extraction of metals [13] and textile manufacturing [14].

The structure of native cellulose is complex and displays unique water adsorption and lipophilic surface area that relate to its complex fibril morphology and the variable accessibility of the hydrophilic groups. The unique hydrophile-lipophile surface characteristics of cellulose are revealed by calorimetric results since its known hydration states are characterized by free and nonfreezing/freezable water. In the case of free water, it is categorized as unbound water in polymers;

while nonfreezing bound water is strongly bound to the accessible hydroxyl groups, freezable bound water has weak interactions with the biopolymer chain. Free water undergoes similar freezing to normal bulk water (below 0°C according to the cooling rate) while freezable bound water solidifies well below the normal freezing point of bulk water.

Cross-linked cellulose (cf. Scheme 1) has been reported [6, 15–18], where complete cross-linking with epichlorohydrin (EP) results in a net loss of one hydroxyl group. In turn, cross-linking alters the hydrophile-lipophile balance of the biopolymer and its physicochemical properties. This is inferred due to changes in the surface functional groups and their accessibility due to structural modification [17]. As well, cross-linking modifies the textural properties (surface area and pore structure) of biopolymers due to *pillaring effects* [18] where the linkers are inferred to serve as bridges (*pillars*) between the biopolymer units (cf. Scheme 1). The textural properties of cellulose and its modified forms are anticipated to have variable thermodynamics and kinetics of hydration due to the role of *pillaring effects* on the biopolymer structure and properties. In an effort to obtain a greater understanding of the role of *pillaring effects* for cross-linked materials, a detailed structural characterization is needed since sparse studies exist for cellulose materials [19–21].

This study was motivated to address the above knowledge gaps through a systematic investigation of the structure-property relationship concerning the role of *pillaring effects* and adsorption-based properties [22–24] for cellulose and its cross-linked forms. Herein, the morphology, structure, conformation, and thermal stability of cellulose materials was studied by a range of spectroscopic (FTIR, XRD, SEM, and solution/solid-state ^{13}C NMR) methods along with thermal analysis. This study highlights several contributions related to cellulose-modified materials: (i) the development of a greater understanding of semicrystalline/amorphous domains of cellulose and cross-linking and (ii) the use of a multi-instrumental study of cellulose and its cross-linked forms to develop an improved understanding of *pillaring effects* and adsorption-based properties. This study represents *a first example* that details the structural characterization of cellulose and its physicochemical properties (thermal stability and hydration) in relation to the role of *pillaring effects.* In turn, a greater understanding of the physicochemical properties of cellulose and its cross-linked forms will advance the science and technology of such materials. We envisage that such biopolymer materials can be synthetically engineered for specialized solid phase extraction (SPE) of steroids or lipids for analytical to biotechnology applications to address chemical separations in complex media such as urine to water/alcohol mixtures [7, 25].

2. Material and Methods

2.1. Materials. Cellulose (medium fibre from cotton linters), epichlorohydrin (EP), HCl, NaOH, potassium bromide (KBr), phenolphthalein (phth), *p*-nitrophenol (PNP), sodium bicarbonate, deuterated dimethyl sulfoxide (DMSO-d_6), and 1-butyl-3-methylimidazolium chloride ($[C4Mim]^+Cl^-$) ionic liquid (IL) was obtained from Sigma-Aldrich Canada Ltd. (Oakville, ON). HPLC-grade acetone was obtained from Fisher Scientific (NJ, USA). All chemicals used were of ACS grade unless specified and were used as received without further purification.

2.2. Synthesis of Cross-Linked Cellulose. The synthesis of cross-linked cellulose was modified from a previous report [6]. In brief, cellulose (2 g) was cross-linked upon dropwise addition over one minute of EP with heating and stirring in 16 mL of 2 M NaOH in a 100 mL round-bottom flask under argon gas at 80°C for 3 h. Requisite volumes of EP ($\rho = 1.18$ g/mL) corresponding to different cellulose to cross-linker feed ratios (varied from low to high, 4 to 0.5) were used in Table 1. The reaction was stirred for 12 h before neutralization with 1 M HCl solution. The product was separated from the supernatant by vacuum filtration and washed with several generous portions of cold Millipore water, followed by drying at ca. 50°C. The dry product was exhaustively washed in a Soxhlet extractor with HPLC-grade acetone for 24 h followed by drying in a vacuum oven at 56°C for 12 h. The material was ground in a mortar and pestle and isolated using a 40-mesh sieve.

2.3. Characterization of Cellulose Materials

2.3.1. FTIR Studies. FTIR spectra were obtained using a Bio-Rad FTS-40 spectrophotometer with powdered samples by mixing with pure spectroscopic grade KBr in a polymer/KBr weight ratio of 1 : 10 followed by grinding in a small mortar and pestle. The diffuse reflectance infrared Fourier transform (DRIFT) spectra were obtained in reflectance mode at 295 K with a resolution of 4 cm^{-1} over the 400–4000 cm^{-1} spectral range. 16 scans were recorded and corrected relative to a background of pure KBr. Quantification of the spectral bands was carried out by integration of the FTIR signatures using OriginPro 2015 software.

2.3.2. Thermogravimetric Analyses (TGA). Thermogravimetric properties of the materials were measured using a Q50 (TA Instruments) operating with a heating rate of 5°C·min^{-1} up to 500°C with a nitrogen carrier gas. Thermograms were integrated using OriginPro 2015 software.

2.3.3. Elemental Analyses. Elemental microanalyses were obtained with a PerkinElmer 2400 CHN Elemental Analyzer. The combustion oven temperature was set above 925°C while the reduction oven was held above 640°C. The instrument was purged with a mixture of pure oxygen and helium gas. The calibration standard was acetanilide. Elemental analysis results of samples were obtained in duplicate with an estimated precision of ±0.3%. The degree of substitution and weight content (%) of EP in the products was estimated from the CHN results.

2.3.4. X-Ray Diffraction (XRD) Studies. XRD spectra of the samples were obtained using a Panalytical Empyrean powder X-ray diffractometer with monochromatic Co-Kα1 radiation and an applied voltage (40 kV) with a fixed current (45 mA). Samples were mounted in a horizontal configuration after evaporation of methanol films, and the XRD patterns were

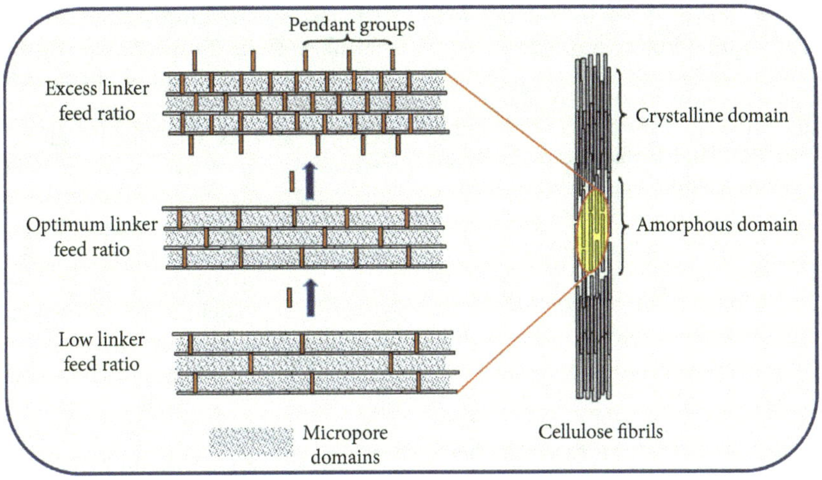

SCHEME 1: Incremental cross-linking ("*pillaring*") of the fibril structure of cellulose by variable cross-linking with epichlorohydrin.

TABLE 1: Synthetic ratios of cellulose and epichlorohydrin (EP) for cross-linked cellulose.

Reaction conditions	CLE-4*	CLE-1*	CLE-0.5*
Weight of EP (g)	0.28**	1.14**	2.28**
Moles of EP	0.00303	0.0123	0.0246
Moles of cellulose	0.0123	0.0123	0.0123
Weight of product (g)	1.40	1.54	1.74
Mole ratio (cellulose: EP)	4.05***	1.00***	0.50***

*Numerical descriptors in the cross-linked polymer refer to the ratio of moles of cellulose monomer $(C_6H_{10}O_5)_n$ to EP. **Calculated using a density value 1.18 g/mL (25°C). ***Ratio of cellulose monomer $(C_6H_{10}O_5)_n$ to EP.

measured in continuous mode over a 2θ range ($2\theta = 7 - 50°$) with a scan rate of $3.2°\cdot\text{min}^{-1}$. The crystallinity index (CrI) was calculated from the intensity ratio of the crystalline peak ($I_{002} - I_{AM}$) and total intensity (I_{002}). Unit cell parameters were obtained using Le Bail refinement fitting based on the unit cell parameters of cellulose 1α, due to similarity between the diffraction patterns of cross-linked cellulose materials and simulated pattern of cellulose 1α [26].

2.3.5. Solid-State ^{13}C NMR Spectroscopy. ^{13}C solid NMR spectra were obtained with a wide-bore (89 mm) spectrometer operating at 8.6 T using an Oxford superconducting magnet system equipped with a 4 mm solid probe with cross polarization and magic angle spinning (CP-MAS) operating at 125.55 MHz. Acquisition parameters were controlled using an Avance DRX-360 console and workstation running TopSpin 1.3 (Bruker BioSpin Corp; Billerica, MA, USA), along with standard pulse programs in the TopSpin1.3 software. Samples were packed in 4 mm outer diameter zirconium oxide rotors capped with Teflon MAS rotor caps. Acquisition was carried out using MAS with a rotational speed of 5 kHz, 2 s recycle delay, and 750 μs cross polarization time. The spectra were referenced externally to adamantane ($\delta = 38.6$ ppm) as an internal standard.

2.3.6. ^{13}C NMR Spectroscopy in Solution. ^{13}C NMR spectroscopy in solution was carried out on a Bruker TopSpin 3.2 (5.0 mm PABBO probe) spectrometer operating at 125.55 MHz. All solution NMR spectra of the cellulose materials were obtained in an ionic liquid/DMSO-d_6 solvent. The samples were prepared by use of 1 g of $[C4Mim]^+Cl^-$ with heating and stirring at 363 K in a 3 mL vial for about 5 minutes. Then, 25 mg of the cellulose sample was added and a clear solution was achieved after 20 min. DMSO-d_6 (80 μL) was added to the IL solution to serve as a deuterium field lock for the spectral acquisition. NMR spectra were acquired at 353 K, where the chemical shifts (δ) of ^{13}C nuclei were measured relative to tetramethylsilane (TMS; $\delta = 0.0$ ppm) as the internal standard. The acquisition parameters were as follows: spectral width of 7560 Hz, acquisition time of 1.48 s, recycle delay of 4 s, and a file size of 22 k data points.

2.3.7. Scanning Electron Microscopy (SEM). The surface morphology of cellulose and its modified forms were studied using scanning electron microscopy (SEM; Model JSM-6010, JEOL/EO). The samples were sputter-coated with gold, and the images were collected under the following instrument conditions: accelerating voltage 10 kV, 9 mm working distance (WD), magnification 3000x, and spot size 50. The dimensions of the fibre domains were estimated by use of ImageJ analysis software.

2.3.8. Nitrogen Adsorption/Desorption Isotherms. The surface area and pore structure properties were determined by nitrogen adsorption using a Micromeritics ASAP 2020 (Norcross, GA) system. Briefly, approximately 1.0 g sample was degassed at the following conditions: an evacuation rate of $5\,\text{mmHg}\cdot\text{s}^{-1}$ in the sample chamber to a stable outgas rate of $<10\,\text{mmHg}\cdot\text{min}^{-1}$. The degassing temperature for the samples was set at ca. 100°C for 72 h. The instrumental parameters were calibrated using alumina (Micromeritics). The BET surface area was calculated from the adsorption isotherm profile using 0.162 nm^2 as the surface area for gaseous molecular nitrogen [27, 28]. The micropore SA was obtained using a t-plot (de Boer method) [29]. The Barret-Joyner-Halenda

(BJH) method was used to estimate the pore volume and pore diameter from the adsorption isotherm profile.

2.3.9. Dye Adsorption Studies. Dye adsorption studies of *p*-nitrophenol and phenolphthalein were used to determine the accessible surface area (SA) and surface accessibility of the hydroxyl groups of cellulose and its cross-linked forms. The accessible SA (m^2/g) was estimated from the adsorption of PNP using (1). Q_m is the monolayer adsorption capacity of the adsorbent-dye system at equilibrium (mol/g), obtained according to (2), N is Avogadro's number ($6.02214 \times 10^{23}\ mol^{-1}$), σ is the cross-sectional molecular area of the dye adsorbate (m^2), and Y is the coverage factor ($Y = 1$ for PNP) [30, 31]. The coverage factor is determined according to the number of adsorbed layers of dye species at the surface of a solid adsorbent. The molecular area (σ) is variable, where it is 52.5 Å^2 for PNP adsorbed in a coplanar orientation and 25.0 Å^2 for an orthogonal arrangement relative to a planar surface [27].

$$SA = \frac{Q_m \times N \times \sigma}{Y}, \quad (1)$$

$$Q_e = \frac{(C_o - C_e) \times V}{m}. \quad (2)$$

Similarly, the accessibility of the hydroxyl groups of the materials was estimated by the decolorization of phenolphthalein (phth) at pH 10.5 upon adsorption [32]. A 50 mM stock solution of phth in ethanol (66 μL) was diluted in 100 mL of sodium bicarbonate buffer (pH 10.5) to yield a 33 μM aqueous solution. A 7 mL aqueous solution containing phth (33 μM) was added to vials containing incremental weights of the cellulose sample. The mixtures were shaken for 24 h, centrifuged (Precision Micro-Semi Micro Centricone, Precision Scientific Co.) at 1550 rpm, and the absorbance of the supernatant was measured using a double-beam spectrophotometer (Varian Cary 100) at 295 ± 0.5 K to monitor the absorbance ($\lambda_{max} = 552$ nm). All measurements were carried out in triplicate.

3. Results and Discussion

Previous studies of cross-linked cellulose polymers have been reported in the open literature [19–21], but there is a limited understanding of how cross-linking (*pillaring*) affects the physicochemical properties of such systems. To investigate the role of *pillaring effects* in cellulose, a multi-instrumental approach was used to evaluate the relationship between structure and physicochemical properties of cross-linked cellulose. To this end, several methods were used that include gas/dye adsorption, TGA, microscopy, and spectroscopy (SEM, UV-vis, FTIR, and NMR) to address this goal. The structure and physicochemical properties of cross-linked cellulose (CLE-**X**; X = 4, 1, and 0.5) were studied where materials with greater EP content are denoted by a smaller X indices (cf. Table 1).

3.1. FTIR Spectroscopy Results. The characterization of functional groups for insoluble cellulose materials in the solid state was achieved using diffuse reflectance infrared Fourier transform (DRIFT) spectroscopy. DRIFT spectral signatures were assigned to chemical bonds or functional groups, [33] (cf. Table S1 in the Supporting Information (SI)) for cellulose and its cross-linked forms. The salient spectral features in Table S1 include a broad band attributed to intermolecular bonded OH groups (cf. Figure 1(a)) (~3000–3600 cm^{-1}), C–H stretching (~2800–3000 cm^{-1}), O–H and C–H bending (~1400–1300 cm^{-1}), and C–O–H and C–O–C asymmetric stretching (~1000–1200 cm^{-1}). The spectral signatures of cellulose and its cross-linked forms are highly overlapped. The band intensities (C–H stretching, C–H bending, and C–O–H and C–O–C asymmetric stretching) increase and become sharper as the linker content of the cellulose material increases, with the exception of CLE-0.5. The IR results provide support that the hydroxyl group of cellulose undergo cross-linking with the epoxide ring of EP, as described in detail elsewhere [34]. The increase in intensity and greater sharpness of these bands can be ascribed to *pillaring effects* of the cellulose fibrils. The *pillaring effects* result in a net decrease of one OH group per EP that undergoes cross-linking. The foregoing is supported by a previous study on cross-linked chitosan materials [5].

Quantitative analysis of these bands (cf. Figure 1(b)) provides additional supporting evidence that amorphization of the macromolecular structure of cellulose occurs as the EP content increases. The variable FTIR band intensity (1427 cm^{-1} and 899 cm^{-1}) indicates that the crystallinity index (CrI) of these materials shows parallel trends to the XRD and CP-MAS NMR results (cf. Figure 2(a) and Table S5 in SI) along with other related studies [35–37]. The amorphous domains of cellulose influences its hydration properties, as reported in a sorption calorimetry study of microcrystalline cellulose and milled cellulose [38]. Herein, it can be inferred that cross-linking of cellulose with EP alters the hydrogen bonding network of the biopolymer through *pillaring effects*, according to the reduced intensity of ^{13}C solid NMR spectral signatures of the amorphous cellulose domains. In the case of ^{13}C solid NMR spectra, the spectral lines for the D-glucose units of the cross-linked cellulose become sharper with increased EP content. The effect relates to greater motional dynamics of the cross-linked framework versus cellulose due to greater crystallinity and extensive hydrogen bonding of the native biopolymer network [39].

The relative similarity between the IR spectral signatures of native cellulose and its cross-linked forms indicates that the basic structural units and the crystalline domains are preserved, in agreement with parallel observations for cross-linked β-cyclodextrin reported elsewhere [40]. The attenuation of IR bands for the cross-linked cellulose are consistent with the net loss of one –OH group for each cross-link formed. In cases where excess EP undergoes cross-linking, as for the CLE-0.5 biopolymer, potential side reactions between EP and nonreacted pendant epoxy groups occur (cf. Scheme 1). This may contribute to secondary effects due to grafting versus cross-linking in the IR spectra [41]. The spectral signatures at about 1037–1200 cm^{-1} (C–O stretching and bending) and 2800–3000 cm^{-1} (C–H stretching vibrations) provide supporting evidence along with the

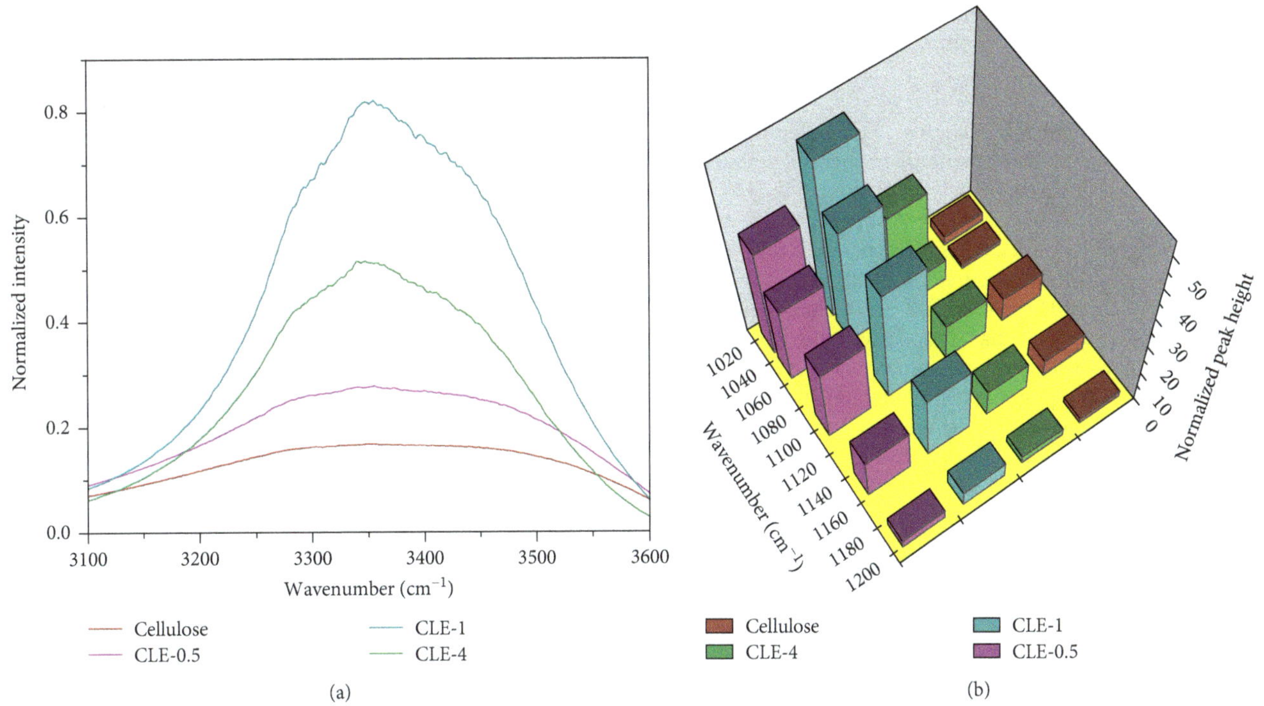

FIGURE 1: FTIR spectral results for cellulose and its cross-linked polymers: (a) –OH spectral region and (b) C–O peak areas.

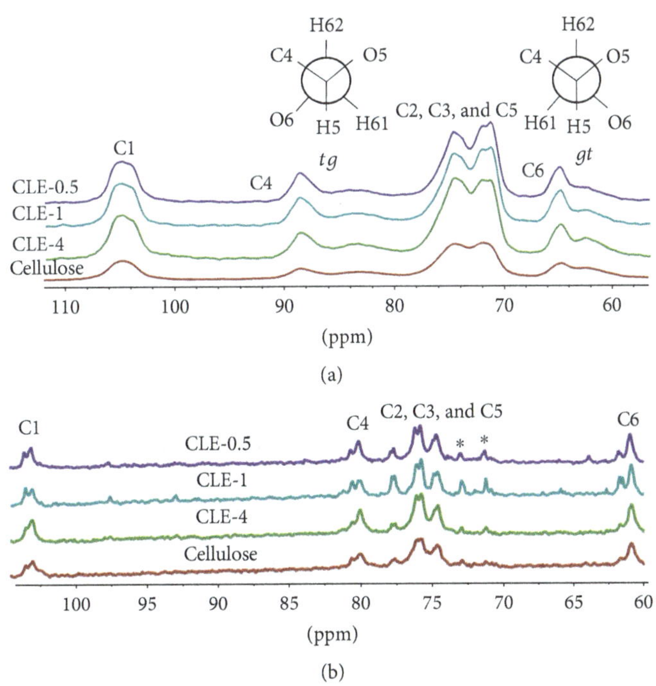

FIGURE 2: ^{13}C NMR spectra of cellulose and its cross-linked forms: (a) solid ^{13}C CP-MAS NMR spectra at 295 K and (b) solution NMR spectra in a IL-DMSO-d_6 solution system at 353 K where the asterisk (*) denotes signatures due to the EP linker unit.

presence of unique ^{13}C NMR lines between 70 and 74 ppm (cf. Figure 2) for the cross-linked reaction products. The variation of the C6 signatures for the spectra in solution of the cross-linked cellulose strongly support that bonding occurs at the primary hydroxyl sites (C6-OH) [35–37]. FTIR studies [35–37] indicate that the IR spectral signatures at about 1037, 1057, 1200, and 1335 cm^{-1} relate to C–O stretching (C6 and/or C3), C–O–H *in-plane* bending at C6, and *in-plane* bending at C2 or C3. An increase in the IR signal intensities of the bands at 2800–3000 cm^{-1} confirm the introduction of C-H groups from EP [41].

3.2. Thermal Gravimetric Analysis (TGA) Results. Cross-linked materials may possess modified thermal stability, as observed by use of thermal analysis methods [42, 43]. TGA profiles provide a relative assessment of the thermal stability since the weight loss profiles versus temperature indicate that thermal decomposition occurs, especially for structurally related materials [40, 44, 45]. The TGA results shown in Table S2 (SI) support that the native and cross-linked cellulose display thermal decomposition events between 200 and 400°C. The thermal event for pristine cellulose is near 334°C, while the cross-linked cellulose has variable temperature stability, as follows: CLE-4 (~325.4 and 359.1°C), CLE-1 (~322.6 and 362.4°C), and CLE-0.5 (~310.1 and 341.3°C).

The TGA results indicate that cross-linking of cellulose results in elevated decomposition temperatures, as compared to pristine cellulose. Shifts in the thermogram profiles provide support that cross-linking and surface grafting of EP alter the thermal stability [6, 18], in accordance with minor changes in the degree of cellulose crystallinity (vide infra). In Table S2, thermal events at 359.1°C, 362.4°C, and 341.3°C relate to the decomposition of the cellulose-EP framework, whereas the decomposition of the EP component occurs at 325.4°C (CLE-4), 332.6°C (CLE-1), and 310.1°C (CLE-0.5). The variable thermal stability relates to changes in the heat capacity due to chemical cross-linking and changes in the textural properties (SA and pore structure) of cellulose. It may also be related to the role of surface grafting that contribute changes in the structure and morphology of cellulose (cf. Scheme 1). In general, greater thermal stability occurs as the cross-linker content increases, in agreement with other cross-linked polysaccharides [6, 18]. An exception is noted for CLE-0.5 which has a stoichiometric excess of cross-linker relative to the equivalent number of available hydroxyl groups of cellulose. As noted above in Section 3.1 and Scheme 1, the formation of an EP-based oligomer or a surface grafted form of cellulose provides an account of the observed effects noted for CLE-0.5 (cf. Scheme S1 in SI). The formation of an EP oligomer agrees with the lower decomposition temperature of CLE-0.5 and corresponds to a thermal event observed for an EP-based homopolymer [46].

3.3. Elemental Analysis. Table S3 (SI) lists the results for the elemental analyses (CHO) of cellulose and its cross-linked forms, along with the related EP content. The results for EP (Table S3) are based on its formula weight after accounting for loss of HCl to enable comparison with the other cellulose materials. An increase in the C (%) content occurs at increased cross-linking, in agreement with other studies for related materials [47, 48]. In Table S3, the identity of the products varies according to the degree of substitution (CLE-4: 0.32, CLE-1: 0.33, and CLE-0.5: 0.39) and EP content: CLE-4: 9.91 wt.%, CLE-1: 10.3 wt.%, and CLE-0.5: 11.9 wt.%. These results are consistent with the trends in elemental analysis and provide further support that cross-linking occurs between cellulose and EP for CLE-4 and CLE-1. By contrast, cross-linking and surface grafting may occur for CLE-0.5.

3.4. X-Ray Diffraction (XRD). Structural changes for cellulose and its cross-linked forms were studied by XRD. The XRD results (Figure 3) of cellulose and its cross-linked forms are shown along with a sample of cellulose that was heat-treated with 2 M NaOH for 3 h. The XRD patterns display similar features with the following Miller indices, ~16° (101), ~19° (10i), ~24° (021), ~26° (002), and ~40° (040). These signatures are characteristic of cellulose I [49] and also concur with FTIR spectral results. Similar diffraction patterns are noted for each material in Figure 3 that parallel the FTIR results in Table S1, indicating that the basic structural units and crystalline domains of cellulose are well preserved. The refined unit cell parameters in Table S4 (SI) were calculated on the basis of known unit cell dimensions of cellulose Iα ($a = 6.717$, $b = 5.962$, $c = 10.400$, and $\alpha = 118.08$, $\beta = 114.80$, and $\gamma = 80.37$) which vary slightly from reported values for Iα [26, 50]. This data supports a one-chain triclinic unit cell structure (P1 space group) with glucosyl linkages and hydroxymethyl groups in the *gauche-trans* (*gt*) and *trans-gauche* (*tg*) conformations with dihedral angles of 180° and −60°, respectively [50]. The XRD spectra of the treated cellulose had a reduced 002 peak with a separation between 101 and 10i relative to cellulose and its cross-linked forms. This effect may relate to the greater accessibility of the amorphous domains of cellulose upon treatment with NaOH, in agreement with previous reports [37, 51]. These results are consistent with ^{13}C CP-MAS NMR spectral results, where no significant changes to the crystalline nature of cellulose are noted with greater cross-linking, in agreement with the CrI index reported herein (cf. Table S5; 70%, CLE-4: 71.9%, CLE-1: 70.2%, and CLE-0.5: 73.9%), along with a similar study [52].

3.5. Solid- and Solution-State ^{13}C NMR Spectroscopy. ^{13}C NMR spectroscopy is a valuable structural tool for the study of interactions of soluble and poorly soluble carbonaceous materials, especially for structurally complex materials with short- and long-range order similar to that of semicrystalline cellulose [40]. NMR studies carried out in the solid state and in solution provide complementary insight on the molecular structure, interactions, geometry, and dynamics of cellulose since the ^{13}C NMR spectral signature reveals information on its local environment and presence of short-range order/disorder within the biopolymer network. ^{13}C NMR spectra obtained in solution yield sharper lines for polymers due to averaging of anisotropic interactions by fast isotropic molecular tumbling to produce well-defined spectral features. Figure 2 illustrates the respective ^{13}C solid CP-MAS and solution-based NMR spectra for cellulose and its cross-linked forms.

In Figures 2(a) and 2(b), the ^{13}C NMR spectra in the solid state and in solution reveal that the resonance lines of cellulose and its cross-linked forms reside over a well-defined spectral region (62–105 ppm), in agreement with other reports on cellulose and related polysaccharides [53–55]. The ^{13}C NMR spectra of cellulose in the solid state and in solution reveal

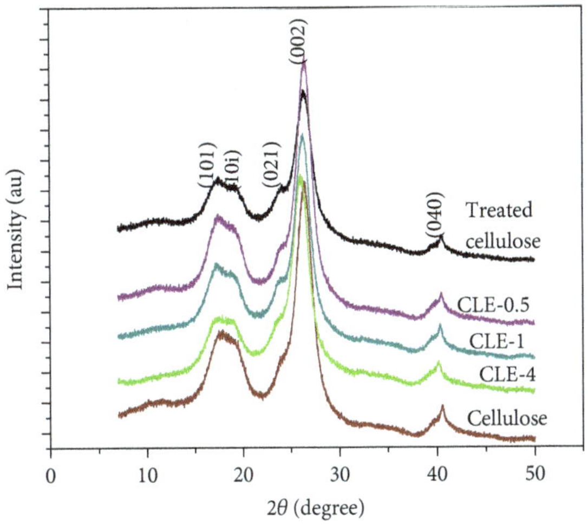

FIGURE 3: XRD patterns of cellulose and its cross-linked materials.

spectral features for C6 (~61.1 ppm), C2 (~74.5 ppm), C3 (~75.6 ppm), C5 (~77.4 ppm), C4 (~80.2 ppm), and C1 (~105 ppm), in good agreement with other NMR studies [54, 55]. The ^{13}C NMR spectral lines for cross-linked cellulose materials show a slight shift versus native cellulose that indicates changes to the local chemical environment of C2-, C3-, and C6-OH groups upon cross-linking. The appearance of new resonance lines from the cross-linker domains confirms the covalent linkages between cellulose and EP. There is further evidence of cross-linking by the variable thermal events in the TGA profiles, intensity change of ^{13}C solid NMR spectra due to structure modification of the amorphous domains, and the FTIR bands at ~1037–1200 cm^{-1} that appear upon cross-linking of cellulose. Related NMR studies for polymers cross-linked with EP reveal new ^{13}C NMR signatures at 69 to 73 ppm, in agreement with the NMR results reported herein. The ^{13}C NMR spectral signatures of modified cellulose bear similar structural features to the polymorphic form of native cellulose (type I) [55, 56] that concur with XRD results in Figure 3. The ^{13}C NMR spectral results are further supported by a shift of the IR band at 1099 cm^{-1} assigned to the native polymorphic form (cellulose I). Evidence of changes in the fine structure of the cellulose backbone is also seen in the solution ^{13}C NMR spectra due to the isotropic conditions of the IL solvent system (Figure 2(b)). The trends are assigned to *pillaring effects* of cellulose due to changes in the motional dynamics of the biopolymer network in the solid NMR spectrum (Figure 2(a)). While the FTIR and SEM results do not show large variation with the level of cross-linking, evidence of cross-linking and *pillaring effects* is noted for cellulose from the NMR spectral results in Figure 2. The ^{13}C NMR signatures (Figure 2(a)) of cellulose (80 to 92 ppm) provide evidence of the crystalline and amorphous biopolymer domains [55, 57–59], along with other ^{13}C NMR signatures ($\delta = 60 - 65$ ppm). The sharper ^{13}C NMR bands at ~92 ppm and ~65 ppm relate to a crystalline domain of cellulose while bands at ~80 ppm and ~60 ppm are due to amorphous domains. A comparison of the ^{13}C CP-MAS NMR spectra of microcrystalline and amorphous cellulose reveals the absence of C4 (~88 ppm) and C6 (~65 ppm) signatures for amorphous cellulose [55], and this may relate to differences in the dynamics and the cross-polarization transfer efficiency due to the morphology of these different materials. The chemical shift values for the C6 nuclei agree with the *gt*- and *tg*-preferred conformation of hydroxymethyl groups, where the dihedral angles are 180° and −60°, respectively, in agreement with other reports and XRD results [60, 61]. Herein, a comparison of the NMR spectra of cellulose and its cross-linked forms (Figure 2(a)) shows attenuated NMR line intensity ($\delta = 80$ to 84 ppm and 62 ppm) attributed to ^{13}C NMR spectral signatures for amorphous cellulose. The NMR results provide supporting evidence that structural changes occur upon cross-linking that influence biopolymer solvation that differ uniquely with native cellulose. The similar crystallinity index estimated by the XRD results for native cellulose and its cross-linked forms suggests that the amorphous domains undergo a cross-linking process while the crystalline domains appear largely unaffected.

3.6. Scanning Electron Microscopy (SEM). SEM provides evidence of the changes in morphology and surface character of polymers due to cross-linking, as shown by the SEM micrographs for cellulose and its cross-linked forms (Figures 4(a)–4(d)). The micrographs reveal variable surface morphology and fibre structure of cellulose when compared with cross-linked cellulose. The SEM results for cellulose (Figure 4(a)) show evidence of fibre bundle assemblies with a slightly roughened surface appearance and attenuated pore structure. By contrast, the SEM results for cross-linked cellulose (Figures 4(b)–4(d)) show greater variation in the surface roughness upon cross-linking. Evidence of fibre bundle disintegration is noted by the mesoporous polymer surface in Figures 4(b)–4(d), as compared with unmodified cellulose. The loss of fibre structure is further evidenced by the formation of small individual particulates with variable shape that appear as semiordered and closely packed structures. A variable fibril structure was observed as the EP content increased, where an exception was noted for CLE-0.5, in agreement with the trend in fibre size. The largest identified fibre in the SEM micrographs (cellulose: $14.5 \pm 0.2\,\mu$m, CLE-4: $11.3 \pm 0.2\,\mu$m, CLE-1: $9.6 \pm 0.1\,\mu$m, and CLE-0.5: $14.4 \pm 0.2\,\mu$m) along with the variable thermal stability (cf. Table S2 in SI) provides support for the change in molecular structure. The SEM image of CLE-0.5 reveals small fibril strands which appear entangled, as evidenced by the formation or exposure of new fibril bundles in accordance with the effects of excess EP and surface grafting of cellulose, as described in Sections 3.1 to 3.5.

The entanglement may relate to self-reaction via cross-linking and/or EP surface grafted cellulose species due to the excess EP cross-linker used in the synthesis, in good agreement with the lower thermal decomposition temperature of CLE-0.5 from TGA studies (cf. Table S2 in SI). A comparison of the SEM images of cellulose and CLE-0.5 reveals that such materials have large fibril bundles where individual strands show slight size variation. Cellulose and CLE-0.5 have larger fibrils (14.4–$14.5 \pm 0.2\,\mu$m) that provide further evidence of cellulose surface grafting, especially for CLE-0.5

due to the use of an excess linker for its synthesis (cf. Figure 4).

3.7. Dye-Based and Nitrogen Adsorption Surface Area Estimates. Dye-based methods for use in solid-solution and gas adsorption isotherms for solid-gas systems provide complementary characterization of textural (surface area (SA) and pore structure) properties of materials from isotherm studies. [62]. However, it should be noted that differences exist for SA estimates from solid-gas isotherms versus solid-solution systems using dye-based methods. The differences may relate to solvent swelling which has been previously reported for starch and cellulose biopolymers [62, 63]. Estimates of SA for cross-linked polymers from adsorption isotherm studies were obtained using PNP in aqueous media and nitrogen gas as the adsorbate probes (Figure 5). The inset in Figure 5 shows that CLE-1 had the largest SA among the cellulose materials and was attributed to the optimum stoichiometric synthetic ratio used for the reaction since the surface accessibility of the –OH groups of unmodified cellulose is about 30%. Among the biopolymers, native cellulose had the lowest SA based on the dye adsorption method (cf. Table 2 and inset of Figure 5), in agreement with complementary results obtained from FTIR, SEM, solution/solid-state ^{13}C NMR spectroscopy, and phth decolorization. On the basis of reactions with surface accessible –OH groups at the amorphous domains of cellulose, EP cross-linking with cellulose yields materials with variable morphology and textural properties (cf. Figures 4 and 5 in [39]), in agreement with studies on EP cross-linked cyclodextrin materials [40].

Gas adsorption/desorption isotherms provide a standard approach for the characterization of textural properties of porous materials [28, 62]. The isotherm for CLE-1 with nitrogen is shown in Figure 5. This system corresponds to a type II isotherm according to the International Union of Pure and Applied Chemistry (IUPAC) adsorbent classification system. The saturation of the monolayer occurs at a lower relative pressure $(p/p^{\circ}) \approx 0.2$ while greater adsorption occurs at $p/p > 0.8$, in agreement with results for other cross-linked polysaccharides [5]. The isotherm displays low gas uptake up to a relative pressure (p/p°) below 0.8, while greater adsorption occurs for $p/p > 0.8$ due to adsorption at the grain boundaries of the material [6]. The BET SA estimate for CLE-1 is relatively low ($1.91\ m^2{\cdot}g^{-1}$), where the average pore width is about 10.1 nm which is similar to unmodified cellulose ($SA = 0.957\ m^2 \cdot g^{-1}$). Similar SA values for cellulose and its cross-linked forms relate to the number (ca. 30%) of accessible –OH groups of cellulose that do not allow for complete pillaring of the cellulose fibrils. This interpretation is supported by evidence of crystalline domains and surface grafting in CLE-0.5, where a stoichiometric excess of the EP cross-linker was used. A similar solvent swelling (%) for CLE-0.5 and C-EP sonication is listed in Table 2. In the case of C-EP, greater surface grafting was reported [39] for sonication-assisted synthesis conditions that further affirm that cross-linking and surface grafting occur for CLE-0.5. The apparent SA obtained by the dye-based method and nitrogen adsorption reveals differences in the estimated values by each of the two methods which are in accordance with solvent swelling effects in water, as outlined herein. Variable SA was reported [64] for never-dried and air-dried commercial softwood cellulose fibres, where it was noted that the mode of drying of cellulose led to morphological changes due to collapse of its biopolymer network structure upon loss of solvent [65].

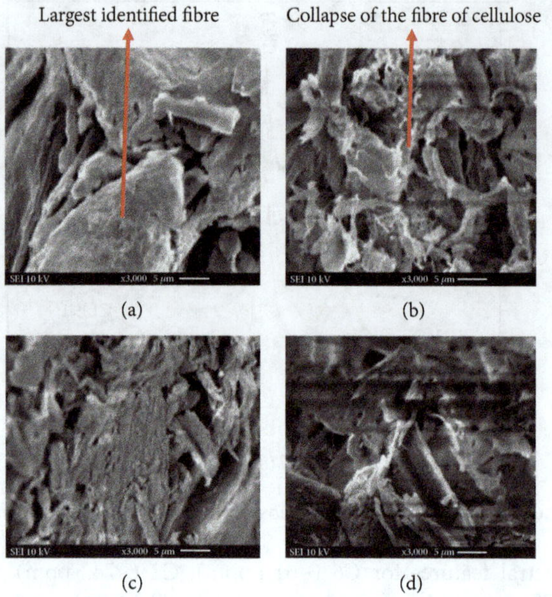

FIGURE 4: SEM micrographs of (a) cellulose, (b) CLE-4, (c) CLE-1, and (d) CLE-0.5 at 295 K.

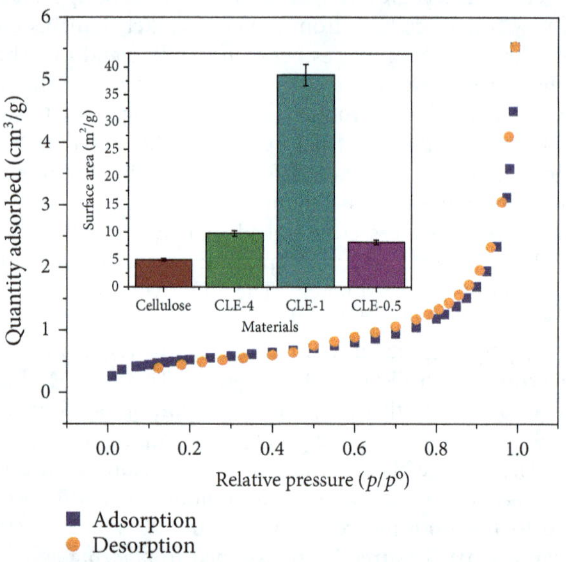

FIGURE 5: Nitrogen adsorption/desorption isotherms of cellulose materials, where the inset provides SA estimates using the dye-based method with PNP.

3.8. Phenolphthalein (phth) Decolorization. Physicochemical properties such as SA and pore structure contribute

TABLE 2: Effect of cross-linking on the physicochemical properties of biomaterials.

Material	Precursor	Q_m (mg/g)	Dye-based SA (m^2/g)	N_2-based SA (m^2/g)	Water swelling (%)	Reference
Native	Cellulose	**2.17***	**4.93***	**0.957**	**161**	[18, 39]
Cellulose		**0.23†**				This work
CLE-0.5	Cellulose	4.33*	9.83*	NA	215	[18]
CLE-2	Cellulose	17.0*	38.6*	NA	205	[18]
CLE-1	Cellulose	NA	NA	1.91	NA	This work
CLE-4	Cellulose	3.60	8.19	NA	297	[18]
C-EP (heating)	Cellulose	1.85†	NA	1.18	260	[39]
C-EP (sonication)	Cellulose	0.82†	NA	0.871	280	[39]
CE-1	Cellulose	72.8*	NA	NA	244	[6]
CE-2	Cellulose	45.2*	NA	NA	205	[6]
CE-3	Cellulose	24.6*	NA	NA	161	[6]
CPL-1	Chitosan	114*	124*	NA	NA	[33]
CPL-2	Chitosan	43.0*	46.7*	NA	NA	[33]
CPL-3	Chitosan	29.2*	31.6*	NA	NA	[33]

* denote values obtained with *p*-nitrophenol (PNP) with Q_m values originally reported in mmol/g. † denote values obtained with 2-naphthoxy acetic acid (NAA). NA: not available.

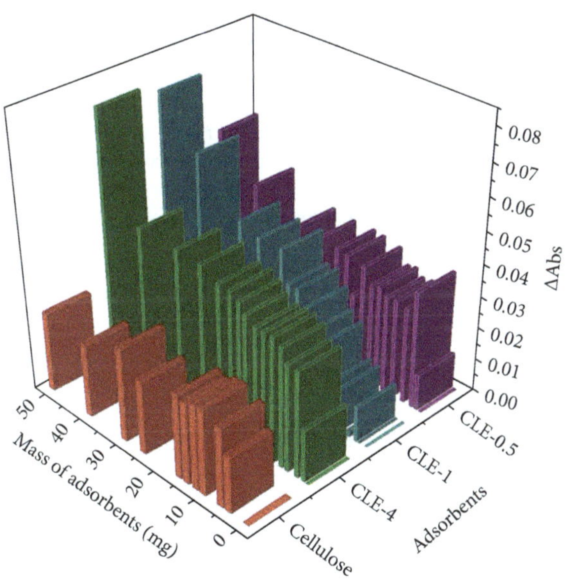

FIGURE 6: Phenolphthalein decolorization in aqueous solution with variable amounts of cellulose and cross-linked polymers at 295 K and pH 10.5.

significantly to the sorption properties of amorphous and nontemplated porous materials [32, 66]. Cross-linking and other surface modification [17, 42, 43, 67] alter the textural properties of polysaccharide materials as revealed by dye adsorption methods. Dye decolorization results for phth with cellulose and its cross-linked forms are shown in Figure 6. The trends reveal that pristine cellulose had the lowest decolorization effect on phth, while greater decolorization effects (ΔAbs) occur for cross-linked cellulose with increased surface accessibility of –OH groups, in agreement with results for such biopolymers [68]. Thus, the relationship between the level of cross-linking and surface accessibility of the –OH groups is supported by the ^{13}C NMR line intensity variation of EP signatures for cross-linked cellulose (cf. Figure 2) and the IR spectral intensity of the –OH groups (cf. Figure 1(a)). According to a previous study [69], decolorization of phth was related to hydrogen bonding interactions that result in pK_a shifts of the bound dye due to complex formation with the biopolymer –OH groups [68]. The –OH group surface accessibility of cellulose and steric effects for its cross-linked forms correlate with ΔAbs, in agreement with a study of cyclodextrin polymers [70]. The hydroxyl groups of cellulose have low surface accessibility (ca. 30%) due to its fibril morphology that results from extensive hydrogen bonding of the biopolymer network. The role of cross-linking is hypothesized to result in pillaring of the fibril structure that changes the textural properties of cross-linked chitosan relative to the unmodified biopolymer [5]. Cross-linking is asserted to introduce defects into the hydrogen bond network that enhance the surface accessibility of the hydroxyl groups due to *pillaring effects* (cf. Scheme 1), as revealed by more effective dye decolorization. Cross-linking of cellulose with EP results in a net loss of one hydroxyl group of cellulose for each cross-link formed (cf. Scheme S1 in SI). At elevated levels of cross-linking, the –OH group surface accessibility of cellulose may reveal steric effects [32, 68], in agreement with changes in morphology noted in the SEM results (Figure 4). In general, the dye adsorption results for cellulose and its cross-linked forms reveal enhanced adsorption properties due to the greater binding site accessibility by solvent and dye adsorbates.

3.9. Comparison with Other Cross-Linked Cellulose Polymers. A comparison of the cross-linked cellulose polymers in this work with other studies is provided in Table 3. Many polymers were prepared in homogenous media, whereas the

Table 3: Summary of studies for cross-linked cellulose polymers.

Reactants	Reaction media	Structural characterization	Ref.
Cellulose, EP	Homogeneous (urea/NaOH solution)	TGA, AFM, and UV-vis spectroscopy	[71]
Cellulose, EP, and NH_4OH	Heterogeneous	FTIR	[72]
Nanocellulose, polyamideamine-EP (PAE)	Homogeneous	SEM and Hg porosimetry	[73]
Cellulose, EP, NH_4OH, and $NaIO_2$	Heterogeneous	XRD, TGA, and FTIR	[74]
CMC, EP	Homogeneous	SEM and FTIR	[75]
Bulk cellulose and EP	Heterogeneous	FTIR, SEM, XRD, TGA, solid/solution ^{13}C NMR, dye decolorization, N_2 adsorption, and dye-based SA estimates	This work

polymers in this study were modified under heterogeneous conditions due to the limited solubility of cellulose in aqueous media. In many cases, the structural properties of the polymers in Table 3 were not extensively characterized, as compared with the complementary methods reported herein that included adsorption isotherms, TGA, SEM, and spectroscopy (UV-vis, FTIR, and NMR).

4. Conclusions

Cellulose was structurally modified by cross-linking with EP at variable mole ratios. Structural characterization was carried out in solution and the solid state by various complementary methods to reveal that cross-linking occurs at the amorphous versus crystalline domain of cellulose. The claim that the cross-linking of cellulose occurred at the amorphous domains is affirmed by similar features in the XRD patterns and FTIR and ^{13}C NMR spectra of the materials. Gas adsorption with nitrogen and dye adsorption (PNP and phth) in aqueous media along with SEM results reveal that cross-linked cellulose has a unique structure and morphology that differs relative to that of pristine cellulose. Cross-linking of cellulose in the amorphous domains led to *pillaring effects* of the biopolymer network (cf. Scheme 1). Evidence of the pillaring of cellulose fibrils was further affirmed by the improved (up to ≈ 8-fold greater; Figure 6) adsorption properties of cross-linked versus native cellulose. *Pillaring effects* contribute to greater surface accessibility of polar functional groups (-OH) and active sites that favour adsorption and hydration properties of the cross-linked biopolymer. A key outcome of this study concerns the role of the amorphous domains of cellulose that undergo cross-linking since the crystalline domains are structurally preserved and do not appear to undergo any appreciable cross-linking reaction. This study will contribute favourably to the valorization and utilization of native cellulose since its synthetic utility and recalcitrant nature is governed by the relative crystallinity (amorphous versus crystalline domains) of the biopolymer. By contrast, the physicochemical properties of cellulose can be tuned via cross-linking of the amorphous cellulose domains in a controlled fashion, where such materials may have utility as advanced SPE materials in biotechnology for lipid fractionation to environmental remediation of petrochemicals in aquatic environments and chemical separations [6, 11, 76].

Conflicts of Interest

The authors declare that they have no conflicts of interest.

Supplementary Materials

Scheme S1: cross-linking reaction between two cellulose repeat units and epichlorohydrin in aqueous solution. Table S1: FTIR band assignments for cellulose and various cross-linked forms. Table S2: TGA results for cellulose and its cross-linked materials. Table S3: elemental analysis results of cellulose and its cross-linked forms. Table S4: unit cell parameters of cellulose materials and cross-linked forms. Table S5: crystallinity index of cellulose and the cross-linked polymers. Table S6: selected physicochemical properties of dyes. *(Supplementary Materials)*

References

[1] X. Qiu and S. Hu, ""Smart" materials based on cellulose: a review of the preparations, properties, and applications," *Materials*, vol. 6, no. 3, pp. 738–781, 2013.

[2] A. Yamazawa, T. Iikura, A. Shino, Y. Date, and J. Kikuchi, "Solid-, solution-, and gas-state NMR monitoring of ^{13}C-cellulose degradation in an anaerobic microbial ecosystem," *Molecules*, vol. 18, no. 8, pp. 9021–9033, 2013.

[3] T. Mekonnen, P. Mussone, H. Khalil, and D. Bressler, "Progress in bio-based plastics and plasticizing modifications," *Journal of Materials Chemistry A*, vol. 1, no. 43, pp. 13379–13398, 2013.

[4] T. Oyama, "Cross-linked polymer synthesis," *Encyclopedia of Polymeric Nanomaterials*, pp. 496–505, 2015.

[5] M. H. Mohamed, I. A. Udoetok, L. D. Wilson, and J. V. Headley, "Fractionation of carboxylate anions from aqueous solution using chitosan cross-linked sorbent materials," *RSC Advances*, vol. 5, no. 100, pp. 82065–82077, 2015.

[6] L. Dehabadi and L. D. Wilson, "Polysaccharide-based materials and their adsorption properties in aqueous solution," *Carbohydrate Polymers*, vol. 113, pp. 471–479, 2014.

[7] L. Dehabadi and L. D. Wilson, "Nuclear magnetic resonance investigation of the fractionation of water–ethanol mixtures

with cellulose and Its cross-linked biopolymer forms," *Energy & Fuels*, vol. 29, no. 10, pp. 6512–6521, 2015.

[8] S. Rimdusit, K. Somsaeng, P. Kewsuwan, C. Jubsilp, and S. Tiptipakorn, "Comparison of gamma radiation crosslinking and chemical crosslinking on properties of methylcellulose hydrogel," *Engineering Journal*, vol. 16, no. 4, pp. 15–28, 2012.

[9] V. K. Thakur and M. K. Thakur, "Processing and characterization of natural cellulose fibers/thermoset polymer composites," *Carbohydrate Polymers*, vol. 109, pp. 102–117, 2014.

[10] Z. Yue-Hong, Z. Wu-Quan, G. Zhen-Hua, and G. Ji-You, "Effects of crosslinking on the mechanical properties and biodegradability of soybean protein-based composites," *Journal of Applied Polymer Science*, vol. 132, no. 5, 2015.

[11] N. Reddy, R. Reddy, and Q. Jiang, "Crosslinking biopolymers for biomedical applications," *Trends in Biotechnology*, vol. 33, no. 6, pp. 362–369, 2015.

[12] K. Vanherck, G. Koeckelberghs, and I. F. J. Vankelecom, "Crosslinking polyimides for membrane applications: a review," *Progress in Polymer Science*, vol. 38, no. 6, pp. 874–896, 2013.

[13] A. S. Ayoub and S. S. H. Rizvi, "An overview on the technology of cross-linking of starch for nonfood applications," *Journal of Plastic Film & Sheeting*, vol. 25, no. 1, pp. 25–45, 2009.

[14] S. Thaseen, "Durable press treatments to cotton, viscose, bamboo and tencel fabrics," *International Journal of Research in Engineering and Technology*, vol. 3, no. 8, pp. 32–35, 2014.

[15] Y. Li, H. Xiao, M. Chen, Z. Song, and Y. Zhao, "Absorbents based on maleic anhydride-modified cellulose fibers/diatomite for dye removal," *Journal of Materials Science*, vol. 49, no. 19, pp. 6696–6704, 2014.

[16] R. J. Moon, A. Martini, J. Nairn, J. Simonsen, and J. Youngblood, "Cellulose nanomaterials review: structure, properties and nanocomposites," *Chemical Society Reviews*, vol. 40, no. 7, pp. 3941–3994, 2011.

[17] I. Udoetok, L. Wilson, and J. Headley, "Quaternized cellulose hydrogels as sorbent materials and pickering emulsion stabilizing agents," *Materials*, vol. 9, no. 8, pp. 645–660, 2016.

[18] I. A. Udoetok, R. M. Dimmick, L. D. Wilson, and J. V. Headley, "Adsorption properties of cross-linked cellulose-epichlorohydrin polymers in aqueous solution," *Carbohydrate Polymers*, vol. 136, pp. 329–340, 2016.

[19] Y.-X. Bai and Y.-F. Li, "Preparation and characterization of crosslinked porous cellulose beads," *Carbohydrate Polymers*, vol. 64, no. 3, pp. 402–407, 2006.

[20] J. A. Seo, J. C. Kim, J. K. Koh, S. H. Ahn, and J. H. Kim, "Preparation and characterization of crosslinked cellulose/sulfosuccinic acid membranes as proton conducting electrolytes," *Ionics*, vol. 15, no. 5, pp. 555–560, 2009.

[21] M. J. Peña, S. T. Tuomivaara, B. R. Urbanowicz, M. A. O'Neill, and W. S. York, "Methods for structural characterization of the products of cellulose- and xyloglucan-hydrolyzing enzymes," *Methods in Enzymology*, vol. 510, pp. 121–139, 2012.

[22] T. Heinze and T. Liebert, "Unconventional methods in cellulose functionalization," *Progress in Polymer Science*, vol. 26, no. 9, pp. 1689–1762, 2001.

[23] M. Ioelovich, "Cellulose as a nanostructured polymer: a short review," *BioResources*, vol. 3, no. 4, pp. 1403–1418, 2008.

[24] F. H. A. Rodrigues, C. Spagnol, A. G. B. Pereira et al., "Superabsorbent hydrogel composites with a focus on hydrogels containing nanofibers or nanowhiskers of cellulose and chitin," *Journal of Applied Polymer Science*, vol. 131, no. 2, 2014.

[25] N. A. Manaf, B. Saad, M. H. Mohamed, L. D. Wilson, and A. A. Latiff, "Cyclodextrin based polymer sorbents for micro-solid phase extraction followed by liquid chromatography tandem mass spectrometry in determination of endogenous steroids," *Journal of Chromatography A*, vol. 1543, pp. 23–33, 2018.

[26] A. D. French, "Idealized powder diffraction patterns for cellulose polymorphs," *Cellulose*, vol. 21, no. 2, pp. 885–896, 2014.

[27] T. Allen, *Particle Size Measurement: Volume 2: Surface Area and Pore Size Determination*, Springer, London, UK, 1997.

[28] K. Sing, "The use of nitrogen adsorption for the characterisation of porous materials," *Colloids and Surfaces A: Physicochemical and Engineering Aspects*, vol. 187-188, pp. 3–9, 2001.

[29] J. C. P. Broekhoff and J. H. de Boer, "Studies on pore systems in catalysts: XI. Pore distribution calculations from the adsorption branch of a nitrogen adsorption isotherm in the case of "ink-bottle" type pores," *Journal of Catalysis*, vol. 10, no. 2, pp. 153–165, 1968.

[30] C. H. Giles, T. H. MacEwan, S. N. Nakhwa, and D. Smith, "786. Studies in adsorption. Part XI. A system of classification of solution adsorption isotherms, and its use in diagnosis of adsorption mechanisms and in measurement of specific surface areas of solids," *Journal of the Chemical Society (Resumed)*, pp. 3973–3993, 1960.

[31] M. M. Lynam, J. E. Kilduff, and W. J. Weber Jr, "Adsorption of p-nitrophenol from dilute aqueous solution: an experiment in physical chemistry with an environmental application," *Journal of Chemical Education*, vol. 72, no. 1, p. 80, 1995.

[32] M. H. Mohamed, L. D. Wilson, and J. V. Headley, "Estimation of the surface accessible inclusion sites of β-cyclodextrin based copolymer materials," *Carbohydrate Polymers*, vol. 80, no. 1, pp. 186–196, 2010.

[33] D. Y. Pratt, L. D. Wilson, and J. A. Kozinski, "Preparation and sorption studies of glutaraldehyde cross-linked chitosan copolymers," *Journal of Colloid and Interface Science*, vol. 395, pp. 205–211, 2013.

[34] S. Eyley and W. Thielemans, "Surface modification of cellulose nanocrystals," *Nanoscale*, vol. 6, no. 14, pp. 7764–7779, 2014.

[35] C. P. Azubuike and A. O. Okhamafe, "Physicochemical, spectroscopic and thermal properties of microcrystalline cellulose derived from corn cobs," *International Journal of Recycling of Organic Waste in Agriculture*, vol. 1, no. 1, pp. 9–16, 2012.

[36] X. Colom and F. Carrillo, "Crystallinity changes in lyocell and viscose-type fibres by caustic treatment," *European Polymer Journal*, vol. 38, no. 11, pp. 2225–2230, 2002.

[37] S. Y. Oh, D. I. Yoo, Y. Shin et al., "Crystalline structure analysis of cellulose treated with sodium hydroxide and carbon dioxide by means of X-ray diffraction and FTIR spectroscopy," *Carbohydrate Research*, vol. 340, no. 15, pp. 2376–2391, 2005.

[38] V. Kocherbitov, S. Ulvenlund, M. Kober, K. Jarring, and T. Arnebrant, "Hydration of microcrystalline cellulose and milled cellulose studied by sorption calorimetry," *The Journal of Physical Chemistry B*, vol. 112, no. 12, pp. 3728–3734, 2008.

[39] I. A. Udoetok, L. D. Wilson, and J. V. Headley, "Ultra-sonication assisted cross-linking of cellulose polymers," *Ultrasonics Sonochemistry*, vol. 42, pp. 567–576, 2018.

[40] D. Y. Pratt, L. D. Wilson, J. A. Kozinski, and A. M. Mohart, "Preparation and sorption studies of β-cyclodextrin/epichlorohydrin copolymers," *Journal of Applied Polymer Science*, vol. 116, no. 5, pp. 2982–2989, 2010.

[41] X. Hou, J. Yang, J. Tang, X. Chen, X. Wang, and K. Yao, "Preparation and characterization of crosslinked polysucrose microspheres," *Reactive and Functional Polymers*, vol. 66, no. 12, pp. 1711–1717, 2006.

[42] A. H. Karoyo and L. D. Wilson, "Nano-sized cyclodextrin-based molecularly imprinted polymer adsorbents for perfluorinated compounds—a mini-review," *Nanomaterials*, vol. 5, no. 2, pp. 981–1003, 2015.

[43] M. H. Mohamed, L. D. Wilson, and J. V. Headley, "Tuning the physicochemical properties of β-cyclodextrin based polyurethanes via cross-linking conditions," *Microporous and Mesoporous Materials*, vol. 214, pp. 23–31, 2015.

[44] M. H. Mohamed, L. D. Wilson, and J. V. Headley, "Design and characterization of novel β-cyclodextrin based copolymer materials," *Carbohydrate Research*, vol. 346, no. 2, pp. 219–229, 2011.

[45] J. Singh, N. S. Mishra, Uma, S. Banerjee, and Y. C. Sharma, "Comparative studies of physical characteristics of raw and modified sawdust for their use as adsorbents for removal of acid dye," *Bioresources*, vol. 6, no. 3, pp. 2732–2743, 2011.

[46] Y. Ren, G. Wu, X. Zhao, X. Liu, and F. Liu, "Effect of poly(epichlorohydrin) on the thermal and mechanical properties of poly(vinyl chloride)," *Journal of Applied Polymer Science*, vol. 118, no. 6, pp. 3416–3424, 2010.

[47] G. Ma, B. Qian, J. Yang, C. Hu, and J. Nie, "Synthesis and properties of photosensitive chitosan derivatives(1)," *International Journal of Biological Macromolecules*, vol. 46, no. 5, pp. 558–561, 2010.

[48] L. Poon, S. Younus, and L. D. Wilson, "Adsorption study of an organo-arsenical with chitosan-based sorbents," *Journal of Colloid and Interface Science*, vol. 420, pp. 136–144, 2014.

[49] J. Bian, F. Peng, X. P. Peng, P. Peng, F. Xu, and R. C. Sun, "Acetic acid enhanced purification of crude cellulose from sugarcane bagasse: structural and morphological characterization," *BioResources*, vol. 7, no. 4, 2012.

[50] Y. Nishiyama, J. Sugiyama, H. Chanzy, and P. Langan, "Crystal structure and hydrogen bonding system in cellulose Iα from synchrotron X-ray and neutron fiber diffraction," *Journal of the American Chemical Society*, vol. 125, no. 47, pp. 14300–14306, 2003.

[51] I. P. Samayam, B. L. Hanson, P. Langan, and C. A. Schall, "Ionic-liquid induced changes in cellulose structure associated with enhanced biomass hydrolysis," *Biomacromolecules*, vol. 12, no. 8, pp. 3091–3098, 2011.

[52] D. Ishimura, Y. Morimoto, and H. Saito, "Influences of chemical modifications on the mechanical strength of cellulose beads," *Cellulose*, vol. 5, no. 2, pp. 135–151, 1998.

[53] F. Aloulou, S. Boufi, and J. Labidi, "Modified cellulose fibres for adsorption of organic compound in aqueous solution," *Separation and Purification Technology*, vol. 52, no. 2, pp. 332–342, 2006.

[54] J. S. Moulthrop, R. P. Swatloski, G. Moyna, and R. D. Rogers, "High-resolution ^{13}C NMR studies of cellulose and cellulose oligomers in ionic liquid solutions," *Chemical Communications*, no. 12, pp. 1557–1559, 2005.

[55] C. D. Tran, S. Duri, A. Delneri, and M. Franko, "Chitosan-cellulose composite materials: preparation, characterization and application for removal of microcystin," *Journal of Hazardous Materials*, vol. 252-253, pp. 355–366, 2013.

[56] R. H. Newman and T. C. Davidson, "Molecular conformations at the cellulose–water interface," *Cellulose*, vol. 11, no. 1, pp. 23–32, 2004.

[57] P. T. Larsson, U. Westermark, and T. Iversen, "Determination of the cellulose Iα allomorph content in a tunicate cellulose by CP/MAS ^{13}C-NMR spectroscopy," *Carbohydrate Research*, vol. 278, no. 2, pp. 339–343, 1995.

[58] P. T. Larsson, K. Wickholm, and T. Iversen, "A CP/MAS ^{13}C NMR investigation of molecular ordering in celluloses," *Carbohydrate Research*, vol. 302, no. 1-2, pp. 19–25, 1997.

[59] Y. Sun, L. Lin, H. B. Deng et al., "Structural changes of bamboo cellulose in formic acid," *BioResources*, vol. 3, no. 2, pp. 297–315, 2008.

[60] F. Horii, A. Hirai, and R. Kitamaru, "Solid-state ^{13}C-NMR study of conformations of oligosaccharides and cellulose," *Polymer Bulletin*, vol. 10, no. 7-8, pp. 357–361, 1983.

[61] Y. Yoneda, K. Mereiter, C. Jaeger et al., "van der Waals versus hydrogen-bonding forces in a crystalline analog of cellotetraose: cyclohexyl 4′-O-Cyclohexyl β-D-cellobioside cyclohexane solvate," *Journal of the American Chemical Society*, vol. 130, no. 49, pp. 16678–16690, 2008.

[62] L. D. Wilson, M. H. Mohamed, and J. V. Headley, "Surface area and pore structure properties of urethane-based copolymers containing β-cyclodextrin," *Journal of Colloid and Interface Science*, vol. 357, no. 1, pp. 215–222, 2011.

[63] L. Dehabadi, F. Fathieh, L. D. Wilson, R. W. Evitts, and C. J. Simonson, "Study of dehumidification and regeneration in a starch coated energy wheel," *ACS Sustainable Chemistry & Engineering*, vol. 5, no. 1, pp. 221–231, 2016.

[64] M. Kimura, Z.-D. Qi, H. Fukuzumi, S. Kuga, and A. Isogai, "Mesoporous structures in never-dried softwood cellulose fibers investigated by nitrogen adsorption," *Cellulose*, vol. 21, no. 5, pp. 3193–3201, 2014.

[65] X. Jie, Y. Cao, J.-J. Qin, J. Liu, and Q. Yuan, "Influence of drying method on morphology and properties of asymmetric cellulose hollow fiber membrane," *Journal of Membrane Science*, vol. 246, no. 2, pp. 157–165, 2005.

[66] G. Crini, "Recent developments in polysaccharide-based materials used as adsorbents in wastewater treatment," *Progress in Polymer Science*, vol. 30, no. 1, pp. 38–70, 2005.

[67] I. A. Udoetok, L. D. Wilson, and J. V. Headley, "Self-assembled and cross-linked animal and plant-based polysaccharides: chitosan–cellulose composites and their anion uptake properties," *ACS Applied Materials & Interfaces*, vol. 8, no. 48, pp. 33197–33209, 2016.

[68] M. H. Mohamed, L. D. Wilson, J. V. Headley, and K. M. Peru, "Thermodynamic properties of inclusion complexes between β-cyclodextrin and naphthenic acid fraction components," *Energy & Fuels*, vol. 29, no. 6, pp. 3591–3600, 2015.

[69] M. Bertau and G. Jörg, "Saccharides as efficacious solubilisers for highly lipophilic compounds in aqueous media," *Bioorganic & Medicinal Chemistry*, vol. 12, no. 11, pp. 2973–2983, 2004.

[70] M. H. Mohamed, L. D. Wilson, D. Y. Pratt, R. Guo, C. Wu, and J. V. Headley, "Evaluation of the accessible inclusion sites in copolymer materials containing β-cyclodextrin," *Carbohydrate Polymers*, vol. 87, no. 2, pp. 1241–1248, 2012.

[71] B. Guo, W. Chen, and L. Yan, "Preparation of flexible, highly transparent, cross-linked cellulose thin film with high mechanical strength and low coefficient of thermal expansion," *ACS Sustainable Chemistry & Engineering*, vol. 1, no. 11, pp. 1474–1479, 2013.

[72] S. Saxena, S. Garg, and A. K. Jana, "Synthesis of cellulose based polymers for sorption of azo dyes from aqueous solution," *Journal of Environmental Research & Development*, vol. 6, no. 3, 2012.

[73] Z. He, X. Zhang, and W. Batchelor, "Cellulose nanofibre aerogel filter with tuneable pore structure for oil/water separation and recovery," *RSC Advances*, vol. 6, no. 26, pp. 21435–21438, 2016.

[74] S. Kumari, D. Mankotia, and G. S. Chauhan, "Crosslinked cellulose dialdehyde for Congo red removal from its aqueous solutions," *Journal of Environmental Chemical Engineering*, vol. 4, no. 1, pp. 1126–1136, 2016.

[75] W. Wei, S. Kim, M.-H. Song, J. K. Bediako, and Y.-S. Yun, "Carboxymethyl cellulose fiber as a fast binding and biodegradable adsorbent of heavy metals," *Journal of the Taiwan Institute of Chemical Engineers*, vol. 57, pp. 104–110, 2015.

[76] M. A. Hubbe, K. R. Beck, W. G. O'Neal, and Y. C. Sharma, "Cellulosic substrates for removal of pollutants from aqueous systems: a review. 2. Dyes," *Bioresources*, vol. 7, no. 2, 2012.

9

A Comparison Study of Antiultraviolet and Sustained Release Properties of Polydopamine/ Avermectin Microcapsule and Microsphere

Zhichuan Shen,[1,2] **Xinhua Zhou**,[1,2] **Huina Qiu**,[1] **Hua Xu**,[1,3] **Huayao Chen**,[1,3] **and Hongjun Zhou**[1,2]

[1]*School of Chemistry and Chemical Engineering, Zhongkai University of Agriculture and Engineering, Guangzhou 510220, China*
[2]*Key Laboratory of Agricultural Green Fine Chemicals of Guangdong Higher Education Institution, Guangzhou 510220, China*
[3]*Guangzhou Key Lab for Efficient Use of Agricultural Chemicals, Guangzhou 510220, China*

Correspondence should be addressed to Xinhua Zhou; cexinhuazhou@163.com and Hongjun Zhou; hongjunzhou@163.com

Academic Editor: Ulrich Maschke

By using dopamine (DA) as the monomer, the model drug avermectin (AVM) was loaded on polydopamine microspheres (AVM/PDAMS) and polydopamine microcapsules (AVM@PDAMC) by the method of impregnation and encapsulation, respectively. The materials' structures were systematically characterized using Fourier transform infrared spectroscopy (FTIR), zeta potential analysis, scanning electron microscopy (SEM), thermogravimetric analysis (TGA), and differential scanning calorimetry (DSC). The comparison of antiultraviolet capability as well as release behaviors under different pH values of the materials were studied. The results showed that a spherical appearance was observed from both materials. The use of AVM/PDAMS and AVM@PDAMC made the decomposition temperature of AVM increase to 235°C and 245°C, respectively. After being exposed to ultraviolet light for 1400 min, the residual ratios of AVM of AVM/PDAMS and AVM@PDAMC were 42% and 54%, respectively. Both AVM/PDAMS and AVM@PDAMC showed acid sensitivity. AVM/PDAMS and AVM@PDAMC took about 13 h and 60 h to reach the release rate of 50% under pH 3. The release process of AVM/PDAMS could be explained by the Weibull model at pH 3, while the release behavior of AVM@PDAMC fitted the Baker–Lonsdale equation. At pH 7 and pH 9, both of the delivery materials followed the Korsmeyer–Peppas model and belonged to the Fick diffusion.

1. Introduction

Pesticides are widely used to control insects, plant diseases, and weeds. However, due to climate conditions and the administration model, most of the applied pesticides decompose before reaching their targets [1]. The pesticides deposited onto plants are easily decomposed or volatilized due to exposure to the sun and are leached into the soil with the rain water [2, 3]. To avoid such degradation and waste, the methods of encapsulation or adsorption of the pesticide with a suitable vehicle are used in this field to provide the properties of controlled release as well as responsiveness to the environment [4, 5].

Several delivery systems have been developed for loading pesticides [6, 7]. Among them, polydopamine (PDA) nanoparticles obtained from the autoxidation of its dopamine (DA) have attracted wide attention in the past few years. For PDA microspheres, in a weak alkaline solution (pH = 8.50), DA monomers can be oxidized to a quinoid structure. Then, the intramolecular cyclization and oxidative rearrangement will take place via the 1,4-Michael addition reaction to form 5,6-dihydroxyanthraquinone. Finally, all these structures will self-polymerize to form the PDA microspheres through physical interaction or covalent bonding [8]. For PDA microcapsules, by using an oil droplet as a soft template, it can enrich the hydroxide ions spontaneously at the oil/water interface and provide it a negative potential so that the in situ emulsion interfacial polymerization can be used to prepare PDA microcapsules [9, 10]. Yu et al. [11] investigated the loading and release behavior of PDA microcapsules on the mitochondrial marker Rh6G and found that the drug release process showed pH responsiveness. Ho and Ding [12] found that the PDA microspheres exhibited pH responsiveness during the controlled release of the anticancer drug camptothecin. Tamanna and Yu [13] loaded the antimicrobial rifampicin on PDA microspheres and found that the system maintained high stability at an acidic pH. This material has garnered important interest in the application of the controllable release of drugs as it can be governed by the responsivity of the molecular chains or the surface charge in reaction to pH [14–17]. Moreover, studies have shown that PDA can absorb ultraviolet light, which is why it is widely used in the preparation of UV-shielding composites [18–24]. Although both systems use only PDA as the carriers, drugs are loaded on the surface of PDA microspheres by means of an adsorption method, while the loading of drugs in PDA microcapsules employ a coating method. The different loading methods inevitably lead to differences in the structure, properties, and drug release behavior. To the best of our knowledge, fewer studies have paid much attention to the difference in UV resistance as well as the controlled release behavior of pesticides between the PDA microspheres and the microcapsules. Hence, the difference between these two systems requires more research.

The biological pesticide avermectin (AVM), a kind of macrocyclic lactone, is commonly used in agriculture by virtue of its insecticide and acaricide activities [25]. On one hand, research has shown that AVM is susceptible to photolysis and oxidation, leading to its instability under exposure to UV light and a short half-life [26]. On the other hand, the structural composition of AVM contains a large number of hydroxyl and aromatic heterocycles, which can be adsorbed onto the surface of PDA or encapsulated by a PDA coating by the formation of π–π stacking and hydrogen bonds [27]. This connection allows the PDA delivery systems to exhibit properties of pH responsiveness and controlled release. Therefore, AVM can be applied as a model drug for studying the difference of release behaviors between the PDA microspheres and the PDA microcapsules as well as their UV resistance.

Herein, AVM was loaded on the PDA microspheres and microcapsules by the method of wet-impregnating adsorption and encapsulation, respectively. The differences in structure, antiultraviolet performance, and release behavior of the two systems were analyzed with a series of characterization tests. The release mechanism was also revealed to provide a useful reference for the field of controlled release of pesticides.

2. Materials and Methods

2.1. Chemicals. 3-Hydroxytyramine hydrochloride (DA) and 1,1,1-tris(hydroxymethyl)-methanamin (Tris) were all obtained from Shanghai Aladdin Biochemical Technology Co. Ltd. (Shanghai, China). Ethanol, 1-butanol, sodium hydroxide, and hydrochloric acid were obtained from Tianjin Damao Chemical Reagent Factory (Tianjin, China). Avermectin (AVM, wt = 97%) was obtained from Henan Kai Rui Biotechnology Development Co. Ltd. All chemicals were of analytical grade and used as received without any further purification.

2.2. Synthesis

2.2.1. Preparation of Polydopamine Microsphere and Its Adsorption for Avermectin. According to the literature [28, 29], 200 mL (0.02 mol/L) of Tris solution (pH = 8.50) was added to a flask to be completely stable at 25°C with 30 min of stirring. Then, 200 mg of DA was added to the solution and the reaction was stirred for another 24 h before being filtered, washed, and dried at 70°C. The samples were denoted as PDAMS. After that, 25 mg of PDA was immersed in a 50 mL ethanol solution with 150 mg of AVM at 30°C under stirring for 24 h before being filtered, washed, and dried at 60°C [30]. The samples were denoted as AVM/PDAMS.

2.2.2. Preparation of Polydopamine Microcapsule and Its Encapsulation for Avermectin. According to the literature [10], 200 mL (0.02 mol/L) of Tris solution (pH = 8.50) and 20 mL of 1-butanol solution with 200 mg AVM were added to a flask to be completely stable at 25°C with 30 min of stirring. Then, 200 mg of DA was added to the solution and the reaction was stirred for another 24 h before being filtered, washed, and dried at 60°C. The samples were denoted as AVM@PDAMC. The preparation of polydopamine microcapsules without AVM was synthesized in the same method and denoted as PDAMC. The above synthesis processes are shown in Scheme 1.

2.3. Testing the Adsorption and Encapsulation Properties. A UV Lambda 365 ultraviolet-visible spectrometer from PerkinElmer (Waltham, MA, USA) was applied to measure the amount of AVM loaded on the PDAMS and PDAMC. A linear regression for the relationship between the solution concentration (C) and absorbance (A) of the AVM standard solutions at different concentrations was performed at $\lambda = 245$ nm to obtain a standard curvilinear equation: $A = 0.028C - 0.0655$; $R^2 = 0.9997$. UV spectroscopy was used to measure the absorbance of this solution before and after the adsorption in AVM ethanol solution.

Loading content (LC) of AVM/PDAMS was calculated by the following equation:

$$\mathrm{LC} = \frac{[(C_0 - C_t) \times V]}{[M_1 + (C_0 - C_t) \times V \times 0.001]}, \quad (1)$$

where C_0 is the original mass concentration (mg/L) of the AVM ethanol solution, C_t is the mass concentration (mg/L) of the AVM ethanol solution at the relevant time interval, V is the volume (L) of the ethanol solution, and M_1 is the mass (g) of the PDAMS.

According to [31], the operational method to calculate the loading content (LC) of AVM@PDAMC is as follows: 50 mg AVM@PDAMC was dispersed in a flask containing

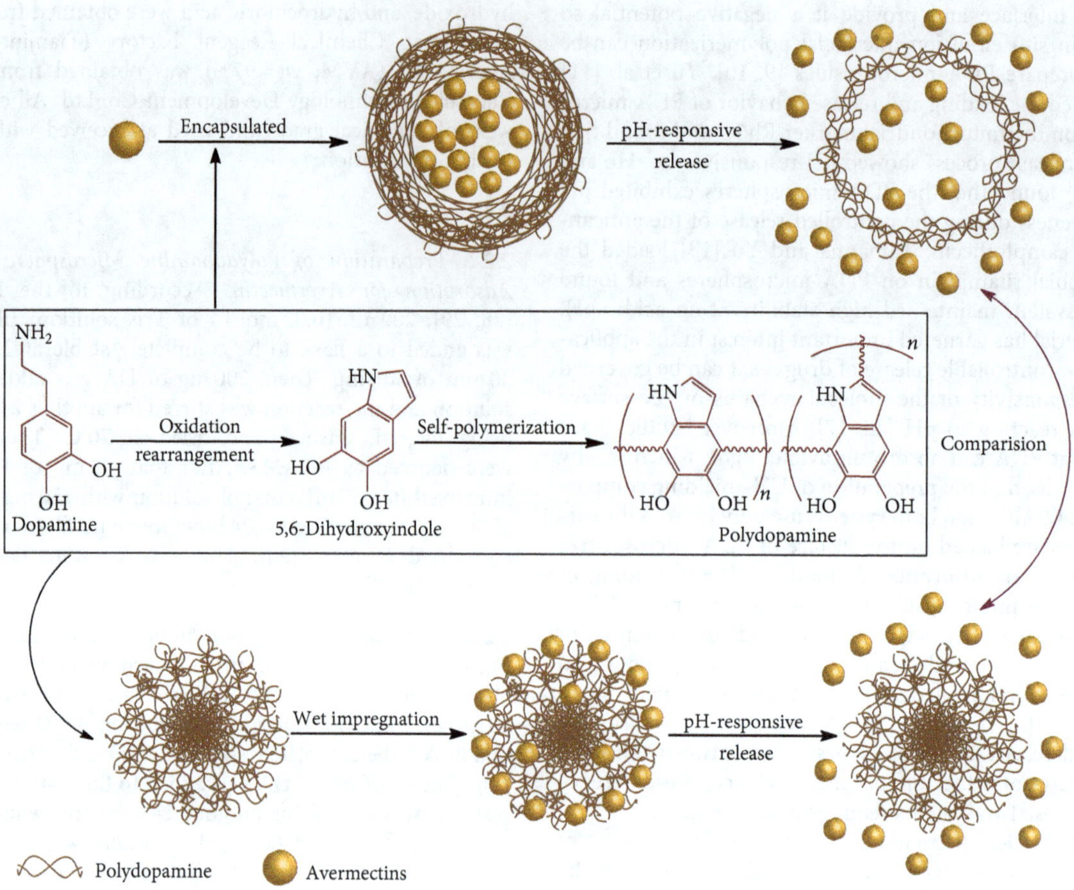

Scheme 1: Schematic illustration of the synthetic process of AVM@PDAMC and AVM/PDAMS.

150 mL of ethanol solution at 25°C under stirring for 3 min. Then, 1 mL of supernatant of the sample solution was transferred using a pipetting gun and an equal volume of the original ethanol solution was added to the flask to replace the withdrawn sample. After that, the reaction was stirred continuously for 24 h and 1 mL of supernatant of the sample solution was also transferred. The absorbance of the above two supernatants was obtained by a UV Lambda 365 ultraviolet-visible spectrometer, and the amount of AVM was calculated by the standard curvilinear equation above.

The following equation was used to calculate the LC of AVM@PDAMC:

$$LC = \frac{[(C_{24} - C_3) \times V]}{M_2}, \tag{2}$$

where C_3 is the adsorption amount (mg/L) of AVM on the outer surface of PDAMC, C_{24} is the encapsulation amount (mg/L) of AVM encapsulated by PDAMC, V is the volume (L) of the ethanol solution, and M_2 is the mass (g) of PDAMC.

2.4. Antiultraviolet Properties of AVM/PDAMS and AVM@PDAMC. For m (mg), AVM/PDAMS or AVM@PDAMC was weighted and dispersed in a quartz Erlenmeyer flask with 50 mL of 40% ethanol solution at 25°C. A UV lamp (365 nm, 300 W) was used to degrade the above samples with a distance of 1 m. At intervals of t, 1 mL of the sample solution was transferred and diluted to 25 mL. An equal volume of the original solution was then added to the quartz Erlenmeyer flask to replace the withdrawn sample. The absorbance of the 25 mL solution was obtained. The absorbance was converted to the concentration with a standard curvilinear equation: $A = 0.0545C + 0.0468$; $R^2 = 0.9994$. Then, the residual amount of AVM was calculated as r_i. A $t - r_i$ curve was drawn to study the photodegradation of AVM. r_i was calculated as follows:

$$r_i = \begin{cases} 1 - \dfrac{[\rho_i \times 0.05]}{[m \times LC]}\ (i = 1), \\ 1 - \left\{ \dfrac{(\rho_i \times 0.05)}{(m \times LC)} + \dfrac{\left(\sum_{i=1}^{i-1} \rho_i \times 0.001\right)}{(m \times LC)} \right\}\ (i = 2, 3, 4 \ldots), \end{cases} \tag{3}$$

where ρ_i is the mass concentration (mg/L) of AVM for each sample.

2.5. Sustained Release Properties of AVM/PDAMS and AVM@PDAMC. For M (mg), AVM/PDAMS or AVM@PDAMC was weighted and dispersed in a conical flask with 50 mL of 40% ethanol solution at 25°C. At intervals of t, 1 mL of the sample solution was transferred and diluted to 25 mL. An equal volume of the original solution was then added to the conical flask to replace the withdrawn sample. A UV Lambda 365 ultraviolet-visible spectrometer was used to detect the absorbance of the 25 mL solution and the detected value was converted to the concentration with a standard curvilinear equation: $A = 0.0545C + 0.0468$; $R^2 = 0.9994$. Finally, the cumulative release amount of AVM was calculated as R_i and a sustained release curve was shown to study the release behavior of AVM/PDAMS and AVM@PDAMC. R_i was calculated as follows [32]:

$$R_i = \begin{cases} \dfrac{(\rho_i \times 0.05)}{(M \times LC)} & (i = 1), \\ \dfrac{(\rho_i \times 0.05)}{(M \times LC)} + \dfrac{\sum_{i=1}^{i-1} \rho_i \times 0.001}{(M \times LC)} & (i = 2, 3, 4 \ldots). \end{cases} \quad (4)$$

2.6. Statistical Analysis. A difference factor (f_1) and a similarity factor (f_2) were proposed to compare the differences and similarities between two in vitro dissolution profiles [33]. f_1 calculates the percent difference between the two curves at each time point, and it is a measurement of the relative error between the two curves. f_2 is a logarithmic reciprocal square root transformation of the sum of the squared error, and it is the measure of the similarity in the percent dissolution between the curves. When $f_1 > 15$ and $f_2 < 50$, there were significant differences between the two curves so that these two factors could be used for further analysis of the differences in sustained release of AVM@PDAMC and AVM/PDAMS. f_1 and f_2 were calculated as follows:

$$f_1 = 100 \left[\frac{\left(\sum_{i=1}^{n} |R_i - R_i'| \right)}{\sum_{i=1}^{n} R_i} \right], \quad (5)$$

$$f_2 = 50 \log \left\{ \left[1 + \left(\frac{1}{n} \right) \sum_{i=1}^{n} \left(R_i - R_i' \right)^2 \right]^{-0.5} \times 100 \right\}, \quad (6)$$

where n is the number of the time points, R_i is the dissolution value of the reference formulation at time I, and R_i' is the dissolution value of the test formulation at time i.

2.7. Characterization. The compositions of the particles were analyzed by a Spectrum 100 Fourier infrared spectrometer (PerkinElmer, Waltham, MA, USA) using the KBr squash technique. Before topographical characterization, samples were put in 10 mL ethanol with ultrasonic dispersion for 15 min to obtain a uniform suspension. Then, the upper suspension was dropped on the platform and dried. Gold particles were sprayed onto the surface of the samples under vacuum, and the samples were characterized by an S-4800 scanning electron microscope(Hitachi, Tokyo, Japan) to observe the surface topography. The zeta potential and particle size of the samples were investigated with a 90Plus PALS Zetasizer Nano ZS (Bruker Corporation, Karlsruhe, Germany) in water at pH = 7. Before investigating, samples were put in 5 mL ultrapure water with ultrasonic dispersion for 15 min to obtain a uniform suspension. A TGA 2 thermogravimetric analyzer (Mettler Toledo, Columbus, OH, USA) was used to analyze the heat stability of the particles over the heating range of 40–800°C and heating rate of 10°C·min^{-1}. A Q200 differential scanning calorimeter (TA Instruments, New Castle, DE, USA) was used to conduct differential scanning calorimetry (DSC) and detect the crystalline degree of avermectin in the particles over a heating range of 20–180°C and a heating rate of 10°C·min^{-1}.

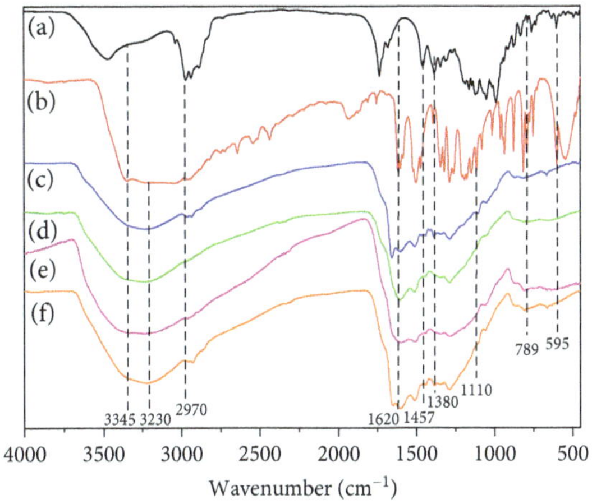

FIGURE 1: FTIR spectra of AVM (a), DA (b), AVM@PDAMC (c), AVM/PDAMS (d), PDAMS (e), and PDAMC (f).

TABLE 1: The particle diameters and zeta potential of PDAMS, PDAMC, AVM/PDAMS, AVM@PDAMC, and AVM.

Materials	Particle diameter (nm)	Polydispersity	Zeta potential (mV)
PDAMS	576.95	0.628	−40.05
PDAMC	583.40	0.540	−39.55
AVM/PDAMS	616.72	0.534	−50.41
AVM@PDAMC	643.64	0.556	−60.11
AVM	—	—	−28.12

3. Results and Discussion

3.1. Characterization. In Figure 1, Fourier transform infrared spectroscopy (FTIR) was adopted to compare the different compositions of DA, AVM, PDAMS, PDAMC, AVM/PDAMS, and AVM@PDAMC. For DA, the bands located at 3345 cm^{-1} and 3230 cm^{-1} were attributed to the stretching vibrations of phenolic amino and phenolic hydroxyl,

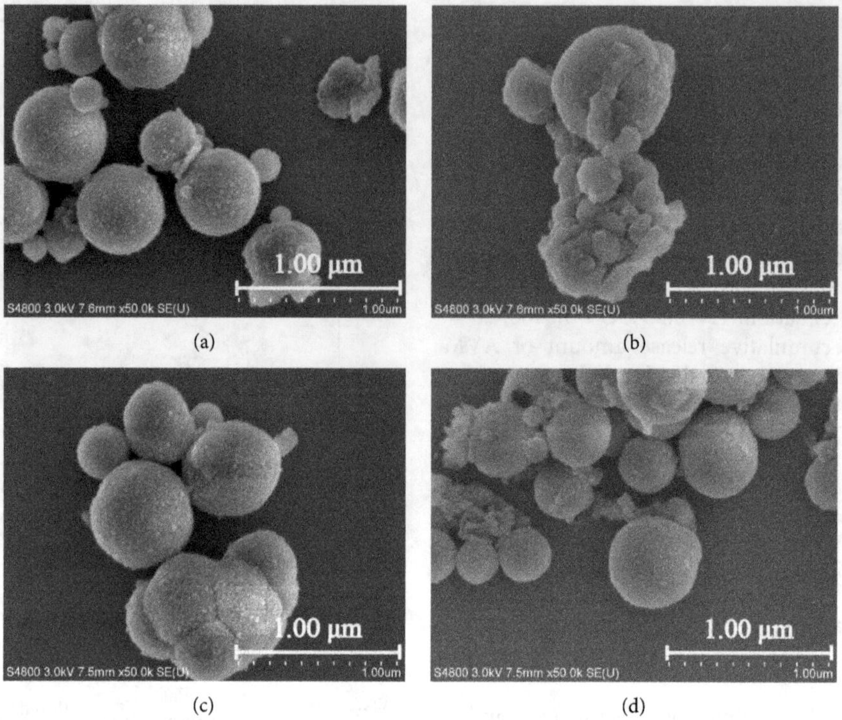

FIGURE 2: SEM images of PDAMS (a), PDAMC (b), AVM/PDAMS (c), and AVM@PDAMC (d).

respectively. The band that appeared at 1620 cm^{-1} was the stretching vibration of the benzene ring and the bending vibration of phenolic amino, respectively. The shear vibration band of the phenolic amino was located at 1520 cm^{-1}. The two bands that appeared at 1384 cm^{-1} and 1110 cm^{-1}belonged to the characteristic peaks of C-O-H and C-O on the phenolic hydroxyl. In comparison with DA, the characteristic peaks of indole and indoline structure were observed in the spectral line of PDAMS, PDAMC, AVM/PDAMS, and AVM@PDAMC, which confirmed the formation of polydopamine [34, 35]. Compared with AVM, the characteristic peaks located at 2970 cm^{-1}, 1457 cm^{-1}, 1380 cm^{-1}, 789 cm^{-1}, and 595 cm^{-1} belonged to C-H, and the heteroaromatic ring of AVM was observed in the spectral lines of AVM/PDAMS and AVM@PDAMC. This phenomenon meant that AVM had been successfully loaded on the PDAMS and PDAMC.

As listed in Table 1, the zeta potential of AVM/PDAMS shifted from −40.05 to −50.41 mV, while the zeta potential of AVM@PDAMC changed from −39.55 mV to −60.11 mV due to the loading of negatively charged AVM. Meanwhile, after loading AVM, the particle sizes of AVM/PDAMS and AVM@PDAMC increased to 39.77 nm and 56.64 nm, respectively. It was obvious that both the zeta potential and particle size of AVM@PDAMC had a greater change than that of AVM/PDAMS. This may be explained by the fact that AVM@PDAMC was loaded with more AVM. The polydispersity of all the samples were higher than 0.3, which was indicative of the wide distribution of nanoparticle size.

Figure 2 depicts the SEM images of PDAMS, PDAMC, AVM/PDAMS, and AVM@PDAMC. All the materials showed a spherical shape and the particle size distribution of the materials was generally between 480 and 560 nm. However, the particle sizes of a small part of the particles were in the range of about 200–300 nm, which may be related to the pH difference during the process of synthesis [12]. Such a phenomenon was inconsistent with the polydispersity in Table 1. In comparison with PDAMC in Figure 2(b), AVM@PDAMC in Figure 2(d) showed a better ballability performance. This phenomenon may be explained as follows: On one hand, AVM with a negative potential was dissolved in 1-butanol making these droplets generate electrostatic repulsion to each other so that they could be more stably dispersed, which was beneficial to the formation of spherical microcapsules. On the other hand, the π–π stacking between the AVM molecules and PDA molecules facilitated the coating of PDA on the surface of 1-butanol droplets. These two factors together made the AVM@PDAMC perform better ballability. Moreover, due to the adhesion of PDA, the phenomenon of adhesive agglomeration was observed among the particles [36]. In addition, the particle size from the SEM images did not corroborate well with DLS. This is because DLS is an intensity-based technique, whereas SEM is number-based, making them fundamentally different [37, 38]. DLS measures the R_H of the dispersed particles, whereas SEM provides the projected surface area based on how much of the incident electrons were reflected by the sample. Hence, the size obtained by SEM is usually smaller than that of DLS [39].

As shown in Figure 3, AVM/PDAMS and AVM@PDAMC were subjected to TGA to obtain thermal stability information. For PDAMS, PDAMC, AVM/PDAMS, and

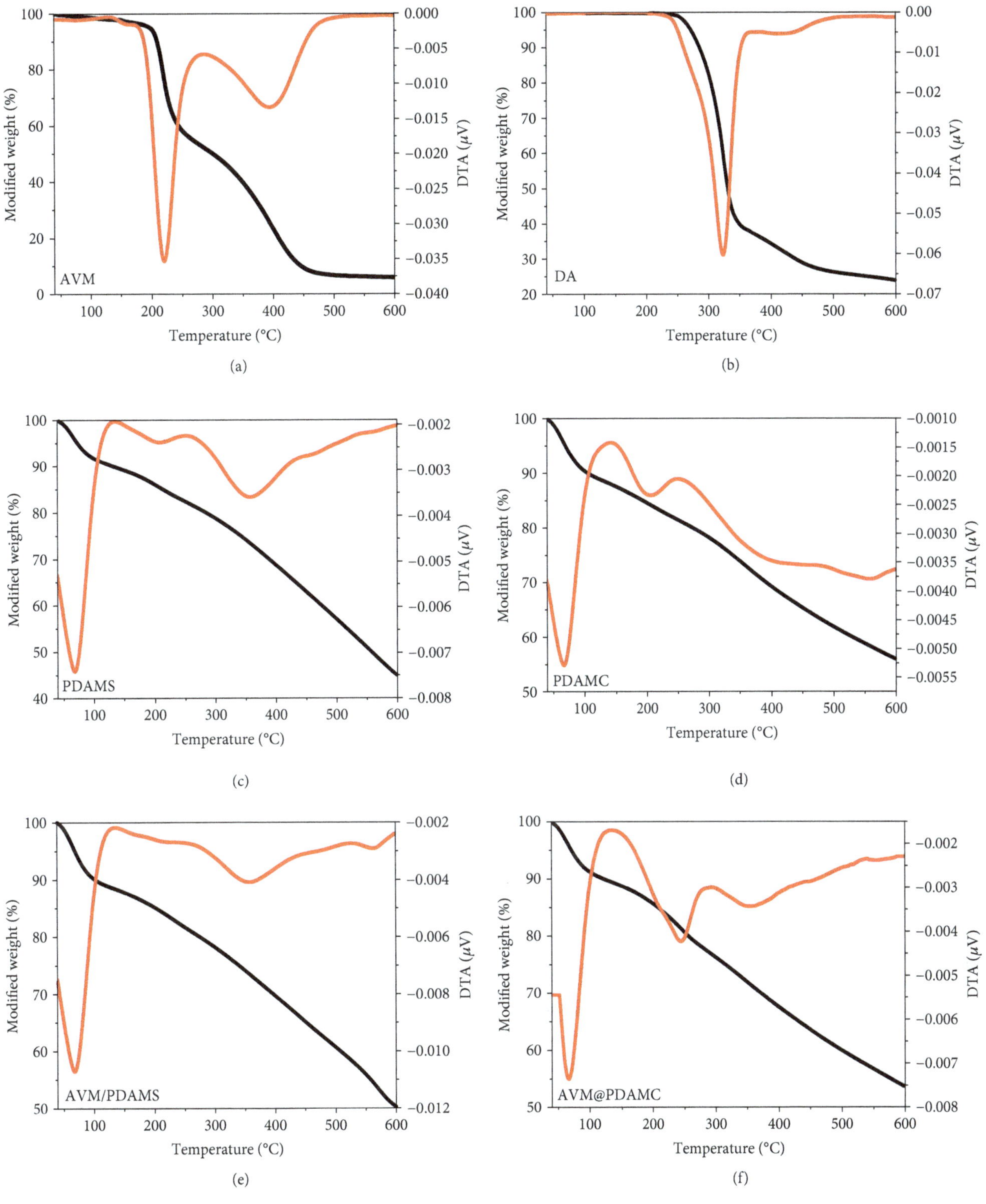

FIGURE 3: TG and DTG curves of AVM (a), DA (b), PDAMS (c), PDAMC (d), AVM/PDAMS (e), and AVM@PDAMC (f).

AVM@PDAMC, the loss in mass from 50°C to 100°C was due to the elimination of residual physically adsorbed water. No sharp peak of thermal decomposition was observed during the subsequent heating process, which indicated that DA molecules were polymerized into PDA through the formation of covalent bonds [40]. Compared to the curves of

AVM, PDAMS, and PDAMC, the loss in mass from 200 to 600°C of AVM/PDAMS and AVM@PDAMC was caused by the deposition of AVM and PDA. The thermal decomposition peak of AVM loaded on PDAMS and PDAMC was not obvious and both of them shifted from 220°C to 235°C and 245°C, respectively, which indicated the improvement in the thermal stability of AVM. Moreover, the thermal decomposition temperature of AVM loaded on PDAMC was higher than that of PDAMS. This means that, in comparison to the direct adsorption effect of PDAMS, the coating effect of PDAMC could better improve the thermal stability of AVM as it avoids direct contact between AVM and the external environment.

Figure 4 exhibits the DSC thermograms of AVM and polydopamine materials. After the load of AVM, the melting peak of AVM/PDAMS increased from 105°C to 108°C, while the melting temperature of AVM@PDAMC shifted from 110°C to 116°C. The increase of the melting temperatures of these two kinds of materials may be due to the higher melting point of AVM (150–155°C, CAS, 71751-41-2) and the π–π stacking effect between AVM and PDA. It also confirmed that AVM was successfully loaded on PDA. The melting temperature change of AVM@PDAMC was greater than that of AVM/PDAMS, indicating that the coating method makes the interaction between AVM and PDA more effective than that of the adsorption method.

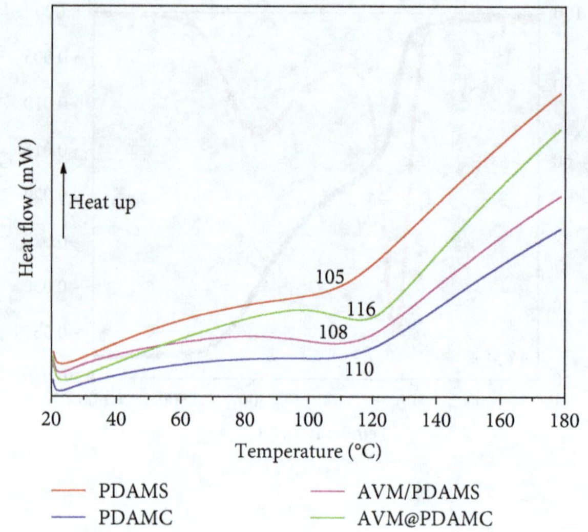

FIGURE 4: DSC curves of PDAMS, PDAMC, AVM/PDAMS, and AVM@PDAMC.

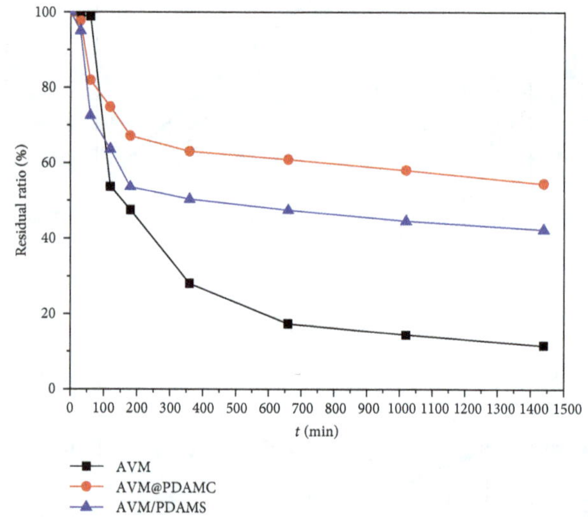

FIGURE 5: The residual ratio of AVM/PDAMS and AVM@PDAMC under irradiation of UV light.

3.2. Study of Antiultraviolet Properties. Figure 5 shows the drug release residual curves of AVM, AVM/PDAMS, and AVM@PDAMC under ultraviolet radiation. After being irradiated with UV light for 200 min for both AVM/PDAMS and AVM@PDAMC, the residual AVM in solution began to reach equilibrium. Moreover, the residual AVM of AVM/PDAMS was less than 50% since 600 min, while that of AVM@PDAMC remained above 50% until 1400 min and the pure AVM had been degraded by 89% at the same time. This comparison on residual AVM between AVM/PDAMS and AVM@PDAMC indicated that the PDA vehicle could improve the UV resistance of AVM, and PDAMC provided significantly better UV resistance than PDAMS. It further demonstrated that when compared to the directed adsorption effect of the outer surface of PDAMS, the coating method could make PDA exert a shielding effect against ultraviolet irradiation and minimize the degradation effect of UV light to AVM.

3.3. Sustained Release Performance Study. The AVM release profiles of AVM/PDAMS and AVM@PDAMC at different pH values are shown in Figure 6. It was clear that both AVM/PDAMS and AVM@PDAMC showed significant pH responsiveness. They all performed the fastest release at pH 3 and the slowest release at pH 7. Moreover, at pH 3, the release of AVM from AVM/PDAMS reached the 50% cumulative release rate (R_i) at 13 h, while that of AVM@PDAMC needed an additional 47 h. At pH 9, AVM released from AVM/PDAMS needed 50 h to reach 50% of R_i, while it took 90 h to make that of AVM@PDAMC reach the same rate. AVM/PDAMS spent 59 h reaching 50% of R_i at pH 7, while AVM@PDAMC did not reach the same R_i until 150 h. In general, AVM@PDAMC achieved a better sustained release effect than AVM/PDAMS.

Theoretically speaking [36], PDA usually carries a positive charge due to the protonation of amino groups under acidic conditions so that the AVM molecule with a negative charge can combine closely with the PDA vehicle. This factor would make the release of AVM from the PDA vehicles perform slower at pH 3 than that at pH 9. However, our experimental result was different. So, the possible sustained release mechanism of our PDA systems could be explained as follows: at pH 7, the AVM with a large amount of heteroaromatic rings and hydroxyl groups could form π–π stacking structures and hydrogen bonds with the phenol structure and amino group of PDA, so that the AVM could be

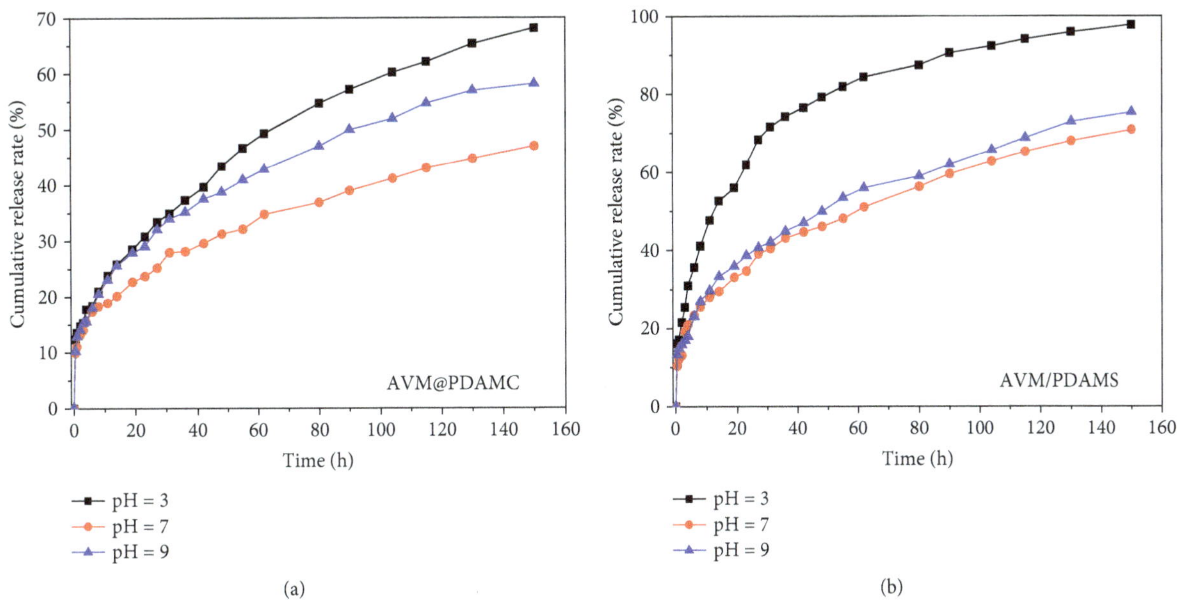

FIGURE 6: Release curves of AVM/PDAMS (a) and AVM@PDAMC (b) under different pH values.

TABLE 2: Statistical variability results for the release kinetic of AVM/PDAMS and AVM@PDAMC under the same pH.

Fit factors	pH = 3	pH = 7	pH = 9
f_1	42.35	30.60	20.21
f_2	26.89	43.99	51.00

TABLE 3: Release kinetic models of drug release.

Release model	Formulation
Zero-order model	$y = K_1 t$
First-order model	$y = K_2[1 - \exp(-K_1 t)]$
Hixson–Crowell model	$y = (K_2 - K_1 t)^3$
Weibull model	$y = 1 - \exp\left[\left(-t^{K_2}\right)/K_1\right]$
Higuchi model	$y = K_1 t^{1/2}$
Baker–Lonsdale model	$(3/2)\left[1 - (1 - y)^{2/3}\right] - y = K_1 t$
Korsmeyer–Peppas model	$y = K_2 t^{K_1}$
Quadratic model	$y = 100\left(K_1 t^2 + K_2 t\right)$
Logistic model	$y = K_3/\{1 + \exp[-K_2(t - K_1)]\}$

adsorbed or encapsulated effectively by PDA [41, 42]. On the contrary, when the release environment was acidic, the π–π stacking and the hydrogen bond were destroyed, and the PDA molecules underwent protonation, which reduced the interaction among the PDA chains, resulting in the stretch of the molecular chain structure and the enlargement of the molecular distance, finally leading to the loose union between AVM and PDA. The influence of this destructive effect exceeded the electrostatic attraction between the protonated PDA and the AVM, which caused a macroscopic phenomenon of accelerating the release of AVM. Under basic conditions, on one hand, the deprotonated PDA with a negative charge showed an electrostatic repulsion to AVM; on the other hand, mutual interaction generated by the π–π stacking and hydrogen bond limited the diffusion of AVM from the PDA vehicle into the solution. These two forces competed with each other, eventually leading to a sustained release rate that was faster than that of the neutral condition, but slower than that of the basic environment. In addition, when compared to the directed adsorption of PDAMS for AVM, PDAMC encapsulated the AVM in its hollow cavity, which meant that the release of AVM was not only controlled by the charge properties and the π–π stacking of PDA, but was also restricted by the barrier effect of the PDA coating [43]. Therefore, the controlled release performance of PDAMC for AVM was better than that of PDAMS.

The above results proved that the systems performed the higher release rate at lower pH and the lower release of pesticide at neutral pH. This fact indicated that the AVM/PDAMS system can be well stored in a neutral environment and activated while sprayed into earth under acidic conditions such as the red clay in Southern China [44], while AVM@PDAMC can also be activated in soil under basic environmental conditions.

3.4. Statistical Analysis. The results of f_1 and f_2 are listed in Table 2. It was clear that at pH 3 and pH 7, the sustained release of AVM/PDAMS and AVM@PDAMC was obviously different, while at pH 9 this was not obvious. This showed that different release environments had different degrees of influence on these two kinds of materials.

3.5. Study of Release Kinetics. In order to further study the release regulars of AVM/PDAMS and AVM@PDAMC,

Table 4: Fitting results for release curves of AVM/PDAMS and AVM@PDAMC.

Release model	pH	AVM@PDAMC K_1	K_2	K_3	R^2	AVM/PDAMS K_1	K_2	K_3	R^2
Zero-order model	3	0.0042	—	—	0.9635	0.0053	—	—	0.7433
	7	0.0024	—	—	0.9148	0.0035	—	—	0.8898
	9	0.0031	—	—	0.8918	0.0040	—	—	0.8734
First-order model	3	0.0227	0.6857	—	0.8667	0.0639	0.8870	—	0.9279
	7	0.0485	0.3953	—	0.7429	0.0498	0.5726	—	0.7971
	9	0.0455	0.4932	—	0.8321	0.0438	0.6307	—	0.8765
Hixson–Crowell model	3	0.0023	0.5963	—	0.9027	0.0020	0.7593	—	0.6685
	7	0.0016	0.5662	—	0.8586	0.0018	0.6413	—	0.8275
	9	0.0018	0.5994	—	0.8297	0.0020	0.6430	—	0.8064
Weibull model	3	15.7661	0.5727	—	0.9538	6.2909	0.5922	—	0.9951
	7	10.8111	0.3710	—	0.9825	7.8666	0.4169	—	0.9861
	9	10.0934	0.4199	—	0.9886	9.1389	0.4799	—	0.9930
Higuchi model	3	0.0606	—	—	0.9666	0.1005	—	—	0.7808
	7	0.0432	—	—	0.8259	0.0627	—	—	0.8367
	9	0.0530	—	—	0.8813	0.0671	—	—	0.9119
Baker–Lonsdale model	3	8.7246×10^{-4}	—	—	0.9935	0.0030	—	—	0.9811
	7	3.4348×10^{-4}	—	—	0.9840	8.2042×10^{-4}	—	—	0.9866
	9	5.5707×10^{-4}	—	—	0.9871	0.0010	—	—	0.9919
Korsmeyer–Peppas model	3	0.4392	0.0785	—	0.9750	0.3072	0.2269	—	0.9717
	7	0.3154	0.0941	—	0.9898	0.3184	0.1350	—	0.9950
	9	0.3380	0.1051	—	0.9938	0.3539	0.1247	—	0.9967
Quadratic model	3	-4.5900×10^{-7}	1.1162×10^{-4}	—	0.8162	-1.1684×10^{-6}	2.2483×10^{-4}	—	0.5998
	7	-4.1335×10^{-7}	8.7622×10^{-5}	—	0.4807	-6.2282×10^{-7}	1.2960×10^{-4}	—	0.5316
	9	-5.0533×10^{-7}	1.0750×10^{-4}	—	0.6004	-6.4022×10^{-7}	1.3630×10^{-4}	—	0.6741
Logistic model	3	38.0052	0.0329	0.7066	0.9934	13.4592	0.0772	0.9055	0.9679
	7	22.0411	0.0354	0.4440	0.9693	19.5788	0.0419	0.6222	0.9699
	9	20.9608	0.0430	0.5287	0.9594	21.3379	0.0467	0.6653	0.9502

several kinetic models (Table 3) were applied to fit the release data [45, 46]. The results are shown in Table 4 and Figures 7 and 8. From R^2, at pH 7 and pH 9, the release processes of both AVM/PDAMS and AVM@PDAMC were more in accordance with the Korsmeyer–Peppas equation, and their value of K_1 was smaller than 0.45, indicating that the release behaviors were controlled by the Fick diffusion. Moreover, their release processes tended to be a one-way release. The result also showed that there was an exponentially decreasing concentration gradient from the center of the vehicle to the surrounding environment during the release of AVM, and the PDA chains were tightly bound to each other [47]. In contrast, at pH 3, the release behavior of AVM@PDAMC fitted the Baker–Lonsdale model and illustrated that AVM@PDAMC maintained its original good spherical appearance without collapse as the model was suitable for the release process of spherical nanodrug delivery systems [47]. AVM/PDAMS was in line with the Weibull model, and $K_2 < 1$, which indicated that the fitting curve had a higher initial slope, reflecting the large release of AVM adsorbed on the surface of PDAMS during the sustained release.

4. Conclusion

In this paper, AVM/PDAMS and AVM@PDAMC were designed and prepared to investigate antiultraviolet performance and sustained release behaviors. The materials all had a spherical appearance. The application of PDA facilitated the improvement of the thermal stability of AVM. The coating effect of PDAMC not only made the thermal stability of AVM more enhanced, but it also improved the loading amount of PDA for AVM. AVM@PDAMC exhibited better UV resistance than AVM/PDAMS, and the formation of the PDA coating improved the UV-shielding effect of the system. At pH 7 and pH 9, the sustained release process of AVM@PDAC and AVM/PDAMS could be described by the Korsmeyer–Peppas equation and followed the Fick diffusion. At pH 3, the Baker–Lonsdale model and the Weibull model could be applied to explain the release behaviors of AVM@PDAMC and AVM/PDAMS, respectively. Under acidic conditions, the release rate of both AVM@pDAMC and AVM/PDAMS increased. The time required for AVM/PDAMS to reach 50% of R_i was 47 h earlier than that of AVM@PDAMC, while at pH 7, the R_i of

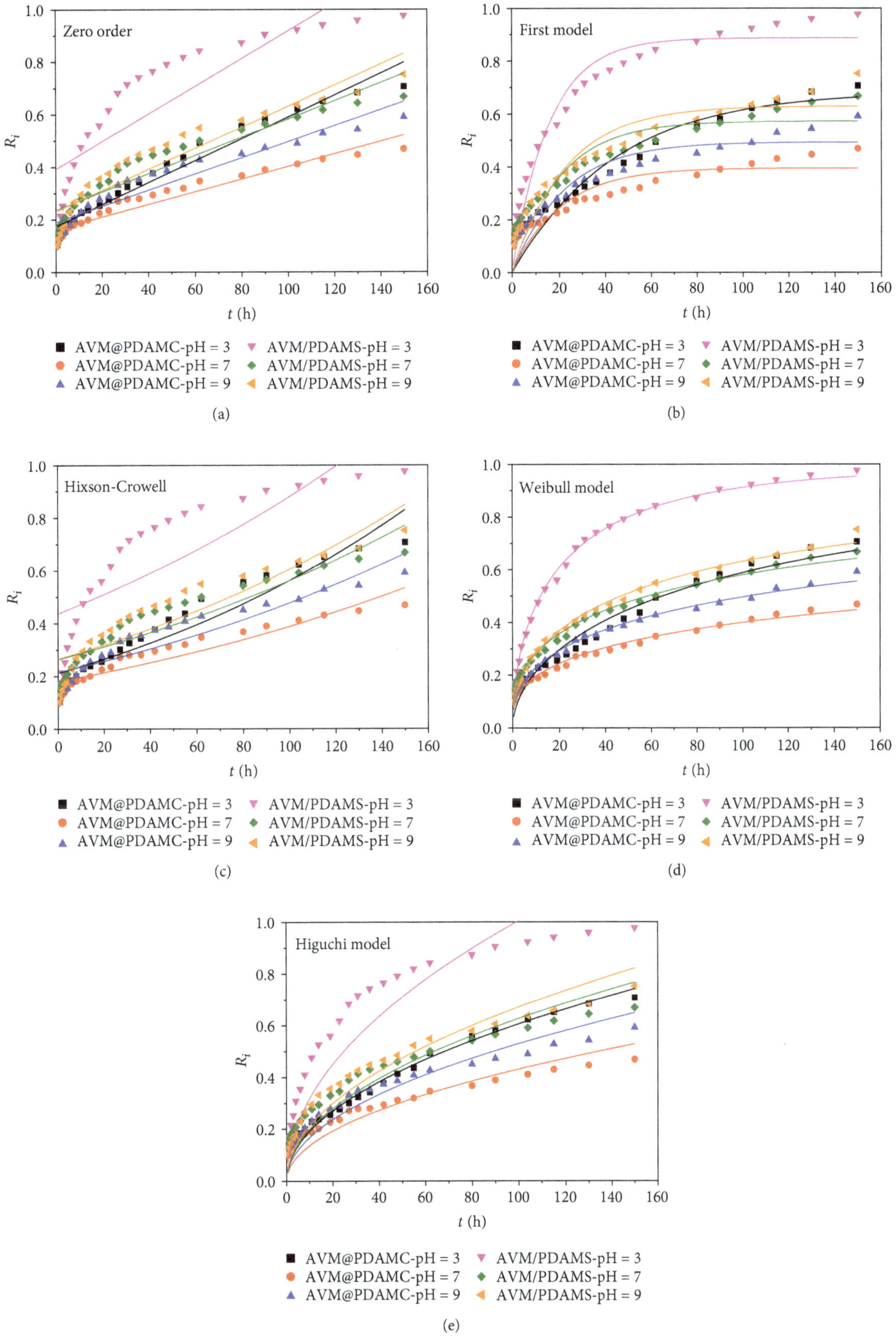

FIGURE 7: Nonlinear fit of release curves of AVM/PDAMS and AVM@PDAMC.

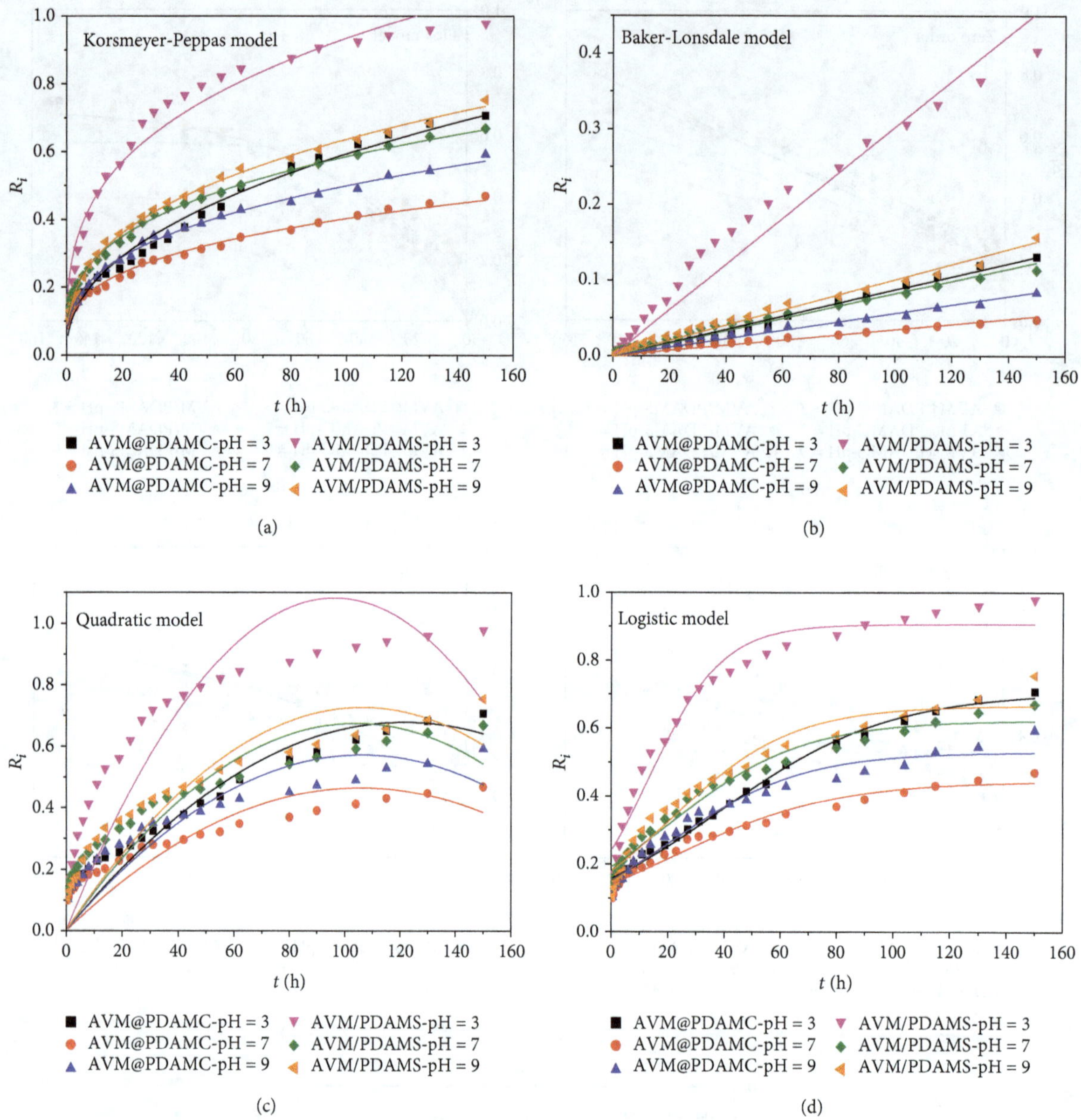

FIGURE 8: Nonlinear fit of release curves of AVM/PDAMS and AVM@PDAMC.

AVM@PDAMC did not reach 50% during the whole process. PDAMC exhibited superior controllability for AVM over PDAMS. The π–π stacking played a key role in the controlled release of AVM from the PDA vehicle and created a competitive relationship with the electrostatic interactions between AVM and PDA. This polydopamine drug delivery system with a controlled release of pesticides has potential application value in the agricultural field with the stimulation caused by the changes of the soil environment or the physiological system of pests.

Conflicts of Interest

The authors declare no competing financial interest.

Authors' Contributions

Zhichuan Shen and Xinhua Zhou conceived and designed the experiments. Huina Qiu performed the experiments. Zhichuan Shen, Hongjun Zhou, and Xinhua Zhou analyzed the data. Hua Xu and Huayao Chen provided the reagents/-materials. Zhichuan Shen, Hongjun Zhou, and Xinhua Zhou wrote/edited the paper.

Acknowledgments

This research was funded by the National Natural Science Foundation of China (Grant no. 21576303), the Natural Science Foundation of Guangdong Province (Grant no.

2016A030313375, 2017A030311003), the Science and Technology Program of the Department of Agriculture of Guangdong Province (Grant no. 2017LM4177), and the Climbing Program for Students in the Universities of Guangdong Province (Grant no. pdjh2017b0255).

Supplementary Materials

Graphic abstract. The polydopamine microcapsule had a better release control property for avermectin than the polydopamine microsphere. At pH 3, the polydopamine microcapsule and polydopamine microsphere showed sensitive release for avermectin and followed the Baker–Lonsdale model and the Weibull model, respectively. The polydopamine microcapsule had a better anti-UV capability for avermectin than the polydopamine microsphere. *(Supplementary Materials)*

References

[1] M. G. Mogul, H. Akin, N. Hasirci, D. J. Trantolo, J. D. Gresser, and D. L. Wise, "Controlled release of biologically active agents for purposes of agricultural crop management," *Resources, Conservation and Recycling*, vol. 16, no. 1-4, pp. 289–320, 1996.

[2] J. Gan, M. Hussain, and N. M. Rathor, "Behaviour of an alginate-kaolin based controlled-release formulation of the herbicide thiobencarb in simulated ecosystems," *Pesticide Science*, vol. 42, no. 4, pp. 265–272, 1994.

[3] M. Margni, D. Rossier, P. Crettaz, and O. Jolliet, "Life cycle impact assessment of pesticides on human health and ecosystems," *Agriculture, Ecosystems and Environment*, vol. 93, no. 1-3, pp. 379–392, 2002.

[4] Z.-Z. Li, S. A. Xu, L. X. Wen et al., "Controlled release of avermectin from porous hollow silica nanoparticles: influence of shell thickness on loading efficiency, UV-shielding property and release," *Journal of Controlled Release*, vol. 111, no. 1-2, pp. 81–88, 2006.

[5] S. He, W. Zhang, D. Li et al., "Preparation and characterization of double-shelled avermectin microcapsules based on copolymer matrix of silica–glutaraldehyde–chitosan," *Journal of Materials Chemistry B*, vol. 1, no. 9, pp. 1270–1278, 2013.

[6] R. Issa, A. Akelah, A. Rehab, R. Solaro, and E. Chiellini, "Controlled release of herbicides bound to poly(oligo(oxyethylene) methacrylate) hydrogels," *Journal of Controlled Release*, vol. 13, no. 1, pp. 1–10, 1990.

[7] N. A. Peppas and R. Langer, "New challenges in biomaterials," *Science*, vol. 263, no. 5154, pp. 1715–1720, 1994.

[8] S. Hong, Y. S. Na, S. Choi, I. T. Song, W. Y. Kim, and H. Lee, "Non-covalent self-assembly and covalent polymerization co-contribute to polydopamine formation," *Advanced Functional Materials*, vol. 22, no. 22, pp. 4711–4717, 2012.

[9] J. K. Beattie, A. M. Djerdjev, A. Gray-Weale et al., "PH and the surface tension of water," *Journal of Colloid and Interface Science*, vol. 422, pp. 54–57, 2014.

[10] J. Cui, Y. Yan, G. K. Such et al., "Immobilization and intracellular delivery of an anticancer drug using mussel-inspired polydopamine capsules," *Biomacromolecules*, vol. 13, no. 8, pp. 2225–2228, 2012.

[11] B. Yu, D. A. Wang, Q. Ye, F. Zhou, and W. Liu, "Robust polydopamine nano/microcapsules and their loading and release behavior," *Chemical Communications*, no. 44, pp. 6789–6791, 2009.

[12] C.-C. Ho and S.-J. Ding, "The pH-controlled nanoparticles size of polydopamine for anti-cancer drug delivery," *Journal of Materials Science: Materials in Medicine*, vol. 24, no. 10, pp. 2381–2390, 2013.

[13] T. Tamanna and A. Yu, "Polydopamine nanoparticle as a stable and capacious nano-reservoir of rifampicin," *International Journal of Medical, Health, Biomedical, Bioengineering and Pharmaceutical Engineering*, vol. 10, no. 2, pp. 35–38, 2016.

[14] Q. Liu, B. Yu, W. Ye, and F. Zhou, "Highly selective uptake and release of charged molecules by pH-responsive polydopamine microcapsules," *Macromolecular Bioscience*, vol. 11, no. 9, pp. 1227–1234, 2011.

[15] K. Kono, K. Kawakami, K. Morimoto, and T. Takagishi, "Effect of hydrophobic units on the pH-responsive release property of polyelectrolyte complex capsules," *Journal of Applied Polymer Science*, vol. 72, no. 13, pp. 1763–1773, 1999.

[16] A. Lamprecht, H. Yamamoto, H. Takeuchi, and Y. Kawashima, "Microsphere design for the colonic delivery of 5-fluorouracil," *Journal of Controlled Release*, vol. 90, no. 3, pp. 313–322, 2003.

[17] A. Stefanache, M. Ignat, C. Peptu, A. Diaconu, I. Stoleriu, and L. Ochiuz, "Development of a prolonged-release drug delivery system with magnolol loaded in amino-functionalized mesoporous silica," *Applied Sciences*, vol. 7, no. 3, p. 237, 2017.

[18] M. d'Ischia, A. Napolitano, V. Ball, C. T. Chen, and M. J. Buehler, "Polydopamine and eumelanin: from structure–property relationships to a unified tailoring strategy," *Accounts of Chemical Research*, vol. 47, no. 12, pp. 3541–3550, 2014.

[19] Y. Liu, K. Ai, and L. Lu, "Polydopamine and its derivative materials: synthesis and promising applications in energy, environmental, and biomedical fields," *Chemical Reviews*, vol. 114, no. 9, pp. 5057–5115, 2014.

[20] K.-Y. Ju, Y. Lee, S. Lee, S. B. Park, and J. K. Lee, "Bioinspired polymerization of dopamine to generate melanin-like nanoparticles having an excellent free-radical-scavenging property," *Biomacromolecules*, vol. 12, no. 3, pp. 625–632, 2011.

[21] K. Shanmuganathan, J. H. Cho, P. Iyer, S. Baranowitz, and C. J. Ellison, "Thermooxidative stabilization of polymers using natural and synthetic melanins," *Macromolecules*, vol. 44, no. 24, pp. 9499–9507, 2011.

[22] J.-J. Feng, P. P. Zhang, A. J. Wang, Q. C. Liao, J. L. Xi, and J. R. Chen, "One-step synthesis of monodisperse polydopamine-coated silver core–shell nanostructures for enhanced photocatalysis," *New Journal of Chemistry*, vol. 36, no. 1, pp. 148–154, 2012.

[23] S. L. Phua, L. Yang, C. L. Toh et al., "Simultaneous enhancements of UV resistance and mechanical properties of polypropylene by incorporation of dopamine-modified clay," *ACS Applied Materials & Interfaces*, vol. 5, no. 4, pp. 1302–1309, 2013.

[24] S. Xiong, Y. Wang, J. Yu, L. Chen, J. Zhu, and Z. Hu, "Polydopamine particles for next-generation multifunctional biocomposites," *Journal of Materials Chemistry A*, vol. 2, no. 20, pp. 7578–7587, 2014.

[25] R. W. Burg, B. M. Miller, E. E. Baker et al., "Avermectins, new family of potent anthelmintic agents: producing organism and fermentation," *Antimicrobial Agents and Chemotherapy*, vol. 15, no. 3, pp. 361–367, 1979.

[26] L. S. Crouch, W. F. Feely, B. H. Arison et al., "Photodegradation of Avermectin B1a thin films on glass," *Journal of Agricultural and Food Chemistry*, vol. 39, no. 7, pp. 1310–1319, 1991.

[27] Y. Li, M. Zhou, Y. Pang, and X. Qiu, "Lignin-based microsphere: preparation and performance on encapsulating the pesticide avermectin," *ACS Sustainable Chemistry & Engineering*, vol. 5, no. 4, pp. 3321–3328, 2017.

[28] Q. Yue, M. Wang, Z. Sun et al., "A versatile ethanol-mediated polymerization of dopamine for efficient surface modification and the construction of functional core-shell nanostructures," *Journal of Materials Chemistry B*, vol. 1, no. 44, pp. 6085–6093, 2013.

[29] J. Yan, L. Yang, M. F. Lin, J. Ma, X. Lu, and P. S. Lee, "Polydopamine spheres as active templates for convenient synthesis of various nanostructures," *Small*, vol. 9, no. 4, pp. 596–603, 2013.

[30] H. Chen, Y. Lin, H. Zhou, X. Zhou, S. Gong, and H. Xu, "Synthesis and characterization of chlorpyrifos/copper(II) Schiff base mesoporous silica with pH sensitivity for pesticide sustained release," *Journal of Agricultural and Food Chemistry*, vol. 64, no. 43, pp. 8095–8102, 2016.

[31] C. Huayao, L. Yueshun, Z. Hongjun, Z. Xinhua, G. Sheng, and X. Hua, "Highly efficient alginate sodium encapsulated chlorpyrifos/copper(II) Schiff base mesoporous silica sustained release system with pH and ion response for pesticide delivery," *RSC Advances*, vol. 6, no. 115, pp. 114714–114721, 2016.

[32] H. Chen, Y. Lin, H. Xu et al., "Preparation of sustained-release chlorpyrifos particles via the emulsification coacervation method and their sustained-release performance," *Journal of Macromolecular Science, Part A*, vol. 54, no. 2, pp. 91–96, 2017.

[33] V. Pillay and R. Fassihi, "Evaluation and comparison of dissolution data derived from different modified release dosage forms: an alternative method," *Journal of Controlled Release*, vol. 55, no. 1, pp. 45–55, 1998.

[34] X. Zheng, J. Zhang, J. Wang, X. Qi, J. M. Rosenholm, and K. Cai, "Polydopamine coatings in confined nanopore space: toward improved retention and release of hydrophilic cargo," *Journal of Physical Chemistry C*, vol. 119, no. 43, pp. 24512–24521, 2015.

[35] N. Nishizawa, A. Kawamura, M. Kohri, Y. Nakamura, and S. Fujii, "Polydopamine particle as a particulate emulsifier," *Polymer*, vol. 8, no. 3, p. 62, 2016.

[36] X. Jia, W. Sheng, W. Li, Y. Tong, Z. Liu, and F. Zhou, "Adhesive polydopamine coated avermectin microcapsules for prolonging foliar pesticide retention," *ACS Applied Materials & Interfaces*, vol. 6, no. 22, pp. 19552–19558, 2014.

[37] X. Zhao, S. Zhu, Y. Song, J. Zhang, and B. Yang, "Thermal responsive fluorescent nanocomposites based on carbon dots," *RSC Advances*, vol. 5, no. 20, pp. 15187–15193, 2015.

[38] K.-H. Kim, H. Xing, J. M. Zuo, P. Zhang, and H. Wang, "TEM based high resolution and low-dose scanning electron nanodiffraction technique for nanostructure imaging and analysis," *Micron*, vol. 71, pp. 39–45, 2015.

[39] S. Bhattacharjee, "DLS and zeta potential—what they are and what they are not?," *Journal of Controlled Release*, vol. 235, pp. 337–351, 2016.

[40] R. Batul, T. Tamanna, A. Khaliq, and A. Yu, "Recent progress in the biomedical applications of polydopamine nanostructures," *Biomaterials Science*, vol. 5, no. 7, pp. 1204–1229, 2017.

[41] J. Fu, Z. Chen, M. Wang et al., "Adsorption of methylene blue by a high-efficiency adsorbent (polydopamine microspheres): kinetics, isotherm, thermodynamics and mechanism analysis," *Chemical Engineering Journal*, vol. 259, pp. 53–61, 2015.

[42] H. Gao, Y. Sun, J. Zhou, R. Xu, and H. Duan, "Mussel-inspired synthesis of polydopamine-functionalized graphene hydrogel as reusable adsorbents for water purification," *ACS Applied Materials & Interfaces*, vol. 5, no. 2, pp. 425–432, 2013.

[43] W. Sheng, W. Li, G. Zhang, Y. Tong, Z. Liu, and X. Jia, "Study on the UV-shielding and controlled-release properties of a polydopamine coating for avermectin," *New Journal of Chemistry*, vol. 39, no. 4, pp. 2752–2757, 2015.

[44] H. Liu, X. Wu, Q. Wang, S. Wang, D. Liu, and G. Liu, "Responses of soil ammonia oxidation and ammonia-oxidizing communities to land-use conversion and fertilization in an acidic red soil of southern China," *European Journal of Soil Biology*, vol. 80, pp. 110–120, 2017.

[45] F. O. Costa, J. J. S. Sousa, A. A. C. C. Pais, and S. J. Formosinho, "Comparison of dissolution profiles of ibuprofen pellets," *Journal of Controlled Release*, vol. 89, no. 2, pp. 199–212, 2003.

[46] P. Costa and J. M. Sousa Lobo, "Modeling and comparison of dissolution profiles," *European Journal of Pharmaceutical Sciences*, vol. 13, no. 2, pp. 123–133, 2001.

[47] M. L. Bruschi, *Mathematical Models of Drug Release*, Elsevier Science, Amsterdam, 2015.

10

Synthesis and Characterization of Conductive Polypyrrole: The Influence of the Oxidants and Monomer on the Electrical, Thermal, and Morphological Properties

Abdirahman Yussuf, Mohammad Al-Saleh, Salah Al-Enezi, and Gils Abraham

Polymeric Products and Customization Program, Petroleum Research Center, Kuwait Institute for Scientific Research, P.O. Box 24885, 13109 Safat, Kuwait

Correspondence should be addressed to Abdirahman Yussuf; ayussuf@kisr.edu.kw

Academic Editor: Hossein Roghani-Mamaqani

Conductive polymer, polypyrrole (PPy), was synthesized by chemical oxidative polymerization technique for a period of four hours at room temperature using pyrrole monomer (mPPy) in aqueous solution. Different oxidants such as ferric chloride ($FeCl_3$) and ammonium persulphate ($N_2H_8S_2O_8$) and surfactant sodium dodecyl sulphate ($C_{12}H_{25}NaO_4S$) were used. The produced PPy samples were characterized by using different techniques such as the electrical resistivity by four probe technique, thermogravimetry analysis (TGA), differential scanning calorimetry (DSC), Fourier transform infrared (FTIR) spectroscopy, X-ray diffraction (XRD), and scanning electron microscopy (SEM). The performance of the oxidants has been investigated and compared. It was found that both oxidants, $FeCl_3$ and $N_2H_8S_2O_8$, have decreased electrical resistivity as a function of temperature, which means increased conductivity. However, $FeCl_3$ has achieved better performance than $N_2H_8S_2O_8$, where it has achieved a lower resistivity of about 60 ohms at room temperature, which indicates higher conductivity of PPy samples with $FeCl_3$ as an oxidant. Similarly, further investigation of $FeCl_3$ oxidant has been conducted by varying its concentration, and its influence on the final properties was reported. It has been observed that the morphology of PPy samples has a significant influence on the conductivity. It was found that 0.1 M and 0.05 M concentrations of $FeCl_3$ oxidant and monomer, respectively, have achieved better thermal stability, which is $FeCl_3$/mPPy ratio of 2 as an optimum value. FTIR and XRD results confirmed the structural formation of polypyrrole from pyrrole monomer during the synthesizing process.

1. Introduction

Recently conductive polymers, particularly polypyrrole (PPy), are gaining momentum for many commercial applications due to their excellent thermal stability, good electrical conductivity, relative ease of synthesis, and environmental stability [1–3]. Generally, conductive polymers are conductive in nature due to the existence of a conjugated electron or alternating single bond and double bond system in its chemical structure, because of this unique chemical property, the attention of many researchers and scientists from different research disciplines both in academics and industries around the world has been attracted [4, 5]. PPy has been actively used in many potential applications such as electronic devices, sensors, batteries, microactuators, anti-electrostatic coatings, and biomedical [6–10], and it has been synthesized and prepared by using different techniques such as electrochemical [11, 12] or chemical oxidation of pyrrole monomer [13, 14] in various organic solvents and in aqueous media.

Due to the wide range of applications for the chemically prepared PPy, properties such as processability, thermal stability, and conductivity are the main properties that require further improvements for future advanced applications. Therefore, in a chemical oxidative polymerization of pyrrole monomer, many researchers have reported using different oxidants for synthesizing PPy. For instance, $FeCl_3$ has been used as an oxidant for synthesizing PPy films in order to

Table 1: All materials used and their concentrations.

Name	Designated	Concentration (M)
Pyrrole monomer	mPPy	0.025, 0.05, 0.1, 0.15
Ferric chloride	$FeCl_3$	0.025, 0.05, 0.1, 0.15
Ammonium persulphate	$(NH_4)_2S_2O_8$	0.1
Sodium dodecyl sulphate	$C_{12}H_{25}NaO_4S$	0.0225

study the detection of gases such as ammonia [15, 16], and the use of ammonium persulphate as an oxidant focuses mainly on conductivity behavior. While Khamlich et al. [17] have used ferric chloride as an oxidant for preparing polypyrrole/graphene nanocomposites, investigating the influence of nanofiller on the final properties. Similarly, Hassanzadeh et al. [18] have utilized $FeCl_3$ as an oxidant for chemical synthesis of polypyrrole nanowire arrays for gas sensing applications. Also, many researchers reported PPy composites, for instance, Barrau et al. [19] have done work on carbon nanotube/epoxy resin with PPy, while Achour et al. [20] and Sangawar and Moharil [21] have reported using polymethyl methacrylate (PMMA) nanocomposite with PPy and polyvinylchloride (PVC) and PMMA-filled PPy composites, respectively. However, the use of different oxidants and pyrrole monomer for pure PPy and tailoring their performance on the final properties such as electrical, thermal, and morphological have not been fully exploited, and there are a few reports available in literature which use sodium dodecyl sulphate as a dopant and comparing the performances of ferric chloride and ammonium persulphate as an oxidizing agent in terms of thermal stability, conductivity, and morphology. Therefore, in this work, the influence of different oxidants and its concentrations on the thermal stability, conductivity, and morphology are investigated.

2. Experimental

2.1. Materials. Pyrrole monomer (mPPy) 98% was supplied by Sigma Aldrich. Anhydrous iron (III) chloride, also called ferric chloride ($FeCl_3$) 98%, and ammonium persulphate ($N_2H_8S_2O_8$) 98% were used as the oxidizing agent, and sodium dodecyl sulphate ($C_{12}H_{25}NaO_4S$), anionic surfactant, was used as the dopant/surfactant. All of them were supplied by Sigma Aldrich as well. All materials were used as received.

2.2. Synthesizing and Sample Preparation. Polypyrrole (PPy) was synthesized by chemical oxidative polymerization technique using pyrrole monomer and different oxidants. The chemical polymerization was carried out in a beaker with 100 ml of distilled water by mixing different molar ratios of pyrrole monomer, an oxidant, and a surfactant. Different concentrations of pyrrole monomer and oxidants were prepared, and as shown in Table 1, the polymerization solution was composed of various amounts of the pyrrole and oxidants and a fixed amount for the surfactant. Since this is an exothermic reaction, the addition of distilled water (cooled) was done slowly. A given volume of pyrrole monomer was quickly added to the distilled water with the required amount of oxidant and surfactant; vigorous magnetic stirring was maintained to facilitate the dispersion of pyrrole. The polymerization reaction was carried out for a period of 4 hrs at a temperature of 25°C. The precipitation of fine black particles was nearly immediate. After the prescribed polymerization time, synthesized PPy was filtered from the solution with filter paper and thoroughly washed with distilled water and ethanol several times; PPy was dried in a vacuum oven at about 40°C overnight.

2.3. Electrical Property Evaluation. The resistivity of PPy samples was measured using a four-point probe method, Model 6220 DC, manufactured by Keithley Instruments Inc. The four-point probe method is the most common technique to measure a semiconductor material's resistivity due to its low demand on sample preparation and high accuracy. It consists of a probe station (four probe tips), an ampere meter, a DC current source, and a voltmeter. Two of the probes are used to source current, and the other two probes are used to measure voltage, which means as current flows between the outer probes, the voltage drop across the inner probe is measured. For the measurement procedure, a sample film with a thickness of about 0.4 cm was prepared from the PPy sample powder by using hydraulic press and applying a pressure of 70 MPa and finally dried and cleaned for testing. The ampere meter, DC current source, and the voltmeter were checked to ensure they were on. The sample was placed on the disk and was aligned so that the spot to be measured was under the probe head. After waiting for a few seconds for the voltage to stabilize, the measurements were taken. When sample testing was completed, old samples were removed and a new sample was inserted for testing. Three measurements were done from each batch of samples.

2.4. Fourier Transform Infrared (FTIR) Spectroscopy Analysis. Fourier transform infrared (FTIR) spectra of the samples were obtained with FTIR spectrometer in the wave number range 400 to 4000 cm^{-1}, and the IR spectra of powder were recorded in the KBr medium at room temperature at a 1 cm^{-1} resolution averaging 15 scans. FTIR investigations were carried out to confirm the polymerization of the monomer during the synthesis process and existing functional groups in the produced polypyrrole conductive polymer.

2.5. Thermal Property Measurement. In order to analyze the thermal stability of the produced PPy samples, TGA was performed by using a DTG-60 instrument, manufactured by Shimadzu in Japan. A sample weight of around 10 mg was heated from 25 to 800°C at a heating rate of 10°C min^{-1} under nitrogen atmosphere. The weight loss of the samples was estimated in accordance with ASTM D 3850-2000. DSC measurements were carried out on DSC-50, manufactured by Shimadzu in Japan. The starting temperature was 25°C and was raised at 10°C min^{-1} to 300°C, then cooled down to room temperature.

2.6. Morphological Measurement. X-ray diffraction (XRD) studies were carried out by using a PAN Analytical, Rontgun Diffractometer System, Model PW3040/60, Netherland, with Cu-Kα (1.5418 Å) radiation operating at 30 kV and 20 mA. The XRD patterns were recorded in the 2θ range 10 to 80°

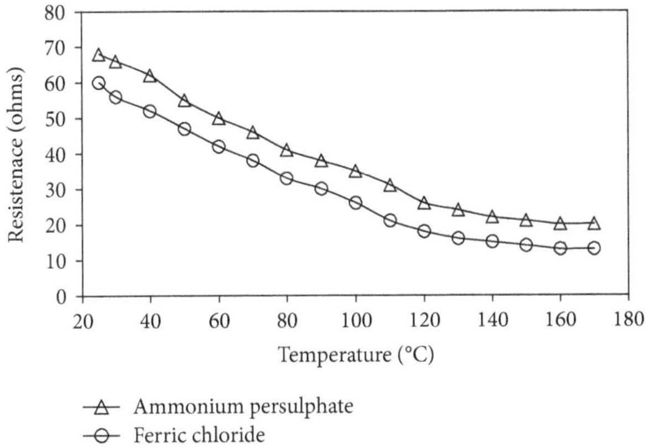

FIGURE 1: Resistance versus temperature for different oxidants (pyrrole 0.05 M and 0.1 M oxidants at 25°C, duration 4 hrs, and surfactant 0.025 M).

with step width 0.02° and step time 1.25 sec. The morphology of the polypyrrole sample was examined using scanning electron microscope (SEM), A TESCAN TS5135MM model.

3. Results and Discussion

3.1. Electrical Property Evaluation. Figure 1 shows the variation of electrical resistivity as a function of temperature of pure PPy samples with different oxidizing agents (ferric chloride and ammonium persulphate). It is observed that as temperature increases, the electrical resistivity decreases for both oxidants, which means increasing conductivity. It is generally expected that increasing temperature will cause higher energy for an electron in the valence band, which jumps to conduction band; hence, conductivity should increase with temperature. It is also possible that thermal curling affects PPy chain alignment, which could cause increasing conjugation length and decreases resistivity of the sample. However, it is observed that samples containing oxidizing agent, ferric chloride ($FeCl_3$), have achieved slightly lower resistance and higher conductivity compared to the other oxidizing agent ammonium persulphate ($N_2H_8S_2O_8$). Similar observation of increased conductivity by using ferric chloride was reported by Chitte et al. [15]. This phenomenon could be attributed due to structural difference between the two oxidizing agents and their interaction with pyrrole monomer during the chemical synthesis and their morphologies. The morphology of the samples has a significant influence on the conductivity. The highest conductivity was obtained for sample containing ferric chloride as an oxidizing agent as shown in Figure 2 with fibrillar morphology structure compared to the spherical morphology for sample containing ammonium persulphate as shown in Figure 3 (see Morphology Analysis). Similar observation was reported by Khadem et al. [22], where conductivity of fibrillar sample was higher than that of spherical samples. This behavior could be attributed to the better electron transfer within the polymer chain for fibrillar structure compared to the spherical-shaped morphology [22].

Other possible reason why ammonium persulphate has lower conductivity could be related to the addition of anionic surfactant and its more interaction with it, which as a result causes decrease in conductivity. Similar observation was reported by Kudoh [23], who found that the decrease in conductivity is due to the presence of surfactant anion adsorbed on the PPy surface. In order to observe the individual effects of the $FeCl_3$ and pyrrole monomer on the resistivity of the PPy samples, the concentration of the reactants was varied one at a time to evaluate their performance. As shown in Figure 4, with increasing $FeCl_3$ concentration, there are continuous decreases in resistivity occurring till 0.1 M. This could be attributed to the formation of carboxyl defects due to overoxidation by $FeCl_3$. Generally, carbonyl defects could cause interruptions in conjugation and as a result act as impediments to charge transfer which eventually reduces conductivity; this is could be due to the overoxidation. Similar observation of overoxidation causing reduction in the conductivity has been reported in [24].

In the case of pyrrole monomer as shown in Figure 5, resistivity decreases with increase in concentration up to 0.05 M, beyond which resistivity increased with increasing pyrrole monomer concentrations. Hence, lower resistivity and higher conductivity were achieved at 0.05 M monomer concentration.

3.2. FTIR Analysis. FTIR investigations were carried out to confirm the polymerization of the monomer to polymer during the synthesis process. Figures 6(a) and 6(b) show the FTIR spectra of synthesized PPy samples from different oxidants such as ferric chloride and ammonium persulphate, respectively. The FTIR spectra of the PPy samples were recorded in the range of 4000 to 500 cm^{-1} to confirm polymerization. The occurrence of a wide peak at 3400 cm^{-1} is assigned to the presence of N-H stretching vibrations of pyrrole ring. The weak band at 2100 cm^{-1} is due to C-H stretching. The characteristic peak at 1600 cm^{-1} corresponds to the C=C stretching, whereas peaks at 1400 cm^{-1} and 1200 cm^{-1} represent C-N and C-H plane deformation bonds, respectively, and the peak at 1100 cm^{-1} is due to C-C stretching. The peaks observed in this study match well with those available in the literature [22, 25, 26] confirming the formation of polypyrrole polymer, as shown in Figure 6. The FTIR results have clearly shown the formation of PPy polymer and its composition structure details.

3.3. Thermal Stability Results

3.3.1. Differential Scanning Calorimetry (DSC) Results. Figure 7 shows DSC results for synthesized PPy for different oxidants, ferric chloride ($FeCl_3$) and ammonium persulphate ($(NH_4)_2S_2O_8$), while Figure 8 shows DSC results for different concentrations of ferric chloride oxidant. Table 2 presents the glass transition temperature (T_g) and melting temperature (T_m) obtained from the DSC analysis for polypyrrole samples of different oxidants used.

Figure 7 shows the comparison of the two oxidants, ferric chloride and ammonium persulphate. As shown in Figure 7, the PPy sample with ferric chloride as an oxidant had a

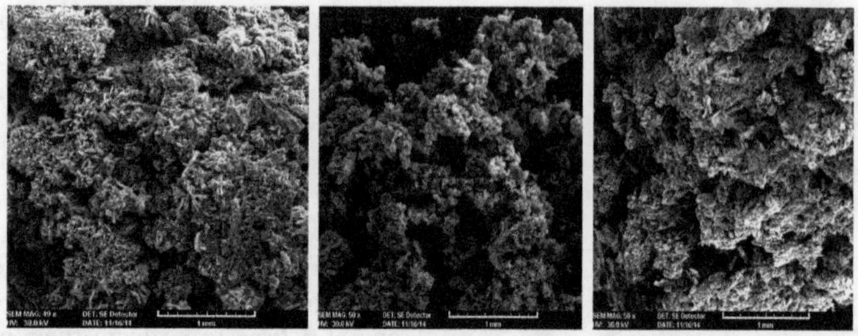

FIGURE 2: Microstructure of PPy samples where ferric chloride was used as an oxidizing agent (oxidant 0.1 M, pyrrole 0.05 M, and surfactant 0.025 M at 25°C and duration of 4 h).

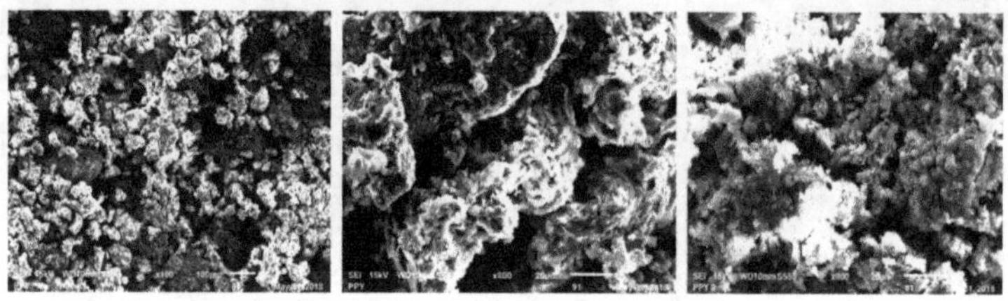

FIGURE 3: Microstructure of PPy samples where ammonium persulphate was used as an oxidizing agent (oxidant 0.1 M, pyrrole 0.05 M, and surfactant 0.025 M at 25°C and duration of 4 h).

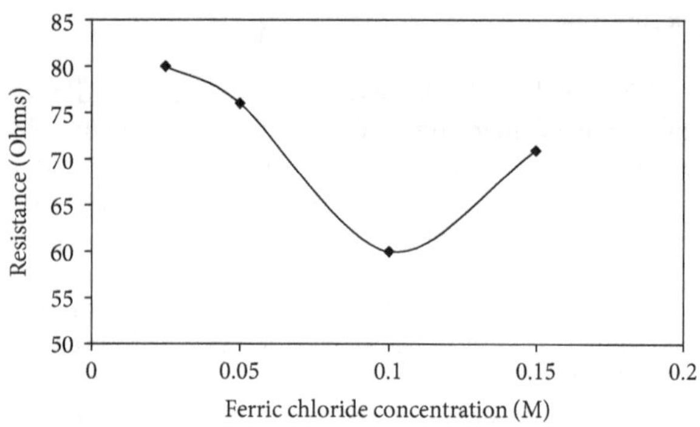

FIGURE 4: Resistance versus different ferric chloride concentrations (pyrrole 0.05 M at 25°C, duration 4 hrs, and surfactant 0.025 M).

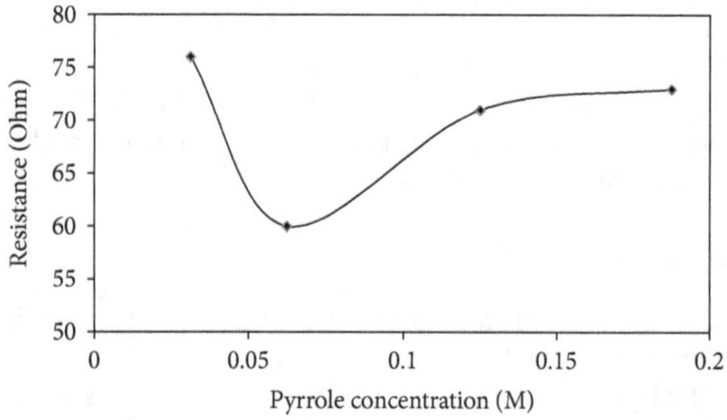

FIGURE 5: Resistance versus different pyrrole concentrations (ferric chloride 0.1 M at 25°C, duration 4 hrs, and surfactant 0.025 M).

higher melting point than the PPy sample with ammonium persulphate as an oxidant and also slightly had a higher glass transition temperature for $FeCl_3$ samples. This could mean ferric chloride as an oxidant has achieved a better and higher thermal stability during synthesis compared to the other oxidant, ammonium persulphate. This could be attributed to the difference in the nature of chemical structure between the two oxidants. It is also believed that the possible reason for this occurrence was due to the low crystallization rate of PPy and its nature of being amorphous material. The DSC results for both oxidants are tabulated in Table 2.

Ferric chloride as an oxidant had a better performance than the other oxidant, ammonium persulphate. Hence, for further investigation, the influence of different concentrations of ferric chloride on the PPy samples in terms of resistivity and thermal stability was studied.

Figure 8 shows the influence of different ferric chloride concentrations on the melting and glass transition temperatures. As shown in Figure 8, the concentration of ferric chloride at 0.1 M achieved slightly a higher melting point of the PPy samples, while the lower concentration had a lower melting and lower glass transition temperatures. It is also important to note that increasing the concentration above 0.1 M had no significant influence on increasing the melting point temperature. All results are tabulated in Table 3.

3.3.2. Thermogravimetric Analysis (TGA). Figure 9 shows the TGA curves for PPy samples with different oxidants (ferric chloride and ammonium persulphate), while Figure 10 shows PPy samples with different concentrations of ferric chloride. The onset temperatures ($T_{20\%}$) of 20% weight loss and ($T_{60\%}$) of 60% weight loss deviation from the baseline T_{20} and T_{60} were used as the indicator of the compound's thermal stability and referred to as $T_{20\%}$ and $T_{60\%}$, respectively. During thermal degradation, the TGA curves display different degradation steps or processes for all of the samples. As shown in Figure 9, the first significant weight loss takes place approximately at temperature between 25 and 100°C due to water evaporation, while the second one starts at 100 and

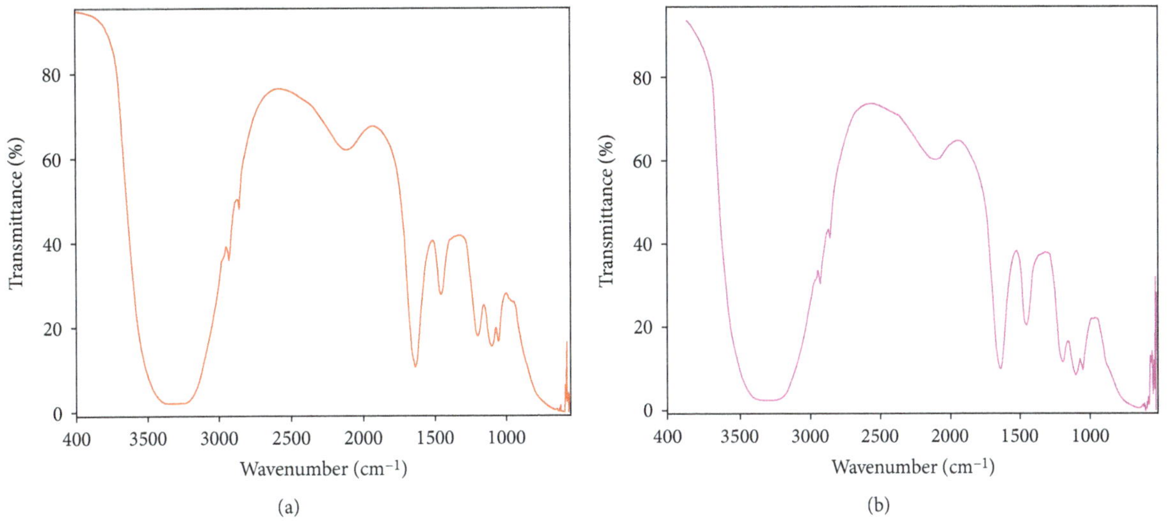

FIGURE 6: The FTIR spectrum of Ppy samples with (a) ferric chloride and (b) ammonium persulphate.

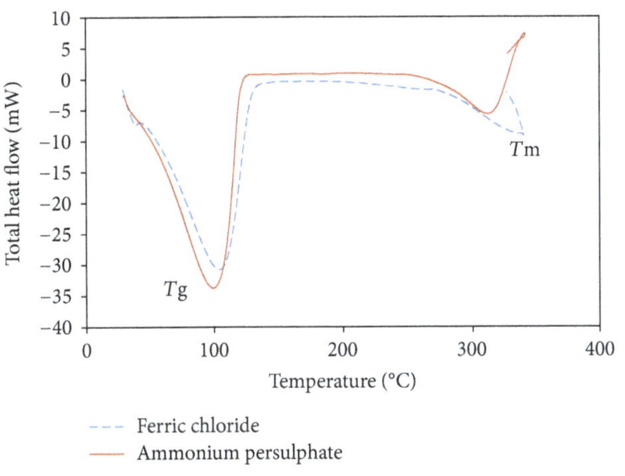

FIGURE 7: DSC results of PPy samples with different oxidants. Pyrrole 0.05 M and 0.1 M oxidants at 25°C, duration 4 hrs, and surfactant 0.025 M.

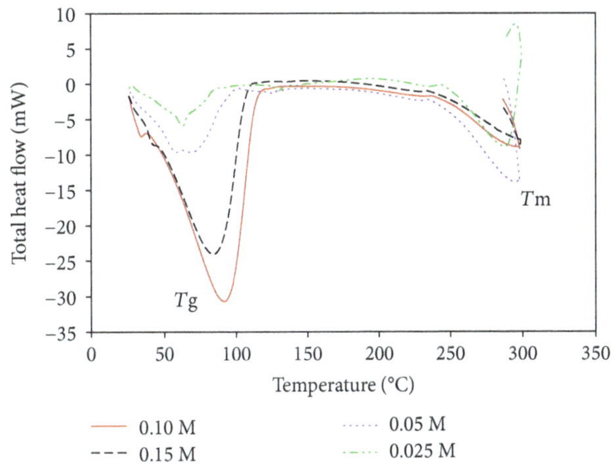

FIGURE 8: DSC results of PPy samples with different ferric chloride concentrations (pyrrole 0.05 M and surfactant 0.025 M at 25°C and duration 4 h).

ends at 180°C for possible impurities in the samples. The main and really mass loss for ferric chloride corresponds to polymer degradation which starts at about 227°C. However, for ammonium persulphate, the main degradation starts at 187°C, which means it decomposes at lower temperature compared to other sample. This could be attributed to chemical structure difference and interaction of monomer and surfactants during synthesis.

Therefore, as shown in Figure 9, PPy samples with ferric chloride oxidant show overall higher thermal stabilities compared to PPy samples with ammonium persulphate oxidant. The data for the variation in these temperatures are summarized in Table 4. As shown in Table 4, the $T_{20\%}$ of PPy sample with ferric chloride is 204°C, where the $T_{20\%}$ of PPy samples with ammonium persulphate is about 83°C. A similar trend was observed for the $T_{60\%}$ of PPy samples, which indicates higher thermal stability for PPy samples with ferric chloride, which is 385°C compared to 200°C for ammonium persulphate.

Figure 10 shows the influence of different ferric chloride concentrations on the sample's degradation and thermal stability. As shown in Figure 10, the ferric chloride concentration at 0.1 M has achieved a better thermal stability for both samples at $T_{20\%}$ and $T_{60\%}$ weight losses compared to

TABLE 2: DSC glass transition and melting temperature results for two oxidants.

Oxidant	Glass transition temperature (T_g) (°C)	Melting point (T_m) (°C)
$FeCl_3$	93	298
$(NH_4)_2S_2O_8$	87	270

Table 3: DSC results for different $FeCl_3$ concentrations.

$FeCl_3$ concentration (M)	Glass transition temperature (T_g) (°C)	Melting point (T_m) (°C)
0.15	85	295
0.1	93	298
0.05	67	289
0.025	63	287

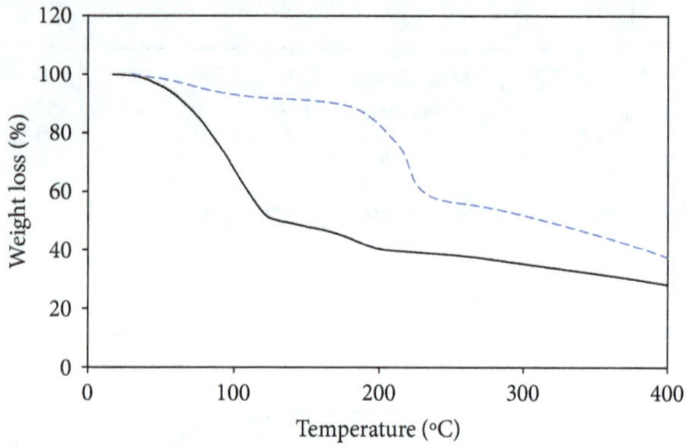

Figure 9: TGA results for PPy samples with different oxidants (pyrrole 0.05 M, oxidant 0.1 M, and surfactant 0.025 M, at 25°C duration 4 h).

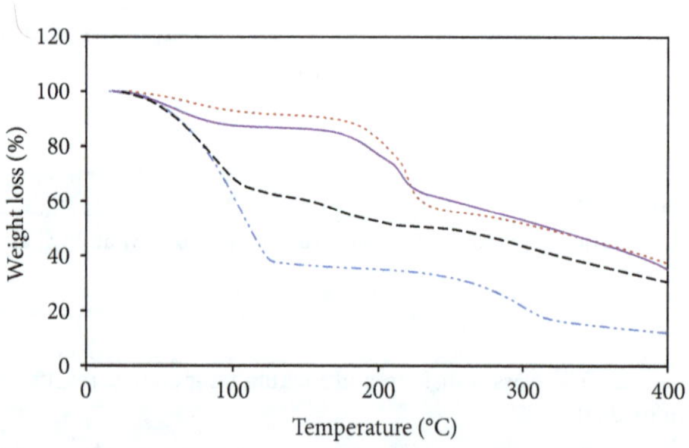

Figure 10: TGA results with different ferric chloride concentrations (pyrrole 0.05 M and surfactant 0.025 M at 25°C and duration of 4 h).

other concentrations, particularly lower concentrations such as 0.05 M and 0.025 M. However, at a higher concentration of 0.15 M, no significant change of the thermal stability was observed. This indicates that 0.1 M concentration of ferric chloride could be the optimum value of achieving good thermal stability of the produced PPy samples.

Table 5 shows $T_{20\%}$ and $T_{60\%}$ for different ferric chloride concentrations. From the TGA data, the thermal degradation temperature increases with increase in concentration of ferric chloride oxidant. For instance, at a concentration of 0.1 M, the highest thermal stability was achieved; however, a slight decrease of thermal stability beyond that concentration was observed. At lower concentrations, such as 0.05 M and 0.025 M, thermal stability decreased drastically, particularly at 0.025 M.

Table 4: TGA results for PPy samples with $FeCl_3$ and $(NH_4)_2S_2O_8$ as oxidants.

Oxidant	$T_{20\%}$ (°C)	$T_{60\%}$ (°C)
$FeCl_3$	202	384
$(NH_4)_2S_2O_8$	83	206

Table 5: TGA results for PPy samples with different concentrations of $FeCl_3$.

$FeCl_3$ concentration (M)	$T_{20\%}$ (°C)	$T_{60\%}$ (°C)
0.15	192	381
0.1	202	384
0.05	80	325
0.025	79	122

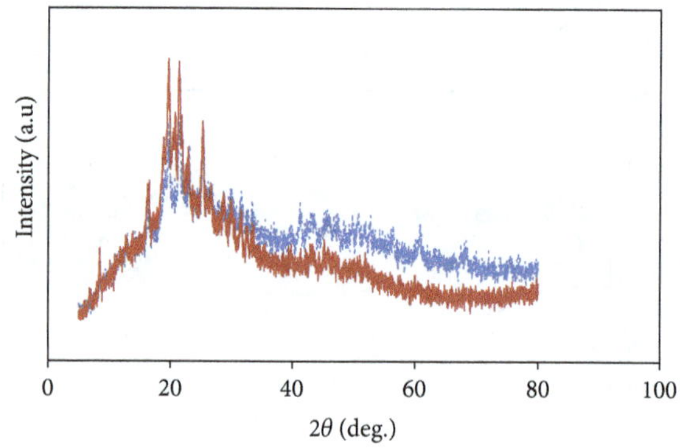

Figure 11: X-ray diffraction of PPy samples (oxidant 0.1 M, pyrrole 0.05 M, and surfactant 0.025 M at 25°C and duration of 4 h).

3.4. Morphological Results

3.4.1. X-Ray Diffraction (XRD) Analysis. Figure 11 shows the X-ray diffraction studies for PPy samples with different oxidants. As shown in Figure 11, the prominent observed peaks and the diffraction in PPy samples with ferric chloride and ammonium persulphate were registered at 2θ, equals 21° and 20°, respectively.

The broad peaks in the region $2\theta = 20°\text{-}21°$ indicate the amorphous nature of the polymer and intermolecular stacking structure. This is in agreement with the previously reported in the literature [15, 22].

3.4.2. Morphology Analysis. The microstructure of polypyrrole samples is shown in Figures 2 and 3 for ferric chloride and ammonium persulphate as oxidizing agents, respectively. The sample has a uniform granular morphology. The amorphous and partially crystalline nature seen in the SEM pictures is reflected in the XRD spectra previously shown.

It was observed that the PPy prepared using a PPy/ferric chloride showed fibrillar morphology in the micrograph, as shown in Figure 2. Similarly, as shown in Figure 3, PPy samples with ammonium persulphate as an oxidant show spherical morphology. Thus, the morphological change of the PPy was strongly dependent on the type of oxidant used during the chemical polymerization process. The morphological structure of the PPy samples has significant influence on the final properties particularly conductivity as shown previously in the section of electrical properties of the samples. It was found that the sample with fibrillar morphology has achieved higher conductivity compared to the sample with spherical morphology.

4. Conclusions

PPy samples were synthesized using different oxidants, and their performance in terms of electrical and thermal properties has been evaluated and compared. It was found that using ferric chloride ($FeCl_3$) as an oxidant had a significant influence on electrical and thermal properties of the produced PPy samples compared to the ammonium sulphate ($N_2H_8S_2O_8$). Similarly, the effects of $FeCl_3$ and pyrrole monomer concentrations on the electrical and thermal stability were examined, and results showed that increasing concentration of $FeCl_3$ up to 0.1 M decreases sample resistivity; similarly, increasing pyrrole monomer concentration decreases resistivity, however, beyond 0.05 M pyrrole concentration resistivity increases, which means at 0.05 M, pyrrole concentration could be optimum concentration to achieve lower resistivity. Hence, higher conductivity of the PPy samples was achieved at concentrations of 0.1 M and 0.05 M for ferric chloride and pyrrole monomer, respectively. It was found that morphological property of the sample plays a significant role in achieving higher conductivity. In terms of thermal property analysis, DSC results have shown that ferric chloride as an oxidant had achieved higher melting point than the PPy sample with ammonium persulphate as an oxidant; also, slightly higher glass transition temperature for samples with ferric chloride was observed. TGA observations showed that the PPy samples with ferric chloride oxidant have higher thermal stability than the samples with ammonium sulphate. It is also shown that TGA result concentration of ferric chloride at 0.1 M had achieved significant thermal stability. Overall result analysis of thermal property indicates that PPy samples with ferric chloride as an oxidant exhibited high thermal stability. FTIR, XRD, and SEM results have shown the formation of polypyrrole polymer during synthesis. All these results indicate the influence of different oxidants which could achieve different performance; therefore, it is possible to engineer the PPy properties according to the requirement and needs for particular applications.

Conflicts of Interest

The authors declare that they have no conflicts of interest.

Acknowledgments

The authors would like to thank the Kuwait Institute for Scientific Research (KISR) for providing the grant that has made this research work possible.

References

[1] G. Wallace, G. M. Spinks, L. A. P. Kane-Maguireand, and P. Teasdale, *Conductive Electroactive Polymers*, CRC Press, New York, USA, 2003.

[2] J. Y. Lee, S. H. Cho, and K. T. Song, *Handbook of Conductive Polymers*, T. A. Skotheim and J. R. Reynolds, Eds., CRC Press, New York, USA, 2007.

[3] P. Chandrasekhar, *Conducting Polymers, Fundamentals and Applications: A Practical Approach*, Springer Publications, New York, USA, 1999.

[4] T.'y. V. Vernitskaya and O. N. Efimov, "Polypyrrole: a conducting polymer; its synthesis, properties and applications," *Russian Chemical Reviews*, vol. 66, no. 5, pp. 443–457, 1997.

[5] R. Ansari, "Polypyrrole conducting electroactive polymers: synthesis and stability studies," *E-Journal of Chemistry*, vol. 3, no. 4, pp. 186–201, 2006.

[6] K. Guerchouche, E. Herth, L. E. Calvet, N. Roland, and C. Loyez, "Conductive polymer based antenna for wireless green sensors applications," *Microelectronic Engineering*, vol. 182, pp. 46–52, 2017.

[7] R. Jain, N. Jadon, and A. Pawaiya, "Polypyrrole based next generation electrochemical sensors and biosensors: a review," *Trends in Analytical Chemistry*, vol. 97, pp. 363–373, 2017.

[8] I. Sapurina, Y. Li, E. Alekseeva et al., "Polypyrrole nanotubes: the tuning of morphology and conductivity," *Polymer*, vol. 113, pp. 247–258, 2017.

[9] B. Yue, C. Wang, X. Ding, and G. G. Wallace, "Electrochemically synthesized stretchable polypyrrole/fabric electrodes for supercapacitor," *Electrochimica Acta*, vol. 113, pp. 17–22, 2013.

[10] N. K. Guimard, N. Gomez, and C. E. Schmidt, "Conducting polymers in biomedical engineering," *Progress in Polymer Science*, vol. 32, no. 8-9, pp. 876–921, 2007.

[11] M. Wysocka-Zolopa and K. Winkler, "Electrochemical synthesis and properties of conical polypyrrole structures," *Electrochimica Acta*, vol. 258, pp. 1421–1434, 2017.

[12] J. M. Pringle, J. Efthimiadis, P. C. Howlett et al., "Electrochemical synthesis of polypyrrole in ionic liquids," *Polymer*, vol. 45, no. 5, pp. 1447–1453, 2004.

[13] Z. D. Kojabad and S. A. Shojaosadati, "Chemical synthesis of polypyrrole nanostructures: optimization and applications for neural microelectrodes," *Materials & Design*, vol. 96, pp. 378–384, 2016.

[14] M. Omastová, M. Trchová, J. Kovářová, and J. Stejskal, "Synthesis and structural study of polypyrroles prepared in the presence of surfactants," *Synthetic Metals*, vol. 138, no. 3, pp. 447–455, 2003.

[15] H. K. Chitte, G. N. Shinde, N. V. Bhat, and V. E. Walunj, "Synthesis of polypyrrole using ferric chloride ($FeCl_3$) as oxidant

together with some dopants for use in gas sensors," *Journal of Sensor Technology*, vol. 1, no. 2, pp. 47–56, 2011.

[16] V. Shaktawat, N. Jain, M. Dixit, N. S. Saxena, K. Sharma, and T. P. Sharma, "Temperature dependence of conductivity of polypyrrole doped with sulphuric acid," *Indian Journal of Pure & Applied Physics*, vol. 46, pp. 427–430, 2008.

[17] S. Khamlich, F. Barzegar, Z. Y. Nuru et al., "Polypyrrole/graphene nanocomposite: high conductivity and low percolation threshold," *Synthetic Metals*, vol. 198, pp. 101–106, 2014.

[18] N. Hassanzadeh, H. Omidvar, and S. H. Tabaian, "Chemical synthesis of high density and long polypyrrole nanowire arrays using alumina membrane and their hydrogen sensing properties," *Superlattices and Microstructures*, vol. 51, no. 3, pp. 314–323, 2012.

[19] S. Barrau, P. Demont, C. Maraval, A. Bernes, and C. Lacabanne, "Glass transition temperature depression at the percolation threshold in carbon nanotube–epoxy resin and polypyrrole–epoxy resin composites," *Macromolecular Rapid Communications*, vol. 26, no. 5, pp. 390–394, 2005.

[20] M. E. Achour, A. Droussi, D. Medine et al., "Thermal and dielectric properties of polypyrrole-poly(methyl methacrylate) nanocomposites," *International Journal of the Physical Sciences*, vol. 6, no. 22, pp. 5075–5079, 2011.

[21] V. S. Sangawar and N. A. Moharil, "Study of electrical, thermal and optical behavior of polypyrrole filled PVC: PMMA thin film thermoelectrets," *Chemical Science Transactions*, vol. 1, no. 2, pp. 447–455, 2012.

[22] F. Khadem, M. Pishvaei, M. Salami-Kalajahi, and F. Najafi, "Morphology control of conducting polypyrrole nanostructures via operational conditions in the emulsion polymerization," *Journal of Applied Polymer Science*, vol. 134, no. 15, 2017.

[23] Y. Kudoh, "Properties of polypyrrole prepared by chemical polymerization using aqueous solution containing $Fe_2\ (SO_4)_3$ and anionic surfactant," *Synthetic Metals*, vol. 79, no. 1, pp. 17–22, 1996.

[24] T. Uyar, L. Toppare, and J. Hacaloğlu, "Spectroscopic investigation of oxidation of *p*-toluene sulfonic acid doped polypyrrole," *Synthetic Metals*, vol. 123, no. 2, pp. 335–342, 2001.

[25] B. Tian and G. Zerbi, "Lattice-dynamics and vibrational-spectra of polypyrrole," *Journal of Chemical Physics*, vol. 92, no. 6, pp. 3886–3891, 2009.

[26] C. R. K. Rao, R. Muthukannan, J. A. Jebin, T. Antony, and M. Vijayan, "Synthesis, and properties of polypyrrole obtained from a new Fe(III) complex as oxidizing agent," *Indian Journal of Chemistry*, vol. 52A, pp. 744–748, 2013.

11

Evaluation of Antifingerprint Properties of Plastic Surfaces Used in Automotive Components

M. P. Luda,[1] N. Li Pira,[2] D. Trevisan,[1] and V. Pau[1]

[1]*Dipartimento di Chimica, Università di Torino, Via P. Giuria 7 10125, Torino, Italy*
[2]*GML, C.R.F. S.C.p.A, Corso Settembrini 40, 10135 Torino, Italy*

Correspondence should be addressed to M. P. Luda; mariapaola.luda@unito.it

Academic Editor: Stefano Turri

The antifingerprint properties of a range of surfaces produced with different technologies (in-mould decoration, in-mould labeling, and painted) were objectively evaluated by depositing on them in standard conditions an artificial fingerprint for direct determination of its visibility. The artificial fingerprint behaves similarly to the real human fingerprints. A classification method is then proposed to classify surfaces on the base of antifingerprint properties by measuring the roughness profile (Ra) and calculating the % variation of gloss (GU 20 and 60°), haze, luminance (L), and diffuse reflectance (R) values after fingerprint deposition. This approach provides an objective and quantitative test method to determine visual antifingerprint properties of coated surfaces, instead of the "easy-to-clean" properties commonly evaluated. The data acquired provides a design guideline for fabricating visually fingerprint-free surfaces by controlling roughness, texture, color, and transparency of surfaces, with the aim of optically masking fingerprints.

1. Introduction

Today, most research is focused on improving easy-to-clean properties and self-cleaning (antistaining, antisoiling) properties of the surfaces used in most industrial sectors [1]. In particular, these properties are very important on surfaces touched and manipulated by the user such as touch screen displays since fingerprints easily stick on them; as a consequence, their optical quality is impaired and information becomes quite difficult to read. For this reason, the fingerprint problem on surfaces is a pressing issue not only in the automotive sector. In most situations, antifingerprint properties are required on surfaces resulting from new technologies such as in-mould decoration (IMD) and in-mould labelling (IML), alternatives of the conventional injection molding process. In IMD, a molten polymer is introduced into the mould cavity where a film carrying the layers of dried paint, decorated with heat-resistant ink, is attached to one side of the mould walls with the paint facing the gate. The paint is transferred to the plastic, and the carrier film is removed when opening the mold. IMD has been employed for the production of various molded products such as automotive interior parts, mobile phone cases, and logo-imprinted plastic products. IML utilizes plastic labels, made with various polymers; the labels become an integral part of the final product, which is delivered as a predecorated item. IML products include picnic ware, mouse mats, or internal automotive components. The development of new effective products with controlled characteristics depends strictly on the knowledge of what governs the antifingerprint properties of the surfaces.

Fingerprints possess amphiphobic properties [2–4], so an appropriate surface roughness and surface chemistry are both crucial factors in the fabrication of fingerprint-free surfaces. Main top-down methods for preparation of superoleophobic plastic surfaces are etching, lithography, and laser processing whilst main bottom-up methods are electrodeposition, hydrothermal method, electrospinning, and sol-gel process [5]. Other bottom-up methods for superoleophobic surfaces have also been reported, such as spray deposition, dip coating, self-assembly, physical vapor deposition (PVD), and the template method. Though all these methods can be applied to fabricate superoleophobic surfaces, few can be used in industrial production because of their high cost and limited production volumes and the inherent complexity of the manufacturing process. So far, many companies carry out applications of hydrophobic and oleophobic coatings to automotive display manufacturers both on plastic and on glass-based displays [6]. The amphiphobic property can also be achieved in a nanoscaled concave structure consisting of cavities, in which the capillary force produced at the liquid-air interface inside the cavity is able to repel the entering liquid (water or oil) [7].

However, imparting antifingerprint properties to surfaces is complex because of fingerprint composition, which depends on the time elapsed since deposition, on the gender, on the age, and on the diet of the person touching the surface [3, 4]. Consequently, successful antifingerprint properties cannot be easily predicted from the resins used or material labels alone.

Currently, the evaluation of the visibility of fingerprints on surfaces is only visual (with the naked eye) so an objective test method is required. To our knowledge, the only method found in literature able to evaluate the behavior of surfaces against fingerprints is the one proposed by Wu et al. [8] who measured the contact angles on smooth and rough surfaces prepared on purpose by sol-gel technique using an artificial fingerprint liquid. This method suggests that a surface with a contact angle above 87° is virtually fingerprint free.

Aside from the fact that the abovementioned method works or not, it is evident that it cannot quantify the visibility of fingerprints; in fact, with equal contact angles, a fingerprint can be more or less visible depending on the optical properties of the surface. An easy-to-clean surface exhibits low wettability by the liquid, and therefore, the correspondent contact angle must be close to 90° or higher. In this paper, a new approach is proposed for an objective classification method; a full characterization of various commercial surfaces for automotive interiors has been performed, and key properties are measured in correlation with the visibility of artificial fingerprints deposited with a standardized method.

2. Material and Methods

2.1. Materials. The 15 commercial surfaces for automotive interiors reported in Table 1(a) were tested. Different technologies of decoration were employed: 5 IMD, 5 IML, and 5 painted surfaces. Some of them were patterned (IMD1, IMD2, IMD3, and IML3), some transparent (IML2, IMD5), and some had a black finishing (IMD4, IML1, IML4, IML5, V1, V2, V3, VN1, and VN3).

Three of the painted surfaces were treated with a physical vapor deposition (PVD) coating layer to hopefully impart antifingerprint properties to the underlying surface (V1, V2, and V3).

The description of each surface with all relevant characteristics is shown in Table 1(a). The reagents for artificial fingerprints were all bought from Sigma-Aldrich and used without changes.

2.2. Methods

2.2.1. Reference Standard Fingerprint Preparation. A standard artificial fingerprint, representing an average human fingerprint, was developed to set up a surface classification based on the visibility of fingerprints in standard conditions.

To replicate the human fingerprint, both artificial sweat and sebum were required. The artificial sweat was prepared by dissolving the compounds listed in Table 2 (left column) in distilled water [9]. The weight % of the constituents listed in Table 2 compared to the complete formula in [9] is ≈80% (% *w*/*w*); this simplification is a good simulation of the real sweat with only 5 constituents easy to retrieve in the market. The complex mix of oils and fatty substances listed in Table 2 (right column) was blended to emulate the natural sebum; all are common oils or easy to recover constituents [10].

The mixture chosen to create the artificial fingerprints was composed of 95% sweat and 5% sebum (% *w*/*w*). An aliquot of nonionic surfactant (Triton X-100) was required (≈2 μl × grams of mixture) to overcome the natural immiscibility between sweat and sebum.

2.2.2. Standard Fingerprint Deposition. The manual stamp for standard deposition was assembled by composing a stainless steel cylindrical support with a nut on the top (dimension: M_{12}) filled with cotton.

A layer of leather used in car interiors was pulled over to fix everything (Figure 1).

The pressure chosen to create the artificial fingerprint, in order to emulate a "soft touch" condition, was ≤3.5 N/cm^2. Since the assembled manual stamp had a contact area equal to 4 cm^2, a 1 kg weight (10 N) was enough.

The deposition procedure was as follows:

(1) Heating the artificial secretion at ≈40°C

(2) Withdrawing 1 μl of heated artificial secretion with a micropipette (20 μl)

(3) Releasing the artificial secretion on the manual stamp and spreading it on the surface with a thin paint brush

(4) Positioning the manual stamp and the weight on the chosen surface

(5) Creating 3 consecutive artificial fingerprints spaced by 20 mm from each other

(6) Washing the manual stamp and the paint brush with distilled water, degreaser, and a cloth

2.2.3. Characterization Techniques. A Kruss DSA30S contact angle machine was used to measure the static apparent contact angle (CA) with distilled water (WCA) and diiodomethane (CH_2I_2CA). To prevent cross-contamination, a new syringe was used for each liquid. A drop of 0.5 μl was used for the static contact angle; the image of the liquid drop

TABLE 1: (a) Main characteristics of the commercial samples and results of the characterization on not imprinted surfaces.

(a)

		Pattern	Appearance	Tech	IR	Ra (nm)	Haze	WCA	CH_2I_2CA	GU 20°	GU 60°	L	R_{550}
Not visible	IML1	Plain, SiO2	Matt	IML	Various	579.80 ± 193.9	12.60 ± 0.2	80.3 ± 2.1	63.3 ± 1.8	17.70 ± 0.1	37.40 ± 0.2	14.40 ± 0.34	1.80 ± 0.05
	IML2	Plain	Trans	IML	Acrylic	275.20 ± 72.2	13.20 ± 0.4	76.5 ± 3.1	52.3 ± 1.6	8.50 ± 0.6	30.60 ± 1.4	15.77 ± 0.37	2.07 ± 0.07
	IML3	Patterned	Matt	IML	Various	138.00 ± 37	4.10 ± 0.5	82.1 ± 1.3	56.4 ± 3.3	3.00 ± 0.4	18.80 ± 2.2	21.88 ± 0.4	3.49 ± 0.1
	IMD1	Patterned	Matt	IMD	Acrylic	207.70 ± 41.2	37.80 ± 1.9	78.2 ± 1.5	54.3 ± 1.5	35.40 ± 2.1	36.80 ± 2.1	39.61 ± 0.37	11.09 ± 0.21
	IMD2	Patterned	Glossy	IMD	Acrylic	84.80 ± 44.2	2.60 ± 0.2	63.6 ± 6.4	51.7 ± 2.6	72.20 ± 2.4	84.20 ± 1.1	7.72 ± 0.37	0.86 ± 0.04
	IMD3	Patterned	Matt	IMD	Acrylic	209.50 ± 25.5	6.70 ± 2.5	82.4 ± 1.8	57.6 ± 1.9	11.10 ± 5.2	24.30 ± 10.2	22.89 ± 1	3.69 ± 0.3
	IMD4	Plain	Matt	IMD	Various	491.40 ± 66.2	11.90 ± 0.6	85.7 ± 1.2	52.9 ± 2.2	6.00 ± 0.9	27.90 ± 2.6	15.45 ± 0.1	1.97 ± 0.02
Visible	V1	Plain	Glossy	P	urethane	41.00 ± 31.4	1.30 ± 0.1	107.10 ± 1.7	60.90 ± 2.6	82.00 ± 0.9	88.20 ± 3.6	1.17 ± 0.16	0.13 ± 0.02
	V2	Plain	Glossy	P	urethane	12.90 ± 3.9	1.60 ± 0.1	108.10 ± 4.2	79.70 ± 5.6	80.10 ± 0.4	89.40 ± 0.3	1.42 ± 0.12	0.16 ± 0.02
	V3	Plain	Glossy	NP	urethane	3.80 ± 0.98	0.20 ± 0.1	88.50 ± 4.0	49.40 ± 4.9	89.10 ± 0.6	93.60 ± 1.1	1.58 ± 0.14	0.18 ± 0.02
	VN1	Plain	Glossy	NP	urethane	11.50 ± 1.9	2.00 ± 0.1	84.20 ± 2.7	50.10 ± 1.2	80.40 ± 0.3	87.80 ± 1.3	0.66 ± 0.14	0.07 ± 0.02
	VN3	Plain	Glossy	P	urethane	6.50 ± 1.4	0.10 ± 0.1	78.20 ± 1.6	41.50 ± 1.3	88.90 ± 0.6	94.90 ± 0.9	1.29 ± 0.25	0.14 ± 0.03
	IML4	Plain	Glossy	IML	Acrylic	5.90 ± 0.65	0.20 ± 0.1	73.10 ± 1.5	53.00 ± 1.8	82.70 ± 3.6	91.70 ± 0.9	0.82 ± 0.12	0.10 ± 0.02
	IML5	Plain	Glossy	IML	Various	4.60 ± 1.8	0.20 ± 0.1	81.40 ± 4.2	60.20 ± 2.3	84.30 ± 6.8	94.50 ± 0.4	3.42 ± 0.36	0.39 ± 0.05
	IMD5	Plain	Trans	IMD	Various	20.10 ± 12.9	4.60 ± 0.4	70.10 ± 2.8	41.30 ± 1.5	77.60 ± 4.3	91.80 ± 1.7	6.49 ± 0.56	0.70 ± 0.07

trans: transparent; IMD: in-mould decoration; IML: in-mould labeling; P: painted surface with PVD treatment; NP: painted surface without treatment.

(b) Roughness characteristics Ra, Rv, and Rp of the commercial sample and their ratios

		Ra (nm)	Rv	Rp	Rt	|Ra/Rv|	Ra/Rp
Visible	V3	3.8	−43.7	42.9	86.6	0.0870	0.0886
	IML5	4.6	−31.3	28.3	59.6	0.1470	0.1625
	IML4	5.9	−38.3	30	68.3	0.1540	0.1967
	VN3	6.5	−60.4	57	117.4	0.1076	0.1140
	VN1	11.5	−54.33	74.87	129.2	0.2117	0.1536
	V2	12.9	−69.6	54.3	123.9	0.1853	0.2376
	IMD5	20.1	−116.13	67.5	183.63	0.1731	0.2979
	V1	41	−86	156.1	242.1	0.4767	0.2627
Not visible	IMD2	84.8	−181.1	352.6	533.7	0.4682	0.2405
	IML3	138	−590.9	362.6	953.5	0.2335	0.3806
	IMD1	207.7	−1655	464.7	2119.7	0.1255	0.4470
	IMD3	209.5	−621.1	942.5	1563.6	0.3373	0.2223
	IML2	275.16	−1175.7	606.3	1782	0.2340	0.4538
	IMD4	491.4	−1726	1024.8	2750.8	0.2847	0.4795
	IML1	579.8	−977	2498.4	3475.4	0.5934	0.2321

Table 2: Components of the artificial sweat and their concentrations in g/l and of artificial sebum and relative % *w/w*.

Artificial sweat		Artificial sebum	
Reagent	g/l	Reagents	% (*w/w*)
Lactic acid	1.26	Palmitic acid	10
NaCl	1.53	Stearic acid	5
Urea	0.60	Coconut oil	15
KCl	0.46	Paraffin wax	10
$NaHCO_3$	0.25	Jojoba oil	15
		Olive oil	20
		Squalene	5
		Cholesterol	5
		Oleic acid	10

was taken by a digital camera. Results are the average of 10 measures carried on each sample and for each liquid.

The surface roughness (Ra) of all the samples was measured by a Bruker Dektak 150 stylus profilometer setting 2 mg of weight, 400 μm of scan length, 0.044 μm of resolution, and 60 sec of scan duration. Rv (the valley depth of the roughness profile below the mean line), Rp (the peak height over the mean line), and Rt (the distance between the highest peak and the deepest valley) were measured as well.

A Varian Cary 500 spectrophotometer was used for direct transmittance (T%) measurements on glass, setting scans ranging from 300 to 800 nm and a scan rate of 600 nm/min.

IR spectra were acquired with a standard PerkinElmer Spectrum 100 (FT-IR) equipped with the ATR sampling system (PerkinElmer Universal ATR).

A Leica DCM8 confocal microscope was used to measure fingerprint thickness using a 20X magnification and an area of 1 cm^2.

Optical characterization was performed with (a) Konica Minolta Rhopoint IQ Glossmeter for the GU and haze parameters, calibrated with the supplied standard, and (b) Konica Minolta 2600d portable spectrophotometer with an integrating sphere for reflectance (*R*) and luminance (*L*) parameters, using an 8 mm aperture and a scanning range from 360 to 740 nm in the SCE mode (specular component excluded). Reflection was measured at two reflection angles: 20° (GU20) and 60° (GU60). Two values were considered for *R*: at 550 nm (R_{550}) and at the maximum difference along all the experimental range (R_{MAX}).

All parameters were measured before the deposition of fingerprints, so on cleaned surfaces and over the fingerprint mark after its deposition; in each case, at least 10 measures were collected (average values and errors are reported).

For any parameter *X* (CA, GU, haze, *R*, and *L*), variation between the value on imprinted and on cleaned surfaces ($\Delta X\%$) was calculated as follows:

$$\Delta X\% = \left[\frac{(X_1 \cdot 100)}{X_0}\right] - 100. \quad (1)$$

According to the theory of error propagation, the new standard deviation σ_X becomes

$$\sigma_X = \pm\left[\frac{\sigma_0}{X_0} + \frac{\sigma_1}{X_1}\right] \cdot \frac{X_1}{X_0}, \quad (2)$$

where $\Delta X\% = \Delta\%$ is the variation of general parameter X, X_1 is the total average value on fingerprint spots, X_0 is the total average value on the cleaned surface, and σ is the standard deviation, being X a ratio between two different average values $X_0 \pm \sigma 0$ and $X_1 \pm \sigma 1$.

An optical microscope Nikon Eclipse LV100 was used to visually compare artificial and real fingerprints.

The principal component analysis (PCA) was carried out with XLSTAT 2015 software, autoscaling the variables according to the following formula:

$$\left[\left(\frac{X1 - \mu 1}{\sigma_1}\right), \left(\frac{X2 - \mu 2}{\sigma 2}\right), \ldots, \left(\frac{Xp - \mu p}{\sigma p}\right)\right], \quad (3)$$

where $X1$ is the value of parameter 1 for the single sample, $\mu 1$ is the average value for parameter 1, and σ is the standard deviation.

3. Result and Discussion

3.1. Comparison of Artificial and Human Fingerprints. Several measurements and comparisons were done on artificial and human fingerprints (deposited by 20 people: 10 men and 10 women) to verify if artificial fingerprints were a good approximation of the real ones.

An artificial fingerprint (the third one of the sequences deposited on the substrate) and a real fingerprint are compared in Figure 2; they look quite similar. The main difference is the pattern due to the leather used in one case and to the papillary crest in the other.

A further comparison was done by measuring transmittance (*T*%) on glass where a human fingerprint and three artificial fingerprints made with different secretion compositions (95 : 5, 80 : 20, and 0 : 100, sweat/sebum *w*%) were deposited (Figure 3). The sebum decreases uniformly the transmittance in the range 400–800 nm: so, the higher is the amount of sebum, the lower is the *T*%. This comparison supports the choice of the 95% (*w*/*w*) sweat and 5% (*w*/*w*) sebum mix as an optimal composition for the artificial fingerprint.

In addition, artificial and human fingerprints exhibit the same trend in modifying the wettability of a surface, as shown by contact angles measured with distilled water on a glass coated with Indio-Tin Oxide (ITO) (Figure 4).

Moreover, human and artificial fingerprints on glass behave similarly also with *R* and haze parameters. Eventually, the thickness measured with a confocal microscope point outs, once more, to a close resemblance between artificial and real fingerprints (Table 3).

3.2. Visibility of the Standard Fingerprint on the Various Surfaces. After the deposition of fingerprints, all samples were observed by 5 people under artificial and natural light at different angles. In accord to their visual perception with

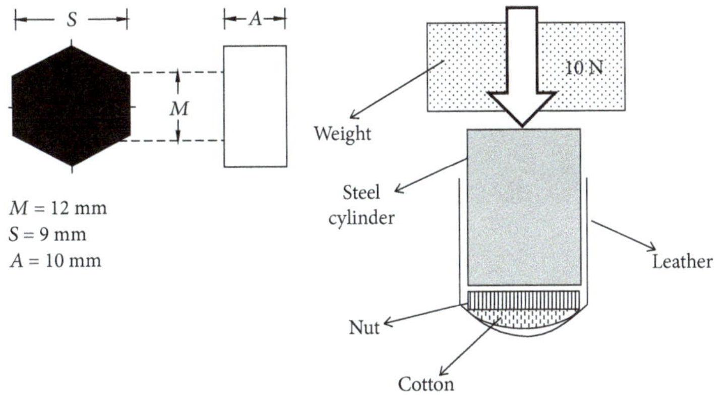

FIGURE 1: (a) View from the above and nut dimensions; (b) side view and stamp configuration.

(a) (b)

FIGURE 2: Images with optical microscope at 20X magnification of third artificial fingerprint (a) and a human fingerprint deposited by a man aging 25 years in low pressure condition (b).

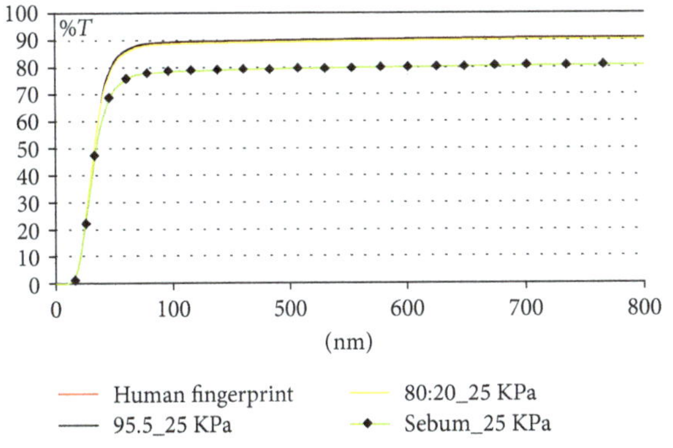

FIGURE 3: T% spectra on glass with human and artificial fingerprints composed by different ratio % sweat/sebum.

the naked eye, samples were divided into two categories: visible or nonvisible, as exemplified in Figure 5. This simple classification represents the customer perception.

There was not disagreement among the observers, and when a fingerprint was only partially visible, the sample was classified in the "visible" category.

The group of observers had no perception of fingerprints in the first seven samples of Table 1(a). The chemical structure of all surfaces was ascertained by ATR spectroscopy. Infrared spectra of the samples are reported in the supplementary materials in Figs S1, S3, and S3. Most of them can be grouped as acrylics or polyurethanes whilst some others pertain to a mixed group (Table 1(a)).

The results of the optical characterization, roughness, and measures of static contact angles on the cleaned surface for the whole set of samples are also reported in Tables 1(a) and 1(b).

The data of Table 1(a) evidence that in both categories, IMD and IML samples can be found; on the contrary, all painted surfaces examined were classified in the "visible" category, no matter if they received or not the antifingerprint treatment.

Apparently, there is no connection between an antifingerprint behavior and the technology of decoration or the polymer of which the surface is made.

On the contrary, glossiness, as measured by the reflection at 20° or 60°, played a major role, since fingerprints were more visible on glossy surfaces than on matt ones. Moreover, the mark was more visible on smooth

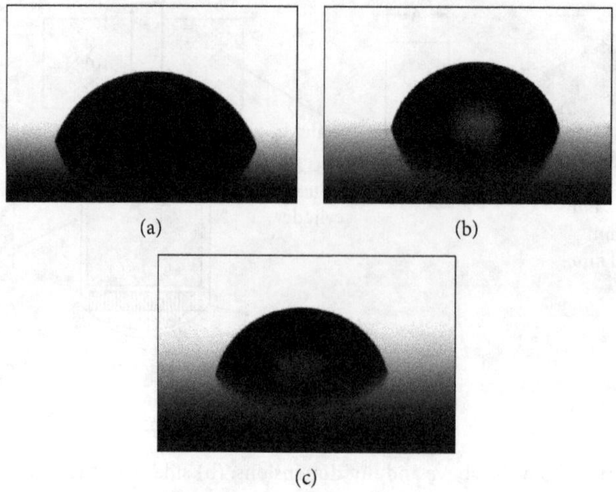

FIGURE 4: Images of water contact angles in different conditions: (a) glass/ITO 60.4 ± 4.2, (b) glass/ITO + real fingerprint 75.7 ± 1.2, (c) glass/ITO + artificial fingerprint 75.6 ± 3.3.

TABLE 3: Average thickness measurements with confocal microscope of artificial and human fingerprint.

Specimen	Average thickness (μm)
Human fingerprint	0.25 ± 0.02
Artificial fingerprint	0.25 ± 0.04

surfaces than on textured samples, no matter if they were glossy or matt.

In samples with high roughness (Ra), haze, L, and R_{550} and low GU, the fingerprint appearance tended to be optically masked; however, it was difficult to predict the limit over which this occurred and a specific cut-off value for each of these properties has to be found.

Apparent static contact angles in water (surface tension 0.073 Nm^{-1}) and in diiodomethane (surface tension 0.051 Nm^{-1}) did not correlate with the visibility of the standard fingerprint; for instance, the only two samples exhibiting low wettability (water contact angle > 90°, V1 and V2) appertained to the visible category. Absolute errors of contact angles measured on cleaned surfaces were relatively low (average values: water ± 3, diiodomethane ± 2.4 corresponding to about 4%). This suggests a sort of macroscopic homogeneity of the surfaces industrially prepared.

The apparent contact angle depends in a complex way on surface energies, roughness, and topography of the samples. Wu et al. [8] measured the contact angles on a series of chemically similar surfaces with different roughness (Ra < 1 μ) using an artificial liquid mimicking the biological one (surface tension 0.02–0.05 Nm^{-1}). By using the Cassie-Baxter approach [11], she found that surfaces with relatively deep valleys were able to trap air under the liquid droplet, improving the nonwetting behavior. Consequently, a new roughness factor for theoretical prediction of antifinger contact angles, Ra/Rv < 0.25, was proposed, as a guideline for an easy-to-clean surface (contact angle > 90°). Smoother surfaces required even deeper valleys to impart antifingerprint property.

In Table 1(b), the full set of roughness parameters for the present commercial samples is reported (Ra < 0.6 μ). As a general trend, Ra/Rv is larger for nonvisible samples (up to 0.59) than for visible ones (up to 0.21). Thus, fingerprints do appear on surfaces with very low Ra/Rv, which could be considered "easy to clean" according to Wu's findings, and fingerprint marks do not appear on surfaces with relatively shallow valleys. However, Wu measures were carried out on a series of homogenous laboratory-prepared surfaces whilst in the present case, surfaces prepared with a number of different industrial techniques and with a large variety of materials (labels, paints, etc.) were investigated, on which other factors play a significant role in the resulting apparent contact angle. In addition, a relatively good negative correlation between Rp (peak value) and $\cos CH_2I_2CA$ was found (R^2 0.7019) for samples of the nonvisible category, suggesting a positive effect of a bumpy structure in finger-free surfaces. Further investigations were carried out on imprinted surfaces to better clarify the reason for appearance of fingerprint marks.

3.3. Imprinted Surface Characterization. In order to correlate the visibility of the fingerprint with the optical properties of the chosen surfaces, the same measurements were performed directly on the artificial fingerprint deposited on the samples. The % differences found in the two sets of measurements (imprinted and not imprinted surfaces) were calculated and reported in Table 4, together with the original value of roughness Ra. Absolute error of contact angles was larger when measured on imprinted surfaces than on cleaned surfaces (average values on cleaned surfaces: water ± 3, diiodomethane ± 2.4; on imprinted surfaces: water ± 4.7, diiodomethane ± 3.1). This suggests that measurements on fingerprint marks were affected by a larger experimental error those than on cleaned surfaces, even if standard deposition of the liquid was performed.

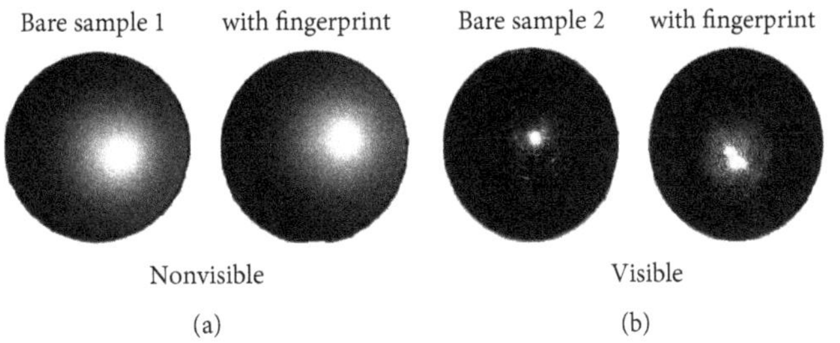

FIGURE 5: IML1 (a) and IML5 (b) views before and after the deposition of the artificial fingerprint.

When fingerprint was visible with the naked eyes, Δhaze, ΔL, ΔR_{550}, and ΔR_{MAX} dramatically increased; on the contrary, ΔGU (both 20° and 60°) and ΔCA tended to decrease. This occurred in all "visible" samples apart from transparent IMD5 where ΔGU slightly increases. All these results demonstrate that fingerprint application makes a surface less brilliant.

"Nonvisible" samples performed somewhat differently: ΔGU and ΔWCA had a contrasted behavior, and Δhaze, ΔL, ΔR, and ΔR_{MAX} were hardly affected; however, ΔCH_2I_2CA decreased suggesting that fingerprint deposition on the substrate occurred in nonvisible cases also but that their appearance was masked by optical diffusion. Haze highly increased only in the patterned glossy IMD2 sample.

Thus, the appearance/nonappearance of fingerprint marks seems to depend on several properties. A comprehensive approach was performed by making a PCA on the data of Table 4 after they were autoscaled (to make measures of different units and range comparable each other).

A relatively good separation between the two classes appears in the score plot (Figure 6(a)); however, the corresponding loading plot (Figure 6(b)) puts in evidence the overlapping contribution of some variables, which therefore gave the same information (ΔGU 20° and ΔGU 60°, ΔL% and ΔR%). At the same time, other variables notably ΔWCA and ΔCH_2I_2CA seem to be less significant despite that contact angles are reported to be crucial to evaluate antifingerprints properties [8, 11]. An explanation could be, as reported in literature [12], that the apparent static contact angles that we measured are less correlated to the quality of the surfaces than the advancing and receding dynamic contact angles.

However, the variation of the static contact angles measured on the fingerprint mark and on the clean surface was under consideration here (ΔWCA and ΔCH_2I_2CA) and not the contact angles themselves. From Table 4, it can be recognized that there is a general decreasing of ΔCA (both in water and in diiodomethane) when measured on the fingerprint mark (visible or not visible) instead of on the cleaned surface. So, even in the case of not visible fingerprint surfaces, some residual material from the finger liquid sticks on the surface making it apparently less hydrophilic. A better knowledge on the oleophobic properties of the investigated surface can be gained by measuring the contact angles by using a probing liquid with a lower surface tension, such as oleic acid (which is one of the main components of the artificial sebum).

All these considerations can explain why the measured static contact angles do not give a significant information in the PCA treatment.

Consequently, a PCA was performed on a reduced number of variables. The results (Figure 7) indicated that the reduction of the variable number did not impair the separation between visible and nonvisible samples.

All "nonvisible" samples were confined in a narrow area of the negative quadrant of the plot where Ra and ΔGU at 20° were dominant, except the IMD2 sample (patterned, glossy). Visible samples were spread over the other quadrants where ΔR_{MAX}, ΔL, and Δhaze were dominant apart for the IMD5 sample (transparent, glossy) which fell close to the zone of "nonvisible" samples.

Such classification is good, but more attention is required for transparent and patterned samples.

3.4. Analytical Method for Antifingerprint Properties. The separation of the two categories of samples (fingerprint visible–nonvisible) was good enough even reducing the number of variables to five (and consequently the number of measures, specifically Ra, Δ% GU 20°, Δ% haze, Δ% R_{MAX}, Δ% L).

A classification method to evaluate and quantify the ability of optically masking fingerprints on chosen surfaces is proposed, based on cut-off values; for most significant parameters, an average value between the limit of each series (visible and not visible) was estimated from the data of Table 4 and taken as cut-off value. The anomalous value for Δhaze % for IMD2 was disregarded in this calculation. The selected cut-off values are reported in Table 5, first row.

Further, the ability to emulate human fingerprints was checked by repeating the measurements of the four parameters (Ra excluded) on four samples using the real human fingerprints of 20 people (10 men and 10 women) in "soft-touch" condition. At least one surface for each technology and two surfaces for each category were selected for this purpose. The proposed cut-off values provided the right classification of the surfaces also when imprinted with human fingerprints (Table 5, right). However, the parameter variation was generally larger with human than with artificial fingerprints, possibly owing to the difficulty to replicate the standard deposition conditions manually (applied pressure) and the possible presence of contaminants.

Eventually, measurements of the five significant parameters were repeated on two samples not previously used in

Table 4: Results on the imprinted surface (all, except Ra) Δ% values.

		Ra (nm)	Δhaze (%)	ΔWCA (%)	ΔCH_2I_2CA (%)	ΔGU 20° (%)	ΔGU 60° (%)	$\Delta L\%^*$ (D65)	$\Delta R\%$ 550 nm	$\Delta R_{MAX}\%$
Not visible	IML1	579.80 ± 193.9	−1.5 ± 0.03	7.3 ± 0.05	−2.7 ± 0.07	−1.5 ± 0.03	−2.5 ± 0.02	−0.3 ± 0.04	−0.7 ± 0.05	4.06 ± 0.13
	IML2	275.20 ± 72.2	4.2 ± 0.05	−5.5 ± 0.08	−10.5 ± 0.11	7.8 ± 0.11	2.2 ± 0.06	0.38 ± 0.07	0.48 ± 0.1	−6.28 ± 0.11
	IML3	138.00 ± 37	4.7 ± 0.2	−3.29 ± 0.1	−8.16 ± 0.1	2.3 ± 0.3	0.1 ± 0.2	−0.2 ± 0.02	−0.4 ± 0.04	−1.13 ± 0.05
	IMD1	207.70 ± 41.2	0.4 ± 0.09	−34.5 ± 0.17	−13.1 ± 0.08	2.9 ± 0.17	−0.3 ± 0.11	−1.06 ± 0.02	−2.7 ± 0.03	−4.2 ± 0.04
	IMD2	84.80 ± 44.2	102.3 ± 0.4	11.2 ± 0.16	−19.3 ± 0.09	−4.2 ± 0.05	−1.9 ± 0.02	2.85 ± 0.08	3.01 ± 0.08	−12.04 ± 0.57
	IMD3	209.50 ± 25.5	4.5 ± 0.8	−6.55 ± 0.1	−9.38 ± 0.[illegible]	5.3 ± 0.95	4.9 ± 0.8	−10.5 ± 0.05	−17.86 ± 0.09	−27.42 ± 0.12
	IMD4	491.40 ± 66.2	−2.6 ± 0.07	−2.57 ± 0.04	−2.65 ± 0.[illegible]	1 ± 0.2	−1.1 ± 0.1	2.48 ± 0.01	3.77 ± 0.02	7.85 ± 0.08
Visible	V1	41.00 ± 31.4	112 ± 0.48	−39.2 ± 0.06	−14.3 ± 0.13	−12.8 ± 0.03	−9 ± 0.07	228.21 ± 0.74	232.3 ± 0.89	2740 ± 4.4
	V2	12.90 ± 3.9	2.5 ± 0.06	−13.3 ± 0.08	−1.5 ± 0.14	−3.9 ± 0.01	−5.4 ± 0.01	23.24 ± 0.34	26.6 ± 0.39	128.6 ± 4.2
	V3	3.80 ± 0.98	606.3 ± 3.8	−8.8 ± 0.08	−9.6 ± 0.07	−4.9 ± 0.02	−2.6 ± 0.02	17.09 ± 0.25	17.05 ± 0.25	157.1 ± 3.6
	VN1	11.50 ± 1.9	26.7 ± 0.17	−4.6 ± 0.06	−9.4 ± 0.05	−4.2 ± 0.02	−1.5 ± 0.04	343.94 ± 1.9	397 ± 2.8	2580 ± 6.5
	VN3	6.50 ± 1.4	1990 ± 2.4	−6.9 ± 0.08	−16.4 ± 0.12	−6.9 ± 0.02	−4.3 ± 0.02	17.47 ± 0.43	22.06 ± 0.53	120 ± 3.1
	IML4	5.90 ± 0.65	170.6 ± 1.3	−20.2 ± 0.13	−10.9 ± 0.07	−4.9 ± 0.1	−3.7 ± 0.02	143.90 ± 0.91	134.7 ± 1.0	1220 ± 11.1
	IML5	4.60 ± 1.8	318.2 ± 2.57	1.6 ± 0.08	−8.5 ± 0.09	−23.6 ± 0.2	−11 ± 0.02	48.25 ± 0.26	46.6 ± 0.29	1520 ± 7.0
	IMD5	20.10 ± 12.9	16.1 ± 0.22	−21.1 ± 0.11	3.1 ± 0.09	2 ± 0.07	0.9 ± 0.03	17.41 ± 0.14	17.33 ± 0.15	95.37 ± 1.1

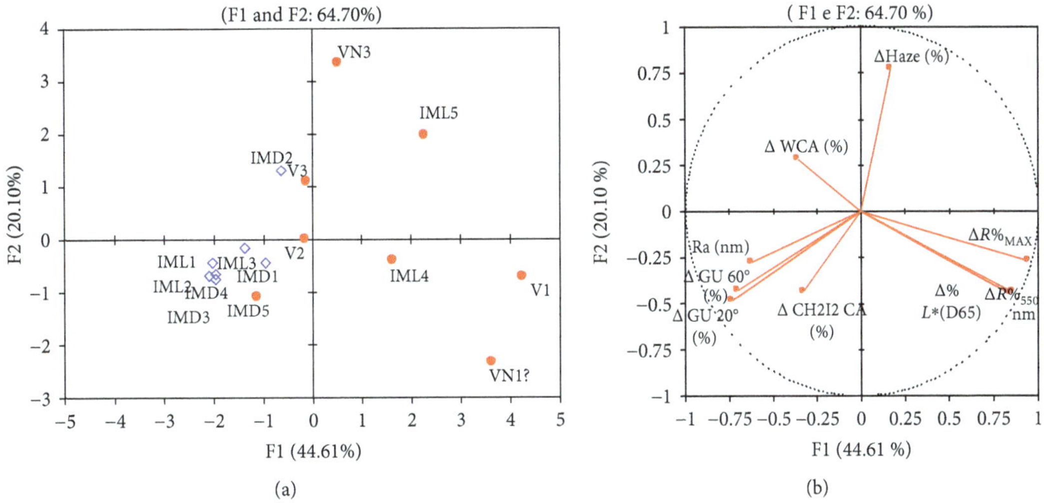

FIGURE 6: Score and loading plots from PCA performed with all variables. Score plot: (a) closed red marks represent "visible" and empty marks represent "nonvisible." Loading plot (b).

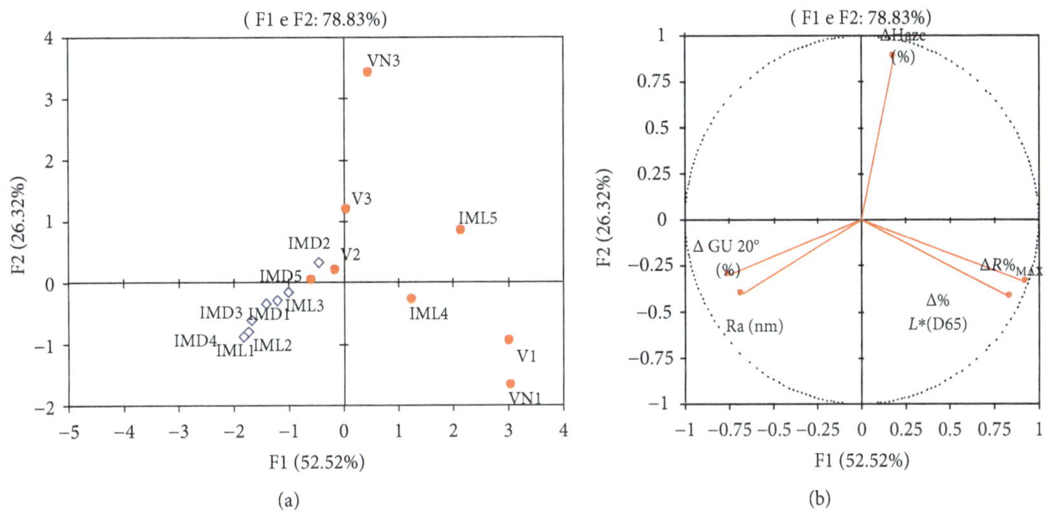

FIGURE 7: Score and loading plots from PCA performed with a reduced number of variables. Score plot: (a) closed red marks represent "visible" and empty marks represent "not visible." Loading plot (b).

TABLE 5: Cut-off values and average results of the four significant parameters on selected surfaces using artificial and human fingerprint.

	Artificial fingerprint				Human fingerprint			
	Δ% haze	Δ% GU 20°	Δ% L	Δ% R_{MAX}	Δ% haze	Δ% GU 20°	Δ% L	Δ% R_{MAX}
Cut-off	10	−2	10	50	10	−2	10	50
IML2	**4.2 ± 0.05**	7.8 ± 0.11	**0.38 ± 0.07**	−6.28 ± 0.11	**5.18 ± 0.1**	2.37 ± 0.1	**−0.08 ± 0.03**	7.88 ± 0.14
IMD1	0.4 ± 0.09	2.9 ± 0.17	**−1.06 ± 0.01**	**−4.20 ± 0.04**	6.24 ± 0.1	14.69 ± 0.1	**−2.87 ± 1**	−5.68 ± 1
VN3	**1990 ± 2.4**	−6.9 ± 0.2	17.47 ± 0.43	120 ± 3.1	**2075 ± 64.8**	−15.59 ± 0.2	94.06 ± 0.7	616 ± 10.7
IML4	170.6 ± 1.3	−4.9 ± 0.1	143.90 ± 0.9	**1220 ± 11.1**	540 ± 5.8	−12.66 ± 0.1	64.75 ± 0.4	**1236 ± 11.8**

order to evaluate the prediction ability of the method. The classification was done by using the proposed cut-off values and then confirmed with naked eye observation (Table 6).

The samples were classified correctly in their respective categories, although some parameters (highlighted in bold) have a dubious value.

4. Conclusions

This work was aimed at elaborating a classification method to objectify the visibility of fingerprints on surfaces, instead of the evaluation of the "easy-to-clean" properties already used.

Table 6: Average results of significant parameters for the selected surfaces and relative classification.

	Ra (nm)	Δhaze (%)	ΔGU 20° (%)	Δ% L	Δ% R_{MAX}	Visibility	Pattern	Appearance	Tech
Cut-off	*50*	*10*	*−2*	*10*	*50*				
VN2	613.07 ± 39.5	**33.33 ± 1.1**	15.15 ± 0.47	−0.87 ± 0.01	−2.42 ± 0.03	Not visible	Plain	Matt	NV
IMD6	47.14 ± 9.9	33.71 ± 0.16	−5.90 ± 0.03	58.27 ± 0.5	716.00 ± 8.14	Visible	Plain	Glossy	IMD

An artificial fingerprint liquid consisting in artificial sebum and artificial sweat (5 : 95% *w/w*) was prepared and deposited on the surfaces in selected standard conditions: both liquid and standard conditions were easily replicable. The similarity of artificial and real human fingerprints was confirmed, and the standard deposition method was shown to eliminate the substantial variability of human fingerprints, offering high reproducibility. The visibility of fingerprints seems to depend on several properties; however, a comprehensive chemometric approach allowed to reduce the significant measurements to 5 properties: the roughness profile (Ra), the % variation of gloss (GU), haze, luminance (L), and diffuse reflectance (R).

This restricted number of variables allowed to classify a surface correctly by using a method based on cut-off values, and the results of the measurements permitted to quantify the antifingerprint optical properties of the surface. The value-added feature of this work is of course the objectification of the visibility of fingerprints on chosen surfaces, instead of the common naked-eye evaluation.

From data acquired, it is evident that rough and matt surfaces promote the optical masking of fingerprints, whilst chemical treatment intended to decrease wettability and reduce adhesion of the fingerprint on the surface was not always successful, confirming that most of commercially treatments available are focused on improving the "easy-to-clean" properties. Glossy surfaces are critical for the point of view of antifingerprint properties; however, an appropriate pattern could be a useful trick in order to hide fingerprints or reduce their visibility. Care should be adopted with transparent surfaces, where roughness and reflectance seem to govern the antifingerprint properties. With future efforts in increasing the number of samples studied, the cut-off values will be more reliable. Moreover, measuring the dynamic contact angles instead of static ones and using a probing liquid with surface tension approaching that of artificial fingerprint liquid will further improve the analysis results, providing a more detailed design guideline for producing visually fingerprint-free surfaces.

Conflicts of Interest

The authors declare that they have no conflicts of interest.

Acknowledgments

The authors wish to thank Dr. M.M. Dugand for the most helpful discussions and for performing confocal analysis. FCA sponsors this work by carrying out optical characterizations and supporting V.P. fellowship.

References

[1] P. Uhlmann, R. Frenzel, B. Voit et al., "Research agenda surface technology: future demands for research in the field of coatings materials," *Progress in Organic Coatings*, vol. 58, no. 2-3, pp. 122–126, 2007.

[2] B. Stoehr, S. McClure, A. Höflich et al., "Unusual nature of fingerprints and the implications for easy-to-clean coatings," *Langmuir*, vol. 32, no. 2, pp. 619–625, 2016.

[3] A. Girod, R. Ramotowski, and C. Weyermann, "Composition of fingermark residue: a qualitative and quantitative review," *Forensic Science International*, vol. 223, no. 1-3, pp. 10–24, 2012.

[4] S. Cadd, M. Islam, P. Manson, and S. Bleay, "Fingerprint composition and aging: a literature review," *Science and Justice*, vol. 55, no. 4, pp. 219–238, 2015.

[5] T. Jiang, Z. Guo, and W. Liu, "Biomimetic superoleophobic surfaces: focusing on their fabrication and applications," *Journal of Materials Chemistry A*, vol. 3, no. 5, pp. 1811–1827, 2015.

[6] B. C. Wilson and D. J. Fiore, "Anti-fingerprint coating applications for automotive touchscreen displays," in *SID Symposium Vehicle Displays & Interfaces*, p. 204, Dearborn, MI, USA, October 2011.

[7] A. Siriviriyanun and T. Imae, "Anti-fingerprint properties of non-fluorinated organosiloxane self-assembled monolayer-coated glass surfaces," *Chemical Engineering Journal*, vol. 246, pp. 254–259, 2014.

[8] L. Y. L. Wu, S. K. Ngian, Z. Chen, and D. T. T. Xuan, "Quantitative test method for evaluation of anti-fingerprint property of coated surfaces," *Applied Surface Science*, vol. 257, no. 7, pp. 2965–2969, 2011.

[9] C. J. Harvey, R. F. LeBouf, and A. B. Stefaniak, "Formulation and stability of a novel artificial human sweat under conditions of storage and use," *Toxicology In Vitro*, vol. 24, no. 6, pp. 1790–1796, 2010.

[10] J. L. Staymates, M. E. Staymates, and G. Gillen, "Evaluation of a drop-on-demand micro-dispensing system for development of artificial fingerprints," *Analytical Methods*, vol. 5, no. 1, pp. 180–186, 2013.

[11] M. Belhadjamor, M. el Mansori, S. Belghith, and S. Mezlini, "Antifingerprint properties of engineering surfaces: a review," *Surface Engineering*, vol. 34, no. 2, pp. 85–120, 2018.

[12] H. B. Eral, D. J. C. M. 't Mannetje, and J. M. Oh, "Contact angle hysteresis: a review of fundamentals and applications," *Colloid and Polymer Science*, vol. 291, no. 2, pp. 247–260, 2013.

12

Viscoelastic and Textural Characteristics of Gels Obtained from Potato Starch Roasted under Several Temperature-Time Conditions

María D. Torres, Francisco Chenlo, and Ramón Moreira

Department of Chemical Engineering, Universidade de Santiago de Compostela, Rúa Lope Gómez de Marzoa, Santiago de Compostela E-15782, Spain

Correspondence should be addressed to María D. Torres; mariadolores.torres.perez@usc.es

Academic Editor: Ulrich Maschke

The influence of roasting conditions on the potato starch (PS) composition, solubility, crystallinity, gel-forming thermal profiles, and texture of the corresponding gels was studied. Thermorheological testing of roasted starches with (RPS) and without (RPSI) the soluble fraction was conducted on a stress-controlled rheometer. Texture profile analysis (TPA) was used to determine the RPS final gel texture. Solubility tests showed equivalent effects of starch roasting for samples treated at 190°C for 8 h (RPS190-8), 210°C for 6 h (RPS210-6), and 230°C for 4 h (RPS230-4). The apparent amylose content of RPS linearly increased with the decreasing degree of crystallinity. Elastic (G') and viscous (G'') properties of RPS depended markedly on apparent amylose content and crystallinity. G'_{gel} values of roasted starches linearly increased with the amylose content suggesting a promotion of the amylose breakdown with roasting temperature. Gels prepared with RPS roasted between 120 and 170°C exhibited intermediate strength and fully thermal reversible features. Roasting between 190 and 210°C favoured strong and fully thermal reversible gels. Although RPS230-4 gels exhibited similar strength to RPS190-8 and RPS210-4, no fully thermal reversible gels were obtained. The soluble fraction removal led to a drop the RPSI gels strength with increasing roasting temperature. No water syneresis was identified for RPS gels during one-week ageing, except for RPS230 gels. Relationships between textural parameters from TPA and viscoelastic gel properties determined by rheology were established.

1. Introduction

Potato starch (PS) is one of the most used biopolymers for industrial starch production, since the extraction procedure is simple in comparison to cereal starches [1]. This starch gains attention by its relatively low price, gluten-free features, abundance, and its renewable and biodegradability characteristics [2]. It is well known that native starches provide viscous, sticky, and cohesive pastes when their aqueous dispersions are heated and gels when above pastes are cooled down [3]. Native starches exhibit thermal decomposition and low shear stress resistance. Modified starches are designed to overcome the shortcomings of native starches, such as loss of viscosity and thickening power upon cooking and storage, syneresis, or retrogradation [4].

Starch modification can be made using physical, chemical, or biological methods [3]. Chemical or enzymatic modifications were extensively studied to extend the native starch applications, although physical modifications throughout the so-called "green chemistry" are currently preferred. The technological potential and industrial utility of native starches are enhanced by adequate modification [2]. Native and modified starches have been used as thickening agents in food processing [5] or in screen printing on textile [6], as glue in the adhesive industry [7], and as binding agents in pharmaceutical [8]. Roasted starchy materials could be an attractive alternative that have gained attention in the last years as the matrix of energy foods by high-performance ultrarunners and athletes [5]. Within nonfood applications, the use of roasted starches can be found in multiple industrial applications such as adhesives, inks, paper, textiles, insecticides, and leather, among others [8, 9]. The roasting is the break down process to form dextrins by heating dry starch, alone or in the presence of an acid catalyst, until a desired conversion degree is obtained [8]. The solubility index has been reported as a simple method to follow the roasting process [10]. The main variables involved in starch roasting are the starch source used, the temperature applied (over the range from 100 to 230°C) and the conversion time (minutes to hours) [5]. Depending on the final use, starch must be roasted in such a way as to ensure that the structure fits the production process [8]. It should be remarked that these starch modifications require present industrial disadvantages from the economical point of view since these require thermal energy (expensive) and long-time processing. Consequently, achieving optimal temperature-time (i.e., energy consumption) operating conditions, depending on the properties required, is an important aspect for the industry.

The knowledge of the thermomechanical behaviour during the preparation of starchy-based gels offers essential information to select the optimum processing conditions to achieve well-defined and tailor-made final products, in particular, for gels from modified starches [5]. The rheology of native starch gels from different sources such as maize and potato starches was extensively studied [11]. Even correlations between gel rheology, structure, and final product quality were found for native starchy gelled systems. Fruit leathers incorporated with roasted sweet potato starch (between 150 and 250°C) have been proposed as a health alternative (low caloric) to sweet fruit snacks made with starch, high fructose syrups, and sugar [12]. In vitro digestibility, crystallinity, rheological, thermal, particle size distribution, and morphological characteristics of toasted ground white and blue maize flours were also comprehensively studied [13]. Nevertheless, the impact of roasting conditions in the absence of any added catalyst on the starch dispersions during gelatinization and on final mechanical gel behaviour needs to be determined [14]. Other starchy materials obtained from toasted ground maize (pinole) are widely consumed by different ethnic groups of Northern Mexico and Southern USA [13]. Latter authors explained that this toasted starchy matrix (about 250–270°C) can be combined with water, spices, or brown sugar; made into a drink and an oatmeal-like paste; or baked to form a more-portable "cake." Given the specific starch functionalization, novelty products are unceasingly being developed and characterized, leading to a noteworthy potato starch value addition.

The effects of roasting of flour or isolated starches in physicochemical, mechanical, and morphological properties have been scarcely addressed. In this context, the main objective of this work is to study the effect of PS roasting conditions on its gel-forming thermal profiles (so-called gel-forming/maturation kinetics), final thermomechanical properties, and thermoreversible properties, which is critically relevant to the industrial processing. A comprehensive research on the thermorheology and texture of native PS and roasted PS (RPS) in a broad range of temperatures was carried out. Rheological testing at small amplitude oscillatory shear (SAOS) and texture profile analysis (TPA) was used to evaluate the system structure evolution and the final gel texture, respectively. Fundamental chemistry (i.e., moisture content, total starch content, apparent amylose ratio, or damage starch), cold solubility, crystallinity, and particle size features of tested starches were also undertaken for further insight.

2. Materials and Methods

2.1. Materials. A commercial potato starch (PS) (CAS 9005-84-9, Panreac Química, S.A., Spain) was selected as a raw material for this research as it is a commonly used reference starch. All reagents used for the different tests were of analytical grade.

2.2. Roasting Process. Commercial PS (moisture content: 24.3 ± 0.3%, dry basis (d.b.)) was dried at 40°C in a vacuum oven before roasting to reduce the moisture content (around 6.1 ± 0.8%, d.b) to values commonly used in roasting starch processes [10]. Then, the starch samples (25 g) placed on a Petri dish in thin layer were roasted in an air convection oven (UFB500, Memmert, Germany) at different temperatures (120, 150, 170, 190, 210, and 230°C), labelled as RPS120, RPS150, RPS170, RPS190, RPS210, and RPS230, respectively, and times (2, 4, 6, 8, 16 and 24 h) to obtain a set of 36 roasted potato starches (RPS). Note here that time will be labelled as "– and the corresponding time" after temperature indication (RPS120-2, RPS120-4, ... , RPS230-24).

2.3. Proximate Physicochemical Properties

2.3.1. Solubility Index. The starch solubility index was obtained following a modified protocol [15]. Starch samples (around 0.2 g, M_0) were dispersed in 10 mL distilled water and vortexed for 1 min. Starch dispersions were kept at room temperature for 30 min. Then, samples were centrifuged at 550 *g* for 15 min. The supernatant was decanted and dried at 110°C until constant weight (M_1). The solubility index of the starch (g/100 g) was determined as $M_1/M_0 \cdot 100$.

2.3.2. Composition. Moisture content was determined according to the standard method (925.10) [16]. Total starch amount, apparent amylose content, and damage starch level were determined using enzymatic test kits (Megazyme Co., Wicklow, Ireland) following standard procedures [17].

2.3.3. Crystallinity Degree. The determination of the crystalline structure of tested potato starches was conducted on a diffractometer (X-ray PW1710, Philips, Netherlands) by means of CuKα radiation (λ: 0.154 nm) operating at 20 mA and 40 kV. The starches were scanned at room temperature through the 2θ (diffraction angle) between 2 and 50° at a fixed scanning rate (4°/min). Roasted potato starch crystallinity evolution as a function of time of roasted potato starches was also determined. The relative degree of crystallinity was calculated from the ratio of diffraction peak and total diffraction area following the method previously detailed [18].

2.3.4. Particle Size Measurements. The particle size distribution and weight average particle diameter (D_w) of the above-mentioned starches were determined following standard sieving procedures [19]. Standard sieves (200, 125, 80, 63, and 40 μm) were employed. The fractions retained in each sieve were separated and weighed.

2.4. Preparation of Samples for Textural and Rheological Measurements. Starches were dispersed in distilled water (30% *w*/*w*, stirring at 3000 rpm for 15 min) on a magnetic stirrer at 25°C following the conditions previously selected for other starchy materials [20]. The selected starch content is in the range of those previously studied for concentrated starch gels (from 20 to 70% *w*/*w*) [21]. In order to provide further insights on the effect of starch water insoluble fraction on the thermorheology of roasted potato starch (RPS) gels, aqueous dispersions made from roasted starches, where the soluble fraction was previously removed, were also prepared. The aqueous RPS dispersions were filtered and separated at 25°C into insoluble (RPSI) and soluble fractions. The soluble fraction was removed, and RPSI was washed several times with distilled water. After drying at 40°C, the corresponding dispersions were prepared as above-indicated.

For rheological measurements, aqueous starch dispersions (PS, RPS, and RPSI) were directly placed into the rheometer measuring system. This allows tracking of the starch gelatinisation in situ, avoiding further gel matrix disturbances. For texture measurements, stirred PS and RPS dispersions (3000 rpm for 15 min) prepared in cylindrical flasks (4.5 cm height and 2.5 cm diameter) were immediately heated up to 80°C as reported elsewhere [22]. Briefly, samples were held at 80°C for 60 min in order to prepare a set of gels. Then, samples were cooled at room temperature for 1 hour prior to being placed in a refrigerator (5°C) for 24 h in order to allow the full maturation of the gels before performing the texture assays.

2.5. Rheology. Rheological measurements (in terms of storage (G′) and loss (G″) moduli) were performed on a stress-controlled rheometer (MCR301, Anton Paar Physica, Austria) using a Couette geometry (cup diameter 26 mm and bob height 25 mm). Aqueous potato starch dispersions were placed into the Couette cup and sealed with paraffin oil to avoid water evaporation during tests. In order to define the linear viscoelastic region, stress sweeps were made on both starch dispersions and gels at different frequencies (0.1, 1, and 100 Hz) over the range of 0.1 to 1000 Pa. All temperature, time, and frequency sweeps were conducted within the linear viscoelastic region.

Potato starch dispersions were heated from 25 to 80°C at a rate of 1°C/min. In order to stabilise the samples, time sweep assays (0.1 Hz) were thereupon conducted at 80°C for 60 min. Samples were cooled down to 25°C (1°C/min). All the above-mentioned tests were performed at 0.1 Hz and 5 Pa. Gel maturation was monitored by time sweeps (1 Hz, 30 Pa) after cooling at 25°C for 60 min. Frequency sweeps (over the range of 0.01 to 100 Hz, 30 Pa) were thereupon conducted at 25°C without disturbing the gel. All these stages were repeated following an experimental protocol to study the thermal reversibility of obtained gels [23].

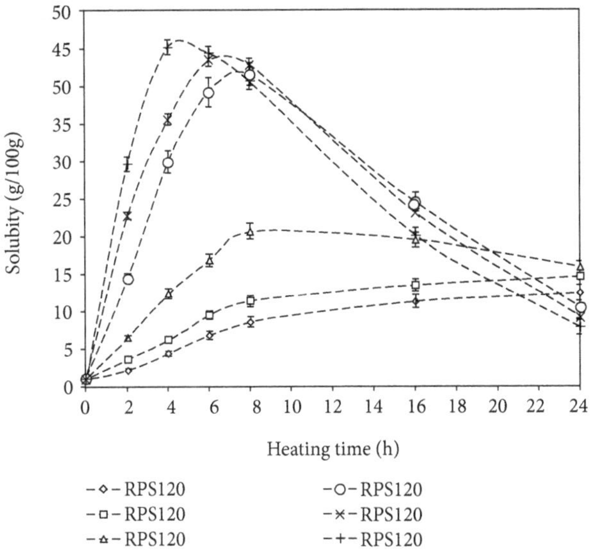

FIGURE 1: Influence of time-temperature roasting conditions on the solubility of potato starch (PS). Note here that in these subsequent plots, error bars are not included whether the uncertainty in data magnitudes is smaller than the symbol size.

2.6. Texture. Texture profile analysis (TPA) measurements were conducted on a texture analyser (TA-XT plus, Stable Micro Systems, UK) with a load cell of 5 kg as reported elsewhere [24]. For this purpose, starch gel samples prepared in cylindrical flasks filled up to 80% (4.5 cm height and 2.5 cm diameter) were submitted to a double penetration using a cylindrical probe P/05R (0.6 mm radius, 2 mm/s crosshead speed, and 5 mm penetration). Experiments were performed 24 h after sample preparation to allow full gel maturation. Before textural testing, samples were equilibrated for about 1 h at 25°C. Note here that hardness (peak force of the first cycle), adhesiveness (negative area of the first bite), springiness (related to elasticity), cohesiveness (ratio between the areas of both peak forces), and chewiness (hardness × cohesiveness × springiness) were determined.

2.7. Syneresis. Starch gel samples, prepared as those used in textural tests, were stored in centrifuge tubes in the fridge (5°C) for 7 days. Each sample was centrifuged (2200 *g* for 15 min), and the percent of water release in relation to the gel weight represented the syneresis [15].

2.8. Statistical Analyses. All tests were performed in triplicate, and data sets were evaluated by means of one-way analysis of variance (ANOVA), using PASW Statistics (IBM SPSS Statistics 22.0.0). A Scheffé test was performed to distinguish means with 95% confidence ($p \leq 0.05$).

3. Results and Discussion

3.1. Physicochemical Features

3.1.1. Solubility. Figure 1 shows the solubility of potato starch after roasting at different temperature and time conditions. RPS120 and RPS150 exhibited the lowest solubility values,

Table 1: Proximate physicochemical features of tested potato starch (PS) (30% *w/w*) roasted at different temperatures (120, 150, 170, 190, 210, and 230°C) for the time corresponding to the highest solubility values for each tested roasting temperature. The corresponding gelatinization peak temperature (T_p) is also displayed.

	PS	RPS120-8	RPS150-8	RPS170-8	RPS190-8	RPS210-6	RPS230-4
Roasting temperature	—	120	150	170	190	210	230
Roasting time, h	—	8	8	8	8	6	4
Initial moisture content, % d.b.	6.1 ± 0.8^{a}	7.1 ± 1.0^{a}	6.8 ± 1.3^{a}	6.6 ± 1.2^{a}	5.8 ± 0.6^{a}	5.5 ± 0.9^{a}	5.3 ± 1.2^{a}
Total starch content, % d.b.	98.7 ± 0.5^{a}	97.4 ± 0.4^{b}	97.1 ± 0.5^{b}	96.7 ± 0.5^{b}	93.5 ± 0.2^{c}	$92.8 \pm 0.4^{c,d}$	92.4 ± 0.3^{d}
Apparent amylose content, %	21.6 ± 0.4^{d}	23.4 ± 0.3^{c}	23.6 ± 0.2^{c}	23.2 ± 0.4^{c}	29.3 ± 0.5^{b}	$29.7 \pm 0.3^{a,b}$	30.6 ± 0.3^{a}
Damaged starch level, %	2.9 ± 0.6^{a}	2.9 ± 0.5^{a}	2.9 ± 0.5^{a}	3.0 ± 0.7^{a}	3.5 ± 0.4^{a}	3.7 ± 0.6^{a}	3.8 ± 0.5^{a}
Damaged/total starch, %	0.029 ± 0.003^{b}	0.030 ± 0.002^{b}	0.030 ± 0.002^{b}	0.031 ± 0.002^{b}	0.037 ± 0.001^{a}	0.040 ± 0.002^{a}	0.041 ± 0.002^{a}
Crystallinity level (30 days), %	22.6 ± 0.9^{a}	$15.1 \pm 0.2^{b,C}$	14.6 ± 0.3^{b}	$14.9 \pm 0.3^{b,C}$	$6.7 \pm 0.3^{c,C}$	$5.4 \pm 0.1^{c,d}$	$5.3 \pm 0.4^{d,C}$
0 day*	—	$17.7 \pm 0.4^{a,B}$	—	$17.2 \pm 0.2^{a,B}$	$10.4 \pm 0.2^{b,B}$	—	$9.8 \pm 0.1^{c,B}$
1 day*	—	$16.2 \pm 0.2^{a,A}$	—	$16.2 \pm 0.4^{a,A}$	$9.5 \pm 0.2^{b,A}$	—	$8.6 \pm 0.3^{c,A}$
3 days*		$15.2 \pm 0.3^{a,C}$	—	$15.1 \pm 0.5^{a,C}$	$6.9 \pm 0.3^{b,C}$	—	$6.1 \pm 0.3^{c,C}$
T_p (°C)	56.2 ± 0.5^{a}	50.5 ± 0.4^{b}	50.1 ± 0.3^{b}	49.6 ± 0.3^{b}	40.1 ± 0.4^{c}	39.9 ± 0.1^{c}	39.7 ± 0.3^{c}

Data are presented as mean ± standard deviation. Data values in a row with different superscript letters are significantly different at the $p \leq 0.05$ level. *Time-crystallinity dependence: measures were performed at different time periods after roasting processing. In this particular case, the significance was evaluated in columns and indicated with different ANOVA superscripts (u, v, and w).

except at 24 h where close values were identified for all samples. The first solubility increase was observed at temperatures between 170 and 190°C, becoming greater at 210 and 230°C. This solubility increase indicates that considerable roasting takes place for RPS190, RPS210, and RPS230. RPS230 gave the highest solubility (g/100 g) values (around 45 ± 0.5) of roasting closely followed by RPS210 and RPS190 (around 42.5 ± 1.0). Potato starch solubility curves featured induction periods in the roasting process, becoming gradually shorter as the roasting temperature increased (from 8 h for RPS120 to 4 h for RPS230). Earlier works [10] found similar induction times for roasted corn starches subjected at temperatures ranging from 150 to 213°C and times below 8 h, although with higher maximum solubility values (up to 80 g/100 g). These authors indicated that the solubility index is as a simple method to follow the roasting process directly related to the dextrin content. The solubility data suggests that the similar effect on starch roasting occurs for RPS190-8, RPS210-6, and RPS230-4. Thus, potato starches featuring the highest solubility values for each tested roasting temperature were selected for further physicochemical and mechanical analyses in order to observe clearer the roasting effect.

Solubility curves of RPS190, RPS210, and RPS230 displayed an initial step increase, and after the maximum the solubility is achieved, the plots showed a downward tendency. This solubility loss was slightly promoted at high temperatures. All RPS solubility curves tend to reach a common solubility value after 24 h (long time) subjected to roasting. This could be explained assuming that larger molecular weight glucans are formed from the fragments produced in the early starch roasting stages [25] or that the nonsolubilisation is owing to the formation of decomposition products and their noncarbohydrate nature polymers [10].

3.1.2. Composition. Composition experiments were carried out in roasted starches after the corresponding times that led to the largest solubility values. Table 1 collects the proximate physicochemical features of tested starches. The average moisture content of the starch before roasting was about 6.1 ± 0.8%, d.b. The average final moisture content (%, d.b.) after roasting was around 2.5 ± 1.1, not exhibiting statistical differences. As expected, total starch content was relatively high in all cases, with a significant decrease with increasing roasting temperature. This result suggests that during roasting, starch is thermally degraded [26]. This is supported by the increase in the damaged/total starch ratio with rising roasting temperature, even though low damaged starch levels were observed in all cases. The determined damaged starch values are in the range of those reported for native starches extracted from gluten-free flours (below 3%) [5]. Apparent amylose content (%) exhibited intermediate values increasing significantly with increasing roasting temperatures. The largest apparent amylose content modifications (around 28%) were identified between RPS170 and RPS190, which suggest that the roasting process is favoured around 190°C [26]. The increase in apparent amylose content after roasting was reported for different starchy cereals and explained because heating promotes the formation of amylodextrins (linear dextrin or short chained amylose) [27].

3.1.3. Crystallinity. The degree of crystallinity (%) of tested starches is presented in Table 1. Crystallinity values statistically decreased with increasing crystallinity measurement times, achieving steady values after three days as evidenced by the negligible variation with those measured after one month. Analysing steady crystallinity values, it was noticed that crystallinity degree significantly decreased with increasing roasting temperature. The largest crystallinity drop was identified between 170 and 190°C (around 36%). This is consistent with the amylose trends above-mentioned, since the crystallinity drop is expected to be related to amylopectin drop (responsible to the amorphous starch features) [25]. Crystallinity values exhibited a linear relationship (slope -0.56 ± 0.05, $R^2 > 0.988$) with amylose content. This result agreed with those reported for native corn starches with different apparent amylose contents (from 0 to 40%), where a linear relationship (slope −1.01) was observed between crystallinity and apparent amylose content, changing crystal type from A to C [24]. In our case, the characteristic B standard pattern of PS (with the strongest diffraction peak at around 17° 2θ, a few small peaks at around 2θ values of 20°, 22° and 23°, and an additional peak at 5° 2θ) evolves towards a more amorphous structure with increasing roasting temperature.

3.1.4. Particle Size. In all cases, the average particle size of tested starches was around $61.2 \pm 2.4\,\mu$m without significant differences among the samples. The most important mass fraction (58.6%) was identified between 80 and 63 μm. The second population (32.1%) was the particles with sizes from 63 to 40 μm, followed by the fraction (9.3%) of particle sizes lower than 40 μm. The above values are consistent with those (over the range from 5 to 100 μm) found for starches isolated from different potato varieties [28] and suggested as suitable for the gel formation [29].

3.2. Thermorheology

3.2.1. Thermal Heating Profiles. Roasted starch samples with the largest solubility values (i.e., RPS120-8, RPS190-8, and RPS230-4) for each roasting temperature were selected to the thermorheological assessment (Figure 2). The thermal heating profile (Figure 2(a)) of aqueous roasted starch dispersions did not show the inflection point corresponding to the initial gelatinization temperature (T_0). This behaviour is usually observed in native starches [29]. Instead, in our work, it was identified as a rapid increase of both G′ and G″ moduli with increasing temperature achieving a peak around 40.1 ± 0.4°C for RPS190-8, which represents the gelatinization peak temperature (T_p), remaining practically invariant above this temperature. G′ was larger than G″ over the whole tested temperature range. Similar trends were obtained for all aqueous RPS dispersions, although the values of T_p were shifted to lower temperatures (from 56.2 ± 0.5°C for PS to 39.7 ± 0.4°C for RPS230-4) with rising roasting temperature, but without a uniform trend. The largest temperature drop (around 9.5 ± 0.3°C) was found between RPS170-8 and RPS190-8, which is consistent with physicochemical achievements, indicating that roasting conditions influence the gelatinization temperatures. The drop in gelatinization temperatures was also reported for roasted oat flours [30]. These authors indicated that this fact could be attributed to the presence of some pregelatinized starch in roasted systems. Here, no significant differences in T_p values were observed below RPS170-8 and above RPS190-8. The T_p values from RPS190-8 to RPS230-4 are similar to the initial gelatinization temperature (T_g) identified for PS and lower than those T_p values reported for different native potato starches (45 ± 1°C) [26]. Another relevant characteristic for RPS is the increase of the thermal stability after T_p (with G′ and G″ moduli almost parallel and constant with the increase of temperature) in comparison to other native starches evaluated at similar [21] and lower (<10%) [22] polymer content. RPS190-8 gels immediately attained G′ and G″ steady values after T_p and consequently exhibited the largest themostability features. This is especially important for potato starches which are usually less thermostable than other commonly used native starches as maize or alternative sources such as chestnut [23].

3.2.2. Thermal Cooling Profiles. The thermal cooling profiles for RPS120-8, RPS190-8, and RPS230-4 gels are presented in Figure 2(b), as representative of tested systems. All samples were stabilised for 60 min at 80°C previously to cooling ramps, exhibiting G′ > G″ with constant values. During the cooling process, the values of both moduli of RPS gels rose for all gels. A moderate reinforcement of the gel structure during cooling was also reported for native maize starches [5] and other native starchy pastes [25]. This is consistent with the physicochemical features observed for roasted starches, since larger roasting temperatures involve lower crystallinity and consequently higher apparent amylose content favouring greater molecular entanglements and the easier formation of a gel network [31]. This behaviour could also justify the lowest reinforcement for systems formulated >RPS190-8, which exhibits the highest apparent amylose content.

3.2.3. Maturation Profiles. The maturation profiles in terms of the viscous and elastic modulus evolution with time for above RPS gels (i.e., RPS120-8, RPS190-8, and RPS230-4) are shown in Figure 2(c). The plot displays the G′ and G″ moduli evolution with time at 25°C, as examples of tested samples. Particularly, RPS190-8 reached dynamic gel network equilibration (G′ and G″ stability) instantaneously with steady values for G′ (around 1.6 10^5 Pa) and G″ (around 2.4 10^4 Pa). No differences (<1 min) in the time maturation were observed for gels formed between RPS190-8 and RPS230-4. Longer times (<10 min, RPS120-8, Figure 2(c)) were required for the structure maturation of gels prepared <RPS170-8, even though the maturation times of all RPS gels were much lower than those (around 30 min) reported for concentrated (between 10 and 40% *w/w*) gels prepared with native starches from different sources [32].

In G′ and G″ magnitudes, the viscoelastic features of RPS gels depended notably on roasting conditions, increasing (around 7 times) the gel strength with rising roasting temperature (G'_{gel} (1Hz) from 6.5 10^4 Pa for RPS120-8 to 3.5 10^5 Pa for RPS230-4) achieving similar tan δ (G″/G′) values in all cases. These moduli values are larger than those reported for gel maturation profiles made from modified potato starch (30% *w/w*) under different gelatinization degrees, where G'_{gel} was in all cases below 5.7 10^4 Pa [23].

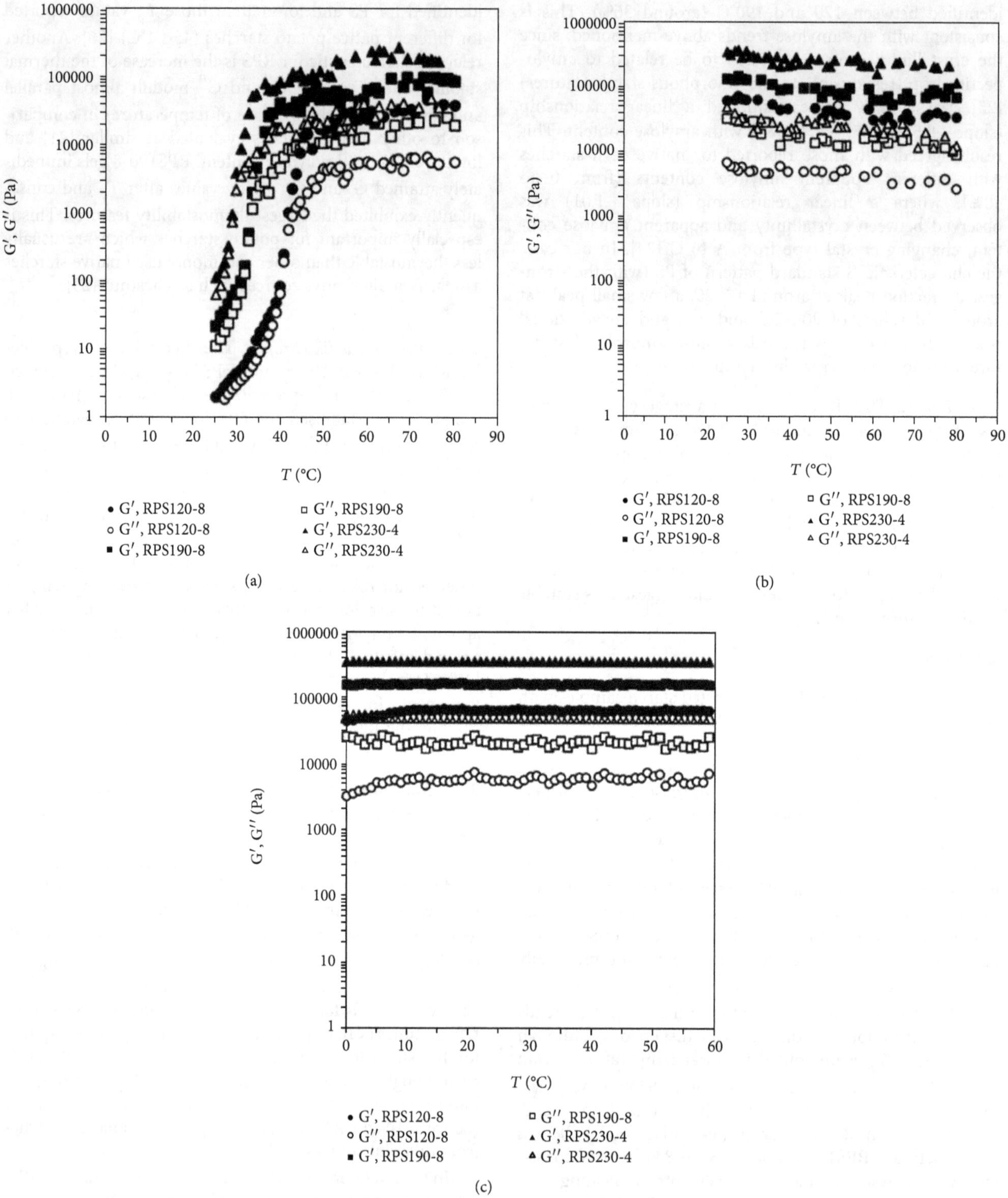

FIGURE 2: Thermorheological features of aqueous dispersions of potato starch (30% *w/w*) roasted at 120°C for 8 h (RPS120-8), 190°C for 8 h (RPS190-8), and 230°C for 4 h (RPS230-4): (a) temperature sweep on heating (from 25 to 80°C at 1°C/min); (b) temperature sweep on cooling (from 80 to 25°C at 1°C/min); and (c) time sweep (at 25°C).

3.2.4. Mechanical Spectra. Figure 3 shows the mechanical spectrum in terms of viscous and elastic modulus evolution with frequency above RPS gel obtained at 25°C, as a representative example. Note here that the mechanical spectra of untreated samples were previously reported for the authors [23]. All gels prepared exhibited a typical gel

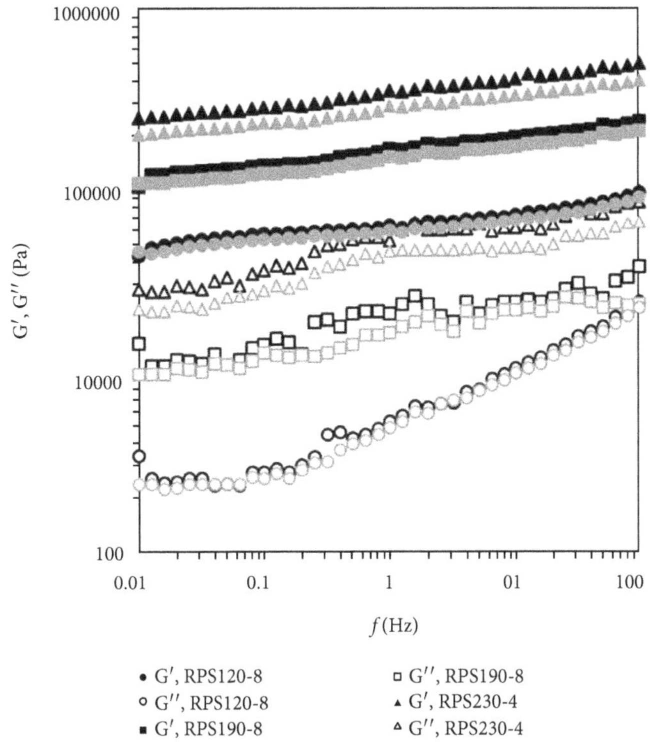

FIGURE 3: Mechanical spectra at 25°C of selected gels prepared from RPS120-8, RPS190-8, and RPS230-4. Note here that grey symbols correspond with the mechanical spectra data collected from the second heating/cooling cycle.

spectrum (G′ > G″ and both modulus frequency independent). This behaviour agrees with the solid characteristics found for other starchy gelled systems such as those made from maize, rice, or chestnut starches at lower starch content (<10% *w/w*) [20, 22]. Concerning strength of RPS gels, the magnitude of G′ and G″ moduli at fixed angular frequency depended greatly on starch roasting temperature. A general strength evaluation of the obtained RPS gels can be given in terms of G'_{gel} (1 Hz) Figure 4(a). G'_{gel} increased with increasing roasting temperature, mainly above RPS190-8. The general behaviour agrees with those results found for gels formed from native corn or rice starches with different apparent amylose contents which occur here with RPS [33], explaining that higher apparent amylose content leads to stronger gels because of their greater molecular entanglements between linear polymer chains and less super globes that are easier to move than long linear chains. In the case of roasted starches above 190°C, the enhancement of the gel strength could be explained because roasting temperature favours the formation of short amylose chains. Synergistic effects on the viscous behaviour of gelled systems of potato starches prepared with fractionated amyloses have been previously reported [34]. A distribution of molecular weights promoted by temperature and acid hydrolysis has been also reported for potato starch and xanthan mixtures [35]. Viscoelastic values for PS and roasted starches from RPS120-8 to RPS170-8 are in the range of those found for food applications such as commercial savoury/sweet gelled systems [22] or nonfood applications such as cosmetics and health personal care products [8]. RPS190-8, RPS210-6, and RPS230-4 exhibited viscoelastic values consistent with those reported for gelled formulations used in biofilm-based products [36].

Further insight on the behaviour of concentrated starch gels is critically relevant in order to improve the processing conditions of in multicomponent systems with high starch content as flour doughs. Figure 4(b) shows the G'_{gel} evolution for gels prepared with RPSI. The mechanical spectrum for RPSI190-8, as an example, is also included, exhibiting weaker gels than those formed with both fractions. It is clear that the RPSI gel strength decreased with increasing roasting temperature. The largest drop (around 45%) is observed between RPSI170-8 and RPSI190-8. The removal of the soluble fraction led to a drop in the apparent amylose content (%, *w/w*) (19.3 ± 0.3 for RPSI120-8, 18.4 ± 0.5 for RPSI150-8, 14.5 ± 0.3 for RPSI170-8, 6.8 ± 0.4 for RPSI190-8, 6.6 ± 0.4 for RPSI210-6, and 5.4 ± 0.4 for RPSI230-4), which can be one of the main factors explaining the gel strength reduction. Note here that the G'_{gel} values exhibited a linear dependence with the apparent amylose content ($R^2 > 0.98$). This behaviour supports the hypothesis that the roasting temperature promotes the breakdown of the apparent amylose because the shorter chains are solubilized and the largest ones remain in the insoluble fraction [33]. Therefore, the behaviour of the RPSI depends only on the apparent amylose content (without influence of apparent amylose modifications). Thus, the formulation of potato starch-

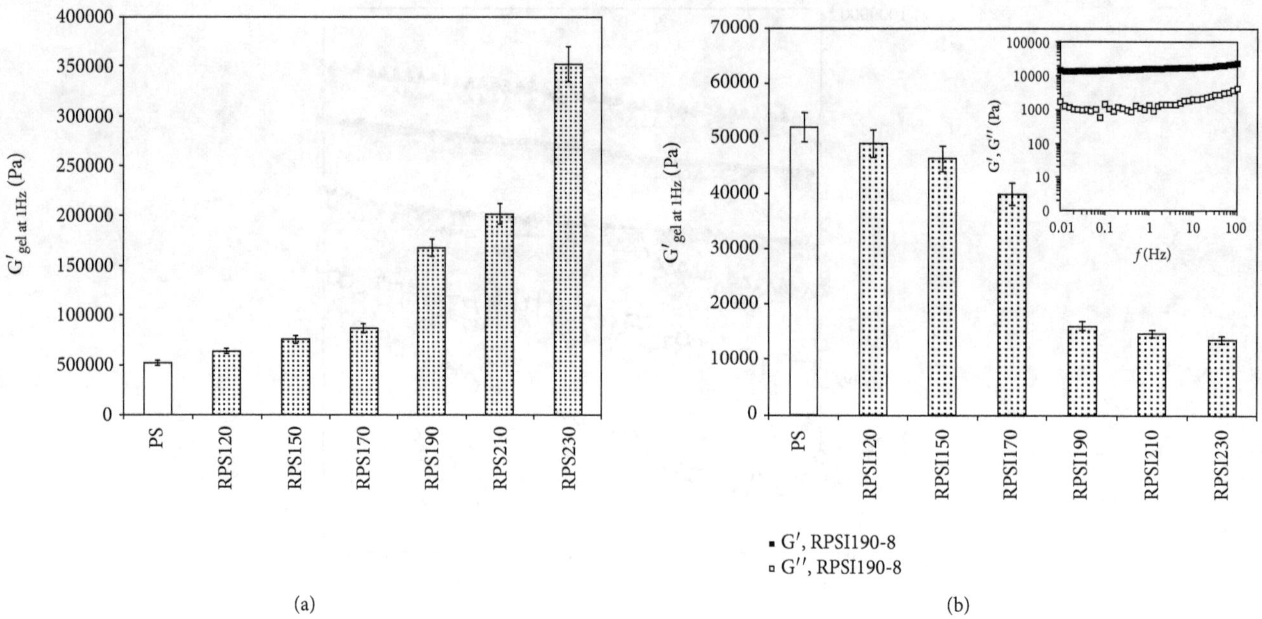

FIGURE 4: Influence of roasting conditions on the mechanical behaviour of the final potato starch gels (G'_{gel} at 1 Hz, Pa) without (a) and with (b) removal of the soluble fraction after starch roasting.

TABLE 2: Texture parameters of tested potato starch gels (30% *w*/*w*) roasted at different temperatures (120, 150, 170, 190, 210, 230°C).

Gels	Hardness (kg)	Adhesiveness (kg s)	Springiness (−)	Cohesiveness (−)	Chewiness (kg)
PS	2.74 ± 0.08^{c}	0.04 ± 0.01^{b}	0.69 ± 0.01^{c}	0.29 ± 0.03^{b}	1.70 ± 0.12^{e}
RPS120-8	$2.78 \pm 0.07^{b,c}$	0.04 ± 0.01^{b}	$0.71 \pm 0.02^{b,c}$	0.39 ± 0.02^{a}	2.32 ± 0.12^{d}
RPS150-8	$2.83 \pm 0.08^{b,c}$	0.04 ± 0.01^{b}	$0.73 \pm 0.01^{b,c}$	0.39 ± 0.02^{a}	$2.43 \pm 0.11^{c,d}$
RPS170-8	2.94 ± 0.09^{b}	$0.05 \pm 0.01^{a,b}$	0.74 ± 0.02^{b}	0.40 ± 0.01^{a}	2.62 ± 0.13^{c}
RPS190-8	3.25 ± 0.06^{a}	0.07 ± 0.01^{a}	0.84 ± 0.04^{a}	0.41 ± 0.02^{a}	3.38 ± 0.08^{b}
RPS210-6	3.32 ± 0.11^{a}	0.08 ± 0.01^{a}	0.88 ± 0.09^{a}	0.42 ± 0.03^{a}	$3.67 \pm 0.09^{a,b}$
RPS230-4	3.38 ± 0.10^{a}	0.08 ± 0.01^{a}	0.93 ± 0.10^{a}	0.42 ± 0.02^{a}	3.98 ± 0.12^{a}

*Data are presented as the mean ± standard deviation. Data values in a column with different superscript letters are significantly different at the $p \leq 0.05$ level.

based gels with a broad viscoelastic features range seems to be feasible combining commonly used separation techniques and roasting conditions. Weak gelling systems with viscoelastic features similar to those obtained in this work can be greatly valuable as basis of formulations for target population with specific requirements, such as puree-based baby or elder food [37].

3.2.5. Thermal Reversibility. Figure 3 shows thermal reversible features for RPS120-8, RPS190-8, and RPS230-4 gels, as an example of the behaviour observed in tested systems. The absence of differences between the mechanical spectra obtained at the end of the first and the second thermal cycle determined the strong thermal reversible characteristics of tested gels. The largest differences were observed for gels made from RPS230-4 (around 7.5% drop in both elastic and viscous moduli), indicating no complete thermal reversibility (Figure 3). Thermal stability characteristics were found for other native starches such as maize or wheat starches [38]; however, the authors are not aware that this rheological feature has been reported for other roasted starches. To sum up, roasting temperatures between 120 and 170°C led to medium strength and fully thermal reversible gels, whereas temperatures between 190 and 210°C favour the formation of strong and fully thermal reversible gels (being the network structures formed in the sample more stable), with no completely thermal reversible gels at 230°C (Figure 3). These outcomes deliver a wide range of roasting conditions in terms of temperature (from 120 to 210°C) and time (from 4 to 8 h) for the potato starch gel processing without jeopardising the stability of the formulation.

3.3. Texture. Table 2 mentions the textural parameters determined for representative RPS gels. The macrostructure behaviour (textural features) was studied for the same selected formulations indicated in the rheological measurements. Hardness (kg) of gels statistically increased with increasing roasting temperatures. No significant variations were observed between RPS120-8 and RPS170-8 and between RPS190-8 and RPS230-4. Similar behaviour was identified for springiness, chewiness, and adhesiveness. No significant differences were noticed for cohesiveness of RPS gels. These trends are consisted with those recently reported for chestnut starch isolated from roasted kernels [39], where starches from roasted chestnuts exhibited higher values of textural parameters when compared with those of fresh or boiled chestnuts. The textural parameters of modified starch gels depend strongly on the processing conditions [25]. Tested RPS gels exhibited larger hardness values (about the double) than those reported for gels prepared with acid-treated sweet potato starches at similar starch content [5], even though the values of textural features for all tested RPS gels were in the range of those reported for commercially offered savoury gelled systems and desserts [22].

The outcomes found from textural tests are related to those obtained from small deformation rheology. Hardness exhibited a linear relationship ($R^2 > 0.98$) with the G'_{gel} (1 Hz) determined in the mechanical spectra for gels formed between RPS120-8 and RPS210-6. On the other hand, RPS230-4 gels featured higher hardness values related to the amylose breakdown above-discussed. Moreover, G'_{gel} (1 Hz) also showed a clear dependence with the ratio of the distances to each peak (D_1/D_2) determined in TPA. An exponential dependence, $G'_{gel} = 289.3\ e^{5.1D1/D2}$ ($R^2 > 0.99$), between both parameters was found. This outcome agrees with those reported for other food materials [15], where springiness (related to D_1/D_2) was correlated with elastic material features.

3.4. Syneresis. The absence of water syneresis observed in the most RPS gels aged during one week at room temperature is a critically relevant feature in gels. Only gels prepared from RPS230-4 exhibited slight water syneresis (below 3% of the total gel). Negligible release of water was considered (<1%) for other concentrated starchy gelled systems as those formulated with sorghum starch [40]. The syneresis absence involves an important industrial advantage to other modified starchy [5, 26] gelling systems where water syneresis arises during the gel formation, suggesting that roasted starches feature more stable structures. Despite the increase of viscoelastic moduli of RPS gels with increasing starch roasting temperature, no gel syneresis was identified. It seems that these modified starches are more hydrophilic, the water is more structural, and then, no free water is available. It was reported [41] that the starch retrogradation rate (related to syneresis) is enhanced by the increasing of apparent amylose content, which is consistent with the results obtained here for RPS systems.

4. Conclusions

To conclude, it should be indicated that physicochemical features, the forming profiles of the gels, and the rheological and textural characteristics of the final gels for potato starch systems strongly depend on the roasting starch conditions. The largest physicochemical, rheological, and textural differences were identified between RPS170 and RPS190 systems. The adequate and almost equivalent solubility effect on starch roasting occurs for RPS190-8, RPS210-6, and RPS230-4. Temperatures below 170°C led to RPS gels with medium strength and fully thermal reversible gel formation, while temperatures between 190 and 230°C promoted strong and fully thermal reversible gel formation, with no fully thermal reversible gels at 230°C. The critically relevant role of the starch soluble fraction was observed in the weakness of RPSI gels with increasing roasting temperature, intimately linked with the apparent amylose content reduction. Linear and exponential dependences were identified between G'_{gel} (1 Hz) and the hardness and springiness ratio (D_1/D_2), respectively. The absence syneresis in RPS gels involves a very important industrial advantage. The T_p values from RPS190-8 to RPS230-4 could allow extending the functional potato starch features, to prepare new types of gelled products such as multicomponent gels or aerated gels where certain limitations (insufficient stability towards processing and environmental conditions or significant syneresis) may occur at large/low temperatures. Overall, this work provided a wide set of potato starches with potential gelling characteristics not only for food applications (PS, < RPS170) such as matrix for savoury/sweet gelatins or pastes, but also for non-food ones (between RPS190 and RPS230), arisen from different roasting starch conditions.

Conflicts of Interest

The authors declare that they have no conflicts of interest.

Acknowledgments

The authors acknowledge the financial support (ED431B) from the Consellería de Cultura, Educación e Ordenación Universitaria de la Xunta de Galicia (Spain). We also would like to thank the use of RIAIDT-USC analytical facilities for X-ray diffraction experimental assays.

References

[1] R. Ezekiel, N. Singh, S. Sharma, and A. Kaur, "Beneficial phytochemicals in potato - a review," *Food Research International*, vol. 50, no. 2, pp. 487–496, 2013.

[2] H. Yu, Y. Cao, Q. Fang, and Z. Liu, "Effects of treatment temperature on properties of starch-based adhesives," *BioResources*, vol. 10, no. 2, 2015.

[3] E. I. Yousif, M. G. E. Gadallah, and A. M. Sorour, "Physico-chemical and rheological properties of modified corn starches and its effect on noodle quality," *Annals of Agricultural Sciences*, vol. 57, no. 1, pp. 19–27, 2012.

[4] J. Waterschoot, S. V. Gomand, E. Fierens, and J. A. Delcour, "Starch blends and their physicochemical properties," *Starch-Starke*, vol. 67, no. 1-2, pp. 1–13, 2015.

[5] M. Witczak, R. Ziobro, L. Juszczak, and J. Korus, "Starch and starch derivatives in gluten-free systems - a review," *Journal of Cereal Science*, vol. 67, pp. 46–57, 2016.

[6] J. G. Cook, *Handbook of Textile Fibres*, Woodhead Publishing Limited, Cambridge, UK, 2009.

[7] M. A. Radeloff and R. H. F. Beck, "Principles of industrial scale chemical starch modification," *Zuckerindustrie*, vol. 137, pp. 222–226, 2012.

[8] J. BeMiller and R. Whistler, *Starch: Chemistry and Technology (Third Edit)*, Elsevier, New York, USA, 2009.

[9] F. F. Dias, "Starch: perspectives and opportunities," *Journal of Scientific and Industrial Research*, vol. 58, pp. 403–413, 1999.

[10] H. C. Srivastava, R. S. Parmar, and G. B. Dave, "Studies on dextrinization. Part I. Pyrodextrinization of corn starch in the absence of any added catalyst," *Starch*, vol. 22, no. 2, pp. 49–54, 1970.

[11] D. Zhang, T. Mu, and H. Sun, "Calorimetric, rheological, and structural properties of potato protein and potato starch composites and gels," *Starch-Stärke*, vol. 69, no. 7-8, article 1600329, 2017.

[12] L. Effah-Manu, I. Oduro, and A. Addo, "Effect of dextrinized sweet potatoes on the physicochemical and sensory quality of infra-red dried mango leather," *Journal of Food Processing & Technology*, vol. 4, no. 5, 2013.

[13] Y. Carrera, R. Utrilla-Coello, A. Bello-Pérez, J. Alvarez-Ramirez, and E. J. Vernon-Carter, "In vitro digestibility, crystallinity, rheological, thermal, particle size and morphological characteristics of pinole, a traditional energy food obtained from toasted ground maize," *Carbohydrate Polymers*, vol. 123, pp. 246–255, 2015.

[14] S. Lagarrigue and G. Alvarez, "The rheology of starch dispersions at high temperatures and high shear rates: a review," *Journal of Food Engineering*, vol. 50, no. 4, pp. 189–202, 2001.

[15] K. Bashir, T. L. Swer, K. S. Prakash, and M. Aggarwal, "Physico-chemical and functional properties of gamma irradiated whole wheat flour and starch," *LWT-Food Science and Technology*, vol. 76, pp. 131–139, 2017.

[16] AOAC, *Official Methods of. Analysis*, Association of Official Analytical Chemistry, Washington, DC, 17th edition, 2000.

[17] AACC, *Approved Methods of Analysis*, American Association of Cereal Chemistry, Eagan, MN, 11th edition, 2010.

[18] P. Correia, L. Cruz-Lopes, and L. Beirão-da-Costa, "Morphology and structure of chestnut starch isolated by alkali and enzymatic methods," *Food Hydrocolloids*, vol. 28, no. 2, pp. 313–319, 2012.

[19] ASAE, *Methods for Determining and Expressing Fineness of Feed Materials by Sieving*, American Society of Agricultural Engineers, St Joseph, MI, 1995.

[20] M. D. Torres, P. Fradinho, A. Raymundo, and I. Sousa, "Thermorheological and textural behaviour of gluten-free gels obtained from chestnut and rice flours," *Food and Bioprocess Technology*, vol. 7, no. 4, pp. 1171–1182, 2014.

[21] A. C. Eliasson and L. Bohlin, "Rheological properties of concentrated wheat starch gels," *Starch*, vol. 34, no. 8, pp. 267–271, 1982.

[22] M. C. Nunes, A. Raymundo, and I. Sousa, "Rheological behaviour and microstructure of pea protein/κ-carrageenan/starch gels with different setting conditions," *Food Hydrocolloids*, vol. 20, no. 1, pp. 106–113, 2006.

[23] M. D. Torres, F. Chenlo, and R. Moreira, "Rheological effect of gelatinisation using different temperature-time conditions on potato starch dispersions: mechanical characterisation of the obtained gels," *Food and Bioprocess Technology*, vol. 11, no. 1, pp. 132–140, 2018.

[24] M. C. Nunes, A. Raymundo, and I. Sousa, "Effect of thermal treatment and composition on the mechanical properties of pea/kappa-carrageenan/starch desserts," in *Gums and Stabilisers for the Food Industry*, P. A. Williams and G. O. Phillips, Eds., vol. 12, pp. 54–64, Royal Society of Chemistry, London, UK, 2004.

[25] S. C. Alcázar-Alay and M. A. A. Meireles, "Physicochemical properties, modifications and applications of starches from different botanical sources," *Food Science and Technology*, vol. 35, no. 2, pp. 215–236, 2015.

[26] A. F. el Sheikha and R. C. Ray, "Potential impacts of bioprocessing of sweet potato: review," *Critical Reviews in Food Science and Nutrition*, vol. 57, no. 3, pp. 455–471, 2015.

[27] I. A. Wani, D. S. Sogi, A. M. Hamdani, A. Gani, N. A. Bhat, and A. Shah, "Isolation, composition, and physicochemical properties of starch from legumes: a review," *Starch-Stärke*, vol. 68, no. 9-10, pp. 834–845, 2016.

[28] H. Wang, B. Zhang, L. Chen, and X. Li, "Understanding the structure and digestibility of heat-moisture treated starch," *International Journal of Biological Macromolecules*, vol. 88, pp. 1–8, 2016.

[29] J. Xing, D. Li, L. Wang, and B. Adhikari, "Temperature thresholds and time-temperature dependence of gelatinization for heat-moisture treated corn starch," *Journal of Food Engineering*, vol. 217, pp. 43–49, 2018.

[30] H. S. Gujral, P. Sharma, and S. Rachna, "Effect of sand roasting on beta glucan extractability, physicochemical and antioxidant properties of oats," *LWT-Food Science and Technology*, vol. 44, no. 10, pp. 2223–2230, 2011.

[31] H. Carrillo-Navas, C. Hernández-Jaimes, R. G. Utrilla-Coello, M. Meraz, E. J. Vernon-Carter, and J. Alvarez-Ramirez, "Viscoelastic relaxation spectra of some native starch gels," *Food Hydrocolloids*, vol. 37, pp. 25–33, 2014.

[32] P. D. Orford, S. G. Ring, V. Carroll, M. J. Miles, and V. J. Morris, "The effect of concentration and botanical source on the gelation and retrogradation of starch," *Journal of the Science of Food and Agriculture*, vol. 39, no. 2, pp. 169–177, 1987.

[33] T. Sasaki, K. Kohyama, and T. Yasui, "Effect of water-soluble and insoluble non-starch polysaccharides isolated from wheat flour on the rheological properties of wheat starch gel," *Carbohydrate Polymers*, vol. 57, no. 4, pp. 451–458, 2004.

[34] J. L. Jane and J. F. Chen, "Effect of amylose molecular size and amylopectin branch chain length on paste properties of starch," *Cereal Chemistry*, vol. 69, pp. 60–65, 1992.

[35] I. G. Mandala and E. D. Palogou, "Effect of preparation conditions and starch=xanthan concentration on gelation process of potato starch systems," *International Journal of Food Properties*, vol. 6, no. 2, pp. 311–328, 2003.

[36] B. W. Peterson, Y. He, Y. Ren et al., "Viscoelasticity of biofilms and their recalcitrance to mechanical and chemical challenges," *FEMS Microbiology Reviews*, vol. 39, no. 2, pp. 234–245, 2015.

[37] M. D. Alvarez and W. Canet, "Dynamic viscoelastic behavior of vegetable based infant purees," *Journal of Texture Studies*, vol. 44, no. 3, pp. 205–224, 2013.

[38] M. Stasiak, M. Molenda, I. Opaliński, and W. Błaszczak, "Mechanical properties of native maize, wheat, and potato starches," *Czech Journal of Food Sciences*, vol. 31, no. 4, pp. 347–354, 2013.

[39] L. Kan, Q. Li, S. Xie, J. Hu, Y. Wu, and J. Ouyang, "Effect of thermal processing on the physicochemical properties of chestnut starch and textural profile of chestnut kernel," *Carbohydrate Polymers*, vol. 151, pp. 614–623, 2016.

[40] H. Singh, N. S. Sodhi, and N. Singh, "Structure and functional properties of acetylated sorghum starch," *International Journal of Food Properties*, vol. 15, no. 2, pp. 312–325, 2012.

[41] Y. Ai and J. L. Jane, "Understanding starch structure and functionality," in *Starch in Food Structure, Function and Applications*, M. Sjöö and L. Nilsson, Eds., Woodhead Publishing Series in Food Science, Technology and Nutrition, UK, 2018.

13

Structural Characterization, Antioxidant Activity, and Biomedical Application of *Astragalus* Polysaccharide Degradation Products

Jian-Min Wang, Xin-Yuan Sun, and Jian-Ming Ouyang

Institute of Biomineralization and Lithiasis Research, Jinan University, Guangzhou 510632, China

Correspondence should be addressed to Xin-Yuan Sun; sunxinyuan1985@163.com and Jian-Ming Ouyang; toyjm@jnu.edu.cn

Academic Editor: Mingqiang Li

To study the antioxidant capacity of *Astragalus* polysaccharides (APS) with different molecular weights, we used hydrogen peroxide to degrade original *Astragalus* polysaccharide (APS0) with an initial molecular weight of 11.03 kDa and obtained three degraded polysaccharides with molecular weights of 8.38 (APS1), 4.72 (APS2), and 2.60 kDa (APS3). The structures of these polysaccharides were characterized by ^{1}H NMR, ^{13}C NMR, FT-IR, and GC/MS. The degradation process did not cause significant changes in the main chain structure of APS. The monosaccharide component of APS before and after degradation was slightly changed. The antioxidant ability *in vitro* (removing hydroxyl and ABTS radicals and reducing ability) and in cells (superoxide dismutase and malondialdehyde generation) of these polysaccharides is closely related to their molecular weight. If the molecular weight of APS is very high or low, it is not conducive to their activity. Only APS2 with moderate molecular weight showed the greatest antioxidant activity and ability to repair human kidney epithelial (HK-2) cells. Therefore, APS2 can be used as a potential antistone polysaccharide drug.

1. Introduction

Radix Astragali is one of the most popular medicinal herbs in China, and it has been used for more than 2000 years. One of its main active ingredients is *Astragalus* polysaccharides (APS), which show antioxidation, antitumor, and antiaging properties and cardiovascular, liver, and kidney protective effects [1, 2].

The polysaccharide property is closely related to its molecular weight, acid group content, monosaccharide composition, glycosidic linkage type, and main chain structure [3–6]. Given the different varieties, qualities, and processing methods, a certain difference exists in its monosaccharide composition and structure [7]. Wang et al. [6] separated APS with a molecular weight of 3.6×10^4 Da from *A. membranaceus* roots by hot water extraction method. The main chain of APS is made up of α–D–$(1 \rightarrow 4)$–Glc and $(1 \rightarrow 6)$–α–D–Glcp, and the branching point is located at O-6. Fu et al. [8] obtained APS with a molecular weight of approximately 3.01×10^5 Da from Mongolian *Astragalus* using low concentration of ethanol for precipitation and gel chromatography for purification. Spectral analysis results of ^{1}H NMR and ^{13}C NMR showed that the APS backbone has a 1,3-linked β-D-Gal residue and the branched portion has β-Glc, 1,6-linked α-Gal; 1,5-linked β-Xyl; 1,4-linked β-Gal; β-D-Gal, 1,2-linked α-Rha; and 1,2,4-linked α-Rha residues.

The molecular weight affects the antioxidant ability of the polysaccharide. In addition, optimal antioxidant ability will vary depending on the type of polysaccharide. Liu et al. [9] reported that *Ganoderma lucidum* polysaccharide (GLP_L1) with low molecular weight (5.2 kDa) has a higher ability to scavenge free radicals, superoxide radicals, and hydrogen peroxide than the component (GLP_L2) with high molecular weight (15.4 kDa). At a concentration of 10 mg/mL, the scavenging rates of GLP_L1 to hydroxyl radical and superoxide anion are close to 75% and 90%, respectively, whereas the scavenging ability of GLP_L2 is only 50% and 60%, respectively. Ma et al. [10] isolated four kinds of polysaccharides, namely, GLP1 (>10 kDa), GLP2 (8–10 kDa), GLP3 (2.5–8 kDa), and GLP4 (<2.5 kDa), from *G. lucidum* by ultrasonic method. At the concentration of 0.5 mg/mL, the reducing ability was in the following order: GLP2 > GLP1 > GLP3 > GLP4, showing that the GLP2 polysaccharide with moderate molecular weight had the best reducing ability. Sheng and

Sun [11] obtained four polysaccharides, namely, CPA-1, CPA-2, CPA-3, and CPA-4, from *Athyrium multidentatum* (Doll.) Ching by H_2O_2 degradation method, and their molecular weights were 14.53, 12.37, 11.55, and 6.40 kDa, respectively. The DPPH scavenging rates of the four polysaccharides with a concentration of 30 μg/mL were 63.35%, 50.23%, 42.45%, and 30.01%, respectively, showing that the polysaccharide with high molecular weight had the greatest scavenging activity.

Free radical-induced oxidative stress is thought to be one of the major causes of kidney disease, including kidney stones [12]. Polysaccharides can remove free radicals in vitro and act as antioxidants to protect living organisms from oxidative damage [13]. Khan [14] certified that free radicals can cause renal epithelial cell damage and allow urine crystals to attach to the damaged epithelial cell layer through simulation of the kidney environment in vitro. This research also proved that renal cell injury induced by free radicals can produce an environment conducive to crystal growth, accelerating the formation of stones. Antioxidant activity means that some antioxidants can protect cells against the damage of reactive oxygen species [15], such as superoxide, singlet oxygen, and hydroxyl radicals. Thus, research and exploring potent natural compounds with antioxidant activities and low cytotoxicity from plants have become very important in the field of biomedicine.

Based on this fact, the effects of the molecular weight of APS on its antioxidant capacity and cell repair ability were studied. The results of this study will provide enlightenment for screening for the optimal active polysaccharides.

2. Experimental Method

2.1. Reagents and Instruments. Astragalus polysaccharides (95% purity) produced by Shaanxi Ciyuan Biology Co. Ltd.; 30% hydrogen peroxide; anhydrous ethanol; trichloroacetic acid (CCl_3COOH); o-phenanthroline; $K_4[Fe(CN)_6]$; ferrous sulfate ($FeSO_4{\bullet}7H_2O$); potassium bromide (KBr); D_2O (99.9%, Sigma); ascorbic acid (Vc); and ferric chloride ($FeCl_3$) were all purchased from Guangzhou Chemical Reagent Factory.

Human kidney proximal tubular epithelial (HK-2) cells were purchased from Shanghai Cell Bank, Chinese Academy of Sciences (Shanghai, China). Dulbecco's modified Eagle's medium (DMEM) and fetal bovine serum were purchased from HyClone Biochemical Products Co. Ltd. (UT, USA). Cell culture plates were purchased from Wuxi NEST Biotechnology Co. Ltd. (Wuxi, China). Malondialdehyde (MDA) kit and superoxide dismutase (SOD) kit were purchased from Jiancheng Institute of Biotechnology (China). Hematoxylin-eosin (HE) kit was purchased from Shanghai Beyotime Biotech Co. Ltd. (Shanghai, China).

The apparatus used in this paper were the following: KQ2200DB Ultrasonic Cleaning Instrument (Kunshan Ultrasonic Instruments Co. Ltd.), Ubbelohde capillary viscometer (0.45 mm, Qihang Glass Instrument Factory, Shanghai, China), conductivity meter (DDS-11A, LEICI, Shanghai, China), ultraviolet-visible spectrophotometer (Cary 500, Varian, Palo Alto, CA, USA), Fourier-transform infrared spectrometer (Equinox 55, Bruker, Karlsruhe, Germany), enzyme mark instrument (Safire2, Tecan, Männedorf, Switzerland), nuclear magnetic resonance (Varian Bruker-600 MHz, Bruker, Germany), 7890A-5975C gas chromatography-mass spectrometer (Agilent, USA), and fluorescence microscope (22DI-E-D282, Leica, Solms, Germany).

2.2. Experimental Methods. We followed the methods of Bhadja et al. [16].

2.2.1. Degradation of Polysaccharides. About 1.2 g original Astragalus polysaccharide (APS0) was weighed and dissolved in 20 mL distilled water. H_2O_2 solution was added to the reaction system—90°C water bath degradation for 2 h. The degradation reaction was cooled at room temperature, at which point the solution pH was adjusted to 7.0 by adding 2 mol/L NaOH solution. The degraded solution was then concentrated to one-third of its original volume at 60°C. The product was precipitated by adding three volumes of anhydrous ethanol. After filtration, the degraded polysaccharide was obtained by drying. The above experimental procedure was repeated by changing the H_2O_2 concentration at 4%, 6%, and 14%; degraded Astragalus polysaccharides with different molecular weights were obtained.

2.2.2. Measurement of Molecular Weights of Polysaccharides. According to the literature [17], the molecular weight was determined by Ubbelohde viscosity method at 25 ± 0.2°C. After determining the drop time T of each polysaccharide in the viscometer, the relative viscosity η_r and the specific viscosity η_{sp} were calculated, where $\eta_r = T_i/T_0$ and $\eta_{sp} = \eta_r - 1$, where T_i and T_0 are the drop time of polysaccharide solution and deionized water. According to the one-point method, the intrinsic viscosity $[\eta] = [2(\eta_{sp} - \ln \eta_r)]^{1/2}/c$, where c is the concentration of the sample to be tested. The molecular weights of each polysaccharide before and after degradation were calculated based on $[\eta]$ values. The relationship between the intrinsic viscosity $[\eta]$ of the polymer solution and the molecular weight M of the polymer can be expressed by the empirical equation of Mark-Houwink: $[\eta] = \kappa M^{\alpha}$, where κ and α are the two parameters of the empirical equation and are constants related to polymer morphology, solvent, and temperature.

2.2.3. Analysis of Carboxylic Group Content. The carboxyl group (–COOH) content of APS was measured by conductometric titration. The conductivity titration curve was plotted using the conductivity value as the Y-axis and the used NaOH volume as the X-axis. The conductivity titration curve can be divided into three parts, a conductivity reduction stage (A), an equilibrium stage (B), and a conductivity increase stage (C). Three tangent lines are constructed from the three-stage curve, and the intersection is a stoichiometric point. The intersection of lines A and B gives the volume of NaOH (V_1) that excessive HCl and $-SO_3H$ consumed; the intersection of lines B and C gives the volume of NaOH (V_2) that excessive HCl and $-SO_3H$ and the –COOH of the APS consumed together; so, $V_2 - V_1$ (platform portion) is the NaOH volume that the –COOH of the APS consumed alone. The $-COOH$ content can be obtained according to the following formula [18]:

$$-COOH(\%) = \frac{C_{NaOH} \times (V_2 - V_1) \times 45/1000}{C_{sample} \times 40/1000} \times 100, \quad (1)$$

wherein C_{NaOH} (mol/L) represents the molar concentration of NaOH, C_{sample} (g/L) represents the mass concentration of the polysaccharide, 45 g/mol is the molar mass of $-COOH$, and 40 mL is the solution volume of polysaccharide. The final value is the average of three parallel experiments.

2.2.4. Fourier-Transform Infrared Spectroscopy (FT-IR) Analysis. A dried polysaccharide sample (2.0 mg) was mixed with KBr (200 mg). After grinding and pressing into KBr pellet, scanning was performed between the ranges of $4000\,cm^{-1}$ to $400\,cm^{-1}$ with a resolution of $4\,cm^{-1}$.

2.2.5. ^1H NMR and ^{13}C NMR Detection of APS. According to the literature [19], 20 mg of fully dried APS1 polysaccharides was weighed and added to a NMR tube containing 0.5 mL of deuterated water (D_2O), which is completely dissolved and placed in the magnetic field of the nuclear magnetic resonance spectrometer for detection.

2.2.6. Monosaccharide Component Detection of APS by GC-MS. According to the literature [20], 10 mg of APS1 polysaccharide was added to a 121°C sealed container containing $2.5\,mol{\cdot}L^{-1}$ trifluoroacetic acid (TFA) (2 mL) for 90 minutes. The solution was concentrated to dryness under reduced pressure, and then the TFA was removed with MeOH to a neutral solution and concentrated to dryness under reduced pressure. The residue was dissolved in 2 mol/L NH_4OH (1 mL) and $1\,mol{\cdot}L^{-1}$ fresh $NaBD_4$ (1 mL). The reaction was carried out at room temperature for 2.5 h and stirring was done at room temperature. Then two drops of acetic acid were added to decompose excess $NaBD_4$ until no bubbles were produced. The solution was concentrated to dryness under reduced pressure. The filtrate was added with MeOH to remove boric acid and dried in vacuo. 1 mL of acetic anhydride was added and acetylated at 100°C for 2.5 h. The acetylated product was extracted with dichloromethane. The organic layer was washed with distilled water, dried, and analyzed by GC-MS. The HP-5MS capillary column ($15\,m \times 250\,\mu m \times 0.25\,\mu m$) was programmed, and the temperature was raised from 135°C to 180°C at 0.5°C/min, then to 190°C at 10°C/min, and up to 310°C at 40°C/min. Helium acts as carrier gas, with a column flow rate of 0.6 mL/min. The acetylated product was identified by debris ions in GC-MS and relative retention times in GC. The structure is identified by peaks and assessed by peak area. Standard monosaccharides (rhamnose, arabinose, fucose, xylose, mannose, glucose, and galactose) are used as reference.

2.2.7. Hydroxyl Radical (•OH) Scavenging Activity of APS with Different Molecular Weights. The •OH scavenging ability of polysaccharide in vitro was detected by H_2O_2/Fe system method [21]. The four different molecular weight polysaccharide solutions ($60\,\mu g/mL$, 1 mL) were incubated with phenanthroline (2.5 mmol/L, 1 mL), ferrous sulfate (2.5 mmol/L, 1 mL), and hydrogen peroxide (20 mmol/L, 1 mL) in phosphate buffer (20 mmol/L, 1 mL, pH 7.4) for 90 min at 37°C. The absorbance measured at 536 nm was designated A_1. The absorbance when hydrogen peroxide (H_2O_2) was replaced with distilled water and polysaccharide solution was A_2 and A_3, respectively. Ascorbic acid (Vc) was used as the positive control group. The ability to scavenge hydroxyl radicals was calculated using the following equation:

$$\text{Scavenging effect}(\%) = \frac{(A_3 - A_1)}{(A_2 - A_1)} \times 100\%. \quad (2)$$

2.2.8. ABTS Radical Scavenging Activity of APS with Different Molecular Weights. ABTS was oxidized to produce a stable blue-green cationic radical $ABTS^+$. The antioxidant reacts with $ABTS^+$ to cause the reaction to fade and initiate a change in the absorbance of a particular wavelength segment to determine the size of the antioxidant to remove ABTS free radicals [22].

The ABTS radical scavenging activity of polysaccharides was performed according to [23] with slight modification. 7 mmol/L ABTS solution was mixed with 2.45 mmol/L potassium persulfate aqueous solution, and then the mixture was incubated in the dark at room temperature for 12–16 h. Then 3 mL $ABTS^+$ solution was added to 1 mL polysaccharide solutions of various concentrations in a test tube. After reacting for 6 min at room temperature, the absorbance was measured at 734 nm.

$$\text{Scavenging effect}(\%) = \left[1 - \frac{(A_1 - A_2)}{A_0}\right] \times 100\%. \quad (3)$$

where A_0 is the control group without polysaccharide, A_1 is the experiment group, and A_2 is the blank group without reagents (the absorbance of polysaccharide solution A_2 was 0).

2.2.9. Reducing Power of APS with Different Molecular Weights. Using the Prussian blue method [24], the polysaccharide sample may reduce the positive Fe^{3+} ions into Fe^{2+} ions in potassium ferricyanide ($K_4[Fe(CN)_6]$) to form Prussian blue and develop at 700 nm. That is, the greater the absorbance at 700 nm is, the stronger is the reduction of polysaccharide [25].

2 mL APS with different molecular weights ($60\,\mu g/mL$) was mixed with 2 mL phosphate buffer solution (PBS) (pH = 6.6) and 2 mL 1% $K_4[Fe(CN)_6]$. The mixture was incubated at 50°C for 20 min. 2 mL 10% trichloroacetic acid was added to the mixture which was then centrifuged for 10 min at 3000 r/min. The supernatant (2 mL) was mixed with 0.5 mL $FeCl_3$ (0.1%, *w*/*v*) solution and 2 mL distilled water. The mixture was shaken well and left to stand for 10 min. Then the absorbance was measured at 700 nm. Phosphate buffer was used as negative control, and Vc was used as positive control for comparison.

Table 1: Physicochemical properties and FT-IR characteristic absorption peaks of original and degraded APS.

	H_2O_2 concentration $C_{(H2O2)}$ (%)	Mean molecular weight (Mr/Da)	-COOH content (%)	Relative intensity of -COOH absorption peak*	Functional group characteristic absorption peak			
					-OH	-COOH	$-CH_2$	Sugar ring
APS0	0	11.03	16.8	1.4	3394.8	1617.1	2927.4	1115.5, 1087.4, 1030.5, 763.5
APS1	4	8.39	16.7	1.2	3393.7	1618.1	2931.3	1106.1, 1082.8, 1022.8, 764.8
APS2	6	4.72	17.2	1.6	3392.7	1616.7	2932.9	1110.8, 1081.9,1023.6, 762.5,
APS3	14	2.60	16.2	1	3396.3	1622.1	2930.8	1113.7, 1082.1, 1027.7, 761.9

*$(100\text{-}T_{APS0}) : (100\text{-}T_{APS1}) : (100\text{-}T_{APS2}) : (100\text{-}T_{APS3})$, where T represents the light transmittance.

2.2.10. Cell Culture. HK-2 cells were cultured in DMEM medium containing 10% fetal bovine serum, 100 U/mL penicillin, and100 μg/mL streptomycin antibiotic with a pH of 7.4 and cultured in a 5% CO_2-humidified environment of 37°C. Cells were passaged by trypsin digestion. Upon reaching 80%–90% confluent, cells were gently blown after trypsinization to form a cell suspension for the following cell experiments.

2.2.11. Superoxide Dismutase (SOD) Activity Detection. SOD activity was assessed using a commercially available kit based on the autooxidation of hydroxylamine. The cell suspension was inoculated in 24-well plates with a cell concentration of 5×10^4 cell/mL and 500 μL per well and was incubated for 24 h in an incubator at 37°C. The experimental model was divided into three groups: (1) control group: only serum-free medium was added, (2) injury group: serum-free medium containing 2.6 mmol/L oxalate was added for 3.5 h, and (3) repair group: 60 μg/mL of APS solutions with different molecular weights was added to the injured cells and repaired for 10 h. At the indicated time points, the treated cells were homogenized in 100 mmol/L Tris-HCl buffer and centrifuged at 10,000 rpm for 20 min and then the SOD activity was determined using assay kits. The absorbance of the supernatant was then measured directly by a microplate reader at 550 nm with a reference wavelength of 600 nm.

2.2.12. Malondialdehyde (MDA) Detection. For lipid peroxidation assay, we used a commercial kit to quantify the generation of MDA according to the manufacturer's protocol. Cell culture and polysaccharide treatment were performed using a method similar to that of SOD measurement. After the repair effect was completed, the cells were harvested by trypsinization and cellular extracts were prepared by sonication in ice-cold buffer (50 mM Tris-HCl, pH 7.5, 5 mM EDTA, and 1 mM DTT). After sonication, lysed cells were centrifuged at 10000 rpm for 20 min to remove debris. The supernatant was subjected to the measurement of MDA levels by detecting the absorbance at 532 nm.

2.2.13. Hematoxylin and Eosin (HE) Staining. Cell suspension with a cell concentration of 5×10^4 cells/mL was inoculated per well in 12-well plates, 1 mL of DMEM containing 10% fetal bovine serum was added per well, and the cells were incubated for 24 h. Cell culture and polysaccharide treatment were performed using a method similar to that of SOD measurement. After the repair was completed, the supernatant was removed by aspiration and washed 2 times with PBS. Then the cells were fixed with 4% polyoxymethylene at room temperature for 15 min. After fixation, the cells were stained with hematoxylin stain for 15 minutes. The cells were then washed with distilled water for 2 min to remove excess stains. Thereafter, the cells were stained with an eosin staining solution for 5 min. The cells were washed with pure water for 2 min to remove excess eosin. After treatment, the cells were observed under a microscope, the nuclei were stained purple, and the cytoplasm was stained pink.

2.2.14. Statistical Analysis. Experimental data were expressed as mean ± SD. The experimental results were analyzed statistically using SPSS 13.0 software. The differences of means between the experimental groups and the control group were analyzed by Tukey's test. $p < 0.05$ indicates significant difference; $p < 0.01$ indicates extremely significant difference.

3. Results

3.1. Degradation of APS. Original *Astragalus* polysaccharide (APS0) with a molecular weight of 11,033 Da was degraded by H_2O_2 with concentrations of 4%, 6%, and 14%. Three kinds of degraded APS, namely, APS1, APS2, and APS3, with the molecular weights of 8376, 4716, and 2600 Da, respectively, were obtained.

3.2. Change in the Contents of the -COOH in APS before and after Degradation. The content of the -COOH in APS was measured by conductivity titration. The results are shown in Table 1. The carboxyl content of three degraded APS was between 16.2% and 17.2%, which only slightly changed (16.8%) before degradation.

The carboxyl group content of APS2 (17.2%) increased slightly. The reason is that hydroxyl radicals produced by the H_2O_2 degradation system break the sugar chain, exposing the carboxyl groups. Thus, high levels of carboxyl were detected in polysaccharides with low molecular weight after degradation [26]. Chang et al. [27] degraded the original polysaccharide with a molecular weight of 1.29×10^2 kDa by ethanol precipitation and obtained two degraded polysaccharides with molecular weights of 60.0 and 52 kDa. The content of uronic acid was increased from 70.8% before degradation to 79.5% and 86.2%.

The carboxyl group content (16.9%) of APS3 decreased slightly, probably because free radicals generated by high concentration of H_2O_2 (14%) led to oxidative decarboxylation of polysaccharides, allowing the carboxyl groups in the polysaccharide chain to be converted into other groups, such as -OH [28].

3.3. Structural Characterization of APS by FT-IR Spectra. Figure 1 shows the FT-IR spectra of four APS. The FT-IR spectra of polysaccharides before and after degradation were similar, and no new peaks appeared, which indicated that the degradation of H_2O_2 did not cause significant effect on the overall structure of polysaccharides.

The absorption peaks of all four APS are shown in Table 1. The strong absorption band at 3392.7 cm^{-1} corresponds to the absorption peak of the stretching vibration of -OH in the polysaccharide. The absorption band at 2932.9 cm^{-1} corresponds to the stretching vibration of C–H [29]. The two distinct absorbance peaks, at about 1616.7 and 1417.9 cm^{-1}, suggested the presence of uronic acids. The absorption peaks at 1113.7 and 1027.7 cm^{-1} indicate that the polysaccharide contains α-glucose residues [30].

As the polysaccharide of each sample has the same mass (2.0 mg), the intensity of the absorption peak can reflect the content of the characteristic functional groups (such as the -COOH) [31]. Compared with the original polysaccharide before degradation, the absorption peaks of the -OH and -COOH at 3408 and 1620 cm^{-1} were enhanced and the peaks of APS2 were the strongest, which was probably due to the exposure of the -OH and -COOH of polysaccharides after degradation, thus increasing their content [32, 33]. The intensity change of the absorption peak of the -COOH is consistent with the change rule of the measured content of the -COOH (Table 1).

3.4. Structural Analysis of APS by ^1H NMR and ^{13}C NMR. The polysaccharides were characterized by ^1H and ^{13}C NMR spectroscopy and typical spectra are shown in Figure 2.

3.4.1. ^1H NMR. Table 2 lists the attribution of each line in the ^1H NMR spectra of APS1 and APS2. The ^1H NMR spectrum of APS1 was similar to that of APS2 (Table 2). For APS2, the signal peaks at δ5.31 and δ4.87 ppm correspond to the chemical shifts of H-1 of (1→4)-α-D-Glcp and (1→6)-α-D-Glcp, respectively. The highly overlapping peak at δ3.42–3.88 ppm belongs to the H-2, H-3, H-4, H-5, and H-6 signals of (1→4)-α-D-Glcp and (1→6)-α-D-Glcp of polysaccharides [34]. The proton signals at δ5.12 and 5.08 ppm belong to the H-1 signal of (1→6)-α-D-Gal and (1→)-α-D-Rha, respectively, of polysaccharide.

3.4.2. ^{13}C NMR Spectrum. Table 3 lists the attribution of the lines in the ^{13}C NMR spectra of APS1 and APS2. The ^{13}C NMR spectrum of APS1 was similar to that of APS2. For APS2, the signal at δ103.7 ppm belongs to the C-1 signal peak of β-D-Glcp of polysaccharide. The signals at δ99.6, δ71.5, δ73.2, δ76.7, δ71.2, and δ60.4 ppm are attributed to C-1, C-2, C-3, C-4, C-5, and C-6 signals of (1→4)-α-D-Glcp, respectively. The signals at δ71.9, δ73.1, δ69.3, δ70.4, and δ65.6 ppm are attributed to C-2, C-3, C-4, C-5, and C6 signals of (1→6)-α-D-Glcp, respectively [34, 35]. The signals at δ100.2, δ72.7, δ72.3, δ81.2, and δ73.3 ppm are

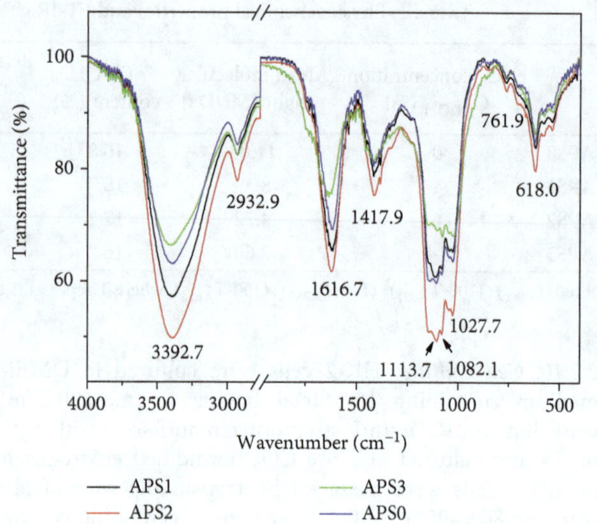

FIGURE 1: Infrared spectrum of APS with different molecular weights.

attributed to C-1, C-2, C-3, C-4, and C-5 signal peaks of (1→4)-GalA, respectively [36]. The signals at δ92.1, δ72.2, δ70.1, δ83.7, and δ69.0 ppm correspond to the C-1, C-2, C-3, C-4, and C-5 signal peaks of (1→)-α-D-Rha, respectively. The signals at δ70.5, δ72.4, and δ63.7 ppm are attributed to the C-4, C-5, and C-6 signal peaks of (1→6)-α-D-Gal, respectively.

3.5. Monosaccharide Composition of APS. The GC spectra of APS1, APS2, APS3, and 7 standard monosaccharides after derivatization were obtained by GC-MS, as shown in Figure 3. The retention time of each monosaccharide peak in APS is consistent with the retention time of standard monosaccharides of glucose, arabinose, rhamnose, and galactose (Figure 3(a)). According to the peak area, the molar ratio of monosaccharides in APS1 can be calculated as 11.4 : 7.8 : 0.7 : 1 (Figure 3(b)). The molar ratio of monosaccharides in APS2 is 13.9 : 8.2 : 1.2 : 1 (Figure 3(c)). The molar ratio of monosaccharides in APS3 is 13.3 : 8.4 : 1.1 : 1 (Figure 3(d)).

3.6. Differences in Scavenging Capacity of •OH Radicals of Polysaccharides with Different Molecular Weights. Among the reactive oxygen free radicals in the body, hydroxyl radicals (•OH) are known as the strongest oxidants, which can induce damage of surrounding biological macromolecules, such as certain proteins, nucleic acids, and unsaturated fatty acids. Such injuries can cause aging, cancer, and other diseases [37].

The •OH scavenging capacity of the four kinds of APS is shown in Figure 4. The •OH scavenging rate increased with the increase in APS concentration from 0.15 mg/mL to 3 mg/mL. The IC_{50} of APS0, APS1, APS2, APS3, and the positive control group (Vc) were 3.1, 2.7, 2.4, 4.2, and 0.06 mg/mL, respectively, which indicated that all APS had the ability to remove •OH radicals and APS2 with moderate molecular weight had the strongest scavenging

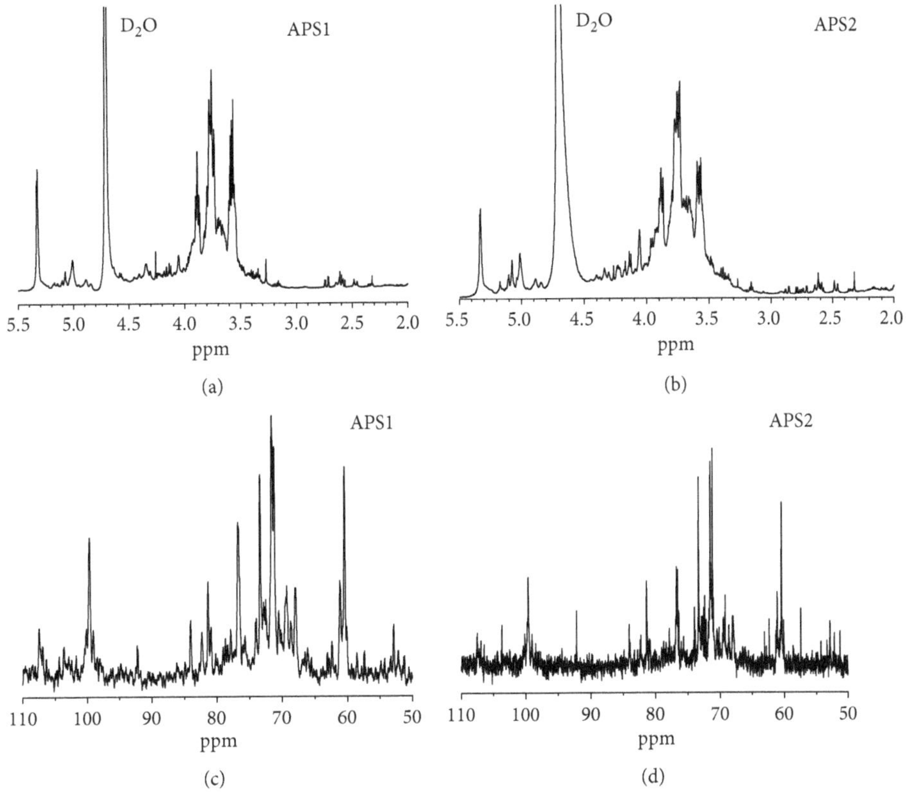

FIGURE 2: Nuclear magnetic resonance spectra of APS1 and APS2. (a) ^{1}H NMR spectrum of APS1; (b) ^{1}H NMR spectrum of APS2; (c) ^{13}C NMR spectrum of APS1; (d) ^{13}C NMR spectrum of APS2.

TABLE 2: Structure characterization of APS1 and APS2 by ^{1}H NMR spectra.

	Composition*	Chemical shifts (δppm) H-1	H-2	H-3	H-4	H-5	H-6
APS1	(1→6)-α-D-Glcp	4.89	3.48	3.63	3.40	3.85	3.64
	(1→4)-α-D-Glcp	5.33	3.55	3.87	3.56	3.73	3.70
	β-D-Glcp	5.18					
	(1→6)-α-D-Gal	5.10					
	(1→)-α-D-Rha	5.08					
	(1→4)-GalA	4.96					
APS2	(1→6)-α-D-Glcp	4.87	3.46	3.63	3.42	3.87	3.63
	(1→4)-α-D-Glcp	5.31	3.53	3.88	3.54	3.76	3.71
	β-D-Glcp	5.19					
	(1→6)-α-D-Gal	5.12					
	(1→)-α-D-Rha	5.08					
	(1→4)-GalA	4.97					

*Gal: galactose; Glcp: glucose; Rha: rhamnose; GalA: glucuronic acid.

TABLE 3: Structure characterization of APS1 and APS2 by ^{13}C NMR spectra.

	Composition*	Chemical shifts (δppm) C-1	C-2	C-3	C-4	C-5	C-6
APS1	β-D-Glcp	103.6					
	(1→4)-α-D-Glcp	99.6	71.5	73.3	76.7	71.1	60.4
	(1→4)-GalA	100.1	72.6	72.4	81.3	73.3	
	(1→6)-α-D-Glcp		71.8	73.0	69.3	70.4	65.7
	(1→)-α-D-Rha	92.2	72.3	70.1	83.7	69.0	
	(1→6)-α-D-Gal				70.5	72.4	63.6
APS2	β-D-Glcp	103.7					
	(1→4)-α-D-Glcp	99.6	71.5	73.2	76.7	71.2	60.4
	(1→4)-GalA	100.2	72.7	72.3	81.2	73.3	
	(1→6)-α-D-Glcp		71.9	73.1	69.3	70.4	65.6
	(1→)-α-D-Rha	92.1	72.2	70.1	83.7	69.0	
	(1→6)-α-D-Gal				70.5	72.4	63.7

*Gal: galactose; Glcp: glucose; Rha: rhamnose; GalA: glucuronic acid.

capacity. At a concentration of 2.5 mg/mL, the scavenging rate of APS2 (63.4%) was twice as high as that of APS3 (31.7%) with the lowest molecular weight and it was also obviously higher than that of APS0 (45.3%) with the highest molecular weight. However, the scavenging rates of the four polysaccharides were all less than that of Vc (97.8%).

3.7. Differences of ABTS Radical Scavenging Capacity of APS with Different Molecular Weights. Four kinds of APS with different molecular weights all had an obvious scavenging effect on ABTS free radicals (Figure 5) and showed a dose-dependent relationship in the concentration range of 0.15–3 mg/mL. At the concentration of 2.5 mg/mL, the scavenging

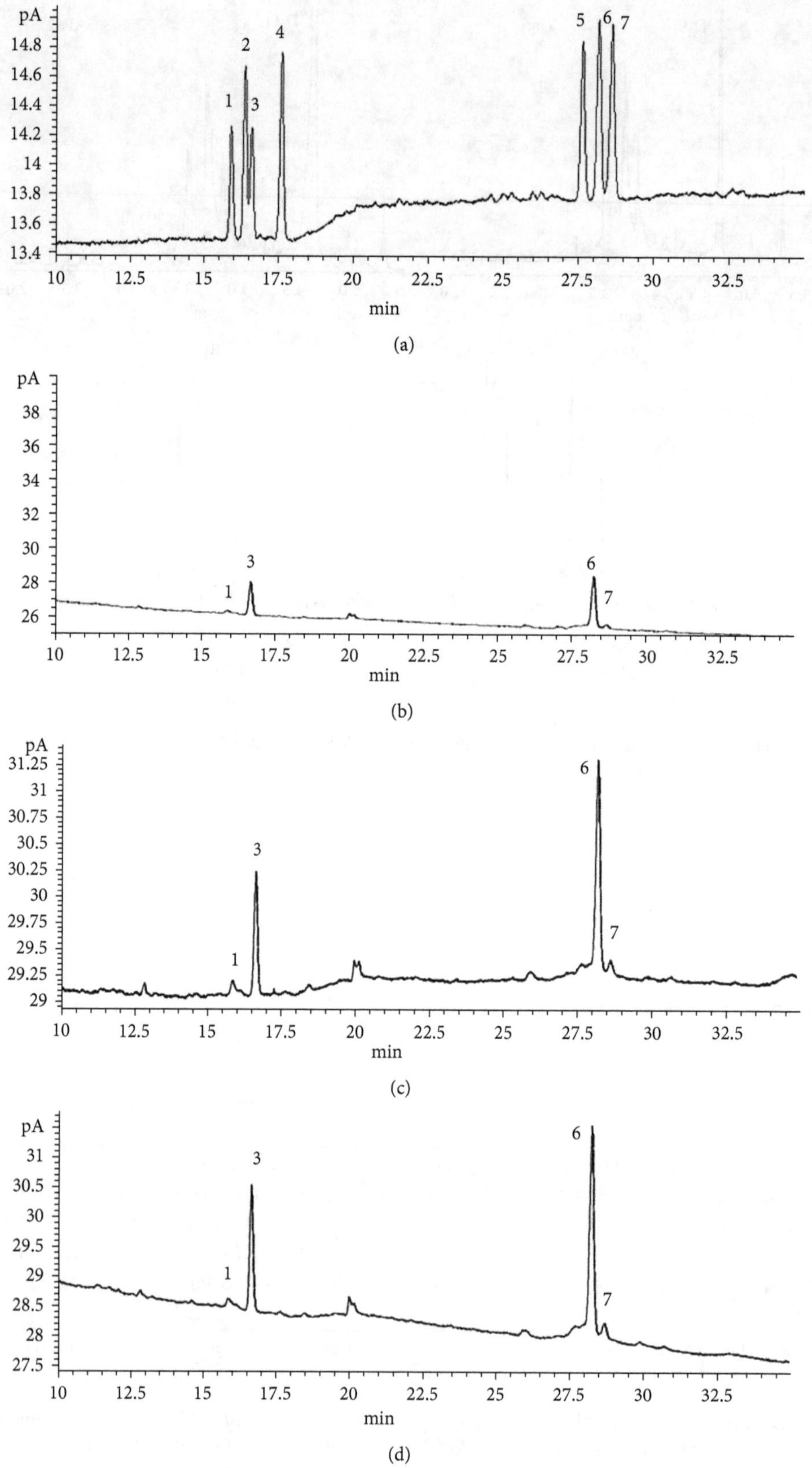

FIGURE 3: Gas chromatography-mass spectrometry (GC-MS) results of APS. (a) Gas chromatogram of standard monosaccharides. (b) Monosaccharide component of APS1. (c) Monosaccharide component of APS2. (d) Monosaccharide component of APS3. 1: rhamnose; 2: fucose; 3: arabinose; 4: xylose; 5: mannose; 6: glucose; 7: galactose.

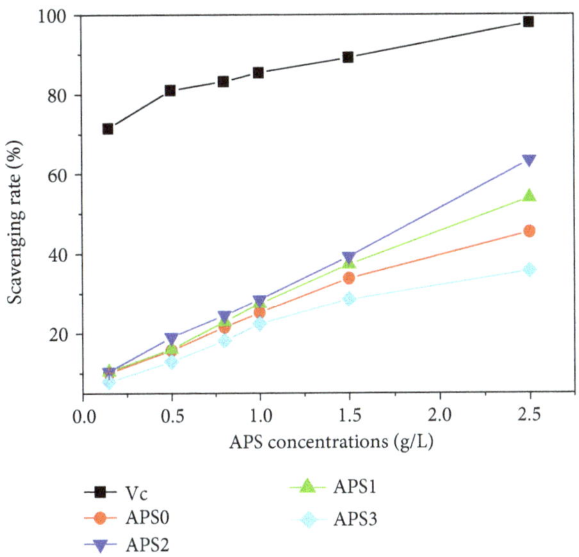

FIGURE 4: Hydroxyl radical scavenging capacity of APS with different molecular weights.

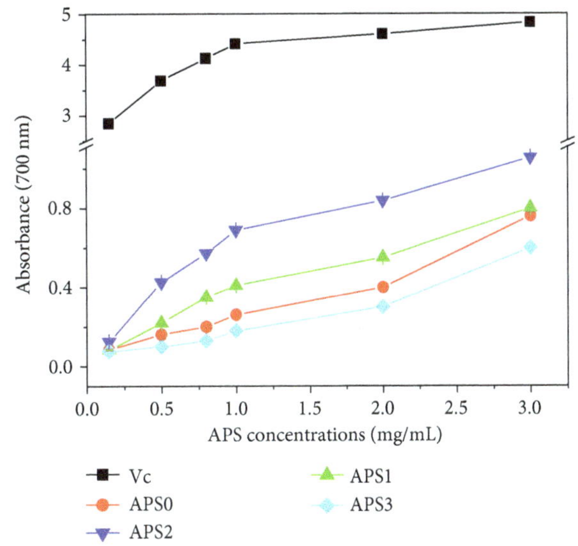

FIGURE 6: Reducing power of APS with different molecular weights.

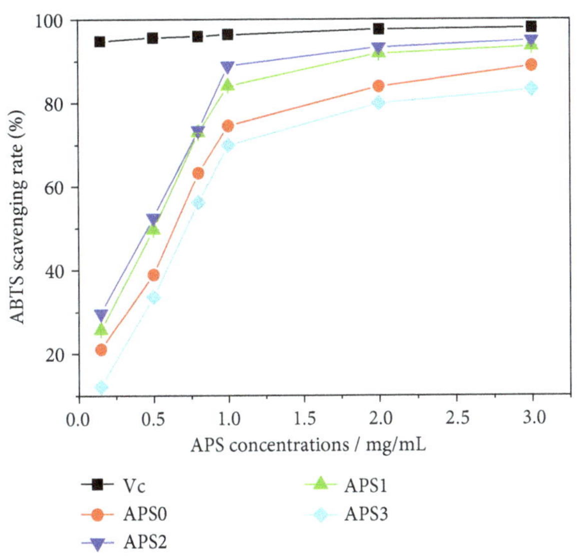

FIGURE 5: ABTS radical scavenging capacity of APS with different molecular weights.

rates of APS0, APS1, APS2, and APS3 were 88.8%, 93.5%, 95%, and 83.2%, respectively, which indicated that APS2 with moderate molecular weight had the highest antioxidant activity and was slightly lower than Vc (98%).

3.8. Differences of Reducing Ability of APS with Different Molecular Weights. The reducing ability of materials in vivo provides hydrogen atoms to destroy the free radical reaction chain, so as to achieve antioxidation. Therefore, the reducing ability is an important indicator of the potential antioxidant capacity of antioxidants.

As shown in Figure 6, the reducing ability of each polysaccharide showed a concentration-dependent manner. When the concentration was 2.5 mg/mL, the ability of the four polysaccharides to reduce Fe^{3+} was in the following order: APS2 (1.06) > APS1 (0.8) > APS0 (0.76) > APS3 (0.6). APS2 had the highest reducing ability.

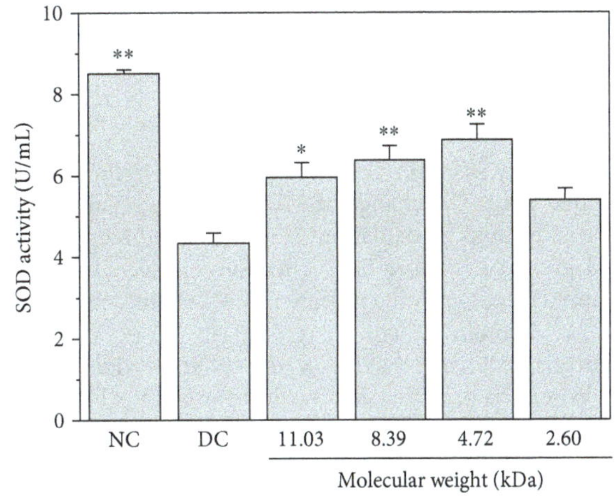

FIGURE 7: Changes of SOD activity after the damaged HK-2 cells were repaired by APS with different molecular weights for 10 h. NC: normal control; DC: damaged control. Compared with the DC group, * indicates $p < 0.05$. ** indicates $p < 0.01$.

3.9. Changes of SOD Activity in Damaged Cells and APS-Repaired Cells. The reduction of SOD activity in organism implies decreased ability to resist free radical-induced damage in organism [38]. Figure 7 shows the changes of SOD activity in damaged HK-2 cells after polysaccharide repair for 10 h. It can be seen that the activity of intracellular SOD increases first and then decreases with the decrease of APS molecular weight. The SOD activity in the control group was 8.5 ± 0.1 U/mL, and it decreased to 4.34 ± 0.25 U/mL after 2.6 mmol/L oxalate damage. After

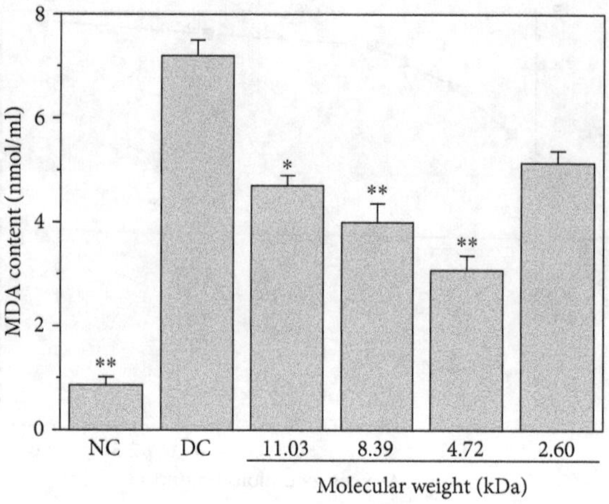

FIGURE 8: Changes of MDA content after the damaged HK-2 cells were repaired by APS with different molecular weights for 10 h. NC: normal control; DC: damaged control. Compared with the DC group, * indicates $p < 0.05$. ** indicates $p < 0.01$.

damaged cells were repaired by APS0, APS1, APS2, and APS3 for 10 h, the SOD activity increased to 5.95 ± 0.38, 6.3 ± 0.36, 6.87 ± 0.37, and 5.39 ± 0.29 U/mL, respectively. APS2 with moderate molecular weight has the strongest ability to enhance SOD activity.

3.10. Changes of MDA Content in Damaged Cells and APS-Repaired Cells. The change of MDA content usually reveals the level of lipid peroxidation *in vivo* and indirectly reflects the degree of cell injury [39]. After the cells were injured by 2.6 mmol/L oxalate, the release of MDA increased from 0.86 ± 0.16 nmol/mL (control group) to 7.2 ± 0.3 nmol/mL (Figure 8), indicating that HK-2 cells suffered oxidative damage. When the damaged cells were repaired by APS0, APS1, APS2, and APS3 for 10 h, the generation content of MDA decreased from 7.2 ± 0.3 nmol/mL to 4.7 ± 0.3, 4.0 ± 0.3, 3.08 ± 0.28, and 5.15 ± 0.23 nmol/mL, respectively. That is, APS inhibit the release of MDA to the outside of the cells. Especially, APS2 with moderate molecular weight has the strongest ability to inhibit MDA release.

3.11. Cell Morphology Repair of APS with Different Molecular Weights in HK-2 Cells. The morphological changes of renal proximal tubule epithelial cells (HK-2) after repair by APS with different molecular weights were observed by hematoxylin and eosin staining (Figure 9). Normal HK-2 grows tightly and is plump. After oxidative damage by 2.6 mmol/L oxalic acid, the inherent form of HK-2 cells was lost. The cell volume was reduced, and nuclei staining was enhanced. Many dense apoptotic bodies were formed. After repair in injured HK-2 cells by APS with different molecular weights, the cell number increased and the cell morphology recovered similar to normal cells in varying degrees. Especially for the APS2-repaired cells, the morphology is the closest to normal cells. By contrast, APS3 with a lower molecular weight and APS0 with a higher molecular weight were less effective than APS2.

4. Discussion

4.1. Chemical Structure of APS. The structure of polysaccharides is the basis for their biological activity. Many factors affect the biological activity of polysaccharides, including molecular weight, acid group content, monosaccharide composition, ligand type, and main chain structure [3–6]. In the present study, the monosaccharide composition and type of sugar residues of APS were analyzed by ^{1}H NMR, ^{13}C NMR, GC-MS, and FT-IR spectroscopy.

From the results of comprehensive analysis (Figures 1–3), α- and β-glycosidic bonds exist in the four kinds of APS. The main structure is that the main chain is composed of $(1 \rightarrow 4)$ connected Glcp and the branch point is located at the C-6 position of $(1 \rightarrow 6)$ connected Glcp, both containing 1,4-linked glucuronic acid fragments (Figure 10). The main structure is basically consistent with the structure illustrated by Li and Zhang [29]. Fu et al. [8] also found that APS mainly consist of glucose, rhamnose, xylose, and galactose, which is inconsistent with our results in that the monosaccharides of APS consisted of rhamnose, arabinose, fucose, sugar, mannose, glucose, and galactose. Even with the same species of polysaccharides, differences will exist in their monosaccharide components due to their origin, processing method, separation process, and purification conditions [27].

The main chain structures of the four polysaccharides were similar from FT-IR results (Figure 1), but the absorption peak intensities of characteristic functional group of the polysaccharides before and after degradation were different (Table 1). This finding is mainly because the hydroxyl radicals produced by H_2O_2 degradation can attack the polysaccharide glycosidic bonds and cleave them, resulting in the change of side chain structure but not the main chain structure in polysaccharides. For example, Tian et al. [40] analyzed the chitosan structure before and after H_2O_2 degradation by infrared spectroscopy and nuclear magnetic resonance. The depolymerization only leads to changes of side groups, and no distinct change takes place in the structures of the main chain.

4.2. Effects of Molecular Weight on the Activity of APS Polysaccharides

4.2.1. Reasons for the Low Activity in Low-Molecular Weight APS. For different types of plant polysaccharides, the range of molecular weights that exhibit the best bioactivity is different. When the molecular weight of APS is very low, an active polymeric structure might not form, resulting in the loss of polysaccharide activity [41]. In low-molecular weight polysaccharide, the unique bond linking method of polysaccharide and their three-dimensional structure (conformation) based on intramolecular hydrogen bond of polysaccharide will be destructed. The carbonyl group in low-molecular weight polysaccharide changed to be in an open chain, whereas in high-molecular weight polysaccharide, it was attached to a ring molecule. The destruction of the ring structure will break the hydrogen bond structure of the polysaccharide molecule, leading to reduced bioactivity.

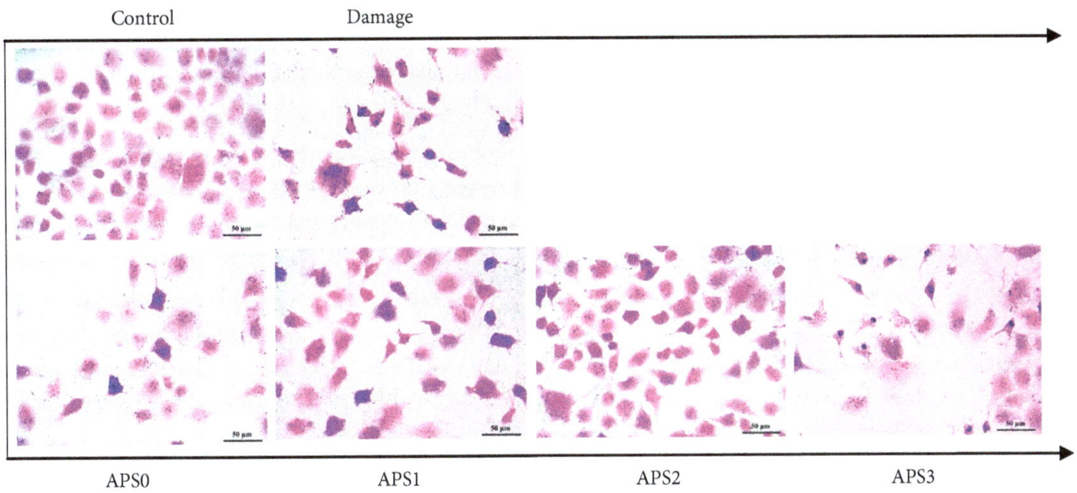

FIGURE 9: Morphological changes after the damaged HK-2 cells were repaired by APS with different molecular weights. Oxalic acid concentration: 2.6 mmol/L, injury time: 3.5 h; polysaccharide concentration: 60 μg/mL, repair time: 10 h.

[→4)-α-D-Glcp→(1→4)-α-D-Glcp→(1→[4)-α-D-Glcp(1→]4)- - α-D-Glcp-(1→]n
6
↑
1
α-D-Glcp

(a)

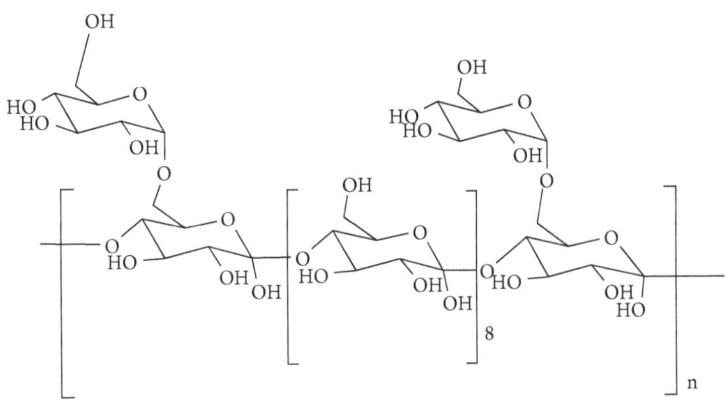

(b)

FIGURE 10: Molecular structure of APS. (a) Chain structure; (b) skeleton structure.

Polysaccharide has two mechanisms of clearing the hydroxyl radical: the first is to remove the already-generated hydroxyl radicals and the second is to inhibit the generation of hydroxyl radicals [42]. The former mechanism is related to that hydroxyl radical can quickly capture the hydrogen atom of the C-H chain of polysaccharide and form H_2O, whereas, the carbon atom of polysaccharide generates a single electron and turns into carbon free radical which then will be oxidized to be a peroxy radical that decomposes harmless products to organism [43]. The latter mechanism is related to the transition of metal ions that catalyze the generation of hydroxyl radicals, so the chelating ability may affect the hydroxyl radicals [44].

Hydroxyl radical can quickly capture the hydrogen atom of the C-H chain of polysaccharide and form H_2O [43]. In addition, the acid group of polysaccharide (such as –COOH) can decrease the production of hydroxyl radicals through the inactivating Fenton reaction. The carboxyl group content has considerable effect on scavenging •OH radicals [45]. APS contained many hydroxyl groups and carboxyl groups in each monomer unit. Polysaccharides with low molecular weights would have more reductive hydroxyl group or carboxyl group terminals (on per unit mass basis) to accept and eliminate the free radicals [46]. The chain structure of APS3 with the lowest molecular weight suffered significant destruction, and its molecular structure was the most loose. Therefore, the ability of APS3 to clear free radicals was the lowest. Chen et al. [47] reported that when the hydroxyl groups from polysaccharide are replaced with sulfuric acid groups, the antioxidant activity of polysaccharides decreases, which also indicates that the reduction in the number of hydroxyl groups will reduce the antioxidant activity of polysaccharides. Lai et al. [48] extracted two molecular weights of mung bean polysaccharides by ethanol precipitation. At a concentration of 0.8 mg/mL, the DPPH free radical scavenging rate (70.2%) of mung bean polysaccharide with low molecular weight (45 kDa) was weaker than that of mung bean polysaccharide (91.6%) with high molecular weight (83 kDa).

4.2.2. Reasons for the Low Activity in High-Molecular Weight APS. APS0 has high molecular weight, molecular volume, and viscosity and relatively low solubility, which affect its biological activity. The biological activity of polysaccharides is closely related to the single-helix structure of the main chain and hydrophilic group (hydroxyl group) located on the surface of the spirochetes [49]. Another reason is because the polysaccharide with high molecular weight has a more compact structure and stronger intramolecular hydrogen bonds, resulting in less exposure of effective active groups. Qi et al. [50] found that the polysaccharide chain has a reduction and a nonreducing end. The greater the molecular weight of polysaccharides is, the lesser the reduction and nonreducing end contents are and the weaker the antioxidant activity is.

Polysaccharide degradation can effectively improve its antioxidant capacity. After degradation of APS0 to the polysaccharide with a suitable molecular weight, the branching degree of the polysaccharide becomes small, providing sufficient spatial extent to form a regular helical structure, which favors the exposure of more hydroxyl groups to the surface of the helical structure and is good for exerting bioactivity.

For example, *Lycium barbarum* polysaccharides with low molecular weight (10.2 kDa) have anticancer activity, whereas those with high molecular weight (6.50×10^3 kDa) have no anticancer activity [51]. Similarly, *Ligusticum chuanxiong* polysaccharide with low molecular weight has a strong reducing ability [52]. At a concentration of 5 mg/mL, the IC_{50} values of reduction capacity of three kinds of *Rhizoma chuanxiong* polysaccharides with molecular weights of 2.83×10^4, 1.23×10^4, and 6.31×10^4 Da are 1.05, 1.1, and 0.84 mg/mL, respectively. However, the immunoregulatory effect of *Opuntia* polysaccharides with high molecular weight (133 kDa) on macrophages is better than that of *Opuntia* polysaccharides with low molecular weight (168 kDa) [53].

4.2.3. APS with Moderate Molecular Weight Has the Strongest Activity. As shown in Figures 4–6, the antioxidant capacity of APS was in the following order: APS2 > APS1 > APS0 > APS3, and APS2 had the strongest antioxidant capacity. APS with a very high (APS0) or low (APS3) molecular weight had weakened antioxidant capacity. The ability of each polysaccharide to repair damaged HK-2 cells also showed the same activity sequence.

A number of studies have also confirmed that polysaccharides with moderate molecular weight have more conducive activity [54, 55]. For example, Im et al. [54] studied the immunomodulatory and antitumor functions of aloe polysaccharides with different molecular weights. The results showed that the inhibitory rate of aloe polysaccharide G2E1DS2 (Mw = 5–400 kDa) with moderate molecular weight on the growth of S180 cells in mice was 91.53%. The inhibition rates of G2E1DS3 (>400 kDa) with the highest molecular weight and G2E1DS1 (<5 kDa) with the lowest molecular weight were only 14.87% and 15.89%, respectively. Fan et al. [55] isolated three polysaccharides (DDP1, DDP2, and DDP3) with molecular weights of 51.5, 26.1, and 6.95 kDa, respectively, from *Dendrobium denneanum* by hot water extraction. At the concentration of 2 mg/mL, the scavenging ability of the three polysaccharides to •OH, DPPH, and ABTS radicals was in the following order: DDP2 > DDP1 > DDP3. Moreover, the antioxidant capacity of DDP2 with moderate molecular weight (26.1 kDa) was the strongest.

Chen et al. [5] obtained three kinds of APS, namely, APS1-1, APS2-2, and APS3-3, by DEAE resin column chromatography, with molecular weights of 112, 98.3, and 20.4 kDa, respectively, and with the uronic acids content of 0.0%, 5.9%, 11.6%, respectively. Their corresponding total antioxidant capacities were 1152.8, 1566.3, and 1580.4 μmol/L, respectively, at the concentration of 2 mg/mL. These findings showed that polysaccharides with a high uronic acid content and a low molecular weight have great antioxidant ability. There are two variables (molecular weight and uronic acid content) in this study, and it is difficult to determine which factor is the most important in affecting antioxidant activity. In our study, we obtained four different molecular weight APS by H_2O_2 degradation method and the differences of their carboxyl content were small (ranging from 16.2% to 17.2%). Therefore, the influence of the carboxyl group content on the antioxidant ability can be avoided. At the same time, the molecular weight of APS used in our study ranged from 2.60 to 1.03 kDa, which is much less than what Chen et al. [5] reported. We also found that only APS with moderate molecular weight showed the greatest antioxidant capacity.

5. Conclusions

Three degraded APS with molecular weights of 8.38, 4.72, and 2.60 kDa were obtained by degradation of primitive APS (APS0, Mw 11.03 kDa) with different concentrations of H_2O_2. The results of ^{1}H NMR, ^{13}C NMR, FT-IR, and GC-MS showed no significant difference in the main chain structure of the four polysaccharides. All four polysaccharides had $(1 \rightarrow 4)$-D-glucose as the main chain and $(1 \rightarrow 6)$-D-glucose linkage as the branch chain. The monosaccharide component of APS before and after degradation was slightly changed. APS with four molecular weights had the ability to scavenge hydroxyl and ABTS radicals and good reducing ability. The antioxidant capacity of APS was in the following order: APS2 > APS1 > APS0 > APS3, that is, the antioxidant effect of APS2 with moderate molecular weight was the best. Moreover, the polysaccharides showed good ability to enhance SOD activity, inhibit MDA release, and recover cell morphology. APS, especially APS2, can be used as potential antioxidative drugs for renal protection.

Conflicts of Interest

The authors declare no competing financial interest.

Acknowledgments

This research work was granted by the National Natural Science Foundation of China (nos. 21701050 and 81670644) and the China Postdoctoral Science Foundation (no. 2017M612837).

References

[1] R. Z. Zhong, M. Yu, H. W. Liu, H. X. Sun, Y. Cao, and D. W. Zhou, "Effects of dietary Astragalus polysaccharide and Astragalus membranaceus root supplementation on growth performance, rumen fermentation, immune responses, and antioxidant status of lambs," *Animal Feed Science and Technology*, vol. 174, no. 1-2, pp. 60–67, 2012.

[2] Y. Fan, Y. Hu, D. Wang et al., "Effects of Astragalus polysaccharide liposome on lymphocyte proliferation in vitro and adjuvanticity in vivo," *Carbohydrate Polymers*, vol. 88, no. 1, pp. 68–74, 2012.

[3] Y. Yu, M. Shen, Q. Song, and J. Xie, "Biological activities and pharmaceutical applications of polysaccharide from natural resources: a review," *Carbohydrate Polymers*, vol. 183, pp. 91–101, 2018.

[4] J. H. Xie, M. L. Jin, G. A. Morris et al., "Advances on bioactive polysaccharides from medicinal plants," *Critical Reviews in Food Science and Nutrition*, vol. 56, Supplement 1, pp. S60–S84, 2015.

[5] R.-Z. Chen, L. Tan, C.-G. Jin et al., "Extraction, isolation, characterization and antioxidant activity of polysaccharides from Astragalus membranaceus," *Industrial Crops and Products*, vol. 77, pp. 434–443, 2015.

[6] K.-p. Wang, J. Wang, Q. Li et al., "Structural differences and conformational characterization of five bioactive polysaccharides from Lentinus edodes," *Food Research International*, vol. 62, pp. 223–232, 2014.

[7] M. Jin, K. Zhao, Q. Huang, C. Xu, and P. Shang, "Isolation, structure and bioactivities of the polysaccharides from Angelica sinensis (Oliv.) Diels: a review," *Carbohydrate Polymers*, vol. 89, no. 3, pp. 713–722, 2012.

[8] J. Fu, L. Huang, H. Zhang, S. Yang, and S. Chen, "Structural features of a polysaccharide from Astragalus membranaceus (Fisch.) Bge. var. mongholicus (Bge.) Hsiao," *Journal of Asian Natural Products Research*, vol. 15, no. 6, pp. 687–692, 2013.

[9] W. Liu, H. Wang, X. Pang, W. Yao, and X. Gao, "Characterization and antioxidant activity of two low-molecular-weight polysaccharides purified from the fruiting bodies of Ganoderma lucidum," *International Journal of Biological Macromolecules*, vol. 46, no. 4, pp. 451–457, 2010.

[10] C.-w. Ma, M. Feng, X. Zhai et al., "Optimization for the extraction of polysaccharides from Ganoderma lucidum and their antioxidant and antiproliferative activities," *Journal of the Taiwan Institute of Chemical Engineers*, vol. 44, no. 6, pp. 886–894, 2013.

[11] J. Sheng and Y. Sun, "Antioxidant properties of different molecular weight polysaccharides from Athyrium multidentatum (Doll.) Ching," *Carbohydrate Polymers*, vol. 108, pp. 41–45, 2014.

[12] P. Wang, X. Jiang, Y. Jiang et al., "In vitro antioxidative activities of three marine oligosaccharides," *Natural Product Research*, vol. 21, no. 7, pp. 646–654, 2007.

[13] K. W. Kim and R. L. Thomas, "Antioxidative activity of chitosans with varying molecular weights," *Food Chemistry*, vol. 101, no. 1, pp. 308–313, 2007.

[14] S. R. Khan, "Hyperoxaluria-induced oxidative stress and antioxidants for renal protection," *Urological Research*, vol. 33, no. 5, pp. 349–357, 2005.

[15] R. Wang, P. Chen, F. Jia, J. Tang, F. Ma, and B. Xu, "Characterization and antioxidant activities of polysaccharides from Panax japonicus C.A. Meyer," *Carbohydrate Polymers*, vol. 88, no. 4, pp. 1402–1406, 2012.

[16] P. Bhadja, C.-Y. Tan, J.-M. Ouyang, and K. Yu, "Repair effect of seaweed polysaccharides with different contents of sulfate group and molecular weights on damaged HK-2 cells," *Polymer*, vol. 8, no. 5, p. 188, 2016.

[17] J.-Y. Yin, S.-P. Nie, Q.-B. Guo, Q. Wang, S. W. Cui, and M.-Y. Xie, "Effect of calcium on solution and conformational characteristics of polysaccharide from seeds of Plantago asiatica L," *Carbohydrate Polymers*, vol. 124, pp. 331–336, 2015.

[18] L. Luo, M. Wu, L. Xu et al., "Comparison of physicochemical characteristics and anticoagulant activities of polysaccharides from three sea cucumbers," *Marine Drugs*, vol. 11, no. 12, pp. 399–417, 2013.

[19] R. Li, W.-c. Chen, W.-p. Wang, W.-y. Tian, and X.-g. Zhang, "Antioxidant activity of Astragalus polysaccharides and antitumour activity of the polysaccharides and siRNA," *Carbohydrate Polymers*, vol. 82, no. 2, pp. 240–244, 2010.

[20] S. Bhatia, P. Rathee, K. Sharma, B. B. Chaugule, N. Kar, and T. Bera, "Immuno-modulation effect of sulphated polysaccharide (porphyran) from Porphyra vietnamensis," *International Journal of Biological Macromolecules*, vol. 57, pp. 50–56, 2013.

[21] J. Jiang, F. Kong, N. Li, D. Zhang, C. Yan, and H. Lv, "Purification, structural characterization and in vitro antioxidant activity of a novel polysaccharide from Boshuzhi," *Carbohydrate Polymers*, vol. 147, pp. 365–371, 2016.

[22] Y. Cai, Q. Luo, M. Sun, and H. Corke, "Antioxidant activity and phenolic compounds of 112 traditional Chinese medicinal plants associated with anticancer," *Life Sciences*, vol. 74, no. 17, pp. 2157–2184, 2004.

[23] R.-B. Xu, X. Yang, J. Wang et al., "Chemical composition and antioxidant activities of three polysaccharide fractions from pine cones," *International Journal of Molecular Sciences*, vol. 13, no. 12, pp. 14262–14277, 2012.

[24] K. I. Berker, B. Demirata, and R. Apak, "Determination of total antioxidant capacity of lipophilic and hydrophilic antioxidants in the same solution by using ferric-ferricyanide assay," *Food Analytical Methods*, vol. 5, no. 5, pp. 1150–1158, 2012.

[25] M. Hassasroudsari, P. Chang, R. Pegg, and R. Tyler, "Antioxidant capacity of bioactives extracted from canola meal by subcritical water, ethanolic and hot water extraction," *Food Chemistry*, vol. 114, no. 2, pp. 717–726, 2009.

[26] Z. Zhao, J. Li, X. Wu et al., "Structures and immunological activities of two pectic polysaccharides from the fruits of Ziziphus jujuba Mill. cv. jinsixiaozao Hort," *Food Research International*, vol. 39, no. 8, pp. 917–923, 2006.

[27] S. C. Chang, B. Y. Hsu, and B. H. Chen, "Structural characterization of polysaccharides from Zizyphus jujuba and evaluation of antioxidant activity," *International Journal of Biological Macromolecules*, vol. 47, no. 4, pp. 445–453, 2010.

[28] J. Tang, J. Nie, D. Li et al., "Characterization and antioxidant activities of degraded polysaccharides from Poria cocos sclerotium," *Carbohydrate Polymers*, vol. 105, pp. 121–126, 2014.

[29] S.-g. Li and Y.-q. Zhang, "Characterization and renal protective effect of a polysaccharide from Astragalus membranaceus," *Carbohydrate Polymers*, vol. 78, no. 2, pp. 343–348, 2009.

[30] J.-Y. Yin, B. C.-L. Chan, H. Yu et al., "Separation, structure characterization, conformation and immunomodulating effect of a hyperbranched heteroglycan from Radix Astragali," *Carbohydrate Polymers*, vol. 87, no. 1, pp. 667–675, 2012.

[31] L. Sun, L. Wang, J. Li, and H. Liu, "Characterization and antioxidant activities of degraded polysaccharides from two marine Chrysophyta," *Food Chemistry*, vol. 160, pp. 1–7, 2014.

[32] M. A. Chaouch, J. Hafsa, C. Rihouey, D. Le Cerf, and H. Majdoub, "Depolymerization of polysaccharides from Opuntia ficus indica: antioxidant and antiglycated activities," *International Journal of Biological Macromolecules*, vol. 79, pp. 779–786, 2015.

[33] C. Q. Qin, Y. M. Du, and L. Xiao, "Effect of hydrogen peroxide treatment on the molecular weight and structure of chitosan," *Polymer Degradation and Stability*, vol. 76, no. 2, pp. 211–218, 2002.

[34] Y. Li, Y.-L. Wang, L. Li et al., "Structural characterization of polysaccharides from the roots of Urtica fissa," *Journal of Asian Natural Products Research*, vol. 11, no. 11, pp. 951–957, 2009.

[35] Z.-Y. Zhu, R.-Q. Liu, C.-L. Si et al., "Structural analysis and antitumor activity comparison of polysaccharides from Astragalus," *Carbohydrate Polymers*, vol. 85, no. 4, pp. 895–902, 2011.

[36] W. Fan, S. Zhang, P. Hao, P. Zheng, J. Liu, and X. Zhao, "Structure characterization of three polysaccharides and a comparative study of their immunomodulatory activities on chicken macrophage," *Carbohydrate Polymers*, vol. 153, pp. 631–640, 2016.

[37] F. Hecht, C. F. Pessoa, L. B. Gentile, D. Rosenthal, D. P. Carvalho, and R. S. Fortunato, "The role of oxidative stress on breast cancer development and therapy," *Tumor Biology*, vol. 37, no. 4, pp. 4281–4291, 2016.

[38] C. M. Muth, Y. Glenz, M. Klaus, P. Radermacher, G. Speit, and X. Leverve, "Influence of an orally effective SOD on hyperbaric oxygen-related cell damage," *Free Radical Research*, vol. 38, no. 9, pp. 927–932, 2009.

[39] H. Chen, H. Yoshioka, G.-S. Kim et al., "Oxidative stress in ischemic brain damage: mechanisms of cell death and potential molecular targets for neuroprotection," *Antioxidants & Redox Signaling*, vol. 14, no. 8, pp. 1505–1517, 2011.

[40] F. Tian, Y. Liu, K. Hu, and B. Zhao, "The depolymerization mechanism of chitosan by hydrogen peroxide," *Journal of Materials Science*, vol. 38, no. 23, pp. 4709–4712, 2003.

[41] S. Alban and G. Franz, "Characterization of the anticoagulant actions of a semisynthetic curdlan sulfate," *Thrombosis Research*, vol. 99, no. 4, pp. 377–388, 2000.

[42] B. Li, S. Liu, R. Xing et al., "Degradation of sulfated polysaccharides from Enteromorpha prolifera and their antioxidant activities," *Carbohydrate Polymers*, vol. 92, no. 2, pp. 1991–1996, 2013.

[43] X. T. Li, Y. K. Zhang, H. X. Kuang et al., "Mitochondrial protection and anti-aging activity of Astragalus polysaccharides and their potential mechanism," *International Journal of Molecular Sciences*, vol. 13, no. 2, pp. 1747–1761, 2012.

[44] J. Wang, Q. Zhang, Z. Zhang, and Z. Li, "Antioxidant activity of sulfated polysaccharide fractions extracted from Laminaria japonica," *International Journal of Biological Macromolecules*, vol. 42, no. 2, pp. 127–132, 2008.

[45] J. Fan, Z. Wu, T. Zhao et al., "Characterization, antioxidant and hepatoprotective activities of polysaccharides from Ilex latifolia Thunb," *Carbohydrate Polymers*, vol. 101, pp. 990–997, 2014.

[46] J. Wang, S. Hu, S. Nie, Q. Yu, and M. Xie, "Reviews on mechanisms ofin vitroantioxidant activity of polysaccharides," *Oxidative Medicine and Cellular Longevity*, vol. 2016, Article ID 5692852, 13 pages, 2016.

[47] Y. Chen, H. Zhang, Y. Wang, S. Nie, C. Li, and M. Xie, "Sulfated modification of the polysaccharides from Ganoderma atrum and their antioxidant and immunomodulating activities," *Food Chemistry*, vol. 186, pp. 231–238, 2015.

[48] F. Lai, Q. Wen, L. Li, H. Wu, and X. Li, "Antioxidant activities of water-soluble polysaccharide extracted from mung bean (Vigna radiata L.) hull with ultrasonic assisted treatment," *Carbohydrate Polymers*, vol. 81, no. 2, pp. 323–329, 2010.

[49] A. Inoue, N. Kodama, and H. Nanba, "Effect of maitake (Grifola frondosa) D-fraction on the control of the T lymph node Th-1/Th-2 proportion," *Biological & Pharmaceutical Bulletin*, vol. 25, no. 4, pp. 536–540, 2002.

[50] H. Qi, T. Zhao, Q. Zhang, Z. Li, Z. Zhao, and R. Xing, "Antioxidant activity of different molecular weight sulfated polysaccharides from Ulva pertusa Kjellm (Chlorophyta)," *Journal of Applied Phycology*, vol. 17, no. 6, pp. 527–534, 2005.

[51] M. Zhang, X. Tang, F. Wang, Q. Zhang, and Z. Zhang, "Characterization of Lycium barbarum polysaccharide and its effect on human hepatoma cells," *International Journal of Biological Macromolecules*, vol. 61, pp. 270–275, 2013.

[52] J.-F. Yuan, Z.-Q. Zhang, Z.-C. Fan, and J.-X. Yang, "Antioxidant effects and cytotoxicity of three purified polysaccharides from Ligusticum chuanxiong Hort," *Carbohydrate Polymers*, vol. 74, no. 4, pp. 822–827, 2008.

[53] I. A. Schepetkin, G. Xie, L. N. Kirpotina, R. A. Klein, M. A. Jutila, and M. T. Quinn, "Macrophage immunomodulatory activity of polysaccharides isolated from Opuntia polyacantha," *International Immunopharmacology*, vol. 8, no. 10, pp. 1455–1466, 2008.

[54] S. A. Im, S. T. Oh, S. Song et al., "Identification of optimal molecular size of modified Aloe polysaccharides with maximum immunomodulatory activity," *International Immunopharmacology*, vol. 5, no. 2, pp. 271–279, 2005.

[55] Y. Fan, X. He, S. Zhou, A. Luo, T. He, and Z. Chun, "Composition analysis and antioxidant activity of polysaccharide from Dendrobium denneanum," *International Journal of Biological Macromolecules*, vol. 45, no. 2, pp. 169–173, 2009.

14

An Optimization Study of Polyacrylamide-Polyethylenimine-Based Polymer Gel for High Temperature Reservoir Conformance Control

Zulhelmi Amir[1,2] **Ismail Mohd Saaid**[1] **and Badrul Mohamed Jan**[2,3,4]

[1]*Petroleum Engineering Department, Universiti Teknologi PETRONAS, 32610 Bandar Seri Iskandar, Perak Darul Ridzuan, Malaysia*
[2]*Department of Chemical Engineering, Faculty of Engineering, University of Malaya, 50603 Kuala Lumpur, Malaysia*
[3]*Center Advanced Frontier Science, Faculty of Science, University of Malaya, 50603 Kuala Lumpur, Malaysia*
[4]*Center for Energy Science, Department of Mechanical Engineering, University of Malaya, 50603 Kuala Lumpur, Malaysia*

Correspondence should be addressed to Ismail Mohd Saaid; ismailsaaid@utp.edu.my

Academic Editor: Shu Seki

This paper presents optimization formulation of organically crosslinked polymer gel for high temperature reservoir conformance control using response surface methodology (RSM). It is always desirable to approach an optimal polymer gel formulation study with adequate performance information related to viscosity and gelation time to minimize excessive water production. In this paper, the effects of polymer and crosslinker concentrations and their influences on gelation time and viscosity were investigated. Central composite design (CCD) was used to determine the optimized organically crosslinked polymer gel formulation. Concentrations of two main raw materials, namely, polyacrylamide (PAM) and polyethylenimine (PEI), were varied in a suitable range. This was to obtain the formulation with the desirable two vital responses, which are viscosity and gelation time. It was found that the results fitted the quadratic equation. Statistically, the quadratic model is reliable and adequate perfectly the variability of the responses obtained from the experimental data. In addition, gelation time and gel viscosity may be controlled by adjusting both polymer and crosslinker concentrations. The optimum formulated organically crosslinked polymer gel with significant desirability factor conditions was achieved at 1.5% *w/v* of PAM and 0.3% *v/v* of PEI.

1. Introduction

The well conformance issue is always associated with poor sweep efficiency during waterflooding [1]. Excessive water production is an indication of low sweep efficiency due to poor well conformance [2]. Generally, a perfect well conformance exhibits a consistent sweep across the whole reservoir, while an imperfect well conformance results in a pocket unrecovered oil left behind [3]. There is no reservoir with ideal conformance in the course of its oil recovery process. Heterogeneous or layered oil reservoir formation with high permeability streak is the common nature of a well with conformance challenges. In this case, injected water flows to a wellbore through its easiest own path in the regions with relatively higher permeability. Thus, oil is bypassed by injected water and leads to lots of unswept oil in the relatively lower permeability region. In many mature reservoirs with long waterflooding history, water cuts can easily rise above 90%, greatly reducing the field profitability [4, 5]. They also lead to active interference in well productivity, early abandonment of the well due to unrecoverable hydrocarbon for mature oilfields, and increase in operating costs to challenge the effect of high produced water [6, 7].

Essential treatments have been introduced to minimize excessive water production to increase hydrocarbon production [8]. As one of the various water management strategies, a polymer gel system has become one of the most pertinent blocking agents to mitigate excessive water production [9–12]. Polymer gel can be deployed as a plug in reducing permeability of the high permeability region. Through the altered permeability, a new batch of injected water may change the direction of the flow, which displaces fluids into unswept "higher permeability" zones where oil is located. It is the least resistance pathways between the injected water source and production wells during the recovery process [5]. Polymer gel is not limited in improving oil recovery, but it also has other beneficial traits over other methods, such as controllable gelation time, adjustable gel strength, the flexibility for pumping without a work-over rig, in-depth penetration, long effective time, low cost, and easy to remove from the wellbore by water recirculation [11, 13–15]. Sorbie and Seright have expounded the concept of gel placement into reservoirs at three stages: injection of low viscosity gelant into the target zones, water injection to displace the low viscosity gel away from the wellbore, and finally shut in well to allow gelation to occur [16]. These polymer gels were designed to propagate through fractures and layers in high permeability and high water saturation zones without damaging productive zones. They help in creating a barrier to prevent water penetration and consequently to reduce the flow of unwanted water in the reservoir [4, 17].

A typical gel system usually composed of two main components, namely, high molecular weight polymer and crosslinker. The crosslinking agent starts to connect itself chemically to two polymer molecules and linking them together by internal or external triggers. Nowadays, more reservoirs with high temperatures are being explored and developed, when exploration of new petroleum fields continues [18]. For elevated temperature application, organically crosslinked polymer gel is preferred over metallically crosslinked because of its thermal stability [17, 19–23]. The covalent bonding between polymer and organic crosslinker provides stable and stronger gels. They have longer gelation times and high thermal stability at high temperatures than coordinated ionic bonding in inorganically crosslinked gel [17, 24]. The most popular commercial gel used in water management is prepared by polyacrylamide (PAM). PAM is readily available with relatively low costs, and it can deliver desirable viscosity and gel strength to block the high permeability fractures or channels. Meanwhile, polyethylenimine (PEI) is commonly studied organic crosslinker and applied in oilfields worldwide in recent years [25]. This paper presents a study on the application of response surface methodology (RSM) in the optimization of the formulation of polymer gel using middle range molecular weight of PAM crosslinked with PEI. The main objective is to obtain the best gel formulation with maximum achievable gelation time and acceptable viscosity.

2. Experimental Approach

2.1. Materials. The used polymer in this study is nonionic polyacrylamide (PAM) that was obtained from Sigma-Aldrich. It is in powder form with a pH of 6 and was used as provided. The molecular weight of PAM is with average 5,000,000–6,000,000 mol. wt. Polyethylenimine (PEI) was used as crosslinker provided by BASF with a pH of approximately 11.7. The molecular weight and active content of PEI were 35,000 mol. wt. and 99%, respectively, as disclosed by the supplier. The chemical structure of PAM and PEI was given in Figure 1. Deionized water was made in our laboratory.

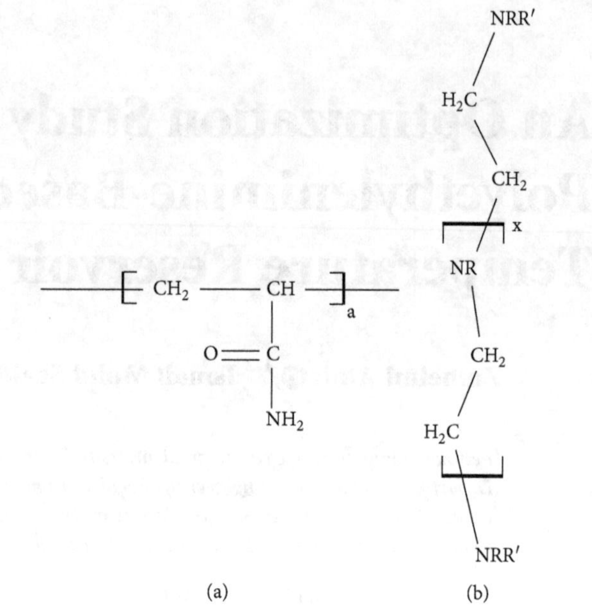

Figure 1: Chemical structure of (a) polyacrylamide and (b) polyethylenimine. Figures from El-Karsani et al. [17].

2.2. Preparation of Polymer Gelant. Polymer gelant in this study was prepared at room temperature. All of the PAM were measured by weight, while PEI were determined by volume. PAM solutions at various concentrations were prepared by mixing initially the polymer powder and deionized water for an hour using the IKA-RW20 digital mixer at 200 rpm to ensure the uniform dissolution of polymer particles. The time and speed of sample mixing were chosen to minimize the shear degradation in PAM samples. Then, various concentrations of PEI were added slowly to the polymer solution. Sequently, hydrochloric acid (HCl) was used to adjust the pH of the gelling solutions to 7 because the gelation process is optimum in neutral condition. The mixture was then stirred further for another 10 minutes.

2.3. Experimental Design and Data Analysis. Statistical design of the experiments with response surface methodology (RSM) was used to formulate the polymer gel systems with desired performance. The RSM is a useful statistical and mathematical method in modelling and optimization studies. To achieve this objective, the statistical software Design Expert version 11.0.3 (Stat-Ease Inc., Minneapolis, USA) was used. In the first step, the steady shear viscosity measurements were designed by central composite design (CCD) with two factors in five levels. The factors are concentrations of PAM and PEI, while the responses are viscosity and gelation time. The factor column in Table 1 presents the range and combination of the independent variable runs of the CCD approach. The experimental design comprises of four factorial points, four axial points at a distance of ±2 from the center, and five replicates of the center point. The two

Table 1: Results of the responses of each experimental run.

Run	Factors		Responses	
	A	*B*	Gelation time (minutes)	Viscosity (mPa.s)
1	2.00	0.5	56.5	478.7
2	2.00	0.5	58.0	504.7
3	2.00	0.2	66.5	533.5
4	1.50	0.3	75.5	286.0
5	2.00	0.5	61.5	487.9
6	3.00	0.5	37.5	947.6
7	2.00	0.5	59.0	531.6
8	2.00	0.5	53.5	470.3
9	2.50	0.3	57.0	733.4
10	2.00	0.9	41.0	620.7
11	1.00	0.5	65.5	88.9
12	1.50	0.7	69.5	346.5
13	2.50	0.7	45.5	736.3

variables were set at five levels which are −2, −1, 0, 1, and 2. The experiments were conducted as per designed plan before the obtained data were plugged into the software. The CCD approach was employed to generate mathematical models from experimental data. The correlation can be used to describe comprehensive relationship between the factors and responses and to predict the gelation time and viscosity of the polymer gel. Furthermore, three-dimensional plots and their respective contour plots were also considered to analyze the relationship. The correlation in CCD is usually shown as a quadratic equation. Then, analysis of variance (ANOVA) was used to analyze statistically the effects of process variables, which is to evaluate the reliability and accuracy of the generated model. To confirm the reliability of the generated models, the R^2 value was calculated. Then, its statistical significance was determined by Fisher's F-test using the same program. In addition, P value (probability of error) with 95% confidence level was used to evaluate the model terms. Finally, in the optimization process, the formulation for polymer gelant was established based on the maximum possible gelation time and acceptable minimum viscosity.

2.4. Viscosity and Gelation Time Measurements. Literature reported some methods applied to determine the gelation kinetics [19, 26, 27]. The methods that are usually used include the bottle test, the sealed tube, the dynamic shear using rheometer, and the static shear using viscometer methods. In this study, the static shear method was used to evaluate the effects of different concentrations of raw materials on the gelation time and viscosity of polymer gel. The steady shear method has been widely used because this technique provides a quantitative measurement of gelation rate, gelation time, and gel viscosity. Moreover, this method is reasonably faster to get the results, accurate, and not very costly compared than the other methods. It has been used by several researchers to determine the gelation time [19, 28]. The viscosity and gelation time were determined from the curves of viscosity against time at high temperatures. The inflection point of the viscosity versus time curve was considered as the gelation point. In the present study, the viscosity and gelation time measurements were taken using a rotational Brookfield Viscometer Model DV-II (AMETEK Brookfield, Massachusetts, USA). This viscometer is equipped with a cylindrical sample holder, a heating jacket, a spindle, and a sealing element. During each test, the gelling solutions were prepared and placed into the sample holder that has been preheated to desired temperature. The volume of the sample was approximately 8 ml. The measurements were taken at temperature of 95°C and ambient pressure. The temperature was controlled by a thermostatic oil bath (Haake CH bath reservoir with Haake F3 temperature controller) connected to the viscometer. Then, the spindle was immersed into the sample, and a steady shear rate of $5.0\,s^{-1}$ was applied for all experiments. The sample was always stirred by the viscometer spindle in the sample cup throughout the experiment. The condition to ensure the temperature throughout the sample was the same. Viscosity data was automatically worked out using a software, and the desired viscosity data was recorded in correspondence to experimental time. Based on the CCD designed matrix, all measuring runs were performed randomly to minimize the effect of unexpected variability in the observed responses. Finally, laboratory tests were carried out in the same conditions based on the generated optimal formulation. This is to validate the mathematical values generated by the software. The error and accuracy between the predicted and experimental values were obtained to determine the precision of the software-generated formula.

3. Results and Discussions

3.1. Gelation Kinetics and Statistical ANOVA Analysis. Measurements of PAM/PEI polymer gel viscosities were performed at 95°C. Figure 2 shows the relationship of PAM/PEI polymer gel viscosities as a function of experimental time. These results have been used for the ANOVA statistical analysis as presented in Table 1. The CCD design in RSM was used to investigate the correlation between the two independent variables, namely, concentrations of polyacrylamide and polyethyleimine, and the two responses, namely, viscosity and gelation time. The analysis predicted and generated mathematical models for each of the responses. These results were used to assess the predicted responses for viscosity and gelation time as a function of polyacrylamide and polyethylenimine concentrations.

3.1.1. Gelation Time. The experimental results as shown in contour and 3D response surface plots (Figure 3) indicate the relationship between gelation time, as a function of polyacrylamide and polyethylenimine concentrations. The gelation time is crucial because it is used in oilfield applications to determine the ability of PAM/PEI polymer gelant to travel deeper through the reservoir. The injection time and depth of propagation of the gel solution into the formation are a function of gelation time. There are different definitions of gelation time reported. Firstly, it is described as the time required to achieve a specific viscosity [29]. Secondly, it is the time required to achieve a specific gel strength [30]. In

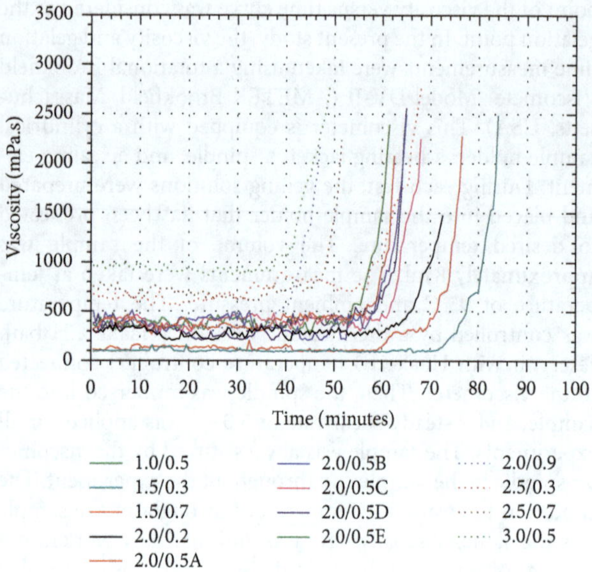

FIGURE 2: Viscosity of PAM/PEI polymer gel against experimental time at various polymer and crosslinker concentrations.

this case, oscillatory viscometer is commonly used to determine the gelation time by measuring the changes of gel strength as a function of time. The maximum slope in the curve of elastic modulus against time can be taken as the gelation period [31]. The last definition of the gelation time is the time required for the elastic and viscous moduli to intersect, and the moduli ratio becomes independent of frequency [32]. In this study, the gelation time is defined as the duration of time required for viscosity to reach the inflection point on the curve of viscosity versus time [21]. The inflection point in the gelation process can be observed as rapidly increasing in viscosity after it starts slowly with no apparent viscosity increment at the beginning. This definition has been used by numerous researches to define the gelation time [26, 28].

It can be seen in Figure 3 that as the polymer concentration increases at the same crosslinker concentration, the gelation time decreases. For instance, when the polymer concentration was less than 2.0%, the gelation time was longer than 60 minutes. Conversely, when the polymer concentration was higher than 3.0%, the gelation time was shorter than 50 minutes. As the polymer concentration or the number of crosslinking sites increases, the rate of gel formation increases, and thereby, the gelation time increases. Similar to the effect of polymer concentration, the gelation time of polymer gel was found to be fairly affected by the initial crosslinker concentrations. Experimental results showed that if the crosslinker system concentration was less than 0.5%, the gelation time would be more than 70 minutes. This is expected because higher PEI concentrations offer more crosslinking opportunities for the polymer. This can also be explained that the available imine groups increased with the increase of PEI content; hence, the crosslinking rate is improved, and consequently, the gelation time shortened. The functional groups of PAM, which are amide groups (-$CONH_2$), will immediately form gel as soon as they react with crosslinker under high temperature. This observation is in agreement with the data reported for organically crosslinked systems [21]. However, it can be observed that the rate of the increase of gelation time by the effect of PEI is less pronounced compared with the effect of PAM. There is also an understanding that delaying gelation is more efficient by retardation of polymer reaction over the crosslinker side [17]. Therefore, PAM is more dominant over PEI in determining the gelation time of the PAM/PEI gel system.

Experimental results for the first response, which is gelation time, were fitted into the equation by the analysis of variance (ANOVA). This equation is a mathematical model that can be used to predict the desired responses at any of the respective factors. ANOVA was used to obtain the model that fits the most to the results. The best equation that fits the results of gelation time is quadratic in nature. The quadratic model to describe the correlation between gelation time (Y_1) as a function of concentration of raw materials in terms of coded factors is expressed as

$$Y_1 = 59.41 - 8.21A - 6.09B - 1.37AB - 1.42A^2 - 0.56B^2, \quad (1)$$

where A represents polyacrylamide concentration and B represents polyethylenimine concentration.

ANOVA analysis was also used to confirm the accuracy of the obtained model. The results of ANOVA analysis show that the model F value given was 8.74, which indicates that the model is significant. It gave indication that the model has only 0.64% chance to occur due to noise. The data also indicates that the mathematical model is significant at 5% confidence level because the value of probability of error (P) is less than 0.05. In addition, based on the results, the obtained model gives a very high R^2 and adjusted R^2 values of 0.8619 and 0.7633, respectively. According to Noordin et al., it is necessary to achieve high R^2 value that is close to 1 [33]. Moreover, having a good agreement with the adjusted R^2 value is also desirable. The high value of R^2 and adjusted R^2 validates the quadratic model is appropriate and satisfactorily fit to the variability of the responses produced from experimental data. The lack of fit F value is 6.31. Even though the lack of fit value relatively has low probability of 5.37% chance to occur due to noise, the lack of fit value implies that it is not significant relative to the pure error. A nonsignificant lack of fit is required as it demonstrates that the model is reliable.

In addition, from the ANOVA results, it is possible to evaluate the signal to noise ratio of the experimental data by assessing the adequate precision (AP). AP value compares the range of the predicted value at the design points [34]. A ratio of more than 4 is anticipated since this value shows the model discrimination. It means the model can be applied to navigate the design space [35]. Based on the results, the response for gelation yielded a high AP value of 9.0485, which is more than 4. This indicates that the predicted model provides an excellent ability to navigate the design space

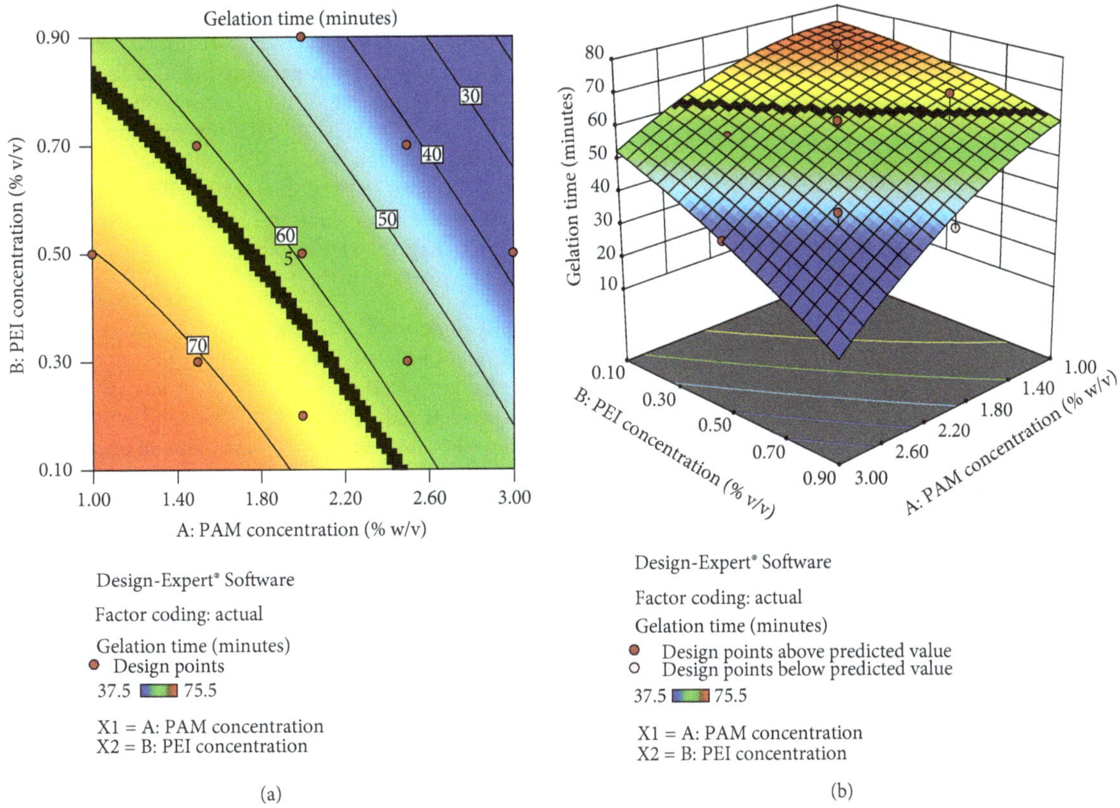

FIGURE 3: Contour and 3D response surface plots from Design Expert for gelation time versus concentration of materials.

designed by the CCD. This is supported by a comparison plot between the predicted and actual values. This comparison plot confirms the accuracy and reliability of the model. Figure 4 presents the diagnostic graph between the predicted and actual values for gelation time. The meaning of the plots is described by the position of the points. Points that are lying above the diagonal line indicate that the values were overestimated, while those below the diagonal line indicate underestimated values [36]. It can be observed that all points are distributed along the diagonal line. The predicted values are in adequate agreement with the actual experimental values. Hence, it could be suggested that the obtained model gives a satisfactory estimation of the predicted values and may be used in predicting the gelation time.

3.1.2. Viscosity. The second response in this study is viscosity of the PAM/PEI based polymer gel at the gelation point. Viscosity is a crucial property of the liquid that manifests the relative motion of the molecules [37]. Basically, liquid flows until stress has diminished and energy has dissipated through the friction. Contradict to the trend of gelation time, the increment of all of the raw material concentrations tend to cause the viscosity of the polymer gel to increase. Viscosity can be influenced by the increment of polymer concentration as can be seen in Figure 5. Based on the results, when the polymer concentration was higher than 2.0%, the viscosity was higher than 400 mPa.s. Similarly, when the polymer concentration was almost 2.5%, the viscosity is more than 800 mPa.s. This is due to the higher polymer concentration that promotes the thickening process in the solutions. The thickening capability of PAM originated from the high molecular weight. In addition, as the polymer loading increase, more polymer chain could stretch in the aqueous solution. The repulsion of the negative electrostatic charges on the polymer chains is the main reason of polymeric stretching in aqueous solution, which consequently causes viscosity to increase [38, 39]. As a result, more crosslinkable sites will be accessible on the polymer backbone, consequently created a denser network of polymer [40]. Moreover, a higher number of polymer molecules lead to higher interactions. Thus, it leads to molecular entanglements between polymer chains [41]. This phenomenon is the reason that upholds higher viscosity for the polymer solutions [39]. On the other hand, the viscosity was found not affected significantly with the increase of crosslinker concentration. Based on the experimental results, polymer concentration seems to have much more influence over crosslinker concentration in determining the viscosity of the gel system [42]. It seems to suggest that a production of lower viscosity of polymer system required lower loading of polymer and crosslinker.

The results from the experimental works were used for the model fitting by ANOVA. Similar to the first response, the quadratic mathematical model is the most suitable for viscosity. The quadratic model to represent the relationship

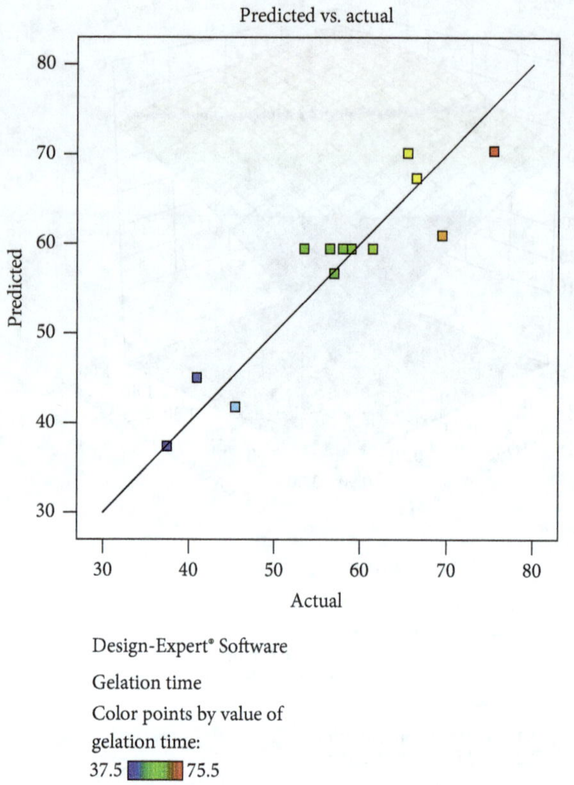

FIGURE 4: Predicted versus actual values plot from Design Expert for gelation time of PAM/PEI polymer gel.

between concentration of raw materials and viscosity (Y_2) in terms of coded factors is as follows

$$Y_2 = 494.88 + 212.88A + 13.82B - 14.4AB + 5.83A^2 + 24.95B^2, \quad (2)$$

where A represents polyacrylamide concentration and B represents polyethylenimine concentration.

From the results of ANOVA analysis, the model F value is 317.51, which means that the model is significant. The probability or P value is established to be lower than 0.05, which implies that the model terms, which are A and A^2 are significant. Values greater than 0.1 indicate the model terms are not significant. The R^2 and adjusted R^2 values are also reasonably near to 1.0, which are 0.9956 and 0.9925, respectively. The generated mathematical equation is well fitted when the difference between R^2 and adjusted R^2 values is less than 0.2. The lack of fit F value of 0.06 indicates that the lack of fit is not significant relative to the pure error. For a lack of fit value this large to occur due to noise, it needs a chance of 97.8%. As mentioned before, nonsignificant lack of fit is desired so that the model can be used to represent the viscosity. The AP ratio obtained in this analysis was 66.78 which was greater than 4. This is desirable and signifies that the model is reliable to navigate the design space. A comparison plot between the predicted and actual values has also been referred to justify the accuracy of the model. Figure 6 illustrates the statistical comparison plot between experimental values and the predicted values for viscosity using the ANOVA generated model. Based on the plot, it is clear that the actual values for gelation time are considered well fitted to predicted values with acceptable variance range. As can be seen, all the experiment design points are lying almost at the same spot and close to the straight line. This suggests that the generated models may provide a good viscosity prediction for PAM/PEI polymer gel.

3.2. Optimization of Polymer Gel Formulation. During the treatment of a deep reservoir, it is important to consider both the effect of gelation time and gelant viscosity when designing the required amount of polymer and crosslinker loading. It should be noted that optimum value of polymer and crosslinker concentration need to be determined. For example, the addition of high crosslinker may cause syneresis [43, 44]. Syneresis must be avoided, especially when gel is employed in naturally fractured reservoirs. Syneresis is caused by overcrosslinking from excessive concentration of crosslinker. Consequently, the water will expulse out of the gel by osmosis and polymer gel dehydrate [45]. Because of this, gel volume reduces, and new flow paths are created, eventually affecting the performances of gel treatment. Conversely, with increasing the polymer dosage, the time required to get a nonflowing polymer gel with a tolerable strength will be shortened. The main objective of this optimization part is to determine the optimum concentrations of raw materials. It supposedly gives the optimized formulation for PAM/PEI polymer gel that is reliable to be applied for high temperature reservoir conformance control. During the optimization process, the polymer gel is setting to have the acceptable lowest viscosity and possible maximum gelation time. A low to very low viscosity of the formulated polymer gel is desirable to obtain good carrying capacity of the fluid and longer propagation of gelant through the reservoir matrix. On the other hand, maximum gelation time is needed for longer propagation and to avoid the gel forming inside the wellbore. Thus, in order to simultaneously optimize all the two factors (concentrations of polyacrylamide and polyethylenimine) correlated to the responses (gelation time and viscosity), a desirability function (d) for multiple responses was used [46]. The values were generated by the Design Expert software when performing the multiple response optimization based on

$$d = \left[\prod_{i=1}^{N} m_i^{r_i}\right]^{1/\sum r_i}, \quad (3)$$

where N represents the number of responses, r_i indicates the importance of particular response, and m_i is the partial desirability function for specific responses.

Generally, the highest value of desirability d represents the optimum conditions that can be achieved from the optimization process [46]. In the optimization process, the gelation time of the PAM/PEI polymer gel was set more important than viscosity. Based on the results, the highest desirability value is 0.898. At this d value, the values of the

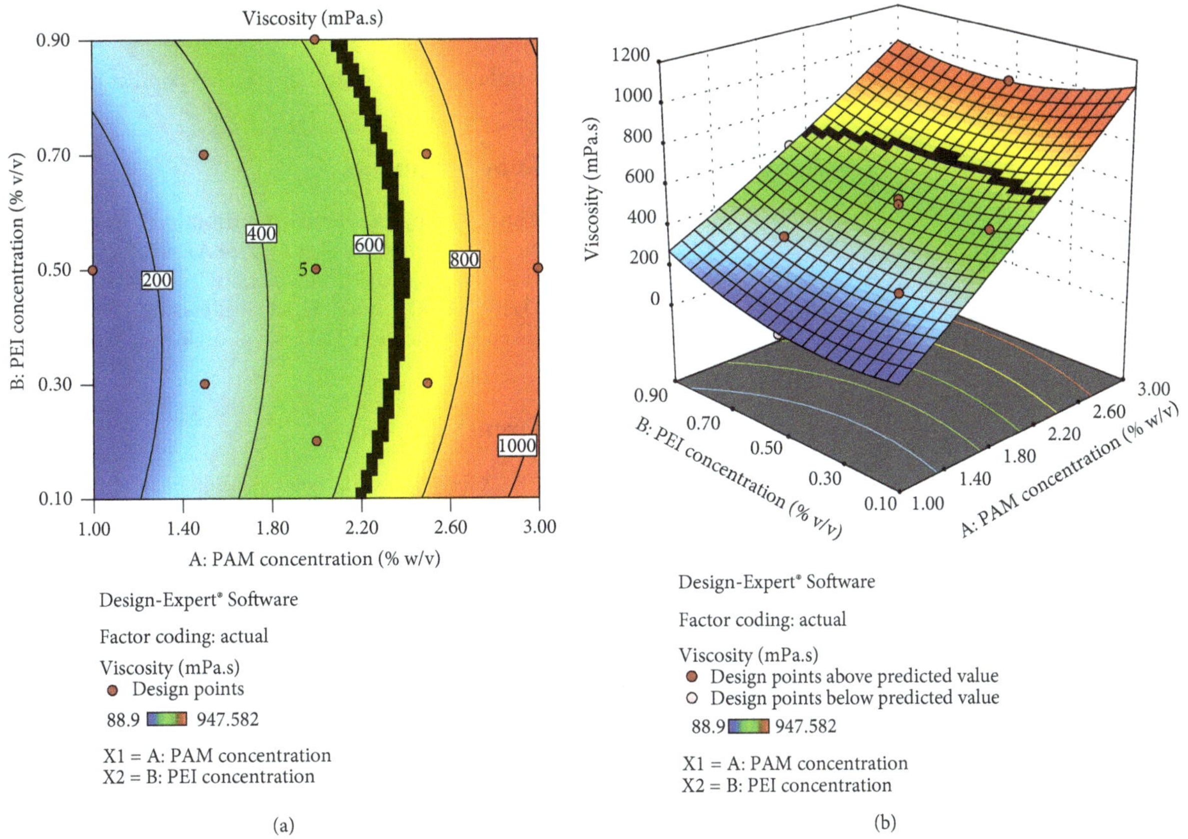

FIGURE 5: Contour and 3D response surface plots from Design Expert for viscosity versus concentration of materials.

partial desirability, m_i, for the two responses, gelation time and viscosity, are 0.86 and 0.23, respectively. This value corresponds to the optimum formulation where the value of factor A (concentration of polyacrylamide) is 1.5% w/v and factor B (concentration of polyethylenimine) is 0.3% v/v. The predicted values for each response parameters, which are gelation time viscosity (Y_1) and viscosity (Y_2), are 70.3 minutes and 284.6 mPa.s, respectively. According to Al-Muntasheri et al. [47], a good range of gelation time for polymer gel should be more than 55 minutes. Since the gelation time for the optimized formulation is about 70 minutes, it is considered good as it is well above 55 minutes. The viscosity on the beginning of the gelation point is relatively within the acceptable range for in situ polymer gel. Then, the d value is the highest achievable desirability; it could considerably represent the optimum condition for the process optimization. Thus, this condition was selected as the optimum concentration of polyacrylamide and polyethylenimine to formulate the PAM/PEI polymer gel.

To validate the optimum process conditions acquired from response surface methodology using CCD, a laboratory experimental test was conducted based on selected optimum conditions. Predicted and experimental conditions were maintained the same for a fair comparison between the two. The test was conducted by formulating the PAM/PEI polymer gel using the optimized formulations as mentioned above. The corresponding gelation time and viscosity were measured as shown in Figure 7 and then was compared to the predicted response values in optimum process condition calculated from the mathematical model. Based on the results, it can be concluded that the two response parameters obtained from the experiment are in good agreement with predicted values. This is due to the high value of calculated accuracy for gelation time and viscosity, which are 97% and 86.5%, respectively. Value of error is also low, which are 2.2 and 38.4 for gelation time and viscosity, respectively. In addition, low value of standard deviation (σ) is shown with values of 1.1 and 19.2 for gelation and viscosity.

4. Conclusions

A polymer gel incorporating polyacrylamide and polyethylenimine has been successfully formulated with the desired properties from RSM. The response surface plots showed that the viscosity of the gelant increases with increasing material concentrations. Gelant viscosity tends to increase with polyacrylamide and polyetylenimine concentrations. Gelation time of the PAM/PEI-based gel also tend to decrease with increasing material concentration. The experimental results for all responses fit the quadratic mathematical model predictions. For gelation time and viscosity, polymer concentration affects these two responses the most. It is concluded that the optimum concentrations for the

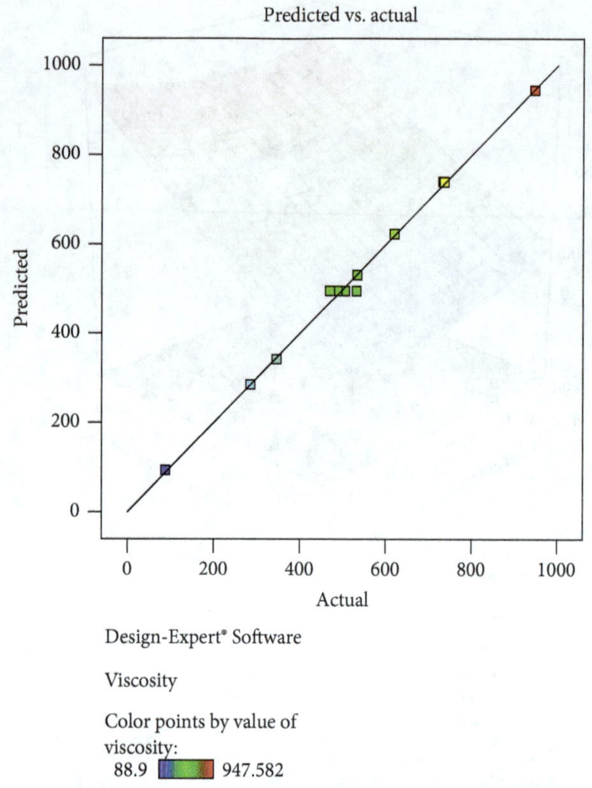

FIGURE 6: Predicted versus actual values plot from Design Expert for viscosity of PAM/PEI polymer gel.

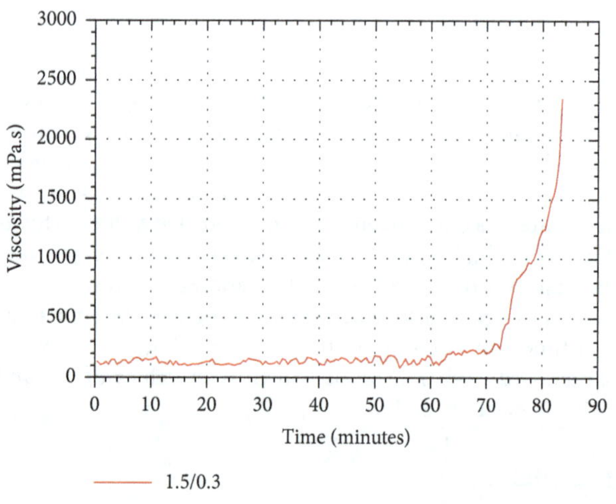

FIGURE 7: Viscosity of PAM/PEI polymer gel against experimental time at optimized polymer and crosslinker concentration.

formulation is achieved at the following conditions where the value of factor A (concentration of polyacrylamide) is 1.5% w/v, and factor B (concentration of polyethylenimine) is 0.3% v/v. This formulation has gelation time of 70.4 minutes and viscosity of 284.6 mPa.s. Furthermore, from the validation experiment, it could be concluded that the two response parameters obtained from both the experiments and predicted by the models were in a good agreement.

Conflicts of Interest

The authors declare that they have no conflict of interest.

Acknowledgments

The authors appreciate the contributions from Universiti Teknologi Petronas (YUTP 0153AA-H05), PETRONAS (GR&T 0153CB-019), University of Malaya (UMRG RP031B-15AFR), and SLAI Fellowship Scheme from Ministry of Education Malaysia and University of Malaya.

References

[1] R. D. Sydansk and L. Romero-Zeron, *Reservoir Conformance Improvement*, Society of Petroleum Engineers Richardson, TX, 2011.

[2] B. Bailey, M. Crabtree, J. Tyrie et al., “Water control,” *Oilfield Review*, vol. 12, pp. 30–51, 2000.

[3] D. Borling, K. Chan, T. Hughes, and R. Sydansk, “Pushing out the oil with conformance control,” *Oilfield Review*, vol. 6, pp. 44–58, 1994.

[4] R. D. Sydansk and G. P. Southwell, “More than 12 years of experience with a successful conformance-control polymer gel technology,” in *SPE/AAPG Western Regional Meeting*, Long Beach, California, June 2000.

[5] R. D. Sydansk and R. S. Seright, “When and where relative permeability modification water-shutoff treatments can be successfully applied,” *SPE Production & Operations*, vol. 22, no. 2, pp. 236–247, 2007.

[6] R. L. Hirsch, R. Bezdek, and R. Wendling, “Peaking of world oil production and its mitigation,” *AICHE Journal*, vol. 52, no. 1, pp. 2–8, 2006.

[7] R. S. Seright, R. H. Lane, and R. D. Sydansk, “A strategy for attacking excess water production,” *SPE Production & Facilities*, vol. 18, no. 3, pp. 158–169, 2003.

[8] R. Seright, “Washout of Cr(III)-acetate-HPAM gels from fractures,” in *International Symposium on Oilfield Chemistry*, Houston, Texas, February 2003.

[9] A. Kabir, “Chemical water & gas shutoff technology—an overview,” in *SPE Asia Pacific Improved Oil Recovery Conference*, Kuala Lumpur, Malaysia, October 2001.

[10] J.-P. Coste, Y. Liu, B. Bai et al., “In-depth fluid diversion by pre-gelled particles. Laboratory study and pilot testing,” in *SPE/DOE improved oil recovery symposium*, Tulsa, Oklahoma, April 2000.

[11] A. Prada, F. Civan, and E. D. Dalrymple, “Evaluation of gelation systems for conformance control,” in *SPE/DOE improved oil recovery symposium*, Tulsa, Oklahoma, April 2000.

[12] G. A. Al-Muntasheri, H. A. Nasr-El-Din, J. Peters, and P. L. J. Zitha, “Investigation of a high temperature organic water shutoff gel: reaction mechanisms,” *SPE Journal*, vol. 11, no. 4, pp. 497–504, 2006.

[13] A. Hamouda and H. Amiri, “Factors affecting alkaline sodium silicate gelation for in-depth reservoir profile modification,” *Energies*, vol. 7, no. 2, pp. 568–590, 2014.

[14] H. Akhlaghi Amiri, A. Hamouda, and A. Roostaei, “Sodium silicate behavior in porous media applied for in-depth profile modifications,” *Energies*, vol. 7, no. 4, pp. 2004–2026, 2014.

[15] D. Perez, F. E. Fragachan, A. R. Barrera, and J. P. Feraud, "Applications of polymer gel for establishing zonal isolations and water shutoff in carbonate formations," *SPE Drilling & Completion*, vol. 16, no. 3, pp. 182–189, 2001.

[16] K. S. Sorbie and R. S. Seright, "Gel placement in heterogeneous systems with crossflow," in *SPE/DOE Enhanced Oil Recovery Symposium*, Tulsa, Oklahoma, April 1992.

[17] K. S. M. El-Karsani, G. A. Al-Muntasheri, and I. A. Hussein, "Polymer systems for water shutoff and profile modification: a review over the last decade," *SPE Journal*, vol. 19, no. 1, pp. 135–149, 2014.

[18] R. Ziegler, "Technology focus: high-pressure/high-temperature challenges (April 2017)," *Journal of Petroleum Technology*, vol. 69, no. 4, pp. 79–79, 2017.

[19] D. Broseta, O. Marquer, N. Blin, and A. Zaitoun, "Rheological screening of low-molecular-weight polyacrylamide/chromium(III) acetate water shutoff gels," in *SPE/DOE improved oil recovery symposium*, Tulsa, Oklahoma, April 2000.

[20] A. Moradi-Araghi, "A review of thermally stable gels for fluid diversion in petroleum production," *Journal of Petroleum Science and Engineering*, vol. 26, no. 1-4, pp. 1–10, 2000.

[21] G. A. Al-Muntasheri, H. A. Nasr-El-Din, and I. A. Hussein, "A rheological investigation of a high temperature organic gel used for water shut-off treatments," *Journal of Petroleum Science and Engineering*, vol. 59, no. 1-2, pp. 73–83, 2007.

[22] K. S. M. El-Karsani, G. A. Al-Muntasheri, A. S. Sultan, and I. A. Hussein, "Gelation of a water-shutoff gel at high pressure and high temperature: rheological investigation," *SPE Journal*, vol. 20, no. 5, pp. 1103–1112, 2015.

[23] R. S. Seright and F. D. Martin, "Impact of gelation pH, rock permeability, and lithology on the performance of a monomer-based gel," *SPE Reservoir Engineering*, vol. 8, no. 1, pp. 43–50, 1993.

[24] J. E. Vasquez, E. D. Dalrymple, I. Abbasy, and L. S. Eoff, "Laboratory evaluation of water swellable materials for fracture shutoff," in *SPE North Africa Technical Conference & Exhibition*, Marrakech, Morocco, March 2008.

[25] J. Morgan, P. Smith, and D. Stevens, "Chemical adaptation and development strategies for water and gas shutoff gels," in *RSC Chemistry in the Oil Industry, 6th International Symposium*, Charllotte Mason College Ambleside, UK, 1997.

[26] J. Vasquez, E. D. Dalrymple, L. Eoff, B. Reddy, and F. Civan, "Development and evaluation of high-temperature conformance polymer systems," in *SPE International Symposium on Oilfield Chemistry*, The Woodlands, Texas, February 2005.

[27] J. Vasquez, F. Civan, T. M. Shaw et al., "Laboratory evaluation of high-temperature conformance polymer systems," in *SPE production and operations symposium*, Oklahoma City, Oklahoma, March 2003.

[28] B. R. Reddy, L. Eoff, E. D. Dalrymple, K. Black, D. Brown, and M. Rietjens, "A natural polymer-based cross-linker system for conformance gel systems," *SPE Journal*, vol. 8, no. 2, pp. 99–106, 2003.

[29] D. S. Jordan, D. W. Green, R. E. Terry, and G. P. Willhite, "The effect of temperature on gelation time for polyacrylamide/chromium (III) systems," *Society of Petroleum Engineers Journal*, vol. 22, no. 4, pp. 463–471, 1982.

[30] R. D. Sydansk, "Acrylamide-polymer/chromium (III)-carboxylate gels for near wellbore matrix treatments," *SPE Advanced Technology Series*, vol. 1, no. 1, pp. 146–152, 1993.

[31] R. Prud'homme and J. Uhl, "Kinetics of Polymer/Metal-Ion Gelation," in *SPE Enhanced Oil Recovery Symposium*, Tulsa, Oklahoma, April 1984.

[32] A. Ponton, S. Warlus, and P. Griesmar, "Rheological study of the sol–gel transition in silica alkoxides," *Journal of Colloid and Interface Science*, vol. 249, no. 1, pp. 209–216, 2002.

[33] M. Y. Noordin, V. C. Venkatesh, S. Sharif, S. Elting, and A. Abdullah, "Application of response surface methodology in describing the performance of coated carbide tools when turning AISI 1045 steel," *Journal of Materials Processing Technology*, vol. 145, no. 1, pp. 46–58, 2004.

[34] M. Khalil, B. Mohamed Jan, and A. Aziz Abdul Raman, "Fractional factorial design optimization of nontraditional completion fluid for perforation with underbalance," *Chemistry and Technology of Fuels and Oils*, vol. 46, no. 5, pp. 340–350, 2010.

[35] Q. K. Beg, V. Sahai, and R. Gupta, "Statistical media optimization and alkaline protease production from Bacillus mojavensis in a bioreactor," *Process Biochemistry*, vol. 39, no. 2, pp. 203–209, 2003.

[36] A. L. Ahmad, S. C. Low, S. R. A. Shukor, and A. Ismail, "Optimization of membrane performance by thermal-mechanical stretching process using responses surface methodology (RSM)," *Separation and Purification Technology*, vol. 66, no. 1, pp. 177–186, 2009.

[37] P. C. Hiemenz and T. P. Lodge, *Polymer Chemistry*, CRC Press, 2007.

[38] R. J. Stokes and D. F. Evans, *Fundamentals of Interfacial Engineering*, John Wiley & Sons, 1997.

[39] D. A. Z. Wever, F. Picchioni, and A. A. Broekhuis, "Polymers for enhanced oil recovery: a paradigm for structure–property relationship in aqueous solution," *Progress in Polymer Science*, vol. 36, no. 11, pp. 1558–1628, 2011.

[40] R. Reichenbach-Klinke, B. Langlotz, B. Wenzke, C. Spindler, and G. Brodt, "Associative copolymer with favorable properties for the application in polymer flooding," in *SPE International Symposium on Oilfield Chemistry*, The Woodlands, Texas, USA, April 2011.

[41] R. P. Wool, "Polymer entanglements," *Macromolecules*, vol. 26, no. 7, pp. 1564–1569, 1993.

[42] K. S. M. ElKarsani, G. A. al-Muntasheri, A. S. Sultan, and I. A. Hussein, "Performance of PAM/PEI gel system for water shut-off in high temperature reservoirs: laboratory study," *Journal of Applied Polymer Science*, vol. 132, no. 17, 2015.

[43] S. L. Bryant, M. R. Rabaioli, and T. P. Lockhart, "Influence of syneresis on permeability reduction by polymer gels," *SPE Production & Facilities*, vol. 11, no. 4, pp. 209–215, 1996.

[44] R. W. Eggert Jr, G. P. Willhite, and D. W. Green, "Experimental measurement of the persistence of permeability reduction in porous media treated with xanthan/Cr(III) gel systems," *SPE Reservoir Engineering*, vol. 7, no. 1, pp. 29–35, 1992.

[45] Q. Ren, H. Jia, D. Yu et al., "New insights into phenol–formaldehyde-based gel systems with ammonium salt for low-temperature reservoirs," *Journal of Applied Polymer Science*, vol. 131, no. 16, 2014.

[46] D. Montgomery, *Design and Analysis of Experiments Fifth Edition*, By John Wiley & Sons, Inc. All rights reserved, 2001.

15

Controlled Depolymerization of Cellulose Fibres Isolated from Lignocellulosic Biomass Wastes

Suh Cem Pang, Lee Ken Voon, and Suk Fun Chin

Resource Chemistry Programme, Faculty of Resource Science & Technology, Universiti Malaysia Sarawak, 94300 Kota Samarahan, Sarawak, Malaysia

Correspondence should be addressed to Suh Cem Pang; scpang@unimas.my

Academic Editor: Qinglin Wu

Various types of lignocellulosic biomass wastes (LBW) had been successfully converted into cello-oligomers with different chain lengths via a controlled depolymerization process. Cellulose fibres isolated from LBW samples were dissolved with room temperature ionic liquid (RTIL) in the presence of an acid catalyst, Amberlyst 15 DRY. The effects of reaction time on the degree of polymerization and yields of water-insoluble cello-oligomers formed were studied. Besides, the yields of water-soluble cello-oligomers such as glucose and xylose were also determined. The depolymerization of cellulose fibres isolated from LBW was observed to follow both second-order and pseudo-second order kinetics under specific conditions. As such, cello-oligomers of controllable chain lengths could be obtained by adjusting the duration of depolymerization process under optimized conditions.

1. Introduction

In recent decades, the production of biofuels from renewable biomass have gained increasing attention of scientists worldwide [1]. However, the use of food resources such as glucose derived from direct extraction of saccharose in sugarcane and sugar beet or through hydrolysis of corn starch for production of biofuels would inevitably compete with food supply [2]. As such, the use of cellulose fibres derived from lignocellulosic biomass wastes (LBW) as the feedstock is considered to be more appropriate than starch and saccharose. Cellulose is the most abundant organic compound on earth, and utilization of cellulose fibres derived from LBW does not compete with food supply [3]. Nevertheless, the presence of extensive intra- and intermolecular hydrogen bond networks, the basicity of glycosidic bonds, and high crystallinity of cellulose have posed great challenges for breaking down cellulose fibres through both chemical or biological processes [4, 5].

In conventional hydrolysis of lignocellulosic biomss materials, mineral acids such as hydrochloric acid, sulfuric acid, or even hydrofluoric acid are being used as liquid catalysts [6]. These liquid catalysts need to be employed in high concentration and/or at high reaction temperature (170–240°C) in order to enhance product yields. Typically, lignocellulosic raw materials are not being completely solubilized in the reaction medium [7]. Although valuable products are obtained from the hydrolysis process, full-scale hydrolyses of cellulose have never been commercially implemented due to issues which include corrosion, high energy demand, catalyst recovery, and degradation of sugars.

In recent years, ionic liquids have been used as solvents and catalysts in various fields of research [8, 9]. Ionic liquids are defined as salts with a melting point below 100°C which exhibit several unique properties such as negligible vapor pressure, extraordinary dissolution properties, wide liquid range, high electrical conductivity, low volatility, and low flammability [10, 11]. Ionic liquids have been reported to dissolve cellulose for up to 25 wt% [12]. This has led to intense research into the conversion of biomass in ionic liquids. Besides, ionic liquids can be easily recovered and recycled by evaporating water or alcoholic residues using vacuum distillation and then purified by passing it through a neutral alumina column [5, 13]. All these unique properties of ionic liquid have led to their uses as the preferred solvent for the depolymerization of cellulose.

The depolymerization of cellulose in ionic liquid occurs by the addition of hydronium ion (H_3O^+) species, which cleavage the 1,4 β-glycosidic bonds to form either cello-oligomers or glucose during the initial stage. However, the presence of H_3O^+ ions can further dehydrate glucose to produce 5-hydroxymethylfurfural, levulinic, and formic acids. These side products are prone to undergo aldol condensation with glucose to form humins which are difficult to be separated from useful products. Such reactions would reduce the efficiency of producing usable feedstock from lignocellulosic biomass. Therefore, it is important to control the depolymerization process such that reaction could be stopped at early stages in order to generate cello-oligomers instead of glucose and to avoid the formation of dehydration and polymeric by-products. However, the production of humins could not be avoided when mineral acid was being used as the catalyst during the hydrolysis of cellulose which occurred instantaneously [14].

Rinaldi and coworkers have reported the use of solid acids such as Amberlyst 15, Amberlyst 35, and Amberlyst 70, as well as Nafion, sulfated zirconate, and zeolite Y as catalysts for the hydrolysis of cellulose dissolved in ionic liquid [15]. The macroreticulated resins functionalized with sulfonic groups (Amberlyst 15, Amberlyst 35) showed good catalytic activity for the depolymerization of cellulose fibres in an ionic liquid. Amberlyst acid resins are highly stable in acidic anion ionic liquid as compared to other basic anion ionic liquids such as 1-butyl-3-methylimidazolium acetate, which leads to rapid decomposition of the catalyst [15]. During the depolymerization of cellulose in ionic liquid, H_3O^+ ions are being progressively released from Amberlyst 15 into ionic liquid within the first hour of contact between cellulose solution and solid catalyst, resulting in an induction period for the formation of glucose [5]. This shows that the depolymerization of cellulose solubilized in ionic liquid in the presence of Amberlyst 15 could afford control of the hydrolysis process during the early stage of cello-oligomer production. However, the study on controlling chain lengths of cello-oligomers is still very lacking. Understanding the mechanisms of the depolymerization process is essential for enabling the production of cello-oligomers with different chain lengths which are suitable for conversion to biofuels or methylcellulose as well as platform chemicals such as surfactants, thickening agents, and glues and as rheology modifiers [16].

Herein, we have reported our findings on controlled depolymerization of cellulose fibres which were isolated from four different types of lignocellulosic biomass wastes including sawdust from wood of *Acacia magnium*, sugarcane bagasse, corn cob, and pith waste of sago (*Metroxyln sagu*). Cellulose fibres were depolymerised by dissolving in ionic liquid in the presence of an acid catalyst, Amberlyst 15 DRY. The yields of cello-oligomers, the mean degree of polymerization (DPw), and the distribution of DPw as well as the major side products dervied from various cellulose fibres of various lignocellulosic biomass wastes were determined. Besides, the kinetics of the depolymerization process of cellulose fibres under specific conditions were investigated.

2. Experiment

2.1. Materials. Four different types of LBW materials were used in this study including wood sawdust, sugarcane bagasse, corncob, and sago pith wastes. The sawdust of wood (*Acacia mangium*) was obtained from the wood research lab of Univerisiti Malaysia Sarawak. Sugarcane bagasse (*Saccharum officinarum* spp.) and corncob (*Earlivee* sweet corn) were collected from a local market, whereas the pith waste of sago (*Metroxylon* sagu) was collected from a local sago starch-processing factory. All lignocellulosic biomass waste samples were washed with deionized water to remove impurities such as sand and dirt, dried in an oven, and than finely ground using a grinder. Sodium hydroxide (NaOH), sodium chlorite ($NaClO_2$), acetic acid, silver nitrate ($AgNO_3$), methanol, and pyridine were purchased from Merck whereas 1-methylimidazole, allyl chloride, dimethyl sulfoxide (DMSO), phenylisocyanate, tetrahydrofuran (THF), and Amberlyst 15 DRY were purchased from Sigma-Aldrich. Potassium sodium tartrate and phenol were purchased from JT Baker. Ultrapurewater (~18.2 MΩ·cm, 25°C) was obtained from the Water Purifying System (ELGA Model Ultra Genetic). All chemicals were used without further purification.

2.2. Extraction of Cellulose Fibres from Lignocellulosic Biomass Wastes. Cellulose fibres were isolated from various lignocellulose biomass wastes based on the method reported by Sun and Cheng [17]. Samples of sawdust, sugarcane bagasse, and corn cob and sago pith waste were first treated with NaOH (12 wt%) at room temperature under stirring (1000 rpm) for 8 h in order to remove the hemicellulose. This procedure was repeated twice. The NaOH treatment had also removed starch granules contained in sago pith waste samples. Iodine test was performed to confirm total removal of starch residues in sago pith waste. In addition, the NaOH-treated samples were delignified in a $NaClO_2$ solution (1 wt%) which had been adjusted to pH 3.58 with acetic acid (10 wt%) and maintained at 70°C for 5 h. Cellulose fibres obtained were washed with deionized water and dried in an oven at 80°C until constant weight.

2.3. Synthesis of 1-Allyl-3-methylimidazolium Chloride (AMIMCl). The room temperature ionic liquid, AMIMCl, was synthesized based on the procedure reported by Zhang et al. with some modifications [18]. 1-Methylimidazole was placed into a three-neck round-bottom flask, and then allyl chloride was added dropwise to the flask at room temperature in the volume ratio of 1:1.2 (*v*/*v*%). The reaction mixture was stirred magnetically under reflux at about 55°C for 7 h. The residual allyl chloride was eliminated using a rotary evaporator, whereas the unreacted 1-methylimidazole was removed by washing thrice with an excess amount of ether. The resulting ionic liquid was stored with a 3A-size molecular sieve in a vacuum desiccator.

2.4. Depolymerization of Cellulose Fibres. A measured volume of ionic liquid, AMIMCl (10 mL), was preheated at 80°C for 30 min to make it less viscous. A sample of cellulose fibres (0.50 g) was then added into the ionic liquid and magnetically stirred until it was completely dissolved. 100 μL of deionized water was added into cellulose solution and stirred for 2 min. Subsequently, the Amberlyst 15 DRY, an acid catalyst (0.05 g, 0.23 mmol $H_3O^+ \cdot g^{-1}$), was added. After the predetermined reaction durations, the solid catalyst was separated by filtering with a 20-mesh sieve and cello-oligomers formed were precipitated by adding water dropwise into the reaction medium. The cello-oligomers were separated into solid and liquid fractions by centrifugation at 6000 rpm. The solid fraction of cello-oligomer samples was washed with ultrapure water until no chloride ions were detected using 0.1 M silver nitrate solution and dried at 60°C in an oven until constant weight and stored in a vacuum desiccator. The liquid fractions of dissovled cello-oligomers were collected separately and stored at 4°C.

2.5. Determination of Degree of Polymerization (DPw). The mean and distribution of DPw of cello-oligomers were determined based on the method reported by Kačík et al. [19]. Generally, phenylisocyanate (1 mL) was added to a cellulose sample (2 mg) dispersed in dry pyridine (6 mL), and the resulting mixture was allowed to react at 110°C for 6 h. Upon completion of reaction, excess phenylisocyanate was decomposed by adding methanol (2 mL). The resulting clear yellow solution was then precipitated in methanol under vigorous stirring. The precipitate was washed with acetic acid (10 wt%) to remove pyridine and then washed with methanol (70 wt%). Finally, the cellulose tricarbanilate (CTC) samples obtained were dried at 60°C in an oven until constant weight. CTC samples were then dissolved in tetrahydrofuran (2 mg/mL) and analyzed by gel permeation chromatography (GPC). Sample solutions were filtered through filters with a mean pore size of 0.20 mm (Whatman, Puradisc 25 NYL). The GPC analysis was performed using a HPLC (Perkin-Elmer FLEXAR) fitted with mixed-bed GPC columns (3 columns, PLgel 10 μM MIXED B, Polymer Laboratories, 7.5 mm ID × 300.0 mm) at 30°C, and stabilized THF was used as the eluent (1.0 mL/min). A UV-Vis detector was used for cello-oligomer detection at the wavelength of 240 nm. Prior to GPC analysis, the system was calibrated using polystyrene standards (Agilent EasiVial, 1.62×10^2 to 7×10^6 g/moL). The mean DPw was calculated by dividing the molecular weight of CTC samples (W) by the monomer equivalent weight of CTC ((1)):

$$\mathrm{DPw} = \frac{\mathrm{W}}{519}. \tag{1}$$

2.6. Determination of Water-Soluble Oligomers. The total reducing sugar (TRS) was quantified using the 3,5-dinitrosalicylic acid (DNS) assay [20]. DNS reagent (2.0 mL) was added to the liquid fraction (1.0 mL) and heated in a boiling water bath for about 10 min. The resulting solution was cooled to room temperature and then diluted with deionized water by a factor of 10. The absorbance of total reducing sugar (TRS) was measured at the wavelenght of 540 nm using a UV-Vis spectrophotometer (Shimadzu Model UV-160A). The concentration of TRS was calculated from the calibration curve of standard glucose solutions.

Glucose and xylose were analysed using high-performance liquid chromatography (HPLC, Shimazu Model LC-20A) equipped with a refractive index detector. The chromatographic separations were carried out using a column (Bio-Rad Aminex HPX-871, 100 × 7.8 mm) and diluted sulfuric acid (5 mM) as the mobile phase. Samples were analysed in isocratic mode at a flow rate of 0.8 mL/min at 65°C and 20 μL injection volume. All detected compounds were identified by comparing retention times with those of corresponding standard reference compounds. Amounts of products formed were quantified by comparing with calibration curves of external authentic samples.

2.7. FTIR Analysis. The chemical structure of cellulose fibres was characterized by Fourier transform infrared spectrometry (FTIR) (Thermo Scientific Nicolet iS10). The dried and ground cellulose sample was mixed thoroughly with potassium bromine (KBr) powder at a ratio of 1:100 (*w*/*w*), before being compressed into pellets. Prior to its use, the KBr powder was dried in an oven to ensure its water content was less than 1 wt%. The FTIR spectrum was generated by scanning at a resolution of 4 cm^{-1} with a total of 64 scans for each sample within the wave number range of 4000–500 cm^{-1}.

2.8. Depolymerization Kinetics of Cellulose Fibres. The depolymerization kinetics of cellulose fibres isolated from various lignocellulosic biomass wastes were investigated based on (2) and (3) [21, 22].

$$\text{Second-order}: \frac{1}{\mathrm{DPw}} = kt + \frac{1}{\mathrm{DPo}}, \tag{2}$$

$$\text{Pseudo-second-order}: \frac{t}{\mathrm{DPw}} = \frac{1}{\mathrm{DPo}}t + \frac{1}{k(\mathrm{DPo})}, \tag{3}$$

where DPo is the mean degree of polymerization values of original cellulose fibres, DPw is the mean degree of polymerization values of cello-oligomers, t is the reaction time of the depolymerization process, and k is the reaction constant.

3. Results and Discussion

3.1. FTIR Spectra. All FTIR spectra showed similar characteristic absorption peaks as compared with those of standard pure cellulose (Figure 1). A strong broad peak observed at 3383–3412 cm^{-1} was attributed to O–H stretching of hydroxyl groups of cellulose with moisture [23]. The absorption bands between 2899 cm^{-1} and 2902 cm^{-1} were due to C–H stretching of methyl and methylene groups [24]. Absorption bands located at 1627–1651 cm^{-1} were due to adsorbed water molecules [25]. Appearances of weak absorption peaks at 1429–1433 cm^{-1} and 1317–1372 cm^{-1} were assigned to bending of CH_2 cellulose [26]. Weak absorption peaks at 1162–1164 cm^{-1} were due to O–H bonds at C2, C3, and C5 of cellulose [25]. Strong peaks observed at about 1061 cm^{-1} were due to C–O stretching [27]. The absorption peak observed at 897 cm^{-1} was due to β-glycosidic linkages between the sugar units [28].

Besides, FTIR spectra were used to confirm the removal of lignin and hemicelluloses from various LBW samples. After strong alkaline treatment, no absorption peak was observed at 1735 cm^{-1}, which was attributed to the complete

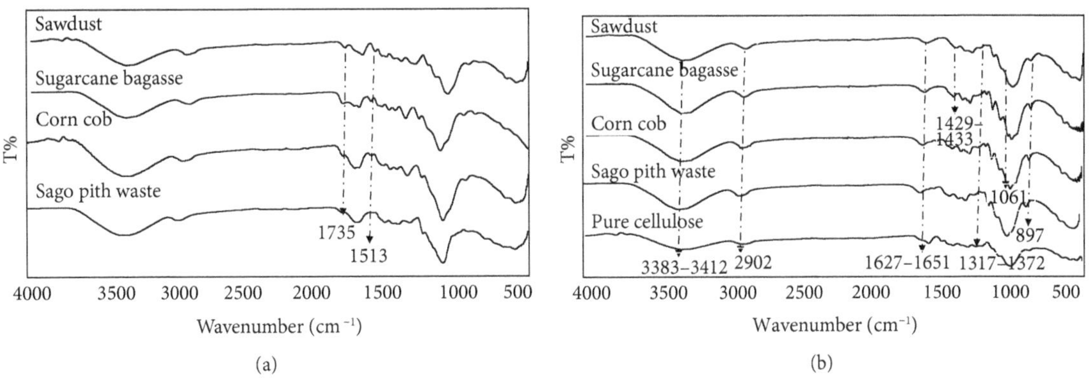

Figure 1: FTIR spectra of various lignocellulosic biomass waste samples (a) before and (b) after alkaline pretreament.

removal of hemicelluloses and/or lignin components [29]. In addition, no absorption peak attributed to the -C=C- stretching of aromatic rings of lignin was observed at 1513 cm^{-1}, thereby confirming that all lignin had been removed from cellulose fibres isolated from various LBW samples (Figure 1(b)) [30].

3.2. Depolymerization of Cellulose Fibres. Cellulose fibres isolated from sawdust, sugarcane bagasse, corn cob, and sago pith waste samples were depolymerized by dissolving in AMIMCl to form 5 wt% of cellulose solution and the subsequent addition of 0.05 g Amberlyst 15 DRY (0.23 mmol H_3O^+) as the catalyst. The schematic reaction mechanism of the depolymerization process is as shown in Figure 2. In this reaction, ion-exchange process occurred between the ionic liquid and acid catalyst resin, resulting in the release of H_3O^+ species which cleavaged the β-glycosidic bonds in the cellulose chains to produce cello-oligomers and other side products such as reducing sugar, glucose, and xylose [5].

Figure 3 shows the mean DPw of cello-oligomers derived from cellulose fibres of various lignocellulosic biomass wastes at different reaction times (5–120 min). The mean DPw values of cello-oligomers were observed to decrease rapidly initially and then reached a constant value after reaction durations of 40 to 80 min.

However, the distribution of DPw values or chain lengths of cello-oligomers produced were observed to be affected by initial DPw values of cellulose fibres used in the depolymerization process. Cellulose fibres derived from sawdust and corn cob with initial DPw values of about 350 were observed to produce cello-oligomers with mean DPw values of 82 and 118, respectively, after a reaction time of about 40 min (Figure 3). However, cellulose fibres derived from sago pith waste and sugarcane bagasse with initial mean DPw values of 564 and 571, respectively, produced cello-oligomers with substantially lower DPw value of 30 and 35 after reaction times of up to 60 min and 100 min, respectively. Cellulose fibres of sago pith waste exhibited a faster rate of depolymerization as compared with that of sugarcane bagasse which could be due to cellulose fibres with comparatively more uniform of DPw distribution.

3.3. Composition of Water-Soluble Oligomers. Generally, the yields of regenerated cello-oligomers obtained via the depolymerization of cellulose fibres of various LBW were observed to decrease with increasing reaction time due to the formation of water-soluble oligomers such as reducing sugar, glucose, and xylose (Figure 4). The major component of water-soluble oligomers produced from different types of cellulose fibres was reducing sugar. Yields of regenerated cello-oligomers and water-soluble oligomers derived from cellulose fibres of sawdust and corn cob after depolymerization for 120 min were 54.3 wt% and 6.15 wt%, and 63.4 wt% and 11.34 wt%, respectively. However, yields of regenerated cello-oligomers from cellulose fibres of sugarcane bagasse and sago pith wastes after depolymerization for 120 min were comparatively lower at 30 wt% and 40 wt%, respectively. Such lower yields could be attributed to comparatively higher percentages of water-soluble oligomers being produced (Figures 4(b) and 4(d)). In addition, cellulose fibres of sugarcane bagasse and sago pith wastes were being depolymerized to cello-oligomers with lower mean DPw values between 30 and 40, respectively (Figure 3). Probably, some shorter-chain cello-oligomers (DPw < 10) were not precipitated, thereby resulting in the comparatively lower yields of cello-oligomers.

3.4. DPw Distribution of Cello-Oligomers. The DPw distribution of cello-oligomers obtained from cellulose fibres of various lignocellulosic biomasses was investigated by gel permeation chromatography (GPC) as shown in Figure 5. All cello-oligomers were observed to exhibit unimodal distribution of DPw at various predetermined reaction durations. Generally, the DPw distribution of cello-oligomers showed a gradual and continuous shift to lower mean DPw values with longer reaction time, indicating that the depolymerization process had caused a chain scission of cellulose molecules, thereby leading to a decrease in mean DPw values with increasing duration of depolymerization. Besides, the width of DPw distribution curves as indicated by the polydispersity index (PDI) was observed to decrease rapidly or gradually, indicating continual reduction in the polydispersity of cello-oligomers with the extent of the depolymerization process (Figure 5(e)).

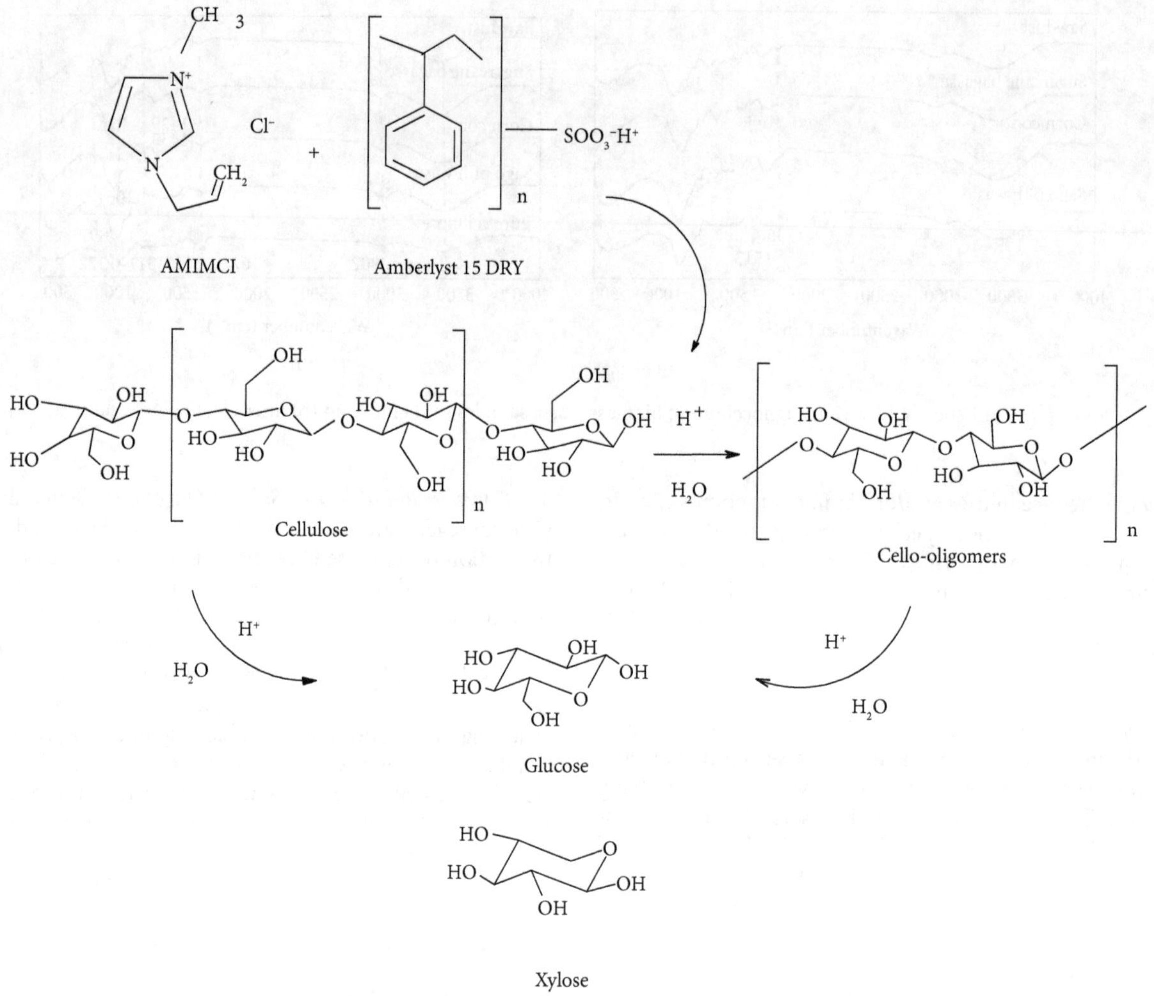

FIGURE 2: Schematic reaction mechanism for the depolymerization process of cellulose fibres.

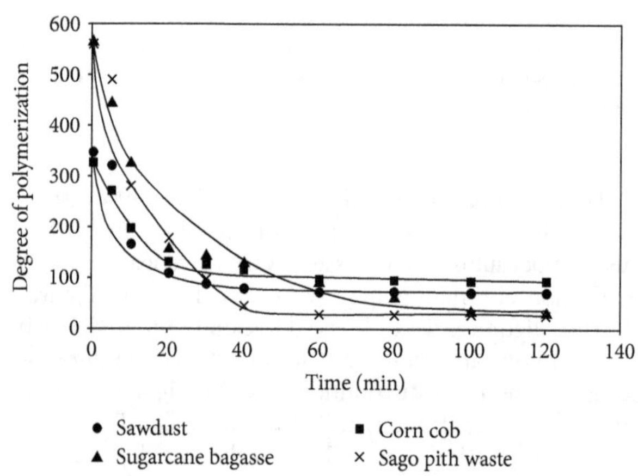

FIGURE 3: Degree of polymerization (DPw) of cello-oligomers derived from cellulose fibres of various lignocellulosic biomass at different reaction times: (•) sawdust (DPw = 351), (▲) sugarcane bagasse (DPw = 571), (■) corn cob (DPw = 329), and (x) sago pith wastes (DPw = 564). Parentheses show the initial DPw values of lignocellulosic biomass samples.

Cello-oligomers derived from cellulose fibres of sawdust were observed to exhibit substantially narrower DPw distribution with its mean peak shifted to lower DPw values after initial depolymerzation duration of 40 min (Figure 5(a)). However, peaks of the mean DPw distribution were observed to have shifted only slightly upon subsequent depolymerization durations between 40 and 120 min. The PDI of cello-oligomers formed was observed to decrease rapidly from 25.5 during the initial 40 min and thereafter decreased very gradually to 14.70 after depolymerization for 120 min.

Similarly, cello-oligomers formed from cellulose fibres of sugarcane bagasse and corn cob samples showed a continual shift in their DPw peaks throughout the depolymerization duration of 120 min (Figures 5(b) and 5(c)). DPw distributions became progressively narrower with increasing depolymerization duration from 10 min to 120 min However, DPI values of cello-oligomers for both sugarcane bagasse and corn cob samples were observed to decrease gradually from 17.6 and 15.3 to 9.78 and 3.21, respectively, after depolymerization for 120 min (Figures 5(e)).

In the case of cello-oligomers formed from cellulose fibres of sago pith waste, DPw peaks were observed to

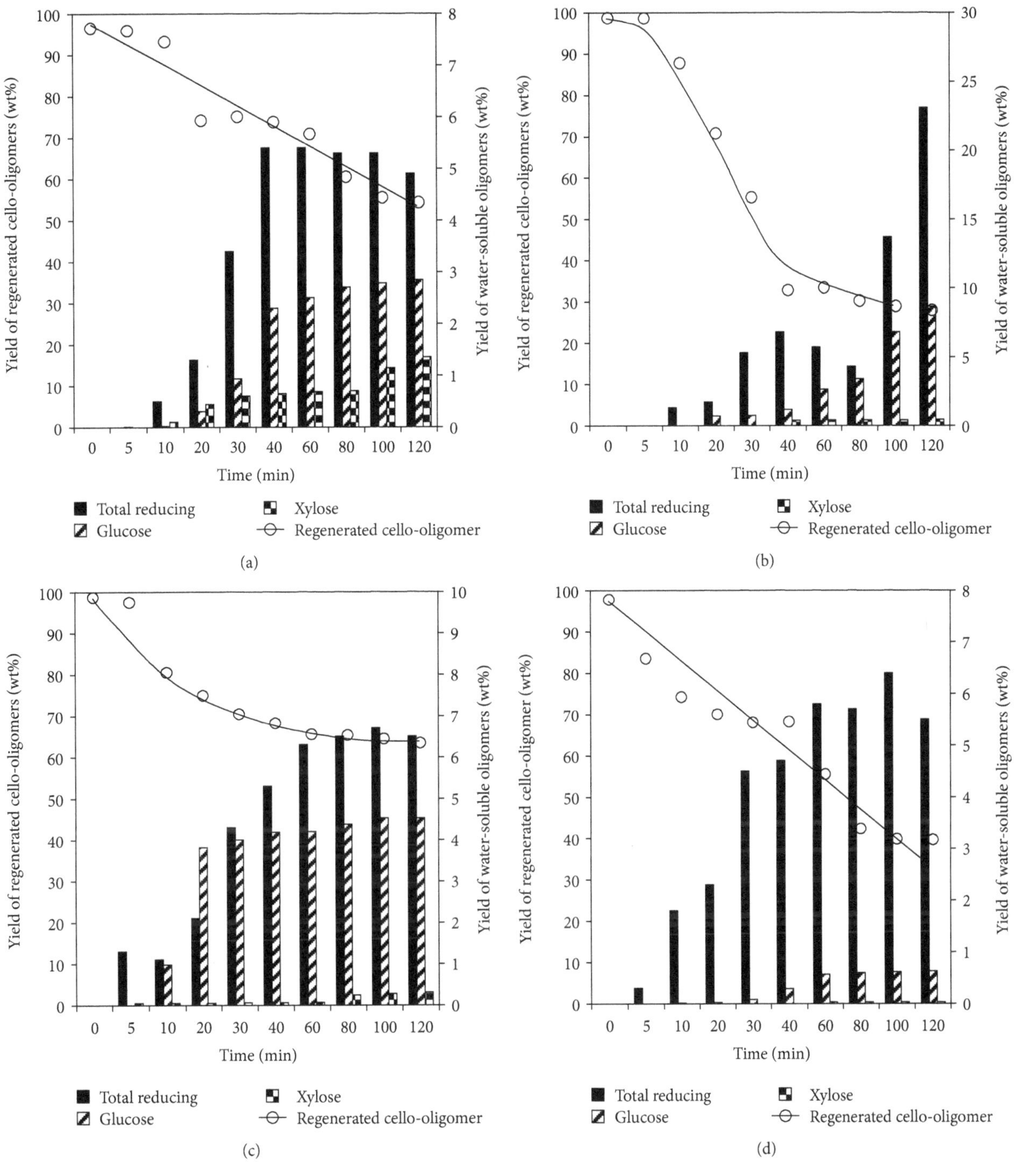

FIGURE 4: Yields of cello-oligomers and water-soluble sugar products derived from depolymerization of cellulose fibres of (a) sawdust, (b) sugarcane bagasse, (c) corn cob, and (d) sago pith wastes at different reaction times (acid catalyst concentration = 0.23 mmol H_3O^+; temperature = 80°C).

shift continuously albeit more rapidly during the intial stage of depolymerization. Correspondingly, the PDI values of cello-oligomers were observed to decrease comparatively more rapidly from 22.4 to 3.02 within the duration of depolymerization.

It is noteworthy that cellulose fibres isolated from both sago pith waste and sugarcane bagasse samples showed comparatively higher initial mean DPw values of 564 and 571, respectively. The depolymerization of such cellulose fibres could have occurred in a controlled fashion with a size-specific preference for the cleavage of longer chains. Hence, cello-oligomers with comparatively lower mean DPw values of 30 and 37, respectively, were obtained after depolymerization for 120 min. However, the lowest PDI

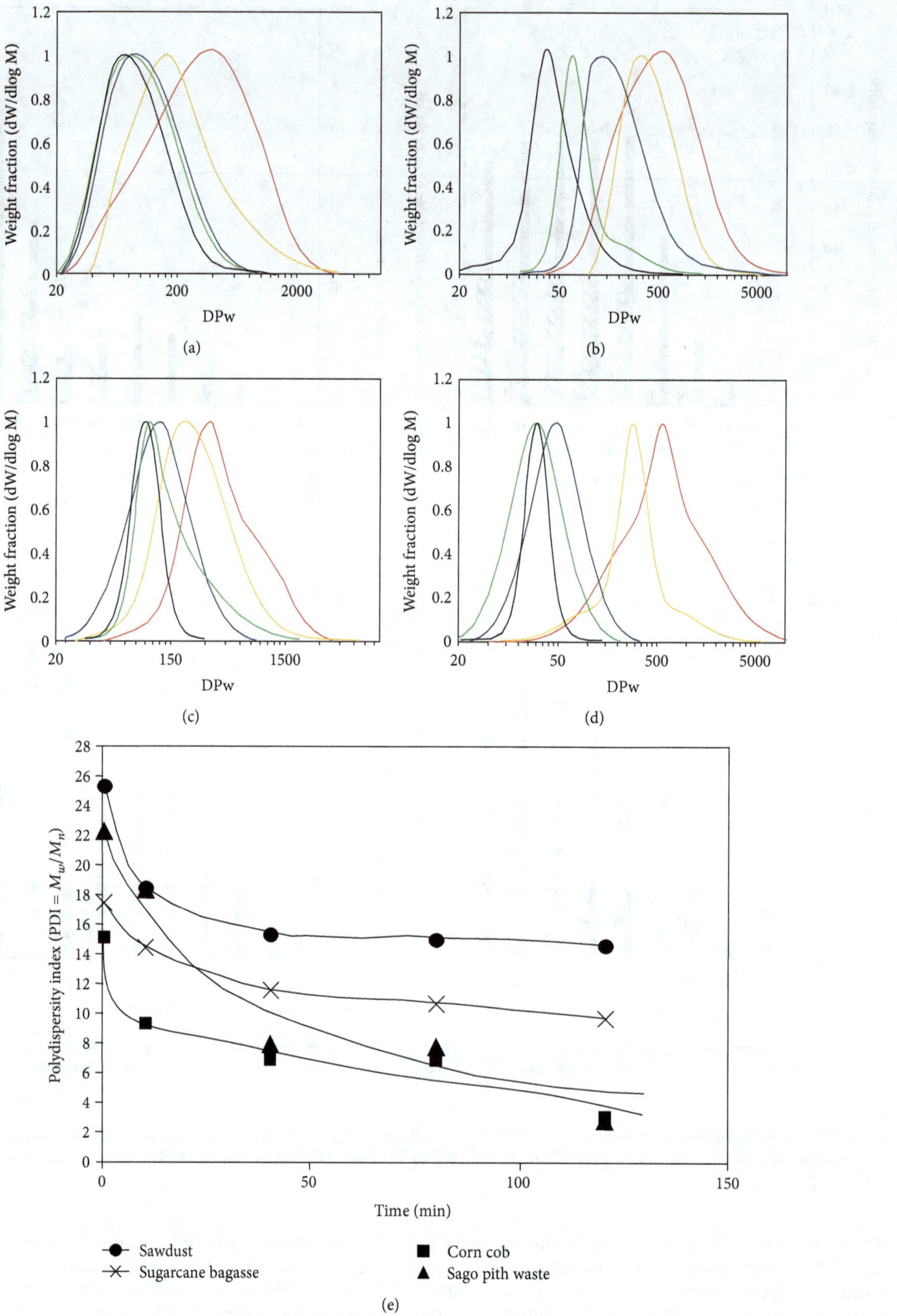

FIGURE 5: DPw distribution of various LBW of the cello-oligomers obtained from cellulose fibres at different reaction times: (a) sawdust, (b) sugarcane bagasse, (c) corn cob, and (d) sago pith wastes and (e) polydispersity index (PDI). (Red line: original cellulose fibres, yellow line: 10 min, blue line: 40 min, green line: 80 min, and black line: 120 min).

Table 1: Mean DPw peak values and polydispersity indices (PDI) for cello-oligomers derived from various types of lignocellulosic biomass wastes.

Source of cellulose fibres	Mean DPw (peak) 0 min	Mean DPw (peak) 120 min	Percentage change of mean DPw ΔDPw (%)	Polydispersity index (PDI) M_w/M_n
Sago pith wastes	564	30	94.7	3.02
Sugarcane bagasse	571	37	93.5	9.78
Sawdust	351	72	79.5	14.70
Corn cob	329	88	78.1	3.21

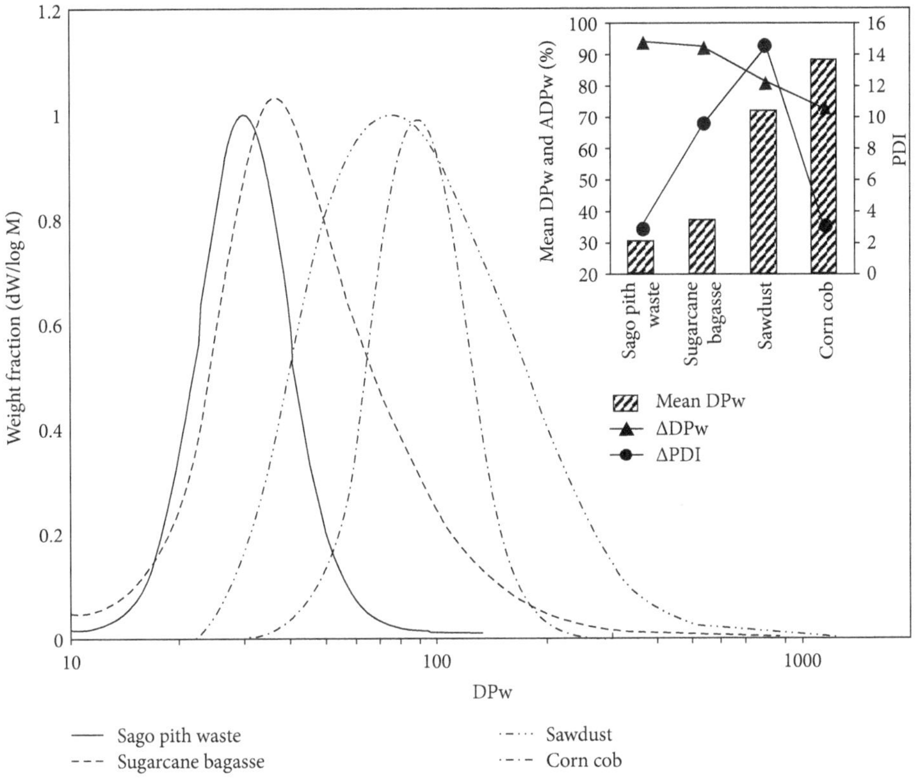

Figure 6: DPw distribution of cello-oligomers obtained from cellulose fibres of various LBW after depolymerization for 120 min. Inset shows the mean DPw value, percentage change of mean DPw (ΔDPw) value, and polydispersity indices (PDI) of cello-oligomers.

values for cello-oligomers formed from sago pith waste and sugarcane baggasse samples were 3.02 and 9.78, respectively (Table 1). The substantial difference in the PDI values could be attributed to the inherent nature and structural differences of cellulose fibres originated from different biomass sources. During depolymerization, hydronium ions from the catalyst enabled the cleavage of longer-chain cello-oligomers to produce cello-oligiomers of shorter chain length while cello-oligomers of low DPw formed water-soluble oligomers.

Figure 6 shows a comparison of the mean DPw distribution of cello-oligomers derived from cellulose fibres of various lignocellulosic biomass wastes after depolymerization for 120 min. The mean DPw peak values of cello-oligomers were observed in the descending order of corn cob, sawdust, sugarcane baggasse, and sago pith waste. In contrast, the width of DPw distributions of cello-oligomers as reflected by their PDI values followed the descending order of sawdust, sugarcane baggasse, corn cob, and sago pith waste (Table 1). Evidently, cello-oligomers derived from cellulose fibres of different biomass sources would possess substantially different DPw distribution and polydispersity which could in turn affect their physical and chemical properties.

3.5. Depolymerization Kinetics of Cellulose Fibres. The depolymerization kinetics of cellulose fibres derived from various types of lignocellulosic biomass wastes under specific conditions were investigated by applying second-order and pseudo-second-order kinetic equations (Figure 7). Depolymerization of cellulose fibres isolated from sugarcane bagasse and sago pith wastes was observed to follow second-order kinetics, whereas depolymerization of cellulose fibres isolated from sawdust and corn cob exhibited pseudo-second-order kinetics. Differences in the rate constants for the depolymerization of cellulose fibres derived from sawdust (1.8×10^{-4} min^{-1}), corn cob (1.6×10^{-3} min^{-1}), sago pith waste

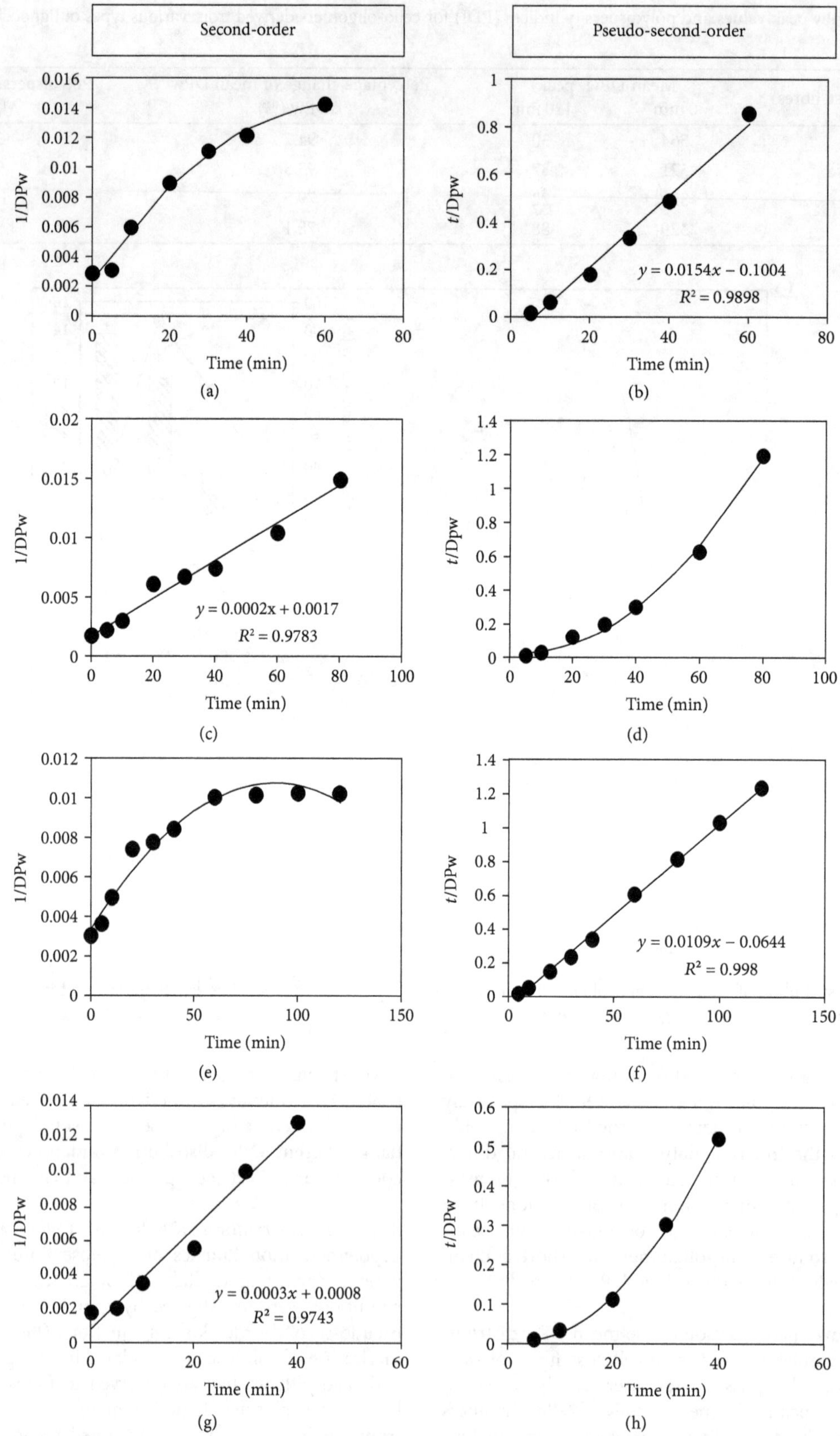

FIGURE 7: Second-order and pseudo-second-order kinetics for the depolymerization of cellulose fibres derived from sawdust (a, b), sugarcane bagasse (c, d), corn cob (e, f), and sago pith wastes (g, h). (Reaction temperature = 80°C; acid catalyst concentration = 0.23 mmol H_3O^+).

(3×10^{-4} min^{-1}), and sugarcane bagasse (2×10^{-4} min^{-1}) could be due to variations in the composition of crystalline phases within their cellulose fibres. As reported by Cheng et al. [31] and Park et al. [32], the crystalline structure of cellulose molecules could affect the hydrolysis rate of cellulose. These results concurred with findings of our previous study that some degree of control could be afforded on the chain lengths of cello-oligomers formed during the depolymerization of cellulose fibres in ionic liquid and in the presence of a solid catalyst. The rate of depolymerization and mean DPw or chain length of cello-ologomers could be controlled by simply varying the reaction time at a fixed acid catalyst concentration and reaction temperature. Henceforth, cello-oligomers of the most desirable chain length could be tailored in order to optimize physicochemical properties for desired applications, as well as yields of desired final products.

4. Conclusion

Cellulose fibres derived from different lignocellulosic biomass wastes such as sawdust, corn cob, sugarcane bagasse, and sago pith wastes were depolymerized under controlled conditions to form cello-oligomers of varying mean DPw values or chain lengths and polydispersity. Cello-oligomers of mean DPw values less than 100 were obtained from all types of lignocellulosic biomass wastes within a depolymerization duration of 120 min. Cello-oligomers with controllable DPw values instead of sugars derived from lignocellulosic biomass could be more ideally suited as the renewable feedstock for sustainable biorefineries in the production of biofuels and biobased functional chemicals. Besides, value-added utilization of such agricultural residues or by-products would therefore provide additional revenues for farmers without any adverse impacts on soil fertility and environment.

Conflicts of Interest

The authors do not have any conflicts of interest.

Acknowledgements

The authors wish to acknowledge the financial support rendered by the Malaysian Ministry of Higher Education (MOHE) via the award of fundamental research grants (Grant nos. FRGS/ST01(01)/967/2013(08) and F07/FRGS/1495/2016), as well as research management and support provided by the Research Innovation and Management Centre (RIMC), Universiti Malaysia Sarawak.

References

[1] H. Rasmussen, H. R. Sørensen, and A. S. Meyer, "Formation of degradation compounds from lignocellulosic biomass in the biorefinery: sugar reaction mechanisms," *Carbohydrate Research*, vol. 385, pp. 45–57, 2014.

[2] D. P. Ho, H. H. Ngo, and W. Guo, "A mini review on renewable sources for biofuel," *Bioresource Technology*, vol. 169, pp. 742–749, 2014.

[3] R. A. Sheldon, "Green and sustainable manufacture of chemicals from biomass: state of the art," *Green Chemistry*, vol. 16, no. 3, pp. 950–963, 2014.

[4] H. Zhao, C. L. Jones, G. A. Baker, S. Xia, O. Olubajo, and V. N. Person, "Regenerating cellulose from ionic liquids for an accelerated enzymatic hydrolysis," *Journal of Biotechnology*, vol. 139, no. 1, pp. 47–54, 2009.

[5] R. Rinaldi, N. Meine, J. vom Stein, R. Palkovits, and F. Schüth, "Which controls the depolymerization of cellulose in ionic liquids: the solid acid catalyst or cellulose ?," *Chemistry Sustainable Materials and Energy*, vol. 3, pp. 266–276, 2010.

[6] M. Yabushita, H. Kobayashi, and A. Fukuoka, "Catalytic transformation of cellulose into platform chemicals," *Applied Catalysis B: Environmental*, vol. 145, pp. 1–9, 2014.

[7] R. Rinaldi and F. Schüth, "Acid hydrolysis of cellulose as the entry point into biorefinery schemes," *Chemistry Sustainable Materials and Energy*, vol. 2, pp. 1096–1107, 2009.

[8] N. V. Plechkova and K. R. Seddon, "Applications of ionic liquids in the chemical industry," *Chemical Society Reviews*, vol. 37, no. 1, pp. 123–150, 2008.

[9] R. D. Rogers, K. R. Seddon, and S. Volkov, Eds., "Green industrial applications of ionic liquids," *Springer Science & Business Media*, vol. 92, pp. 456–478, 2012.

[10] J. Wilkes, "Properties of ionic liquid solvents for catalysis," *Journal of Molecular Catalysis A: Chemical*, vol. 214, no. 1, pp. 11–17, 2004.

[11] H. Olivier-Bourbigou, L. Magna, and D. Morvan, "Ionic liquids and catalysis: recent progress from knowledge to applications," *Applied Catalysis A: General*, vol. 373, no. 1-2, pp. 1–56, 2010.

[12] A. Brandt, J. Gräsvik, J. P. Hallett, and T. Welton, "Deconstruction of lignocellulosic biomass with ionic liquids," *Green Chemistry*, vol. 15, no. 3, pp. 550–583, 2013.

[13] W. Z. Li, M. T. Ju, Y. N. Wang, L. Liu, and Y. Jiang, "Separation and recovery of cellulose from Zoysia japonica by 1-allyl-3-methylimidazolium chloride," *Carbohydrate Polymers*, vol. 92, no. 1, pp. 228–235, 2013.

[14] I. Van zandvoort, Y. Wang, C. B. Rasrendra et al., "Formation, molecular structure, and morphology of humins in biomass conversion: influence of feedstock and processing conditions," *Chemistry Sustainable Materials and Energy*, vol. 6, pp. 1745–1758, 2013.

[15] R. Rinaldi, R. Palkovits, and F. Schüth, "Depolymerization of cellulose using solid catalysts in ionic liquids," *Angewandte Chemie International Edition*, vol. 47, no. 42, pp. 8047–8050, 2008.

[16] L. K. Voon, S. C. Pang, and S. F. Chin, "Regeneration of cello-oligomers via selective depolymerization of cellulose fibers derived from printed paper wastes," *Carbohydrate Polymers*, vol. 142, pp. 31–37, 2016.

[17] Y. Sun and J. Cheng, "Hydrolysis of lignocellulosic materials for ethanol production: a review," *Bioresource Technology*, vol. 83, no. 1, pp. 1–11, 2002.

[18] H. Zhang, J. Wu, J. Zhang, and J. He, "1-Allyl-3-methylimidazolium chloride room temperature ionic liquid: a new and powerful nonderivatizing solvent for cellulose," *Macromolecules*, vol. 38, no. 20, pp. 8272–8277, 2005.

[19] F. Kačík, D. Kačíková, M. Jablonský, and S. Katuščák, "Cellulose degradation in newsprint paper ageing," *Polymer Degradation and Stability*, vol. 94, no. 9, pp. 1509–1514, 2009.

[20] F. Liu, R. K. Kamat, I. Noshadi et al., "Depolymerization of crystalline cellulose catalyzed by acidic ionic liquids grafted onto sponge-like nanoporous polymers," *Chemical Communications*, vol. 49, no. 76, pp. 8456–8458, 2013.

[21] K. J. Vetter, *Electrochemical Kinetics: Theoretical Aspects*, Elsevier, New York, NY, USA, 2013.

[22] D. Robati, "Pseudo-second-order kinetic equations for modeling adsorption systems for removal of lead ions using multi-walled carbon nanotube," *Journal of Nanostructure in Chemistry*, vol. 3, no. 1, p. 55, 2013.

[23] M. Ibrahim, A. Nada, and D. E. Kamal, "Density functional theory and FTIR spectroscopic study of carboxyl group," *Indian Journal of Pure & Applied Physics*, vol. 43, pp. 911–917, 2005.

[24] H. Wang, B. Li, and B. Shi, "Preparation and surface acid-base properties of porous cellulose," *BioResources*, vol. 3, pp. 3–12, 2007.

[25] B. Bodirlau and C. A. Teaca, "Fourier transform infrared spectroscopy and thermal analysis of lignocelluloses fillers treated with organic anhydrides," *Romania Journal Physics*, vol. 54, pp. 93–104, 2007.

[26] E. Smidt, K. Böhm, and M. Schwanninger, *The Application of FT-IR Spectroscopy in Waste Management*, INTECH Open Access Publisher, 2011.

[27] Y. Sun, L. Lin, H. Deng et al., "Structure of bamboo in formic acid," *BioResources*, vol. 3, pp. 297–315, 2008.

[28] S. Y. Oh, D. I. Yoo, Y. Shin et al., "Crystalline structure analysis of cellulose treated with sodium hydroxide and carbon dioxide by means of X-ray diffraction and FTIR spectroscopy," *Carbohydrate Research*, vol. 340, no. 15, pp. 2376–2391, 2005.

[29] W. Chen, H. Yu, Y. Liu, Y. Hai, M. Zhang, and P. Chen, "Isolation and characterization of cellulose nanofibers from four plant cellulose fibers using a chemical-ultrasonic process," *Cellulose*, vol. 18, no. 2, pp. 433–442, 2011.

[30] M. Sain and S. Panthapulakkal, "Bioprocess preparation of wheat straw fibers and their characterization," *Industrial Crops and Products*, vol. 23, no. 1, pp. 1–8, 2006.

[31] G. Cheng, P. Varanasi, C. Li et al., "Transition of cellulose crystalline structure and surface morphology of biomass as a function of ionic liquid pretreatment and its relation to enzymatic hydrolysis," *Biomacromolecules*, vol. 12, no. 4, pp. 933–941, 2011.

[32] S. Park, J. O. Baker, M. E. Himmel, P. A. Parilla, and D. K. Johnson, "Cellulose crystallinity index: measurement techniques and their impact on interpreting cellulase performance," *Biotechnology for Biofuels*, vol. 3, no. 1, pp. 10–17, 2010.

16

A New Route to Fabricate Multifunctional and Multistage Composite Nanoparticle

Xin Wang, Pin Chen, and Xiaohong Hu

School of Material Engineering, Jinling Institute of Technology, Nanjing 211169, China

Correspondence should be addressed to Xiaohong Hu; hxh@jit.edu.cn

Academic Editor: Ming-Guo Ma

The focus of nanoparticle research is on exploring its application in all kinds of field. Among these, multifunctional nanoparticle attracts increasing interest due to its fittest property and adjustable property. Herein, a multifunctional and multistage nanoparticle considering the advantage of both nanogel and pH-responsive property has been designed and synthesized in the research. The composite nanoparticle was obtained by in situ processing and polymerization technique using acetylated β-cyclodextrin and gelatin as materials. Gelatin was first encapsulated into Ac-β-CD nanoparticle in order to investigate optimal fabrication conditions of W/O/W technique. The results showed that the nanoparticle had monodisperse characteristic and coarse spherical morphology, which was influenced by factors such as PVA concentration and water/oil ratio. In further step, two-phase composite nanoparticle could be obtained by combined W/O/W technique and in situ polymerization using optimal preparative parameters of W/O/W technique. Two-phase structure could be confirmed by TEM images and DLS results. Fabrication temperature had no effect on the diameters of composite nanoparticle, but influenced the encapsulated efficiency of nanogel. Finally, composite nanoparticle showed quick pH response property at mild acid medium and no obvious cytotoxicity.

1. Introduction

Nanoparticle is defined as a small sphere with nanoscale size, which attracts increasing interest due to its small-size effect in scientific research [1–3]. From one aspect, small-size effect brought not only large specific surface area but also flexible and adjustable properties [1–3]. From other aspect, nanoparticle is a bridge between bulk material and atomic structure, which offers a method to understand or establish the relationship between bulk material and atomic structure. To date, all kinds of metal, inorganic, and polymeric nanoparticles have been synthesized and prepared using gas condensation, attrition, chemical precipitation, microemulsion method, pyrolysis, and hydrothermal synthesis [2]. Among these kinds of nanoparticles, stimuli-responsive nanoparticles including pH-sensitive, temperature-sensitive, enzyme-sensitive, photo-sensitive, and redox-sensitive have been intensively investigated in the application of biomedical, nanotechnology, information fields [4–9]. On account of mild acid environment in lesion locations and in endosomal/lysosomal compartments of cells, low pH-triggered nanoparticles have been used to deliver drugs, growth factors, and bioactive molecules into targeted tissue or cells [10–13]. Recently, quick response polysaccharide materials to mild acid environment had been synthesized and prepared to pH-responsive nanofiber and nanoparticle [10, 14]. Among these, pH-sensitive acetylated β-cyclodextrin (Ac-β-CD) nanoparticle had been fabricated and introduced into injectable hydrogel [15]. The effectiveness and efficiency of Ac-β-CD nanoparticle had been confirmed by previous research [15].

However, nanoparticle with single function cannot satisfy multiaspect requirements like synergic delivery for two molecules with completely different properties, different response requirements, and different cell endocytosis properties in biomedical field. Since nanogel is another widely used nanoscale material with similar environment to physiological environment, it possesses a large number of advantages like good biocompatibility, similar mechanic properties to soft tissue, and less immune response [16, 17]. On account of merits of both nanoparticle and nanogel, nanogel composite pH-sensitive nanoparticles were designed using double

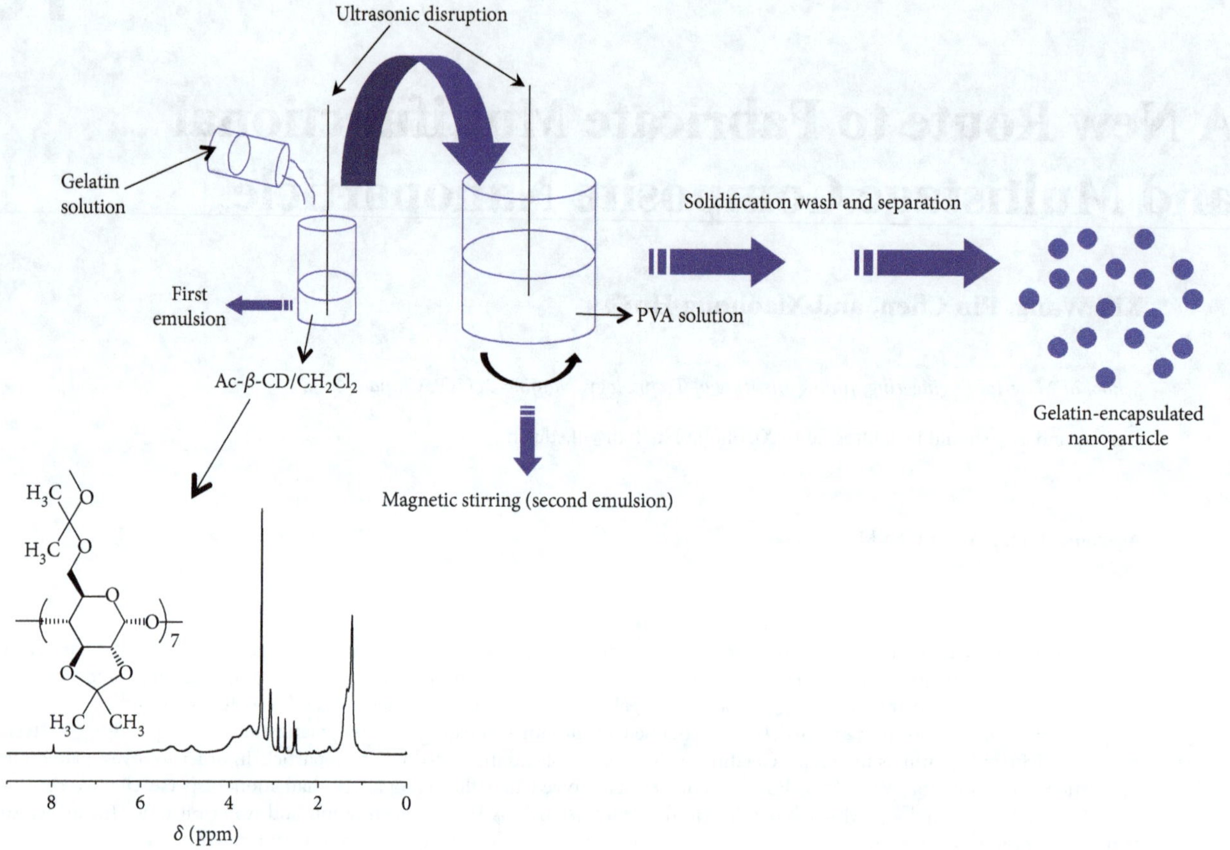

FIGURE 1: Schematic illustration to show the formation of gelatin-encapsulated nanoparticle.

emulsion method in the research. Since hydrogel is hydrophilic-crosslinked polymer network, hydrophilic natural polymer with crosslinkable points is an ideal material for nanogel synthesis. Combined with biocompatibility, biodegradability, and availability, crosslinkable gelatin, which was synthesized in our previous research, is a good choice for nanogel preparation. In order to explore the optimal conditions for double emulsion method, gelatin-encapsulated Ac-β-CD nanoparticles were first prepared and characterized. Then, two-phase composite pH-sensitive nanoparticle was fabricated using similar method. Finally, structure-property information of gelatin nanogel composite Ac-β-CD nanoparticle was investigated in the research. The aim of research is to provide a novel composite nanomaterial for future biomedical application.

2. Experiment Section

2.1. Materials. β-Cyclodextrin (β-CD), 2-methoxypropene, pyridinium 4-toluenesulfonate, dimethyl sulfoxide (DMSO), dichloromethane (DCM), acrylic acid (AA, purified through vacuum distillation), gelatin, ammonium persulfate (APS), and N,N,N$'$,N$'$-tetramethylethylenediamine (TEMED) were purchased from Shanghai Chemical Industries Co. Ltd (China). Poly(vinyl alcohol) (PVA, 87–89 mol% hydrolyzed, CPS: 4.6–5.4), 1-ethyl-3-(3-dimethylaminopropyl)carbodiimide hydrochloride (EDC), and N-hydroxysuccinimide (NHS) were obtained from Aladdin (Shanghai) Co. Ltd. (China). Trypsin, Dulbecco's modified Eagle's medium (DMEM), and fetal calf serum (FBS) were purchased from Gibco. Fluorescein diacetate (FDA) and 3-(4,5-dimethyl) thiazol-2,5-dimethyl tetrazolium bromide (MTT) were got from Sigma. All other chemicals and reagents were used as received.

2.2. Synthesis of G-AA and Ac-β-CD. Acrylic acid-modified gelatin (G-AA) and acetylated β-cyclodextrin (Ac-β-CD) were synthesized according to previous reports [15]. Briefly, G-AA was obtained by EDC/NHS catalyzed coupling reaction and Ac-β-CD was synthesized by acetalization. In the protocol of G-AA synthesis, carboxyl group was first activated by EDC/NHS with molar ratio of 1 : 1.2 : 0.5 (pH 4.0–4.5) for 30 min, then activated AA was reacted with gelatin in water solution (pH 5.0–5.5) for 10 h at room temperature. Finally, G-AA was obtained by dialyzing and freeze-drying. In the protocol of Ac-β-CD synthesis, 50 mM β-CD was reacted with 2-methoxypropene with molar ratio of 1 : 40 under catalysis of 5 mM pyridinium p-toluene sulfonate in anhydrous anaerobic DMSO environment at 30°C for 1 h. Then, Ac-β-CD was obtained by precipitation and filtration and lyophilized. The structure of Ac-β-CD was characterized by ${}_1$H NMR (Figure 1), which was assigned as follows: the chemical shifts from 1.0 ppm to 2.0 ppm are attributed to the protons of C$\underline{H_3}$-C at 1 and 1$'$ positions, the

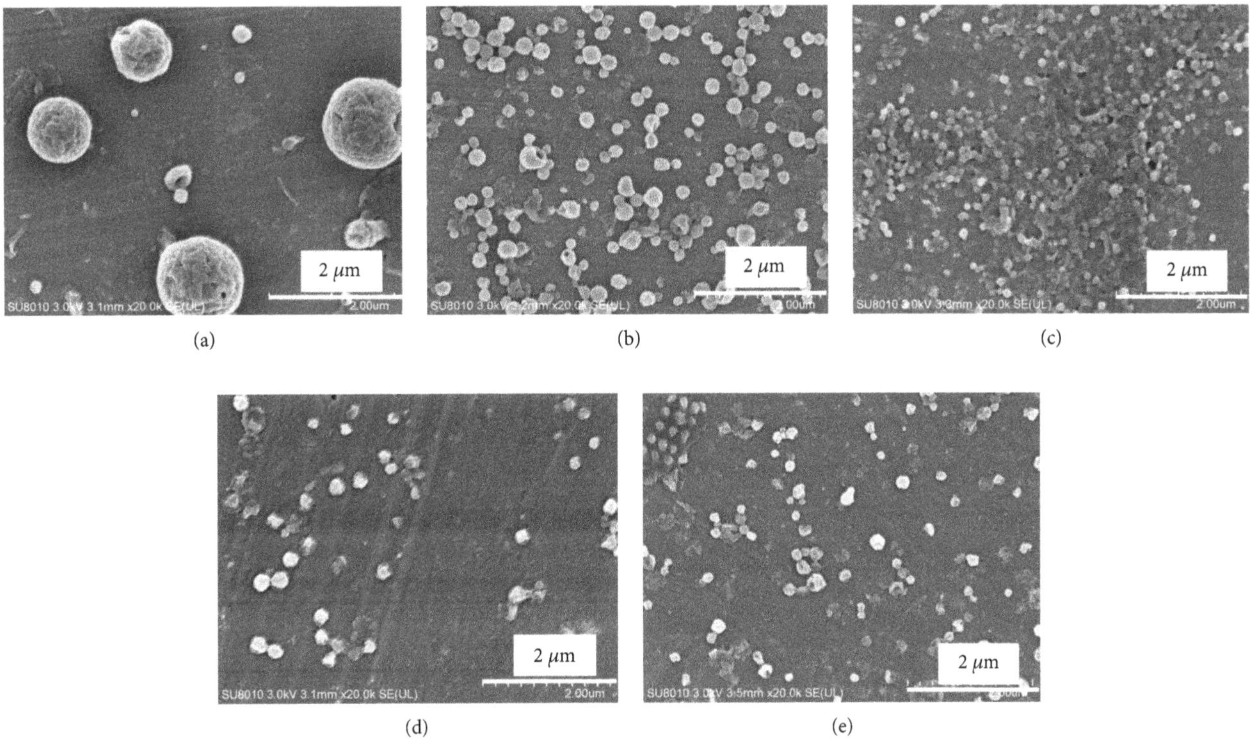

FIGURE 2: SEM images of Nano1 (a), Nano2 (b), Nano3 (c), Nano4 (d), and Nano5 (e).

chemical shift at 3.0 ppm is attributed to the protons of $\underline{CH_3}$-O at 2 position, the chemical shifts from 3.2 ppm to 4.0 ppm are attributed to the protons of β-CD at 3, 4, and 5 positions, respectively, and the chemical shifts at 4.6 ppm and 5.4 ppm are attributed to the protons of 6 position [14].

2.3. Preparation and Characterization of Gelatin-Encapsulated Nanoparticle. Gelatin-encapsulated nanoparticles were prepared by double microemulsion method. Briefly, 200 μL of 10% gelatin solution was first emulsified via probe sonication (Scientz, JY92-II) into 1 mL of 10% *w*/*v* Ac-β-CD/DCM solution, which was further emulsified into 6 mL of 1% *w*/*v* PVA aqueous solution. The obtained emulsion was immediately added into 20 mL of 0.3% *w*/*v* PVA solution to evaporate DCM under magnetic stirring. Finally, nanoparticles were collected by centrifugation (14,000 rpm, 10 min) after 10 h. The collected nanoparticles were washed several times with basic water, lyophilized, and characterized by dynamic light scattering (DLS, nano ZS), scanning electron microscope (SEM, S8100). For DLS test, intensity average was been given to evaluate nanoparticle size. The gelatin encapsulated in nanoparticle was determined by BCA assay. Briefly, 20 μL of dissolved nanoparticle solution was added to one well of a 96-well culture plate, into which 200 μL BCA (agent A/agent $B = 50/1$) solution was added. After 30 min at 60°C, the absorbance at 570 nm was recorded on a microplate reader (Tecan M200), which was used to calculate the gelation concentration by referring to a calibration curve. Encapsulation efficiency of gelatin in nanoparticle was calculated by encapsulated gelatin amount*100%/original gelatin amount.

2.4. Fabrication and Characterization of Two-Phase Composite Nanoparticle. Two-phase composite nanoparticles were also fabricated by double microemulsion method. Briefly, G-AA was dissolved in PBS to obtain 10% *w*/*v* G-AA solution, into which 50 mM APS was added to form homogeneous solution. 200 μL above solution was emulsified into 1 mL of 10% *w*/*v* Ac-β-CD/DCM solution, which was further emulsified into 6 mL of 1% *w*/*v* PVA aqueous solution containing 50 mM TEMED. The obtained emulsion was immediately added into 20 mL of 0.3% *w*/*v* PVA solution to evaporate DCM under magnetic stirring at different temperatures. At the same time, G-AA was crosslinked by radical polymerization to form nanogel structure. Finally, nanoparticles were collected by centrifugation (14,000 rpm, 10 min) after 10 h. The collected nanoparticles were washed several times with basic water, lyophilized, and characterized by dynamic light scattering (DLS, nano ZS), scanning electron microscope (SEM, S8100), and transmission electron microscope (TEM, Tecnai 12) at 200 kV. For DLS test, intensity average was been given to evaluate nanoparticle size. Moreover, transparency at 400 nm for nanoparticle solution in different mediums as a function of time was detected by UV spectroscopy (Cary 50). The gelatin encapsulated in nanoparticle was also determined by the abovementioned BCA assay.

In vitro cultured MCF-7 cells (95% humidified atmosphere + 5% CO_2, 37°C) were used to preliminarily evaluate the cytotoxicity of two-phase composite nanoparticle, which were detached using 0.25% trypsin in PBS before use. In the protocol of cytotoxicity evaluation, 100 μL of 200 μg/mL nanoparticle DMEM suspension was added into each well

Table 1: Fabrication condition and characterizations of gelatin-encapsulated nanoparticles.

Entry	Preparative parameters		Characterization				
	PVA conc. (w/v, %)	Water/oil ratio of the first emulsion	Diameter (nm, SEM)	Hydrodynamic diameter (DLS)		Zeta potential (mV)	Encapsulation efficiency (%)
				Eff. diam. (nm)	PDI		
Nano1	1	1 : 5	324 ± 180	313	0.091	−10.7 ± 4.6	1.8 ± 0.3
Nano2	3	1 : 5	194 ± 60	253	0.150	−12.1 ± 7.3	1.1 ± 0.1
Nano3	5	1 : 5	98 ± 14	191	0.092	−9.3 ± 11.5	1.4 ± 0.4
Nano4	3	1:10	179 ± 58	266	0.111	−14.0 ± 9.4	1.1 ± 0.1
Nano5	3	1 : 20	150 ± 63	267	0.227	−12.0 ± 8.1	1.0 ± 0.1

Table 2: Fabrication condition and characterizations of biphase nanoparticles.

Entry	Preparation temperature (°C)	Characterization				
		Hydrodynamic diameter				Encapsulation efficiency (%)
		In PBS		In acid		
		Eff. diam. (nm)	PDI	Eff. diam. (nm)	PDI	
Nano20	20	199	0.316	245	0.191	12.0 ± 3.1
Nano40	40	195	0.297	253	0.150	14.1 ± 2.9

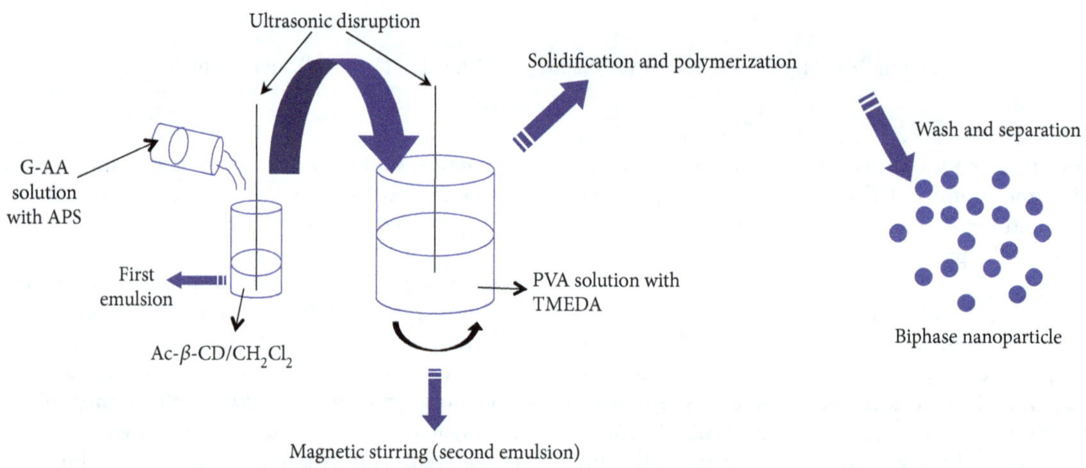

Figure 3: Schematic illustration to show the formation of biphase nanoparticle.

of 96-well culture plate, into which 100 μL cell suspension containing 2000 cells was subsequently added. After 24 h, MTT assay and cell images were used to characterize cytoviability. In MTT assay, 20 μL MTT was incubated with live cells in each well for 4 h to form formazan pigment, which was further dissolved by 200 μL DMSO. Cytoviability was estimated according to the relative absorbance at 490 nm by a microplate reader (Multiskan FC). In addition, viable cells were stained by 5 μg/mL FDA for 10 min, washed with PBS for several times, and observed by fluorescence microscopy (IX73).

2.5. Statistical Analysis. Data were analyzed using the t-test for differences. Results are reported as means ± standard deviation. The significant level was set at $p < 0.05$.

3. Results and Discussion

3.1. Gelatin-Encapsulated Nanoparticle. The detailed structural information of Ac-β-CD and G-AA was described in our previous research [14, 15]. Gelatin-encapsulated Ac-β-CD nanoparticle was fabricated by double water-in-oil-in-water (W/O/W) technique, which was illustrated in Figure 1. Detailedly, gelatin aqueous solution was encapsulated into Ac-β-CD/DCM solution to form the first emulsion, which was further emulsified into water with emulsifier (PVA) to form nanoparticle after DCM was evaporated. Finally, gelatin-encapsulated Ac-β-CD nanoparticle was obtained by wash and separation. Since the influence of preparation conditions on Ac-β-CD nanoparticle's (single O/W technique) structure property had been investigated previously, the optimal preparation conditions like 10% Ac-β-CD

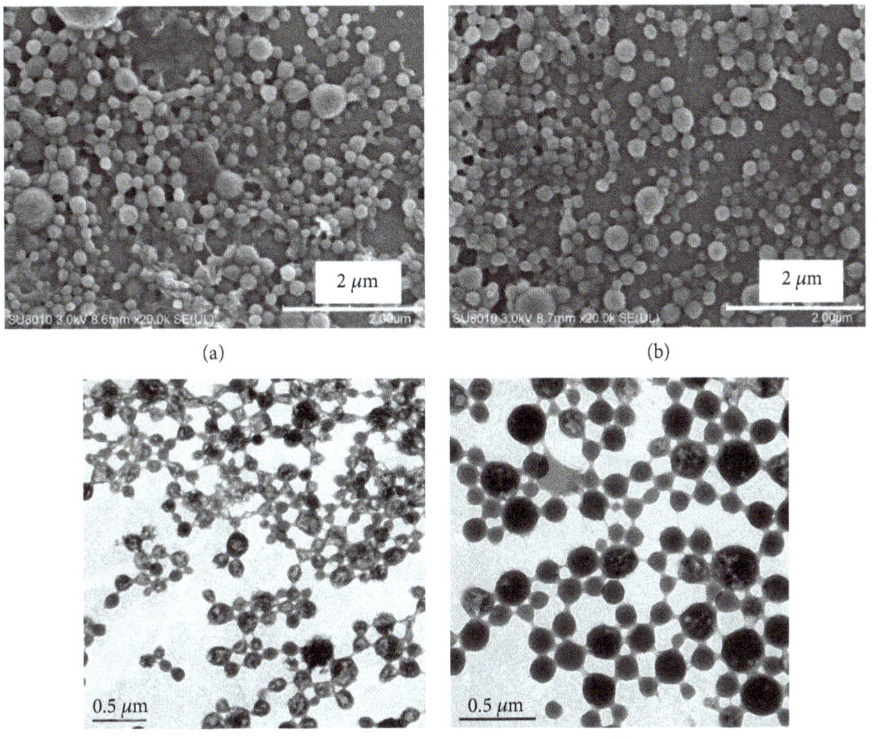

(a) (b) (c) (d)

FIGURE 4: SEM images of Nano20 (a) and Nano40 (b). TEM images of Nano20 (c) and Nano40 (d).

solution and 1 : 6 oil/water ratios for the second emulsion were chosen and fixed in the research. Moreover, preparative parameters like PVA concentration and water/oil ratio of the first emulsion were studied in Figure 2 and Table 1 in the work. All formed nanoparticles exhibited monodisperse characteristic, nearly spherical morphology, and coarse surface, confirmed by DLS results and SEM image (Table 1 and Figure 2). Detailedly, either statistic diameter from SEM images or effective diameter from DLS results decreased with the increase of PVA concentration. However, nanoparticle showed not regular sphere with many defects with higher PVA concentration (Figure 2(c)) and lower water/oil ratio (Figures 2(d) and 2(e)). Additionally, when PVA concentration was higher, nanoparticles seemed to stick together (Figure 2(c)), which might be attributed to unwashed PVA. The water/oil ratio of the first emulsion showed little effect on the diameter of nanoparticle either from SEM images or from DLS results. Moreover, zeta potential of all nanoparticles was from −9.3 mV to −14.0 mV (Table 2). No significant difference was detected for zeta potential of nanoparticle for different preparative parameters. Encapsulation efficiency of inner material is another index to characterize effectiveness for functional nanoparticle. BCA results showed that gelatin could be encapsulated into nanoparticle using this method. But unfortunately, gelatin encapsulation efficiency was as low as 1.0–2.0% *w*/*v*. When PVA concentration was low in preparation, gelatin encapsulation efficiency was higher than other nanoparticles, which might be due to less wash time in preparation. These results indicated that gelatin inside nanoparticle could diffuse into outside aqueous solution.

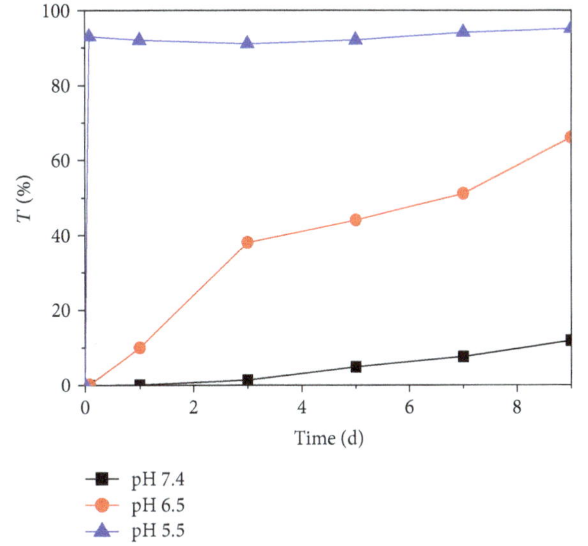

FIGURE 5: Transparency at 400 nm of biphase nanoparticle solutions with different pH values.

3.2. Two-Phase Composite Nanoparticle. As far as nanoparticle size, morphology, and probable emulgator remaining from abovementioned results, preparative parameters of Nano2 as an optimal condition for W/O/W technique were chosen for further investigation. Since crosslinkable gelatin had been verified to be an effective material for hydrogel

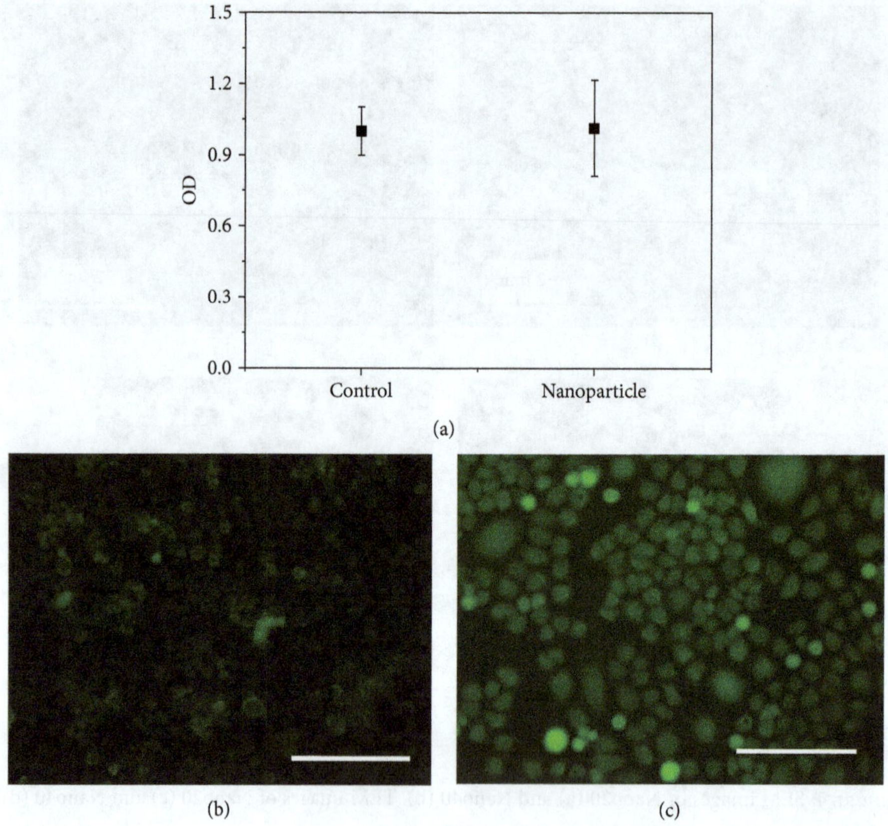

FIGURE 6: (a) Cytoviability of MCF-7 cells control and cocultured with Nano40. Stained MCF-7 cells observed by fluorescence microscopy after cultured 1 d (b, c). MCF-7 cells control (b). Cells cocultured with 100 μg/mL Nano40 (c). The scale is 200 μm.

fabrication with functional properties, it was encapsulated in nanoparticle for in situ nanogel fabrication just as illustrated in Figure 3. In order to make gelatin crosslinked, redox initiation system (APS/TEMED) was used and dissolved in the first and second emulsions. During the process of nanoparticle fabrication, in situ polymerization was taken place simultaneously. Finally, uniformly distributed nanoparticles were obtained regardless of fabrication temperature, which was confirmed by SEM images (Figures 4(a) and 4(b)) and TEM images (Figures 4(c) and 4(d)). In particular, biphase structure was clearly seen in TEM image, which indicated that nanogel embedded in nanoparticle had been successfully formed. Moreover, fabrication temperature seemed to have little influence on the size and morphology of nanoparticles either from SEM images or from TEM images. Hydrodynamic diameters of composite nanoparticle in PBS and in acid were detected by DLS, respectively, which was shown in Table 2. Since Ac-β-CD could be dissolved in acid solution according to previous research, diameter in acid was considered to be the diameter of inner nanogel due to degradation of Ac-β-CD. It was found that the effective diameter of composite nanoparticle was around 200 nm and the effective diameter of nanogel was around 250 nm regardless of fabrication temperature. Moreover, the diameter of nanogel is a little bigger than that of composite nanoparticle, which might be ascribed to the extension of hydrophilic nanogel network in water without hydrophobic Ac-β-CD outsider layer. Furthermore, encapsulated efficiency of nanogel was about 12–14% w/v, which was much larger than that of abovementioned uncrosslinked gelatin. In addition, encapsulated efficiency of nanogel for Nano40 is a little higher than that for Nano20 for the reason that higher fabrication temperature resulted in fast polymerization rate to hinder gelatin molecule diffusion.

Transparency of nanoparticle solution was used to track the degradation process of nanoparticle based on the fact that solution became clear after Ac-β-CD was degraded and dissolved in solution (Figure 5). In PBS (pH 7.4), transparency of nanoparticle solution at 400 nm nearly kept unchanged during the first 3 d, then slowly increased up to 10% in the following 6 d, indicating relative stability of nanoparticle in that solution. In pure water (pH 6.5), transparency of nanoparticle solution at 400 nm gradually increased up to 60% with prolonged time, indicating the continuous degradation of nanoparticle in 9 d. In mild acid solution (pH 5.5), transparency of nanoparticle solution at 400 nm rapidly increased to 90% in the first day, indicating quickly degradation of nanoparticle. The results confirmed the quick pH response property at mild acid medium. The detailed degradation mechanism had been discussed in our previous research. Briefly, degradation of Ac-β-CD nanoparticle was due to structural change of Ac-β-CD. Synthesized Ac-β-CD containing linear acetal groups (LA) and cyclic acetal groups (CA) is sensitive to low pH environment, which was confirmed by $_1$H NMR spectrum.

In view of abovementioned properties, Nano40 was chosen for further cytotoxicity evaluation. Nano40 was cocultured with MCF-7 cells to evaluate its preliminary biocompatibility in vitro. As shown in Figure 6(a), the viability of cells cultured with Nano40 was a little higher than TCPs control, but without significant difference. Fluorescence microscopy images confirmed that live cells lived on the plate homogeneously (Figures 6(b) and 6(c)), with or without Nano40. These results confirmed that the Nano40 nanoparticle could support cell survival.

4. Conclusion

Gelatin-encapsulated Ac-β-CD nanoparticle with monodisperse characteristic, nearly spherical morphology, and coarse surface could be easily fabricated by double water-in-oil-in-water (W/O/W) technique. The diameters of nanoparticles decreased with increase of PVA concentration, but were not influenced by water/oil ratio of the first emulsion. But nanoparticle showed sticky and irregular sphere shape with many defects when PVA concentration was higher and water/oil ratio was lower. Zeta potential of the nanoparticles was from −9.3 mV to −14.0 mV. Further two-phase composite nanoparticle could be obtained by combined W/O/W technique and in situ polymerization. Two-phase structure could be confirmed by TEM images. The effective diameter of composite nanoparticle was around 200 nm, and the effective diameter of nanogel was around 250 nm regardless of fabrication temperature. Encapsulated efficiency of nanogel was about 12–14%, and higher fabrication temperature resulted in higher encapsulated efficiency. Composite nanoparticle showed quick pH response property at mild acid medium and no obvious cytotoxicity.

Conflicts of Interest

The authors declare that there are no competing interests regarding the publication of this paper.

Acknowledgments

This study is financially supported by Natural Science Foundation of Jiangsu Province (BK20171113), Qing Lan Project, the Scholarship of Jiangshu Government, Six Talent Peaks Project in Jiangsu Province (JY-071).

References

[1] A. L. Armstead and B. Li, "Nanotoxicity: emerging concerns regarding nanomaterial safety and occupational hard metal (WC-Co) nanoparticle exposure," *International Journal of Nanomedicine*, vol. 11, pp. 6421–6433, 2016.

[2] D. Bobo, K. J. Robinson, J. Islam, K. J. Thurecht, and S. R. Corrie, "Nanoparticle-based medicines: a review of FDA-approved materials and clinical trials to date," *Pharmaceutical Research*, vol. 33, no. 10, pp. 2373–2387, 2016.

[3] H. Wang, J. Yu, X. Lu, and X. He, *Nanomedicine*, vol. 11, no. 2, pp. 103–106, 2016.

[4] M. R. Preiss and G. D. Bothun, "Stimuli-responsive liposome-nanoparticle assemblies," *Expert Opinion on Drug Delivery*, vol. 8, no. 8, pp. 1025–1040, 2011.

[5] A. R. Town, M. Giardiello, R. Gurjar et al., *Nanoscale*, vol. 9, no. 19, pp. 6302–6314, 2017.

[6] Y. Zhao, K. Thorkelsson, A. J. Mastroianni et al., "Small-molecule-directed nanoparticle assembly towards stimuli-responsive nanocomposites," *Nature Materials*, vol. 8, no. 12, pp. 979–985, 2009.

[7] J. Dong, Y. Wang, J. Zhang et al., "Multiple stimuli-responsive polymeric micelles for controlled release," *Soft Matter*, vol. 9, no. 2, pp. 370–373, 2013.

[8] S. Chen, F. Jiang, Z. Cao, G. Wang, and Z.-M. Dang, "Photo, pH, and thermo triple-responsive spiropyran-based copolymer nanoparticles for controlled release," *Chemical Communications*, vol. 51, no. 63, pp. 12633–12636, 2015.

[9] Z. Cao, H. Wu, J. Dong, and G. Wang, "Quadruple-stimuli-sensitive polymeric nanocarriers for controlled release under combined stimulation," *Macromolecules*, vol. 47, no. 24, pp. 8777–8783, 2014.

[10] E. M. Bachelder, E. N. Pino, and K. M. Ainslie, "Acetalated dextran: a tunable and acid-labile biopolymer with facile synthesis and a range of applications," *Chemical Reviews*, vol. 117, no. 3, pp. 1915–1926, 2017.

[11] B. Wang, P. Liu, B. Shi, J. Gao, and P. Gong, "Preparation of pH-sensitive dextran nanoparticle for doxorubicin delivery," *Journal of Nanoscience and Nanotechnology*, vol. 15, no. 4, pp. 2613–2618, 2015.

[12] Y. Wang, H. Chen, Y. Liu et al., *Biomaterials*, vol. 34, no. 29, pp. 7181–7190, 2013.

[13] Z. Zare-Akbari, H. Farhadnejad, B. Furughi-Nia, S. Abedin, M. Yadollahi, and M. Khorsand-Ghayeni, "PH-sensitive bionanocomposite hydrogel beads based on carboxymethyl cellulose/ZnO nanoparticle as drug carrier," *International Journal of Biological Macromolecules*, vol. 93, no. Part A, pp. 1317–1327, 2016.

[14] X. Hu, S. Chen, X. Gong, Z. Gao, X. Wang, and P. Chen, "Synthesis and preparation of biocompatible and pH-responsive cyclodextrin-based nanoparticle," *Journal of Nanoparticle Research*, vol. 19, no. 3, p. 109, 2017.

[15] B. Chen and X. Hu, "An injectable composite gelatin hydrogel with pH response properties," *Journal of Nanomaterials*, vol. 2017, Article ID 5139609, 7 pages, 2017.

[16] M. Molina, M. Asadian-Birjand, J. Balach, J. Bergueiro, E. Miceli, and M. Calderon, "Stimuli-responsive nanogel composites and their application in nanomedicine," *Chemical Society Reviews*, vol. 44, no. 17, pp. 6161–6186, 2015.

[17] A. Sharma, T. Garg, A. Aman et al., "Nanogel—an advanced drug delivery tool: current and future," *Artificial Cells, Nanomedicine and Biotechnology*, vol. 44, no. 1, pp. 165–177, 2014.

17

Deterministic Approaches for Simulation of Nitroxide-Mediated Radical Polymerization

Mariano Asteasuain [1,2]

[1]*Planta Piloto de Ingeniería Química (PLAPIQUI), Universidad Nacional del Sur (UNS) – CONICET, 8000 Bahía Blanca, Argentina*
[2]*Departamento de Ingeniería Química, Universidad Nacional del Sur (UNS), 8000 Bahía Blanca, Argentina*

Correspondence should be addressed to Mariano Asteasuain; masteasuain@plapiqui.edu.ar

Academic Editor: Hossein Roghani-Mamaqani

Since its development in the last decades, controlled radical polymerization (CRP) has become a very promising option for the synthesis of polymers with controlled structure. The design and production of tailor-made materials can be significantly improved by developing models capable of predicting the polymer properties from the operating conditions. Nitroxide-mediated polymerization (NMP) was the first of the three main variants of CRP to be discovered. Although it has lost preference over the years against other CRP alternatives, NMP is still an attractive synthesis method because of its simple experimental implementation and environmental friendliness. This review focuses on deterministic methods employed in mathematical models of NMP. It presents an overview of the different techniques that have been reported for modelling NMP processes in homogeneous and heterogeneous media, covering from the prediction of average properties to the latest techniques for modelling univariate and multivariate distributions of polymer properties. Finally, an outlook of model-based design studies of NMP processes is given.

1. Introduction

Controlled radical polymerization (CRP), also called reversible-deactivation radical polymerization (RDRP) according to the IUPAC recommendation [1], has emerged in the last decades as a very promising option for the synthesis of polymers with controlled structure. In CRP systems, the high reactivity of the polymer radicals is regulated by the addition of an agent which establishes a balance between active and temporarily inactive (dormant) chains. This dynamic equilibrium is strongly shifted towards the dormant chains. As a consequence, the number of active chains is very small which reduces the interactions between them. For this reason, the overall termination effect is decreased and all chains grow on average at the same rate. CRP systems are distinguished by a linear increase of molecular weight with conversion and by a low amount of dead polymer chains (~1–10%).

There are two general approaches to establishing the necessary balance between active and inactive species: the first one is based on a reversible termination (deactivation), while the second one is based on a reversible (degenerative exchange) transfer. In both cases, the polymer radical propagates a few times during activation periods, before is converted back to the inactive state. The initiation rate, the contribution of the termination reactions, and the dynamics of the activation/deactivation exchange influence the polymer dispersity. As the exchange rate increases, which means that fewer monomer units are added to the growing active chains per activation period, the polymer dispersity decreases. Therefore, CRP is effective in controlling molecular weights. Besides, the dynamic equilibrium between active and dormant radicals spans the lifetime of the living polymer chains from fractions of seconds to several hours. This allows manipulating the chain structure during the polymerization reaction. In this way, it is possible to synthesize complex molecular structures, such as comb or star polymers or copolymers with well-defined structure (block, gradient, or branched copolymers) [2, 3].

The three best-known and most effective variants of CRP are atom transfer radical polymerization (ATRP), nitroxide-mediated polymerization (NMP), and reversible addition-fragmentation chain transfer polymerization (RAFT).

1.1. Atom Transfer Radical Polymerization. The ATRP technique is based on the rapid capture of the propagating radicals in a deactivation reaction. Effective ATRP catalysts ($X - Mt^s/L$) are generally organo-metallic species formed by a transition metal capable of changing their oxidation number, a complexing ligand L, and a counterion X which can form a covalent or ionic bond with the metal centre. The capture of the active radicals is performed by the counterion upon its release after the reduction of the transition metal. Since the activation in ATRP is a bimolecular process, the inactive species D_n formed by the polymer radical and the counterion is inherently stable and can only be activated sporadically, according to the activity of the transition metal that acts as a catalyst. The characteristic kinetic step of ATRP processes is

$$R_n^\bullet + X - Mt^{s+1}/L \underset{k_{\text{act}}}{\overset{k_{\text{desact}}}{\rightleftarrows}} D_n + Mt^s/L. \tag{1}$$

The reduction of the transition metal complex allows the release of the X atom ($X - Mt^{s+1}/L + e \rightarrow X + Mt^s/L$), which captures the propagating radical forming the dormant species D_n. The success of control on the polymerization is in the rate of periodic activation of the dormant species [4].

1.2. Nitroxide-Mediated Polymerization. In NMP, the reaction is also controlled through the reversible termination of the active chains. Here, the mediating agent is a nitroxide that traps the propagating radical, forming a stable radical. The dissociation step of this reversible termination releases the active radical from the dormant species, and this radical can then propagate or terminate with another radical before being deactivated again. The reaction is strongly displaced towards the inactive species. Therefore, most of the polymer chains in the reaction medium are temporarily dormant, and the concentration of radicals capable of propagating or terminating is several orders of magnitude lower than that of conventional radical polymerization [5]. The distinctive reaction of these polymerizations technique is

$$R_n^\bullet + X \underset{k_{\text{act}}}{\overset{k_{\text{desact}}}{\rightleftarrows}} D_n. \tag{2}$$

Here, species X is the nitroxide mediator which, upon trapping the radical, forms the dormant species D_n with n monomer units. The slow thermal dissociation of the dormant species provides a low concentration of radicals, which allows the termination step to be kept to a minimum.

1.3. Reversible Addition-Fragmentation Chain Transfer. RAFT polymerization differs from the two above in that control over the growth of the polymer chains is achieved by a degenerative chain transfer process. Basically, what happens is that a single active radical site is shared among many chains so that these molecules participate in fewer reactions. In these systems, as in conventional radical polymerization, a concentration of radicals in the pseudostationary state is established through the initiation and termination processes. Through bimolecular transfer processes, a minimal number of growing radicals undergoes a chain interchange with dormant species through an intermediate adduct. Control over molecular weight and dispersity is provided by chain transfer agents (CTAs) which exchange the active radical centre between the growing chains. Good control requires that the exchange is rapid compared to propagation [6]. The distinctive kinetic step is

$$R_n^\bullet + TR_m \underset{k_f}{\overset{k_a}{\rightleftarrows}} R_nTR_m \underset{k_a}{\overset{k_f}{\rightleftarrows}} TR_n + R_m^\bullet. \tag{3}$$

The T species acts as a chain transfer agent by transferring the radical centre from the chain $R_n^\bullet$ with n monomer units to the chain $R_m^\bullet$ with m units. In an intermediate step, a radical adduct of 2 branches, R_nTR_m, is generated. The success of this polymerization is given by a rapid exchange, that is, a large $K_{\text{exch}} = k_a/k_f$, to keep the concentration of active radicals low [4].

The three main variants of CRP, ATRP, NMP, and RAFT have their own advantages and disadvantages [7]. Although NMP was initially restricted to a limited number of monomers, the development of new nitroxides and alkoxyamines has allowed using this technique with almost all monomers, with exception of vinyl acetate and vinyl chloride. This can be achieved with few nitroxides such as SG1 (and the BlocBuilder MA alkoxyamine) or TIPNO, which are commercially available [5]. Since the development of the different CRP techniques, NMP lost rapidly the preference against RAFT and ATRP because of several advantages of the latter two, such as the capability of polymerizing more monomers and having more convenient reaction temperatures. However, NMP still remains as an attractive synthesis method because of its thermal activation mechanism, a monocomponent control system, a very simple purification of the product, and no environmental issues.

The development of CRP has made it possible to control key polymer properties such as the molecular weight distribution (MWD), the copolymer composition distribution (CCD), or the length distribution of comonomer sequences (SLD). This feature is outstanding for achieving the production of polymers with prespecified properties [8, 9]. Still, the manipulation of the chain structure is a difficult task since the molecular properties have a strong dependence on the operating conditions, which results in complex interactions between process variables. Thus, the design and production of tailor-made materials can be significantly improved by developing models capable of predicting the polymer properties from the operating conditions [10, 11]. The present review focuses on deterministic methods employed in mathematical models of NMP. It presents an overview of the different techniques that have been reported for modelling NMP processes in homogeneous and heterogeneous media, covering from the prediction of average properties to the latest techniques for modelling univariate and multivariate distributions of polymer properties. Finally, an outlook of model-based design studies of NMP processes is given.

2. Deterministic Methods for the Simulation of NMP

Mathematical modelling is undoubtedly a very valuable tool in polymer science. It helps to gain understanding in the kinetic mechanisms of polymer processes and also complements and may partially substitute expensive and time-consuming laboratory experiments. Besides, the model-based design is becoming a new paradigm in polymer processes [12, 13]. Polymer applications are every day more sophisticated and specific. New synthesis methods like CRP allow producing polymers with controlled, complex microstructures. These include the MWD, the CCD, chain functionality, and topology. The polymer chain microstructure determines several properties of the material. Therefore, the combination of experiments and mathematical models allows designing processes for the production of polymers with precise, tailored characteristics [14].

Most of the mathematical techniques used in the simulation of polymer processes can be grouped into two major approaches: deterministic and stochastic methods. Stochastic approaches are mainly represented by the Monte Carlo (MC) method. The majority of the applications of this method use the kinetic Monte Carlo (KMC) technique proposed by Gillespie [15, 16]. A KMC simulation stochastically follows the evolution in time of a full set of molecules of a sample of the reaction system, based on the probabilities of the individual reaction steps of the kinetic mechanism. A KMC simulation tracks explicitly and individually each molecule in the ensemble of chains. Therefore, it can provide extremely detailed information about the polymer microstructure, which is generally not available with deterministic solvers. Besides, the implementation of these methods is simple. On the other hand, due to the statistical nature of the simulation, the reproducibility and reliability of the results depend on the sample size, which needs to be large enough to guarantee accurate outputs. The problem is that large samples require long computational times and large storage capacities. However, the improvement of the computational efficiency of MC implementations is nowadays an area of active research, which includes parallelization, code optimization, and hybrid deterministic-MC approaches [17–20]. Recent reviews summarize relevant works using MC methods in polymer science [21, 22]. In NMP, these methods have been used to predict copolymer SLD [10, 11, 23, 24], chain functionality and the full MWD [25], kinetics of branching [26], and bivariate MWD-CCD in continuous processes with arbitrary residence time distributions [27] or to track the exact position of functional comonomers in the copolymer chains [28]. Besides, the model-based design of copolymer synthesis by NMP using MC models has also been performed [11, 28, 29]. The other variants of CRP have also been studied using MC models [20, 30–41].

On the other hand, deterministic methods are based on the solution of the differential-algebraic system of population balance equations (PBEs) that represent the polymer system. PBEs are balance equations that describe the evolution in specific properties, called internal coordinates, of the population of polymer molecules. For instance, it is usual to characterize copolymer molecules in a copolymerization reaction by the content of each comonomer in the chain. Therefore, a PBE for this system may involve a differential balance equation for the time evolution of the concentration of species $P_{n,m}$, a copolymer molecule with n units of monomer 1 and m units of monomer 2 (the internal coordinates). PBEs are derived from a kinetic mechanism of the polymer process. Deterministic methods provide unique results, as opposed to the random output of stochastic methods, and generally demand shorter computing times and less storage capacity. Besides, they are usually more suitable for identification and optimization purposes where it is necessary to count with smooth, differentiable structures.

Deterministic methods have a long history in polymer reaction engineering [42]. In particular, there are articles that review deterministic approaches used in mathematical models of the different variants of CRP [12, 43–45]. The present review deepens on deterministic methods employed in mathematical models of NMP. It starts with models describing polymerization rate and average properties only. Then, it discusses techniques for modelling distributions of polymer properties, like the MWD and the SLD. Afterward, it examines complex techniques for predicting multivariate distributions. This is followed by a review of NMP models in heterogeneous media and hybrid deterministic-stochastic methods. Finally, an outlook of model-based design studies of NMP processes is given.

2.1. Average Properties and Polymerization Rate. PBEs constitute a set of equations that is, in principle, infinite in number, because the internal coordinates are distributed properties that theoretically may take values from zero/one to infinity. Since a balance equation must be set for every possible value of the internal coordinates in order to achieve a complete description of the polymer system, an unbounded system of equations is obtained. For instance, in the case of the population, $P_{n,m}$, a mathematical model would need a balance equation for every possible value of n and m in order to get the full distribution of chain lengths. Although it is sometimes possible to assume a maximum value for a distributed property, the resulting number of equations is still usually large.

Taking advantage that some of their averages can sometimes represent the property distributions, there are techniques that allow limiting the number of model equations. The method of moments, originally proposed by Bamford and Tompa [46], is the most popular method for modelling averages of distributed polymer properties. Considering that polymer species are described only by their chain length, the following equations define the number average molecular weight ($\overline{M_n}$), the weight average molecular weight ($\overline{M_w}$), and the dispersity ($Đ$) of a homopolymer in a NMP in terms of the leading moments of the MWD:

$$\overline{M_n} = M_{\text{Mon}} \frac{\lambda_1^R + \lambda_1^P + \lambda_1^D}{\lambda_0^R + \lambda_0^P + \lambda_0^D}, \tag{4}$$

$$\overline{M_w} = M_{\text{Mon}} \frac{\lambda_2^R + \lambda_2^P + \lambda_2^D}{\lambda_1^R + \lambda_1^P + \lambda_1^D}, \tag{5}$$

$$Đ = \frac{\overline{M_w}}{\overline{M_n}}. \tag{6}$$

In these equations, M_{Mon} is the monomer molecular weight, and λ_a^R, λ_a^P, and λ_a^D are the moments of order a ($a = 0,1,2$) of the MWD expressed in molar concentration, of propagating radicals, dead polymer, and dormant radicals, respectively. Besides, (7) expresses the monomer conversion (C) in terms of the first-order moments:

$$C = \frac{\lambda_1^R + \lambda_1^P + \lambda_1^D}{\lambda_1^R + \lambda_1^P + \lambda_1^D + [M]}, \tag{7}$$

where $[M]$ is the monomer molar concentration.

The moments used above are defined as

$$\lambda_a^I = \sum_{n=0}^{\infty} n^a [I_n], \tag{8}$$

where $[I_n]$ is the concentration of a polymer species I ($I = R, P, D$) of chain length n.

The method of moments, as applied in the modelling of average molecular weights and conversion, consists in transforming the univariate PBEs of the polymer species (chain length as the only internal coordinate) into balance equations for the moments of their chain length distributions. This procedure replaces the infinite equations of the chain length distribution with just nine equations for the moments of order zero, one, and two of the MWDs of propagating radicals, dead polymer, and dormant radicals. This method is simple and computationally inexpensive [47]. Its drawback, however, is that it loses information on the full shape of the MWD. The process of transforming the PBEs into the moment equations is not complex, but it can be tedious and lengthy for some cases. Recently, Mastan and Zhu [47] presented a tutorial on the use of the method of moments in different polymer systems.

The method of moments has been used extensively to study CRP systems. Examples on the application of this method to ATRP and RAFT can be found elsewhere [12, 48–54]. In particular, several works employed the method of moments for the simulation of NMP processes. One of the pioneering works was presented by Veregin et al. [55]. Assuming a kinetic mechanism consisting only of activation/deactivation equilibrium and propagation reactions, as well as constant nitroxide and living radical concentrations, they obtained an analytical solution for the average molecular weights and dispersity. In spite of the simplicity of the kinetic mechanism, they obtained a good match with experimental data. Later, more comprehensive representations were proposed. Butté et al. [56] presented a moment-based model suitable for handling NMP and ATRP processes. The model predicted polymerization rate and average properties and was validated against experimental data. Zhu [50] developed a kinetic model for NMP processes using the method of moments. This model predicted conversion, average molecular weights, and dispersity. He used the model for investigating the effect of operating variables, such as the concentration of initiator, nitroxide, and monomer, on the process performance. Also, he investigated the influence of the rate constants of initiation, propagation, termination, transfer, and equilibrium between propagating and dormant radicals, highlighting the significant effect of the transfer to monomer reaction in the broadening of the MWD. Bonilla et al. [57] presented a moment-based model for the NMP of styrene. They considered a detailed kinetic mechanism that included chemical initiation, reversible nitroxyl ether decomposition, monomer dimerization, thermal initiation, propagation, the equilibrium between propagating and dormant radicals, alkoxyamine decomposition, rate enhancement, transfer to monomer and dimer, and conventional termination. Nondimensional model equations improved the numerical behavior in the parameter estimation. The model predictions showed good agreement with experimental data. Kruse et al. [58] modeled the NMP of styrene predicting average molecular weights and conversion. Using an Evans-Polanyi description of the activation energy ($E = E_0 + \alpha\Delta H_R$), the model was fit to experimental data of the reaction at 87°C to obtain parameters ΔH_R of the decoupling reaction of dormant radicals and E_0 of the propagation/depropagation reactions. They compared the fit with different side reactions, such as transfer to monomer, monomer thermal initiation, transfer to polymer, and the reaction between active radicals and nitroxide to form a hydroxy amine. They found that the monomer thermal initiation was critical for obtaining a good fit. On the other hand, the effect of transfer to polymer and the reaction between active radicals and nitroxide were found to be negligible. Belincanta-Ximenes et al. [59] used the model developed by Bonilla et al. [57] to simulate polymerization rate, average molecular weights, and concentration of polymeric species in the NMP of styrene over a range of operating conditions. They also performed a parameter sensitivity analysis showing the effects of temperature, activation/deactivation equilibrium constant, and initial concentrations of nitroxide and initiator. The simulated profiles of conversion, number average molecular weight, and dispersity agreed well with experimental data. Roa-Luna et al. [60] updated the model by Bonilla et al. [57] including the diffusion-controlled (DC) effects on bimolecular radical termination, propagation, and activation/deactivation reactions. They employed free volume theory to model the DC effects. The authors found that DC termination enhanced the living behavior of the system, whereas DC propagation, DC activation, and DC deactivation worsened it. By comparing model predictions with experimental data, the overall conclusion was that the slight improvement in model performance by the inclusion of the DC effects did not justify the increased model complexity. A later work by this group [61] postulated that side reactions between the initiator and the nitroxide might reduce the effective level of alkoxyamine produced in the early stages of the reaction. Based on the previous model by Bonilla et al. [57], these side reactions were modelled by introducing a controller efficiency factor that determined the fraction of nitroxide able to produce dormant species. This model refinement improved the model predictions, particularly those of the average molecular weights. Asteasuain et al. [62] used the method of moments to predict average molecular properties in the NMP of styrene in tubular reactors. They obtained a good comparison with experimental data.

Copolymerization processes increase the complexity of the mathematical model. In addition to chain length, copolymer composition and the length of the monomer sequences are important properties to be considered. The extension of the method of moments to copolymerization reactions uses double-index moments to compute the average molecular properties. The balance equations for these moments are derived from PBEs that include two internal coordinates, the number of units of each comonomer in the chain. For these bivariate PBEs, the double-index moments of order a, b may be defined as

$$\lambda_{a,b}^{I} = \sum_{n=0}^{\infty} \sum_{m=0}^{\infty} n^a m^b [I_{n,m}], \tag{9}$$

where $[I_{n,m}]$ is the molar concentration of a copolymer species I with n units of monomer 1 and m units of monomer 2.

Copolymer systems require being careful in the definition of the average weights in terms of the molar concentration of the polymeric chains because there is not a straightforward relationship between chain length and mass as in homopolymers [63]. Based on the appropriate definitions, the following equations express the average molecular weights, in an NMP, in terms of the leading double-index moments:

$$\overline{M_n} = \frac{M_{\text{Mon1}}\left(\lambda_{1,0}^R + \lambda_{1,0}^P + \lambda_{1,0}^D\right) + M_{\text{Mon2}}\left(\lambda_{0,1}^R + \lambda_{0,1}^P + \lambda_{0,1}^D\right)}{\lambda_{0,0}^R + \lambda_{0,0}^P + \lambda_{0,0}^D}, \tag{10}$$

$$\overline{M_w} = \frac{(M_{\text{Mon1}})^2\left(\lambda_{2,0}^R + \lambda_{2,0}^P + \lambda_{2,0}^D\right) + 2M_{\text{Mon1}}M_{\text{Mon2}}\left(\lambda_{1,1}^R + \lambda_{1,1}^P + \lambda_{1,1}^D\right) + (M_{\text{Mon2}})^2\left(\lambda_{0,2}^R + \lambda_{0,2}^P + \lambda_{0,2}^D\right)}{M_{\text{Mon1}}\left(\lambda_{1,0}^R + \lambda_{1,0}^P + \lambda_{1,0}^D\right) + M_{\text{Mon2}}\left(\lambda_{0,1}^R + \lambda_{0,1}^P + \lambda_{0,1}^D\right)}. \tag{11}$$

The dispersity is calculated using (6) as previously. The conversion can also be expressed in terms of the double-index moments as

$$C = \frac{\left(\lambda_{1,0}^R + \lambda_{1,0}^P + \lambda_{1,0}^D\right) + \left(\lambda_{0,1}^R + \lambda_{0,1}^P + \lambda_{0,1}^D\right)}{\left(\lambda_{1,0}^R + \lambda_{1,0}^P + \lambda_{1,0}^D\right) + \left(\lambda_{0,1}^R + \lambda_{0,1}^P + \lambda_{0,1}^D\right) + [M_1] + [M_2]}. \tag{12}$$

Equations (10), (11), and (12) involve 18 moments, those of order (2,0), (1,1), (0,2), (1,0), (0,1), and (0,0) for each of the three polymeric species. This is twice the number of moments than in homopolymerization models.

The CCD is also an important property of copolymers. Number and weight average values of this distribution can also be expressed in terms of the double-index moments [64]:

$$F_{1,n} = \frac{\lambda_{1,0}^R + \lambda_{1,0}^P + \lambda_{1,0}^D}{\left(\lambda_{1,0}^R + \lambda_{1,0}^P + \lambda_{1,0}^D\right) + \left(\lambda_{0,1}^R + \lambda_{0,1}^P + \lambda_{0,1}^D\right)}, \tag{13}$$

$$F_{1,w} = \frac{M_{\text{Mon1}}\left(\lambda_{2,0}^R + \lambda_{2,0}^P + \lambda_{2,0}^D\right) + M_{\text{Mon1}}\left(\lambda_{1,1}^R + \lambda_{1,1}^P + \lambda_{1,1}^D\right)}{M_{\text{Mon1}}\left(\lambda_{2,0}^R + \lambda_{2,0}^P + \lambda_{2,0}^D\right) + (M_{\text{Mon1}} + M_{\text{Mon2}})\left(\lambda_{1,1}^R + \lambda_{1,1}^P + \lambda_{1,1}^D\right) + M_{\text{Mon2}}\left(\lambda_{0,2}^R + \lambda_{0,2}^P + \lambda_{0,2}^D\right)}, \tag{14}$$

where $F_{1,n}$ and $F_{1,w}$ are the number and weight average copolymer composition of monomer 1 in the copolymer, respectively.

Zhang and Ray [65] used this approach in the development of a comprehensive model for CRP in tank reactors. The model was generic and valid for reversible capping

TABLE 1: Kinetic steps of a nitroxide-mediated copolymerization in terms of the reacting polymer chains.

Step	Equation
Capping and uncapping	$X + R^{\bullet i}_{n,m} \underset{k_{\mathrm{act},i}}{\overset{k_{\mathrm{desact},i}}{\rightleftharpoons}} D^{i}_{n,m}, \quad i = 1, 2.$
Propagation (homo- and cross-propagation)	$R^{\bullet i}_{n,m} + M_j \xrightarrow{k_{pij}} R^{\bullet j}_{n+2-j,m+j-1}, \quad i = 1, 2, j = 1, 2.$
Transfer to monomer	$R^{\bullet i}_{n,m} + M_j \xrightarrow{k_{trmij}} R^{\bullet j}_{2-j,j-1} + P_{n,m}, \quad i = 1, 2, j = 1, 2.$
Termination by combination (homo- and cross-termination)	$R^{\bullet i}_{n,m} + R^{\bullet j}_{r,q} \xrightarrow{k_{tcij}} P_{n+r,m+q}, \quad i = 1, 2, j = 1, 2.$
Termination by disproportionation (homo- and cross-termination)	$R^{\bullet i}_{n,m} + R^{\bullet j}_{r,q} \xrightarrow{k_{tdij}} P_{n,m} + P_{r,q}, \quad i = 1, 2, j = 1, 2.$

$R^{\bullet i}_{n,m}$ is a polymer radical with n units of monomer 1 and m units of monomer 2, with a monomer M_i final unit; $D^{i}_{n,m}$ is a dormant radical with n units of monomer 1 and m units of monomer 2, with a monomer M_i final unit; $P_{n,m}$ is a dead polymer chain with n units of monomer 1 and m units of monomer 2.

CRP systems (i.e., NMP and ATRP). It was validated using experimental data of NMP of styrene and atom transfer radical copolymerization of styrene and n-butylacrylate. Polymerization in batch, semibatch, and a series of continuous tank reactors was analyzed, getting insight into the strengths and weaknesses of the control of the polymer architecture of each operation mode. These authors later extended this work to analyze polymerizations in tubular reactors [66]. Fortunatti et al. [67] used the double-index moment method to predict average molecular weights and copolymer composition in the nitroxide-mediated copolymerization of styrene and α-methyl styrene.

The SLD, that is, the distribution of the length of sequences of each kind of monomer along the polymer chains, is also a significant copolymer property. MC methods are very well suited for giving a detailed insight into this distribution. These methods can predict the explicit sequence of monomers along each copolymer chain in an ensemble of chains [10]. Deterministic models, on the other hand, are in general less exhaustive. They can predict the SLD of the ensemble of chains as a whole, but not of the individual chains. Deterministic models of the SLD are usually set up by formulating a parallel kinetic mechanism that considers the sequences of monomers as the reacting species [68]. For instance, Table 1 shows an extract of a conventional NMP kinetic mechanism, and Table 2 shows the corresponding parallel kinetic mechanism for modelling the SLD [67].

TABLE 2: Kinetic steps of a nitroxide-mediated copolymerization in terms of the monomer sequences.

Step	Equation
Capping and uncapping	$X + S^{\bullet i}_{n} \underset{k_{\mathrm{act},i}}{\overset{k_{\mathrm{desact},i}}{\rightleftharpoons}} S^{Ii}_{n}, \quad i = 1, 2.$
Homopropagation	$S^{\bullet i}_{n} + M_i \xrightarrow{k_{pii}} S^{\bullet i}_{n+1}, \quad i = 1, 2.$
Cross-propagation	$S^{\bullet i}_{n} + M_j \xrightarrow{k_{pij}} S^{di}_{n} + S^{\bullet j}_{1}, \quad i = 1, 2, j = 1, 2, j \neq i.$
Transfer to monomer	$S^{\bullet i}_{n} + M_j \xrightarrow{k_{trmij}} S^{di}_{n} + S^{\bullet j}_{1}, \quad i = 1, 2, j = 1, 2.$
Homotermination by combination	$S^{\bullet i}_{n} + S^{\bullet i}_{m} \xrightarrow{k_{tcii}} S^{di}_{n+m}, \quad i = 1, 2.$
Cross-termination by combination	$S^{\bullet i}_{n} + S^{\bullet j}_{m} \xrightarrow{k_{tcij}} S^{di}_{n} + S^{dj}_{m}, \quad i = 1, 2, j = 1, 2, j \neq i.$
Termination by disproportionation	$S^{\bullet i}_{n} + S^{\bullet j}_{m} \xrightarrow{k_{tdij}} S^{di}_{n} + S^{dj}_{m}, \quad i = 1, 2, j = 1, 2.$

$S^{\bullet i}_{n}$ is a sequence of n units of monomer M_i with an active radical centre. This sequence is positioned at an end of an active radical; S^{Ii}_{n} is a sequence of n units of monomer M_i temporarily inactivated by the nitroxide. This sequence is positioned at an end of a dormant radical; S^{di}_{n} is a dead sequence of n units of monomer M_i. This sequence may be positioned at any place along an active, dormant, or dead polymer chain.

The kinetic mechanism of Table 1 leads to the usual PBEs for $R^{\bullet i}_{n,m}$, $D^{i}_{n,m}$, and $P_{n,m}$, from which the MWD and its averages can be obtained. The parallel kinetic mechanism of Table 2 leads to PBEs for the monomer sequences $S^{\bullet i}_{n}$, S^{Ii}_{n}, and S^{di}_{n}, which define the copolymer SLD. It should be noted that this approach regards the different monomer sequences as entities disconnected from the polymer chains to which they belong, so no tracking of the monomer sequences along single chains is available. Only one internal coordinate, the sequence chain length, characterizes the monomer sequences. Therefore, number and weight average sequence length, as well as its dispersity, can be computed with equations analogous to (4), (5), and (6). The only difference is that SLD averages are chain length averages, not mass averages, and therefore, the corresponding equations are not multiplied by the monomer molecular weight. Following this approach, Ye and Schork [69] developed a moment-based model that described the MWD and the SLD. This model was able to compute the average properties of the MWD and of the SLD. They compared model data with experimental information. The model was used to describe the conventional and NMP of styrene and 4-methylstyrene. Fortunatti et al. [67] used the same methodology to compute the averages of the copolymer SLD in the nitroxide-mediated copolymerization of styrene and α-methyl styrene.

The computation of the SLD moments provides an alternative to the calculation of copolymer composition, without resorting to double-index moments. Since monomer sequences are composed of a single monomer, the copolymer composition can be computed as the ratio of the 1st-order moment of the SLD of one monomer and the sum of the 1st-order moments of the SLD of both comonomers [69], that is,

$$F_{1,n} = \frac{\lambda_1^{S^{\bullet 1}} + \lambda_1^{S^{l1}} + \lambda_1^{S^{d1}}}{\left(\lambda_1^{S^{\bullet 1}} + \lambda_1^{S^{l1}} + \lambda_1^{S^{d1}}\right) + \left(\lambda_1^{S^{\bullet 2}} + \lambda_1^{S^{l2}} + \lambda_1^{S^{d2}}\right)}. \tag{15}$$

Another approach to modelling copolymer reactions is the so-called "pseudo-kinetic rate constants method" [70]. This method defines "pseudo-kinetic" rate constants in terms of the true kinetic constants and the mole fractions of polymer radicals and monomers in the system. The application of the "pseudo-kinetic" constants to the reaction rate expressions of the copolymerization system allows reducing these rate expressions to those of a homopolymerization. Therefore, methods for homopolymerization models, like the one-index moments, can be applied to compute the copolymer average properties. Hernández-Ortiz et al. [71] used this approach to model the nitroxide-mediated copolymerization with cross-linking of vinyl/divinyl monomers. The model assumed that polymer radicals contained only one active radical centre and one dormant radical centre (monoradical assumption). Predictive capabilities included polymerization rate, average molecular weights, gelation point, the evolution of sol and gel weight fractions, cross-link density, and copolymer composition. The model was validated with experimental data of TEMPO-mediated copolymerization of styrene with divinylbenzene. DC effects were assessed and found unimportant. Later [72], they extended their model to consider a more realistic multiradical approach, which took into account the possibility that polymer radicals contained more than one radical centre. In this model, triple-index moments were applied to PBEs whose internal coordinates were the chain length, the number of active centres, and the number of dormant centres. They also obtained a good comparison with experimental data using this model. Recently [73], the group presented a new work in this line that included a different nitroxide and an extended set of experimental information.

Several authors used the commercial software Predici [74] to develop models of NMP systems. Predici is a simulation package for the modelling and dynamic simulation of polymer processes. It employs the Galerkin h-p method to compute the complete MWD without a prior assumption of its shape. Even though the full MWD is available, many works in which Predici was used reported only the prediction of the MWD averages, such as $\overline{M_n}$ and $\overline{M_w}$. These works presented an analysis of reaction steps, in comparison with experimental data and understanding of the influence of operating conditions [75–84]. For instance, Gigmes et al. [75] studied the influence of several kinetic parameters and reaction steps on the livingness and controlled character of a generic NMP. Nicolas et al. [76] presented a comprehensive model of the SG1-mediated copolymerization of methyl methacrylate with small amounts of styrene. They used the model to gain insight into the complex mechanism of the copolymerization and to analyze the influence of different operating conditions. Greszta and Matyjaszewski [77] used a Predici model of the NMP of styrene mediated by TEMPO to find the relevant kinetic steps of this polymerization and to estimate the equilibrium constant of the exchange between active and dormant species using experimental data. Benoit et al. [78] developed a model of the NMP of styrene and of n-butyl acrylate using DEPN as a mediating agent. This model showed good agreement with experimental data. Guillaneuf et al. [79] used a Predici model of the NMP of methyl methacrylate mediated by SG1 to show that the penultimate effect was crucial for the accurate description of this system. Fu et al. [80] developed a model of the high-temperature semibatch NMP of styrene. They studied DC effects, showing that the gel effect was not required to improve the fit of the simulation of polymerizations leading to low molecular weight material. Guillaneuf et al. [81] synthesized a new nitroxide, a dialkylated α-hydrogenated linear compound, and the corresponding 1-phenylethyl alkoxyamine, which yielded enhanced polymerization rate in the NMP of styrene. They used a Predici model of the system to study the effect of temperature and alkoxyamine concentration on the kinetics and on the livingness and controlled character of the polymerization. Payne et al. [82] presented a model of the high-temperature NMP of styrene in batch and CSTRs. Using activation/deactivation coefficients estimated from the batch operation, they analyzed the operation in the CSTR finding explanations for experimental information about the molecular weight of the product. Roa-Luna et al. [84] used a model developed in Predici of the NMP of styrene to study the importance of DC effects. Using a conjunction of model results and experimental data, they concluded that DC effects were weak in this system. Hernández-Meza et al. [83] developed a model in Predici of the NMP using hydroxy-TEMPO as mediating agent and initiated by microwave irradiation. Comparing model outputs of average properties with experimental data, they concluded that microwave-activated production of free radicals from monomer molecules was the most probable initiation mechanism.

2.2. Molecular Weight Distribution and Other Distributions of Polymer Properties

2.2.1. Molecular Weight Distribution. Prediction of the full MWD and other distributions of polymer properties is less frequent than the computation of average values because it demands to solve a more complex mathematical problem, the solution of the theoretically infinite set of PBEs. Different approaches have been reported in the literature for modelling univariate, in most cases the MWD, and multivariate distributions of polymer properties. Some of them have been applied to NMP systems. Usually, authors have resorted to some sort of transformation of the PBEs. One of these is the transformation using probability generating function (pgf). The pgf technique [85, 86] is a powerful method that allows modelling the MWD easily and efficiently in complex polymer systems, without any a priori assumption on the MWD shape. PBEs are transformed into the pgf domain, then the transformed balances are solved for the pgf transform of the MWD, and finally, this transform is numerically inverted to recover the MWD. Asteasuain et al. [62] applied this approach to calculate the full MWD in the NMP of styrene in tubular reactors. The model provided a successful comparison against experimental data. Dias and Costa [87] presented a different transformation-inversion approach to model the full MWD, which uses generating functions of the PBEs. They calculated the MWD in linear and nonlinear CRP, including NMP. They showed that under simplified conditions, it was possible to obtain analytical solutions for the MWD. They also presented solutions based on the numerical inversion of the generating functions for more realistic CRP systems.

Other approaches involve the solution of the original set of PBEs. Naturally, different strategies are applied to reduce its infinite dimensionality. First, they all truncate the chain length dimension at a maximum value beyond which the concentration of polymer chains is assumed zero. Then, different strategies are applied to further reduce the size and complexity of the system of equations. Bentein et al. [88] used a method composed by the integration of moment equations and of modified PBEs. The modification involved the application of the quasi-steady-state approximation (QSSA) to the PBEs of macroradicals and a coarse-graining discretization of the dormant and dead polymer balances. In this discretization, the chain length interval was divided into classes, and the properties of the elements of a class were assigned to all the elements of that class. Convergence was checked by comparing the average properties computed using the method of moments and calculated from the MWD. The authors used this approach to get insight into the complex interplay between reaction rates and polymer properties in the SG1- and the TEMPO-mediated bulk polymerization of styrene at 396 K.

Also, Saldívar-Guerra et al. [89] analyzed the calculus of the full MWD by direct integration of PBEs in addition polymerization. They stated that direct integration was feasible in reasonable computing times using standard computers for a considerable number of realistic systems, including NMP. Apart from the truncation of the maximum chain length, they applied the QSSA to remove the stiffness of the system of DAE. Then, explicit solvers, which are computationally less demanding than implicit ones, were used to solve the remaining differential equations. As an example, the NMP of styrene was presented in this work. In a later work by this group, Zapata-González et al. [90] summarized the methodology for the application of the QSSA in CRP and other polymer systems. They illustrated specific features of the methodology by application to academically and industrially relevant systems of controlled radical polymerization (RAFT and NMP cases) and coordination catalysis polymerization.

2.2.2. Sequence Length Distribution. Some authors also addressed the prediction of the full SLD in CRP of copolymer systems using deterministic models [67, 91–93]. They all computed the SLD by formulating a model for the sequences of monomers as reacting species, as was discussed above. In the case of the SLD, it is generally feasible to solve the sequences PBEs by direct integration straightforwardly because the maximum sequence length usually takes small values. Regarding NMP systems in particular, Fortunatti et al. [67] used this approach to predict the SLDs of styrene and α-methyl styrene in the copolymerization of these monomers. Aguiar et al. [92] used this method to develop a model for the nitroxide-mediated radical copolymerization of styrene-divinylbenzene. This model included the cyclic propagation (cyclization) reactions. Monomer sequences, in this model, are those that connect a radical centre and a pendant double bond present in the same polymer chain. They achieved good agreement between the model predictions and experimental data for solution- and suspension-controlled copolymerizations.

2.2.3. Multivariate Distributions. Often, an appropriate characterization of a polymer demands simultaneous information on more than one property distribution, for instance, the joint MWD-CCD in copolymer systems or the MWD-branching distributions in branched polymers. The computation of these joint distributions involves the solution of multivariate PBEs, in which more than one internal coordinate is used to describe the polymer molecules. For instance, the prediction of the MWD-CCD requires the formulation of bivariate PBEs whose internal coordinates are the number of units of each comonomer in the chain. The solution of multivariate PBEs is a highly complex problem, for which seldom solution approaches have been developed. Reported methods include the numerical fractionation technique, the fixed pivot technique, or the moment distribution method [94–96]. Regarding applications to NMP processes, one of the few reported works is the one by Fortunatti et al. [67]. These authors developed an extension of the pgf method to bivariate distributions [97, 98]. In this method, the bivariate PBEs are transformed using 2D pgfs leading to balance equations for the 2D pgf of the bivariate distribution. After solving this system of equations, the 2D pgf transforms are inverted to obtain the bivariate distribution. In their work [67], Fortunatti et al. calculated the MWD-CCD in the nitroxide-mediated copolymerization of styrene and α-methyl styrene in tubular and semibatch reactors. They analyzed the influence of different feeding policies and temperature profiles on the polymer properties.

2.3. NMP in Heterogeneous Media. CRP techniques have been extensively studied in homogeneous systems. In addition, over the last few years, there has been great interest in adapting these techniques to aqueous heterogeneous systems because they constitute a good alternative for large-scale production. The reasons are many: aqueous dispersed systems are environmentally friendly, show very good heat transfer, better process control, ease of mixing, flexibility, and ease of handling of the final product [99, 100].

CRP in heterogeneous systems presents some complications that have hampered its straightforward dissemination. Among these difficulties, one could mention partitioning of species between aqueous and organic phases, the exit of radicals from polymer particles, poor colloidal stability, and interactions of species with other components of the recipe. The first attempts to adapt CRP to aqueous dispersed systems were carried out in emulsion. They were mostly unsuccessful because the emulsions did not present enough colloidal stability and had mass-transfer limitations. Further studies focused on miniemulsions, where unlike emulsion systems, there is no need of diffusion of reactants through the aqueous phase from monomer droplets to micelles. Miniemulsions are also able to host the polymerization of highly water-insoluble monomers and to form particles containing additives, such as dyes or pigments [100, 101].

Some of the earliest studies on NMP in aqueous dispersed systems were authored by Ma et al. [102–104]. They developed two mathematical models that described the nitroxide-mediated miniemulsion polymerization of styrene initiated by alkoxyamine initiators [103] and by the water-soluble initiator potassium persulfate [102, 104]. The models were based on moment equations and included mechanisms describing reactions in the aqueous and organic phases, particle nucleation, the entry and exit of oligomeric radicals, and the partitioning of nitroxide and styrene between the aqueous and organic phases. Compartmentalization effects were neglected, so a droplet-by-droplet approach was not necessary. Instead, they considered a pseudobulk system and formulated the usual PBEs for the organic and aqueous phases. Model predictions included monomer conversion and average molecular weights. The model outputs matched very well the experimental data. In another work by this group [105], Ma et al. studied specifically the interfacial mass transfer of TEMPO in miniemulsion polymerization. For this purpose, they developed a model consisting of two differential equations that described the concentration of TEMPO in the aqueous and organic phases. The model was used to examine how the diffusivity of TEMPO in the aqueous and organic droplet phases, the average droplet diameter, and the nitroxide partition coefficient influenced the time required for the nitroxide to reach phase equilibrium under nonsteady-state conditions.

In order to apply the pseudobulk approximation, it is necessary that the number of chains per particle is relatively large. This approximation works well for many systems [106]. However, if the number of radicals per particle is low, the compartmentalization effect needs to be taken into account. The well-known Smith-Ewart equations [107] have been traditionally used for this situation. The basic Smith-Ewart equation describes the dynamic evolution of the fraction of particles with $i \geq 0$ radicals, N_i, as

$$\frac{dN_i}{dt} = \rho(N_{i-1} - N_i) + k_d[(i+1)N_{i+1} - iN_i] + c[(i+2)(i+1)N_{i+2} - i(i-1)N_i], \quad (16)$$

where ρ, k_d, and c are the rate frequencies of radical entry, desorption, and overall termination, respectively. This is a chemical master equation that, unlike rate equations, is valid for small values of the property i (i.e., $i = 0,1,2,\ldots$). The basic Smith-Ewart equation can be modified to add more information about the radicals within a particle, such as the chain lengths of one (singly distinguished particle) or two (doubly distinguished particle) of them [108, 109]. This allows the formulation of PBEs describing the length of the polymer chains. When formulating the mathematical model, Smith-Ewart equations are used to describe compartmentalized species, while conventional PBEs are used for the non-compartmentalized species, whose concentration within the particles is high enough. The resulting system of equations can be solved or postprocessed using standard techniques to obtain the full MWD or its averages, such as the method of moments and direct integration.

Different models of NMP in miniemulsion employ modified Smith-Ewart equations. Butté et al. [110] were the first to study compartmentalization effects in NMP in miniemulsion. They used Smith-Ewart equations that were modified to account for the number of living radicals and nitroxide molecules within the particle. Charleux [111] also reported a model of NMP of styrene in miniemulsion using SG1 based on the Smith-Ewart equations, but she considered compartmentalization of living radicals only. Zetterlund and coworkers carried out a number of works on NMP in dispersed media. Zetterlund and Okubo [112, 113] used a model based on Smith-Ewart equations that accounted for the number of living radicals and nitroxides in the particles, to investigate compartmentalization effects in NMP of styrene mediated by TEMPO in miniemulsion at 125°C. They studied the effects of particle size and initial nitroxide concentration on the compartmentalization. The main process variables that were predicted were conversion and reaction rates. Their compartmentalization model, however, was not able to compute the MWD or its averages. In a later work [114], they compared the performance of two mediating agents, TEMPO and TIPNO, on the compartmentalization and livingness of the system. They found that the nature of the nitroxide influenced the minimum particle size beyond which compartmentalization could not be neglected. Subsequently, Zetterlund [115] reviewed and also presented new results on compartmentalization and nitroxide partitioning effects. In this work, they implemented a model in Predici to predict the partitioning of the nitroxide between phases. Using the same model, Zetterlund also studied the NMP of butyl acrylate in miniemulsion mediated by TEMPO [116]. He mentioned that TEMPO-mediated acrylate polymerization is known to be problematic and speculated that this is mainly related to the excessive accumulation of free TEMPO with increasing conversion. The author found out that the maximum particle size for which compartmentalization effects were important was larger for this system than for styrene polymerization. He also postulated that there is a narrow window of particle size of about 110–170 nm for which the miniemulsion polymerization of butyl acrylate is faster than in bulk.

Bentein et al. [117] developed a model of the miniemulsion polymerization of styrene mediated by SG1 at 396 K valid up to high conversion which took into account a targeted chain length and particle diameter. They used three-dimensional modified Smith-Ewart equations that counted compartmentalized macroradicals, initiator radicals, and nitroxide radicals. They used the singly distinguished particle approach to developing PBE that allowed using the method of moments to predict average molecular weights and dispersity. Later, Van Steenberge et al. [118] increased in one the dimensions of the Smith-Ewart equations of the model by Beintein et al. [117] to simultaneously calculate the time evolution of the conversion, number average chain length, dispersity, end-group functionality (EGF), and short-chain branching (SCB) content for the miniemulsion NMP of n-butyl acrylate initiated by poly(nBuA)-(N-tert-butyl-N-(1-diethylphosphono-2,2-dimethylpropyl). They concluded that backbiting could not be neglected for an accurate description of the system, despite the low short-chain branching and that the small loss of end-group functionality at conversions was mainly caused by chain transfer to monomer.

García-Leal et al. [119] studied the NMP of styrene using TEMPO for the production of high molecular weight polystyrene in a mass-suspension process. The main characteristic of the mass-suspension polymerization is that a bulk prepolymerization is first carried out, followed by the suspension of the prepolymer at intermediate conversion in a continuous phase until total conversion. This reduces the inter- and intraparticle mass transfer that occurs at low to moderate conversion in a classic suspension polymerization and permits a better control of the particle size distribution [119]. The authors assumed that the system kinetics behaved as in a conventional polymerization and used the classical PBEs of NMP processed by the method of moments to predict average molecular properties and conversion.

2.4. Hybrid Methods. A KMC simulation provides extremely detailed information of the polymer microstructure because it tracks the evolution in time of a full set of molecules of a sample of the reaction system. However, it usually requires long computational times and large storage capacities. Therefore, a number of authors have been working on the development of hybrid stochastic-deterministic solvers [17, 120, 121]. These solvers try to substitute parts of the stochastic simulation by deterministic techniques in order to accelerate the computation time. Regarding applications to NMP systems, Özçam and Teymour [17] developed a hybrid deterministic-probabilistic method that builds chains one-by-one or chain-by-chain and was named "Chain-by-Chain Monte Carlo" method. A deterministic component of the algorithm, based on the method of moments, is used to obtain information about the global concentration of polymer species and of small molecules, such as monomer or initiator. This information is passed to the MC part of the algorithm. They tested this hybrid model on the synthesis of styrene/methyl methacrylate linear gradient copolymers via NMP and methyl methacrylate/methyl acrylate linear hyperbolic gradient copolymers via ATRP. Results were compared with a conventional KMC model and with a deterministic method-of-moments model and confirmed that full information regarding the microstructure of chains can be obtained using the hybrid method with reduced simulation times and smaller sample sizes.

2.5. Model-Based Design of NMP Processes. Model-based design plays a fundamental role in polymer reaction engineering. Polymers are "process products": the microstructure of the material, and therefore, its application properties are strongly dictated by the conditions of the manufacturing process. Model-based design allows for a faster exploration of the search space, reduced uncertainty, and better, faster, and safer design and operation decisions through a deeper understanding of processes. The controlled growth of polymer chains in CRP processes makes them particularly suitable for synthesizing polymers with tailor-made molecular architecture. Model-based design is a very helpful tool in obtaining the specific operating policies required to achieve the desired molecular structure [14]. Deterministic modelling methods are particularly suitable for model-based design applications due to their fast running times, in comparison with stochastic approaches, and the availability of derivatives for gradient-based optimization solvers. Nevertheless, the model-based design of NMP processes using MC models has also been performed [11, 28, 29].

Different authors performed model-based design studies of NMP processes using deterministic models. Faliks et al. [122] applied an optimal control methodology to the optimization of an NMP of styrene in a plug flow reactor. The main purpose of the work was to show the benefits of optimal control in NMP reaction engineering, for which a typical case study was selected. The optimization problem aimed at finding the best nitroxide radical flux along the length of the reactor, with the objective of matching a target chain length and keeping conversion high. The mathematical model of the process was based on moment equations and predicted conversion and average molecular properties. The authors found that it was possible to reduce the reaction time significantly for a given conversion. However, this came at the expense of a modest increase in dispersity.

Lemoine-Nava et al. [123] used a moment-based model to find optimal operating policies for a simulated industrial reactor of NMP of styrene. They posed a dynamic optimization problem to find time-dependent optimal operating policies that maximized conversion and minimized dispersity.

Asteasuain et al. [124] used a model of the NMP of styrene in tubular reactors based on the pgf technique [62], for the optimization of the process aiming at producing tailor-made MWDs. Optimization variables included the feed rates of styrene, initiator (BPO), and TEMPO at the reactor inlet and lateral feeds, the location of the lateral feeds, and the operating temperature. They showed that it was possible to obtain bimodal and trimodal MWDs, with predetermined peak locations, through proper manipulation of the operating conditions. The authors highlighted that the pgf technique used to model the full MWD presented the advantage that it allowed tailoring the size of the mathematical model as a function of the particular details of the MWD that would be specified in the optimization problem. This permitted to reduce significantly the running time of the optimization tasks.

Zitlalpopoca-Soriano et al. [125] optimized grade transitions in the NMP of styrene in tubular reactors. They investigated the effect of the reactor temperature, monomer feed rate, and cooling water flow rate as manipulated variables for the grade transition. The reactor model involved moment equations that could predict average molecular properties and conversion. The DAE optimization problem was solved using a simultaneous approach wherein the differential and the algebraic variables were fully discretized leading to a large-scale nonlinear programming (NLP) problem. The resulting optimization problem was solved using an interior point algorithm capable of handling large-scale NLPs.

When it comes to process design, optimization objectives are often conflictive. For instance, polymers synthesized by CRP (e.g., NMP) are usually required to have the narrowest possible MWD and highest end-group functionality (EGF). This goal is usually contradictory to the economic objective of minimizing the reaction time for a given conversion. These conflicting objectives are best dealt with by means of multi-objective optimization (MOO) techniques. Recently, Fierens et al. [126] showed the benefits of applying MOO techniques to NMP of styrene initiated by MAMA-SG1 in a semibatch reactor. For this purpose, they coupled a moment-based model of the process with the nondominated sorting genetic algorithm-II (NSGA-II) in FORTRAN code to obtain the Pareto set of optimal solutions. The Pareto set is a collection of optimal solutions in which none of the objective functions can be improved in value without degrading some of the other ones. The design objectives of this study were the minimization of reaction time and of the dispersity and the maximization of the EGF. Process optimization variables included the temperature path, initial reactor charge, and semibatch feeding flows.

Bifurcation analysis is a widely known technique for studying the operability of chemical processes. It helps to assess steady-state multiplicity in chemical reactors and identify stable and unstable operating regions. A few works reported the use of this technique in NMP processes. Lemoine-Nava et al. [127] studied the nonlinear behavior of the NMP of styrene in a CSTR. They selected the cooling water flow rate, feed-stream temperature, cooling water feed temperature, monomer feed-stream concentration, and residence time as the bifurcation parameters. They found input multiplicities, disjoint bifurcations, and isola and hysteresis behaviors, which made the reactor prone to operability and control problems. Later, Zitlalpopoca-Soriano et al. [128] extended this analysis to the NMP of styrene in a tubular reactor. They studied the effect of the cooling flow rate, the feed-stream temperature, the cooling water temperature, the feed-stream monomer concentration, the flow rate of monomer feed, diameter, and global heat transfer. They used moment-based models [57, 125] in both works.

Another very useful application of mathematical models to the process development is the model-based design of experiments. This technique uses process models to design experiments in an optimal way in order to maximize the information that can be extracted from them and minimize the experimental effort required to estimate model parameters. Nabifar et al. [129] demonstrated the benefits of the Bayesian design methodology in complex polymerization processes using an NMP system as a case study. Scott et al. [130] carried out a model-based D-optimal design of experiments in the NMP of styrene. They used a full, moment-based model developed previously by the group [57], as well as a refined model which included new model parameters and revised kinetic steps, and a reduced model for faster simulation runs.

3. Conclusions

Mathematical modelling is a very helpful tool in polymer reaction engineering. It helps to gain understanding in the kinetic mechanisms of polymer processes, complements expensive and time-consuming laboratory experiments, and aids in obtaining the specific operating policies required to achieve the desired molecular structure. This applies especially to CRP, such as NMP, because these synthesis routes are particularly meant to manufacture polymers with controlled, precise molecular structures. This manuscript presents a review that deepens on the main deterministic modelling techniques employed for the simulation of NMP processes. It is described that most of the reported models focus on the prediction of average properties, such as average molecular weights and conversion. The method of moments is an overwhelming most selected option for this task, although commercial packages like Predici have also been employed. Other works applied more sophisticated techniques for modelling distributions of polymer properties, including recent developments for the calculation of multivariate distributions. A description of mathematical models for the simulation of NMP in heterogeneous systems is then presented, which shows that Smith-Ewart equations have been most used for analyzing compartmentalization effects in living radicals and dormant species. Finally, the manuscript gives an outlook of model-based design studies of NMP processes.

Conflicts of Interest

The author declares that there is no conflict of interest regarding the publication of this paper.

Acknowledgments

The author wishes to thank the National Research Council of Argentina (CONICET) and the Universidad Nacional del Sur (UNS) for the financial support.

References

[1] A. D. Jenkins, R. G. Jones, and G. Moad, "Terminology for reversible-deactivation radical polymerization previously called "controlled" radical or "living" radical polymerization (IUPAC Recommendations 2010)," *Pure and Applied Chemistry*, vol. 82, no. 2, pp. 483–491, 2009.

[2] K. Matyjaszewski and K. A. Davis, *Statistical, Gradient, Block and Graft Copolymers by Controlled/Living Radical Polymerizations*, Springer, Berlin, Heidelberg, 2002.

[3] J.-F. Lutz, B. Kirci, and K. Matyjaszewski, "Synthesis of well-defined alternating copolymers by controlled/living radical

polymerization in the presence of Lewis acids," *Macromolecules*, vol. 36, no. 9, pp. 3136–3145, 2003.

[4] K. Matyjaszewski and J. Spanswick, "Controlled/living radical polymerization," *Materials Today*, vol. 8, no. 3, pp. 26–33, 2005.

[5] J. Nicolas, Y. Guillaneuf, C. Lefay, D. Bertin, D. Gigmes, and B. Charleux, "Nitroxide-mediated polymerization," *Progress in Polymer Science*, vol. 38, no. 1, pp. 63–235, 2013.

[6] G. Moad, "Mechanism and kinetics of dithiobenzoate-mediated RAFT polymerization - status of the dilemma," *Macromolecular Chemistry and Physics*, vol. 215, no. 1, pp. 9–26, 2014.

[7] W. A. Braunecker and K. Matyjaszewski, "Controlled/living radical polymerization: features, developments, and perspectives," *Progress in Polymer Science*, vol. 32, no. 1, pp. 93–146, 2007.

[8] M. Zhang and W. H. Ray, "Modeling of "living" free-radical polymerization with RAFT chemistry," *Industrial & Engineering Chemistry Research*, vol. 40, no. 20, pp. 4336–4352, 2001.

[9] J. Liu, L. Tao, J. Xu, Z. Jia, C. Boyer, and T. P. Davis, "RAFT controlled synthesis of six-armed biodegradable star polymeric architectures via a 'core-first' methodology," *Polymer*, vol. 50, no. 19, pp. 4455–4463, 2009.

[10] L. Wang and L. J. Broadbelt, "Explicit sequence of styrene/methyl methacrylate gradient copolymers synthesized by forced gradient copolymerization with nitroxide-mediated controlled radical polymerization," *Macromolecules*, vol. 42, no. 20, pp. 7961–7968, 2009.

[11] L. Wang and L. J. Broadbelt, "Model-based design for preparing styrene/methyl methacrylate structural gradient copolymers," *Macromolecular Theory and Simulations*, vol. 20, no. 3, pp. 191–204, 2011.

[12] Y. N. Zhou and Z. H. Luo, "State-of-the-art and progress in method of moments for the model-based reversible-deactivation radical polymerization," *Macromolecular Reaction Engineering*, vol. 10, no. 6, pp. 516–534, 2016.

[13] T. Bott, W. Gerlinger, and J. Barth, "Tailor made products by smart processes," *Macromolecular Reaction Engineering*, vol. 9, no. 5, pp. 396–400, 2015.

[14] X. Li, S. Liang, W.-J. Wang, B.-G. Li, Y. Luo, and S. Zhu, "Model-based production of polymer chains having precisely designed end-to-end gradient copolymer composition and chain topology distributions in controlled radical polymerization, a review," *Macromolecular Reaction Engineering*, vol. 9, no. 5, pp. 409–417, 2015.

[15] D. T. Gillespie, "A general method for numerically simulating the stochastic time evolution of coupled chemical reactions," *Journal of Computational Physics*, vol. 22, no. 4, pp. 403–434, 1976.

[16] D. T. Gillespie, "Exact stochastic simulation of coupled chemical reactions," *The Journal of Physical Chemistry*, vol. 81, no. 25, pp. 2340–2361, 1977.

[17] D. Demirel Özçam and F. Teymour, "Chain-by-chain Monte Carlo simulation: a novel hybrid method for modeling polymerization. Part I. Linear controlled radical polymerization systems," *Macromolecular Reaction Engineering*, vol. 11, no. 1, 2017.

[18] J. Weng, X. Chen, Z. Yao, and L. T. Biegler, "Parallel Monte Carlo simulation of molecular weight distribution and chemical composition distribution for copolymerization on a graphics processing unit platform," *Macromolecular Theory and Simulations*, vol. 24, no. 5, pp. 521–536, 2015.

[19] E. Pintos, C. Sarmoria, A. Brandolin, and M. Asteasuain, "Modeling of RAFT polymerization processes using an efficient Monte Carlo algorithm in Julia," *Industrial & Engineering Chemistry Research*, vol. 55, no. 31, pp. 8534–8547, 2016.

[20] M. Drache and G. Drache, "Simulating controlled radical polymerizations with mcPolymer—a Monte Carlo approach," *Polymer*, vol. 4, no. 3, pp. 1416–1442, 2012.

[21] A. L. T. Brandão, J. B. P. Soares, J. C. Pinto, and A. L. Alberton, "When polymer reaction engineers play dice: applications of Monte Carlo models in PRE," *Macromolecular Reaction Engineering*, vol. 9, no. 3, pp. 141–185, 2015.

[22] D. Meimaroglou and C. Kiparissides, "Review of Monte Carlo methods for the prediction of distributed molecular and morphological polymer properties," *Industrial & Engineering Chemistry Research*, vol. 53, no. 22, pp. 8963–8979, 2014.

[23] A. S. Cho and L. J. Broadbelt, "Stochastic modelling of gradient copolymer chemical composition distribution and sequence length distribution," *Molecular Simulation*, vol. 36, no. 15, pp. 1219–1236, 2010.

[24] L. Wang and L. J. Broadbelt, "Kinetics of segment formation in nitroxide-mediated controlled radical polymerization: comparison with classic theory," *Macromolecules*, vol. 43, no. 5, pp. 2228–2235, 2010.

[25] M. Drache, "Modeling the product composition during controlled radical polymerizations with mono- and bifunctional alkoxyamines," *Macromolecular Symposia*, vol. 275-276, no. 1, pp. 52–58, 2009.

[26] Y. Reyes and J. M. Asua, "Revisiting chain transfer to polymer and branching in controlled radical polymerization of butyl acrylate," *Macromolecular Rapid Communications*, vol. 32, no. 1, pp. 63–67, 2011.

[27] T. Lemos, P. A. Melo, and J. C. Pinto, "Stochastic modeling of polymer microstructure from residence time distribution," *Macromolecular Reaction Engineering*, vol. 9, no. 3, pp. 259–270, 2015.

[28] S. K. Fierens, S. Telitel, P. H. M. van Steenberge et al., "Model-based design to push the boundaries of sequence control," *Macromolecules*, vol. 49, no. 24, pp. 9336–9344, 2016.

[29] S. K. Fierens, D. R. D'hooge, P. H. M. Van Steenberge, M.-F. Reyniers, and G. B. Marin, "MAMA-SG1 initiated nitroxide mediated polymerization of styrene: from Arrhenius parameters to model-based design," *Chemical Engineering Journal*, vol. 278, pp. 407–420, 2015.

[30] M. Al-Harthi, J. B. P. Soares, and L. C. Simon, "Dynamic Monte Carlo simulation of atom-transfer radical polymerization," *Macromolecular Materials and Engineering*, vol. 291, no. 8, pp. 993–1003, 2006.

[31] M. A. Al-Harthi, J. K. Masihullah, S. H. Abbasi, and J. B. P. Soares, "Dynamic Monte Carlo simulation of ATRP in a batch reactor," *Macromolecular Theory and Simulations*, vol. 18, no. 6, pp. 307–316, 2009.

[32] M. Drache, G. Schmidt-Naake, M. Buback, and P. Vana, "Modeling RAFT polymerization kinetics via Monte Carlo methods: cumyl dithiobenzoate mediated methyl acrylate polymerization," *Polymer*, vol. 46, no. 19, pp. 8483–8493, 2005.

[33] S. K. Fierens, P. H. M. Van Steenberge, M.-F. Reyniers, G. B. Marin, and D. R. D'Hooge, "How penultimate monomer unit

effects and initiator influence ICAR ATRP of *n*-butyl acrylate and methyl methacrylate," *AICHE Journal*, vol. 63, no. 11, pp. 4971–4986, 2017.

[34] P. Ganjeh-Anzabi, V. Hadadi-Asl, M. Salami-Kaljahi, and H. Roghani-Mamaqani, "A new approach for Monte Carlo simulation of RAFT polymerization," *Iranian Journal of Chemistry and Chemical Engineering*, vol. 31, no. 3, pp. 75–84, 2012.

[35] M. Najafi, V. Haddadi-Asl, M. Salami-Kalajahi, and H. R. Mamaghani, "Application of the Monte Carlo simulation method to the investigation of the effect of chain-length-dependent bimolecular termination on ATRP," *e-Polymers*, vol. 9, no. 1, 2009.

[36] M. Najafi, H. Roghani-Mamaqani, V. Haddadi-Asl, and M. Salami-Kalajahi, "A simulation of kinetics and chain length distribution of styrene FRP and ATRP: chain-length-dependent termination," *Advances in Polymer Technology*, vol. 30, no. 4, pp. 257–268, 2011.

[37] M. Najafi, H. Roghani-Mamaqani, M. Salami-Kalajahi, and V. Haddadi-Asl, "A comprehensive Monte Carlo simulation of styrene atom transfer radical polymerization," *Chinese Journal of Polymer Science*, vol. 28, no. 4, pp. 483–497, 2010.

[38] M. Najafi, H. Roghani-Mamaqani, M. Salami-Kalajahi, and V. Haddadi-Asl, "An exhaustive study of chain-length-dependent and diffusion-controlled free radical and atom-transfer radical polymerization of styrene," *Journal of Polymer Research*, vol. 18, no. 6, pp. 1539–1555, 2011.

[39] K. A. Payne, D. R. D'hooge, P. H. M. van Steenberge et al., "ARGET ATRP of butyl methacrylate: utilizing kinetic modeling to understand experimental trends," *Macromolecules*, vol. 46, no. 10, pp. 3828–3840, 2013.

[40] C. Toloza Porras, D. R. D'Hooge, P. H. M. Van Steenberge, M.-F. Reyniers, and G. B. Marin, "A theoretical exploration of the potential of ICAR ATRP for one- and two-pot synthesis of well-defined diblock copolymers," *Macromolecular Reaction Engineering*, vol. 7, no. 7, pp. 311–326, 2013.

[41] A. Mahjub, H. Mohammadi, M. Salami-Kalajahi, and M. T. Angaji, "Simulation of reversible chain transfer catalyzed polymerization (RTCP): effect of different iodide based catalysts," *Journal of Polymer Research*, vol. 19, no. 3, p. 9740, 2012.

[42] C. Kiparissides, "Polymerization reactor modeling: a review of recent developments and future directions," *Chemical Engineering Science*, vol. 51, no. 10, pp. 1637–1659, 1996.

[43] M. A. Al-Harthi, "Highlight on the mathematical modeling of controlled free radical polymerization," *International Journal of Polymer Science*, vol. 2015, Article ID 469353, 12 pages, 2015.

[44] E. Mastan, X. Li, and S. Zhu, "Modeling and theoretical development in controlled radical polymerization," *Progress in Polymer Science*, vol. 45, pp. 71–101, 2015.

[45] R. B. Grubbs, "Nitroxide-mediated radical polymerization: limitations and versatility," *Polymer Reviews*, vol. 51, no. 2, pp. 104–137, 2011.

[46] C. H. Bamford and H. Tompa, "On the calculation of molecular weight distributions from kinetic schemes," *Journal of Polymer Science*, vol. 10, no. 3, pp. 345–350, 1953.

[47] E. Mastan and S. Zhu, "Method of moments: a versatile tool for deterministic modeling of polymerization kinetics," *European Polymer Journal*, vol. 68, pp. 139–160, 2015.

[48] M. Salami-Kalajahi, V. Haddadi-Asl, P. Ganjeh Anzabi, and M. Najafi, "Dithioester-mediated RAFT polymerization: a kinetic study by mathematical modelling," *Iranian Polymer Journal*, vol. 20, pp. 459–478, 2011.

[49] M. Safajou-Jahankhanemlou and M. Salami-Kalajahi, "Modeling of reversible chain transfer catalyzed polymerization by moment equations method," *Iranian Journal of Chemistry and Chemical Engineering*, vol. 32, no. 4, pp. 59–67, 2013.

[50] S. Zhu, "Modeling stable free-radical polymerization," *Journal of Polymer Science Part B: Polymer Physics*, vol. 37, no. 18, pp. 2692–2704, 1999.

[51] D. R. D'Hooge, M.-F. Reyniers, and G. B. Marin, "Methodology for kinetic modeling of atom transfer radical polymerization," *Macromolecular Reaction Engineering*, vol. 3, no. 4, pp. 185–209, 2009.

[52] Y.-N. Zhou and Z.-H. Luo, "Copper(0)-mediated reversible-deactivation radical polymerization: kinetics insight and experimental study," *Macromolecules*, vol. 47, no. 18, pp. 6218–6229, 2014.

[53] Y.-N. Zhou and Z.-H. Luo, "Insight into the ATRP rate controlling ability of initiator structure: micromolecular, macromolecular, and immobilized initiators," *Journal of Polymer Science Part A: Polymer Chemistry*, vol. 52, no. 15, pp. 2228–2238, 2014.

[54] A. R. Wang and S. Zhu, "Modeling the reversible addition–fragmentation transfer polymerization process," *Journal of Polymer Science Part A: Polymer Chemistry*, vol. 41, no. 11, pp. 1553–1566, 2003.

[55] R. P. N. Veregin, P. G. Odell, L. M. Michalak, and M. K. Georges, "Molecular weight distributions in nitroxide-mediated living free radical polymerization: kinetics of the slow equilibria between growing and dormant chains," *Macromolecules*, vol. 29, no. 10, pp. 3346–3352, 1996.

[56] A. Butté, G. Storti, and M. Morbidelli, "Kinetics of "living" free radical polymerization," *Chemical Engineering Science*, vol. 54, no. 15-16, pp. 3225–3231, 1999.

[57] J. Bonilla, E. Saldívar, A. Flores-Tlacuahuac, E. Vivaldo-Lima, R. Pfaendner, and F. Tiscareño-Lechuga, "Detailed modeling, simulation, and parameter estimation of nitroxide mediated living free radical polymerization of styrene 1*," *Polymer Reaction Engineering*, vol. 10, no. 4, pp. 227–263, 2002.

[58] T. M. Kruse, R. Souleimonova, A. Cho, M. K. Gray, J. M. Torkelson, and L. J. Broadbelt, "Limitations in the synthesis of high molecular weight polymers via nitroxide-mediated controlled radical polymerization: modeling studies," *Macromolecules*, vol. 36, no. 20, pp. 7812–7823, 2003.

[59] J. Belincanta-Ximenes, P. V. R. Mesa, L. M. F. Lona, E. Vivaldo-Lima, N. T. McManus, and A. Penlidis, "Simulation of styrene polymerization by monomolecular and bimolecular nitroxide-mediated radical processes over a range of reaction conditions," *Macromolecular Theory and Simulations*, vol. 16, no. 2, pp. 194–208, 2007.

[60] M. Roa-Luna, M. P. Díaz-barber, E. Vivaldo-Lima, L. M. F. Lona, N. T. McManus, and A. Penlidis, "Assessing the importance of diffusion-controlled effects on polymerization rate and molecular weight development in nitroxide-mediated radical polymerization of styrene," *Journal of Macromolecular Science, Part A*, vol. 44, no. 2, pp. 193–203, 2007.

[61] A. R. Murari, S. Contant, L. M. F. Lona, E. Vivaldo-Lima, N. T. McManus, and A. Penlidis, "Modeling insights on the TEMPO mediated radical polymerization of styrene," *Journal of Macromolecular Science, Part A*, vol. 48, no. 9, pp. 681–687, 2011.

[62] M. Asteasuain, M. Soares, M. K. Lenzi et al., ""Living" free radical polymerization in tubular reactors. I. Modeling of the complete molecular weight distribution using probability generating functions," *Macromolecular Reaction Engineering*, vol. 1, no. 6, pp. 622–634, 2007.

[63] N. A. Dotson, R. Galván, R. L. Laurence, and M. Tirrell, *Polymerization Process Modeling*, VCH Publishers, Inc., New York, NY, USA, 1996.

[64] A. Keramopoulos and C. Kiparissides, "Development of a comprehensive model for diffusion-controlled free-radical copolymerization reactions," *Macromolecules*, vol. 35, no. 10, pp. 4155–4166, 2002.

[65] M. Zhang and W. H. Ray, "Modeling of "living" free-radical polymerization processes. I. Batch, semibatch, and continuous tank reactors," *Journal of Applied Polymer Science*, vol. 86, no. 7, pp. 1630–1662, 2002.

[66] M. Zhang and W. H. Ray, "Modeling of "living" free-radical polymerization processes. II. Tubular reactors," *Journal of Applied Polymer Science*, vol. 86, no. 5, pp. 1047–1056, 2002.

[67] C. Fortunatti, C. Sarmoria, A. Brandolin, and M. Asteasuain, "Theoretical analysis of nitroxide-mediated copolymerization of styrene and α-methyl-styrene under different operating policies and reactor designs," *Macromolecular Reaction Engineering*, vol. 8, no. 4, pp. 260–281, 2014.

[68] D. J. Arriola, *Modeling of addition polymerization systems, [Ph.D. thesis]*, University of Wisconsin-Madison, 1989.

[69] Y. Ye and F. J. Schork, "Modeling of sequence length and distribution for the NM-CRP of styrene and 4-methylstyrene in batch and semi-batch reactors," *Macromolecular Reaction Engineering*, vol. 4, no. 3-4, pp. 197–209, 2010.

[70] H. Tobita and A. E. Hamielec, "Kinetics of free-radical copolymerization: the pseudo-kinetic rate constant method," *Polymer*, vol. 32, no. 14, pp. 2641–2647, 1991.

[71] J. C. Hernández-Ortiz, E. Vivaldo-Lima, L. M. F. Lona, N. T. McManus, and A. Penlidis, "Modeling of the nitroxide-mediated radical copolymerization of styrene and divinylbenzene," *Macromolecular Reaction Engineering*, vol. 3, no. 5-6, pp. 288–311, 2009.

[72] J. C. Hernández-Ortiz, E. Vivaldo-Lima, and A. Penlidis, "Modeling of network formation in nitroxide-mediated radical copolymerization of vinyl/divinyl monomers using a multifunctional polymer molecule approach," *Macromolecular Theory and Simulations*, vol. 21, no. 5, pp. 302–321, 2012.

[73] A. J. Scott, A. Nabifar, J. C. Hernández-Ortiz, N. T. McManus, E. Vivaldo-Lima, and A. Penlidis, "Crosslinking nitroxide-mediated radical copolymerization of styrene with divinylbenzene," *European Polymer Journal*, vol. 51, pp. 87–111, 2014.

[74] M. Wulkow, "Computing in technology. Predici," May 2017, http://www.cit-wulkow.de/products/predici-polymerization.

[75] D. Gigmes, D. Bertin, C. Lefay, and Y. Guillaneuf, "Kinetic modeling of nitroxide-mediated polymerization: conditions for living and controlled polymerization," *Macromolecular Theory and Simulations*, vol. 18, no. 7-8, pp. 402–419, 2009.

[76] J. Nicolas, L. Mueller, C. Dire, K. Matyjaszewski, and B. Charleux, "Comprehensive modeling study of nitroxide-mediated controlled/living radical copolymerization of methyl methacrylate with a small amount of styrene," *Macromolecules*, vol. 42, no. 13, pp. 4470–4478, 2009.

[77] D. Greszta and K. Matyjaszewski, "Mechanism of controlled/ "living" radical polymerization of styrene in the presence of nitroxyl radicals. Kinetics and simulations," *Macromolecules*, vol. 29, no. 24, pp. 7661–7670, 1996.

[78] D. Benoit, S. Grimaldi, S. Robin, J. P. Finet, P. Tordo, and Y. Gnanou, "Kinetics and mechanism of controlled free-radical polymerization of styrene and *n*-butyl acrylate in the presence of an acyclic β-phosphonylated nitroxide," *Journal of the American Chemical Society*, vol. 122, no. 25, pp. 5929–5939, 2000.

[79] Y. Guillaneuf, D. Gigmes, S. R. A. Marque, P. Tordo, and D. Bertin, "Nitroxide-mediated polymerization of methyl methacrylate using an SG1-based alkoxyamine: how the penultimate effect could lead to uncontrolled and unliving polymerization," *Macromolecular Chemistry and Physics*, vol. 207, no. 14, pp. 1278–1288, 2006.

[80] Y. Fu, M. F. Cunningham, and R. A. Hutchinson, "Modeling of nitroxide-mediated semibatch radical polymerization," *Macromolecular Reaction Engineering*, vol. 1, no. 2, pp. 243–252, 2007.

[81] Y. Guillaneuf, J.-P. Lamps, J.-M. Catala, D. Gigmes, and E. Drockenmuller, "Improving the control of styrene polymerization at 60 °C using a dialkylated α-hydrogenated nitroxide," *Journal of Polymer Science Part A: Polymer Chemistry*, vol. 50, no. 18, pp. 3750–3757, 2012.

[82] K. A. Payne, J. Debling, P. Nesvadba, M. F. Cunningham, and R. A. Hutchinson, "NMP of styrene in batch and CSTR at elevated temperatures: modeling experimental trends," *European Polymer Journal*, vol. 80, pp. 186–199, 2016.

[83] J. J. Hernández-Meza, G. Jaramillo-Soto, P. R. García-Morán, J. Palacios-Alquisira, and E. Vivaldo-Lima, "Modeling of polymerization kinetics and molecular weight development in the microwave-activated nitroxide-mediated radical polymerization of styrene," *Macromolecular Reaction Engineering*, vol. 3, no. 2-3, pp. 101–107, 2009.

[84] M. Roa-Luna, A. Nabifar, N. T. McManus, E. Vivaldo-Lima, L. M. F. Lona, and A. Penlidis, "Effect of the addition of inert or TEMPO-capped prepolymer on polymerization rate and molecular weight development in the nitroxide-mediated radical polymerization of styrene," *Journal of Applied Polymer Science*, vol. 109, no. 6, pp. 3665–3678, 2008.

[85] M. Asteasuain, A. Brandolin, and C. Sarmoria, "Recovery of molecular weight distributions from transformed domains. Part II. Application of numerical inversion methods," *Polymer*, vol. 43, no. 8, pp. 2529–2541, 2002.

[86] M. Asteasuain, C. Sarmoria, and A. Brandolin, "Recovery of molecular weight distributions from transformed domains. Part I: application of pgf to mass balances describing reactions involving free radicals," *Polymer*, vol. 43, no. 8, pp. 2513–2527, 2002.

[87] R. C. S. Dias and M. R. P. F. N. Costa, "Calculation of CLD using population balance equations of generating functions: linear and non-linear ideal controlled radical polymerization," *Macromolecular Theory and Simulations*, vol. 19, no. 6, pp. 323–341, 2010.

[88] L. Bentein, D. R. D'Hooge, M.-F. Reyniers, and G. B. Marin, "Kinetic modeling as a tool to understand and improve the nitroxide mediated polymerization of styrene," *Macromolecular Theory and Simulations*, vol. 20, no. 4, pp. 238–265, 2011.

[89] E. Saldívar-Guerra, R. Infante-Martínez, E. Vivaldo-Lima, and A. Flores-Tlacuahuac, "Returning to basics: direct integration of the full molecular-weight distribution equations

in addition polymerization," *Macromolecular Theory and Simulations*, vol. 19, no. 4, pp. 151–157, 2010.

[90] I. Zapata-González, E. Saldívar-Guerra, A. Flores-Tlacuahuac, E. Vivaldo-Lima, and J. Ortiz-Cisneros, "Efficient numerical integration of stiff differential equations in polymerisation reaction engineering: computational aspects and applications," *The Canadian Journal of Chemical Engineering*, vol. 90, no. 4, pp. 804–823, 2012.

[91] A. Zargar and F. J. Schork, "Design of copolymer molecular architecture via design of continuous reactor systems for controlled radical polymerization," *Industrial & Engineering Chemistry Research*, vol. 48, no. 9, pp. 4245–4253, 2009.

[92] L. G. Aguiar, M. A. D. Gonçalves, V. D. Pinto, R. C. S. Dias, M. R. P. F. N. Costa, and R. Giudici, "Development of cyclic propagation kinetics for modeling the nitroxide-mediated radical copolymerization of styrene–divinylbenzene," *Macromolecular Reaction Engineering*, vol. 8, no. 4, pp. 282–294, 2014.

[93] A. Zargar and F. J. Schork, "Copolymer sequence distributions in controlled radical polymerization," *Macromolecular Reaction Engineering*, vol. 3, no. 2-3, pp. 118–130, 2009.

[94] J. Chen, X. Zhao, L. Zhang, Z. Cheng, and X. Zhu, "Reversible addition-fragmentation chain transfer polymerization of vinyl acetate under high pressure," *Journal of Polymer Science Part A: Polymer Chemistry*, vol. 53, no. 12, pp. 1430–1436, 2015.

[95] A. Krallis, D. Meimaroglou, and C. Kiparissides, "Dynamic prediction of the bivariate molecular weight-copolymer composition distribution using sectional-grid and stochastic numerical methods," *Chemical Engineering Science*, vol. 63, no. 17, pp. 4342–4360, 2008.

[96] I. Kryven and P. D. Iedema, "A novel approach to population balance modeling of reactive polymer modification leading to branching," *Macromolecular Theory and Simulations*, vol. 22, no. 2, pp. 89–106, 2013.

[97] M. Asteasuain and A. Brandolin, "Mathematical modeling of bivariate polymer property distributions using 2D probability generating functions, 1 - numerical inversion methods," *Macromolecular Theory and Simulations*, vol. 19, no. 6, pp. 342–359, 2010.

[98] A. Brandolin and M. Asteasuain, "Mathematical modeling of bivariate distributions of polymer properties using 2D probability generating functions. Part II: transformation of population mass balances of polymer processes," *Macromolecular Theory and Simulations*, vol. 22, no. 5, pp. 273–308, 2013.

[99] M. F. Cunningham, "Controlled/living radical polymerization in aqueous dispersed systems," *Progress in Polymer Science*, vol. 33, no. 4, pp. 365–398, 2008.

[100] J. K. Oh, "Recent advances in controlled/living radical polymerization in emulsion and dispersion," *Journal of Polymer Science Part A: Polymer Chemistry*, vol. 46, no. 21, pp. 6983–7001, 2008.

[101] J. Qiu, B. Charleux, and K. Matyjaszewski, "Controlled/living radical polymerization in aqueous media: homogeneous and heterogeneous systems," *Progress in Polymer Science*, vol. 26, no. 10, pp. 2083–2134, 2001.

[102] J. W. Ma, M. F. Cunningham, K. B. McAuley, B. Keoshkerian, and M. Georges, "Nitroxide mediated living radical polymerization of styrene in miniemulsion—modelling persulfate-initiated systems," *Chemical Engineering Science*, vol. 58, no. 7, pp. 1177–1190, 2003.

[103] J. W. Ma, J. A. Smith, K. B. McAuley, M. F. Cunningham, B. Keoshkerian, and M. K. Georges, "Nitroxide-mediated radical polymerization of styrene in miniemulsion: model studies of alkoxyamine-initiated systems," *Chemical Engineering Science*, vol. 58, no. 7, pp. 1163–1176, 2003.

[104] J. W. Ma, M. F. Cunningham, K. B. McAuley, B. Keoshkerian, and M. K. Georges, "Model studies of nitroxide-mediated styrene miniemulsion polymerization - opportunities for process improvement," *Macromolecular Theory and Simulations*, vol. 12, no. 1, pp. 72–85, 2003.

[105] J. W. Ma, M. F. Cunningham, K. B. McAuley, B. Keoshkerian, and M. K. Georges, "Interfacial mass transfer in nitroxide-mediated miniemulsion polymerization," *Macromolecular Theory and Simulations*, vol. 11, no. 9, pp. 953–960, 2002.

[106] M. Wulkow and J. R. Richards, "Evaluation of the chain length distribution in free-radical emulsion polymerization—the compartmentalization problem," *Industrial & Engineering Chemistry Research*, vol. 53, no. 18, pp. 7275–7295, 2014.

[107] W. V. Smith and R. H. Ewart, "Kinetics of emulsion polymerization," *The Journal of Chemical Physics*, vol. 16, no. 6, pp. 592–599, 1948.

[108] G. Lichti, R. G. Gilbert, and D. H. Napper, "Molecular weight distribution in emulsion polymerizations," *Journal of Polymer Science: Polymer Chemistry Edition*, vol. 18, no. 4, pp. 1297–1323, 1980.

[109] A. Butté, G. Storti, and M. Morbidelli, "Evaluation of the chain length distribution in free-radical polymerization, 2. Emulsion polymerization," *Macromolecular Theory and Simulations*, vol. 11, no. 1, pp. 37–52, 2002.

[110] A. Butté, G. Storti, and M. Morbidelli, "Pseudo-living polymerization of styrene in miniemulsion," *Dechema-Monographien*, vol. 134, pp. 497–507, 1998.

[111] B. Charleux, "Theoretical aspects of controlled radical polymerization in a dispersed medium," *Macromolecules*, vol. 33, no. 15, pp. 5358–5365, 2000.

[112] P. B. Zetterlund and M. Okubo, "Compartmentalization in nitroxide-mediated radical polymerization in dispersed systems," *Macromolecules*, vol. 39, no. 26, pp. 8959–8967, 2006.

[113] P. B. Zetterlund and M. Okubo, "Compartmentalization in TEMPO-mediated radical polymerization in dispersed systems: effects of macroinitiator concentration," *Macromolecular Theory and Simulations*, vol. 16, no. 3, pp. 221–226, 2007.

[114] P. B. Zetterlund and M. Okubo, "Compartmentalization in NMP in dispersed systems: relative contributions of confined space effect and segregation effect depending on nitroxide type," *Macromolecular Theory and Simulations*, vol. 18, no. 4-5, pp. 277–286, 2009.

[115] P. B. Zetterlund, "Nitroxide-mediated radical polymerization in dispersed systems: compartmentalization and nitroxide partitioning," *Macromolecular Theory and Simulations*, vol. 19, pp. 11–23, 2010.

[116] P. B. Zetterlund, "Nitroxide-mediated radical polymerization of butyl acrylate using TEMPO: improvement of control exploiting nanoreactors?," *Macromolecular Reaction Engineering*, vol. 4, no. 11-12, pp. 663–671, 2010.

[117] L. Bentein, D. R. D'hooge, M. F. Reyniers, and G. B. Marin, "Kinetic modeling of miniemulsion nitroxide mediated polymerization of styrene: effect of particle diameter and nitroxide partitioning up to high conversion," *Polymer*, vol. 53, no. 3, pp. 681–693, 2012.

[118] P. H. M. Van Steenberge, D. R. D'hooge, M.-F. Reyniers, G. B. Marin, and M. F. Cunningham, "4-Dimensional modeling strategy for an improved understanding of miniemulsion NMP of acrylates initiated by SG1-macroinitiator," *Macromolecules*, vol. 47, no. 22, pp. 7732–7741, 2014.

[119] E. García-Leal, M. A. Ordaz-Quintero, E. Saldívar-Guerra, M. E. Albores-Velasco, and E. Vivaldo-Lima, "Nitroxide-mediated controlled radical styrene polymerization via a mass-suspension process," *Macromolecular Reaction Engineering*, vol. 7, no. 12, pp. 699–712, 2013.

[120] C. Schütte and M. Wulkow, "A hybrid Galerkin–Monte-Carlo approach to higher-dimensional population balances in polymerization kinetics," *Macromolecular Reaction Engineering*, vol. 4, no. 9-10, pp. 562–577, 2010.

[121] H. Gao, L. J. Broadbelt, I. A. Konstantinov, and S. G. Arturo, "Acceleration of kinetic Monte Carlo simulations of free radical copolymerization: a hybrid approach with scaling," *AICHE Journal*, vol. 63, no. 9, pp. 4013–4021, 2017.

[122] A. Faliks, R. A. Yetter, C. A. Floudas, Y. Wei, and H. Rabitz, "Optimization of living polymerization through distributed control of a nitroxide radical," *Polymer*, vol. 42, no. 5, pp. 2061–2065, 2001.

[123] R. Lemoine-Nava, A. Flores-Tlacuahuac, and E. Saldívar-Guerra, "Optimal operating policies for the nitroxide-mediated radical polymerization of styrene in a semibatch reactor," *Industrial & Engineering Chemistry Research*, vol. 45, no. 13, pp. 4637–4652, 2006.

[124] M. Asteasuain, M. Soares, M. K. Lenzi et al., ""Living" radical polymerization in tubular reactors, 2 – process optimization for tailor-made molecular weight distributions," *Macromolecular Reaction Engineering*, vol. 2, no. 5, pp. 414–421, 2008.

[125] A. G. Zitlalpopoca-Soriano, E. Vivaldo-Lima, and A. Flores-Tlacuahuac, "Grade transition dynamic optimization of the living nitroxide-mediated radical polymerization of styrene in a tubular reactor," *Macromolecular Reaction Engineering*, vol. 4, no. 8, pp. 516–533, 2010.

[126] S. Fierens, D. D'hooge, P. van Steenberge, M.-F. Reyniers, and G. Marin, "Exploring the full potential of reversible deactivation radical polymerization using Pareto-optimal fronts," *Polymer*, vol. 7, no. 4, pp. 655–679, 2015.

[127] R. Lemoine-Nava, A. Flores-Tlacuahuac, and E. Saldívar-Guerra, "Non-linear bifurcation analysis of the living nitroxide-mediated radical polymerization of styrene in a CSTR," *Chemical Engineering Science*, vol. 61, no. 2, pp. 370–387, 2006.

[128] A. G. Zitlalpopoca-Soriano, E. Vivaldo-Lima, and A. Flores-Tlacuahuac, "Bifurcation analysis of a tubular reactor for nitroxide-mediated radical polymerization of styrene," *Macromolecular Reaction Engineering*, vol. 4, no. 9-10, pp. 599–612, 2010.

[129] A. Nabifar, N. T. McManus, E. Vivaldo-Lima, and A. Penlidis, "A sequential iterative scheme for design of experiments in complex polymerizations," *Chemical Engineering & Technology*, vol. 33, no. 11, pp. 1814–1824, 2010.

[130] A. J. Scott, M. D. Hazlett, E. Vivaldo-Lima, and A. Penlidis, "D-Optimality in model-based experimental designs: applications in NMRP of styrene," *Macromolecular Reaction Engineering*, vol. 9, no. 3, pp. 205–227, 2015.

Processing and Properties of MDF Fibre-Reinforced Biopolyesters with Chain Extender Additives

Armin Thumm,[1] Damien Even,[2] Pierre-Yves Gini,[3] and Mathias Sorieul[1]

[1]*Scion, Private Bag 3020, Rotorua 3046, New Zealand*
[2]*Cabot Performance Materials, 78 Rue Prévochamps, Pepinster B-4860, Belgium*
[3]*Safran Nacelles, Route du Pont 8, 76700 Gonfreville l'Orcher, France*

Correspondence should be addressed to Mathias Sorieul; mathias.sorieul@scionresearch.com

Guest Editor: Simona Zuppolini

Biopolyesters are a way to improve natural fibre composite sustainability. This study explores, for the first time, the potential of using medium density fibreboard (MDF) fibres to reinforce four biobased and biodegradable polyester matrices to create a fully "green composite." Added at 30 wt %, MDF fibres did not improve the strength of the injection moulded NFCs and this deficiency was investigated by measuring fibre length, viscosity, and molecular weight of the matrices. Compared to other lignocellulosic fibres, the use of MDF fibres led to a molecular weight reduction of biopolyesters during processing. This effect was particularly striking for PLA. The addition of a chain extender enhanced the molecular weight of PLA and improved its processability. The tensile strength increase was correlated to a reduction of fibre pull-out, enabling the MDF fibre to fulfil its expected reinforcement role within the biopolyester composite.

1. Introduction

Reinforced plastic composites are a class of materials in which fillers of high modulus are added to a polymer matrix to increase its modulus. The market of composites using natural fillers as reinforcement is expanding. In the emerging field of plastic composites reinforced with biobased fillers, plant lignocellulosic materials are the most widely used. The main advantages of lignocellulosic fillers over traditional inorganic fillers are their low cost, lightweight, renewability, abundant availability, and biodegradability [1]. Composites containing lignocellulosic material can be divided into two categories depending on the morphology of the filler; wood plastic composites (WPC) contain wood flour, and natural fibre composites (NFC) are reinforced with bast/leaf or wood fibres. Wood fibres are one of the most promising lignocellulosic fibre types; they have the advantage of being nonseasonal and of consistent quality. Moreover, unlike wood flour, their high aspect ratio contributes to the strength and stiffness of the composite material when used as a filler [2, 3]. Fibre morphology and physicochemical properties have a strong effect on their reinforcing properties within the thermoplastic matrix [4]. Wood fibre characteristics mainly depend on the fibre origin and refining method. They are extracted from softwood or hardwood, and the main techniques used for the fibrillation/individualisation process are mechanical (e.g., thermomechanical pulps), chemical (e.g., kraft), or a combination of the two (e.g., chemi-thermomechanical pulp) [5].

MDF (medium density fibreboard) fibres are thermomechanical fibres where high temperatures around 170°C lead to cleavage of the fibres along the middle lamella, exposing a comparatively lignin-rich, nonpolar, fibre surface [6–8]. The typical MDF fibre from *Pinus radiata* is approximately 1.5 mm long and 34 μm wide, with an aspect ratio of 44/1 [9–11]. The large production plants devoted to MDF and the relatively low production cost of MDF fibre make it an attractive industrial choice as a filler for NFC.

The NFC market is projected to be US$ 5.8 billion globally by 2021 [12]. Therefore, it is not surprising to observe the burgeoning of wood fibre products derived from pulping processes, appearing for the reinforcement of thermoplastic materials. These products include NCell™ (GreenCore Composites Inc., Canada); reSound NF (PolyOne, USA); CreaMix (CreaFill Fibers Corp., USA); Kareline® (Kareline Natural Composites, Finland); and ForMi (UPM, Finland).

The typical ranges of tensile strength, tensile modulus, and elongation at the break of commercial NFCs range between 12 and 47 MPa, 0.8 and 5.6 GPa, and 1.4 and 5.4%, respectively (Table 1).

Due to their attractive mechanical properties and convenience for compounding, the petrochemical-based polyolefins, polyethylene (PE) and polypropylene (PP), are the primary resins currently used in NFCs [4]. However, the resulting products, a mix of synthetic and biobased materials, are only partially biodegradable and difficult to recycle.

Driven by the increasing environmental impact of traditional plastic and biobased and biodegradable plastics have attracted extensive interest [12, 14, 15]. The creation of a fully biodegradable NFC enables single use or medium term applications and easier end-of-life management. The use of biodegradable polymers in composites has been limited by their higher production costs, slow crystallization, poor stability, narrow processing windows, and incompatibility between hydrophobic polymer and hydrophilic wood fibre [4]. Nevertheless, biopolyesters are more polar in nature than PP or PE [4], which could make them more suitable for use with natural fibres. This feature leads to a better dispersion of the wood-based fillers in a biopolyester matrix compared to PP [13]. In contrast to the widespread use of specific compatibilisers, such as maleated PP (MA-PP) in polyolefins, their use with biopolyester (MA-PHA [16] and MA-PLA [17]) is still limited and noncommercial [18]. Unfortunately, no trends in the effectiveness of compatibilisers on PLA-based NFCs have been identified. However, as for PP [19], fibres with high cellulose content have good interfacial adhesion with PLA [20] without the need of a compatibiliser. The considerable variation in the mechanical properties seen in studies suggests that the grade of PLA, origin of wood, and processing methods are the main factors influencing the performance of the resultant composites [21]. The factors controlling fibre-resin interfacial properties are considered to be the strength of the interfacial adhesion [22] or the mechanical interlocking [23] between a wood and PLA. The results can be summarized as follows: depending on their origin and mode of preparation, wood fillers have mixed effects on tensile strength. An increase in wood filler content in PLA-based NFCs improves the stiffness and decreases the elongation at break [13].

While a large number of natural fibres, including wood fibre and nanocellulose, have been compounded with PLA [24], to our knowledge, no published study has investigated the use of MDF fibres to reinforce bioplastics.

The four main criteria taken into consideration for the selection of a NFC are physical characteristics, mechanical performance, environmental friendliness, and manufacturing cost [25]. The automotive sector is one of the main drivers for the industrial uptake of NFC. For this sector especially, the cost competitiveness and reliable sourcing are crucial. Therefore, MDF fibres have the potential to be an ideal reinforcing agent for NFCs. This study investigates the possibility of creating an economically competitive, fully green NFC using MDF fibres as the reinforcing agent of biopolyester matrices. Peltola et al. [13] hypothesise that the lignin present in thermomechanically pulped (TMP) spruce fibres improves the fibre/PLA interface. Therefore, MDF fibres with a higher lignin content than TMP fibres and surface coated with lignin should be ideal to reinforce a PLA matrix.

Table 1: Range of tensile properties of NFCs made with a biodegradable polymer compared to commercial PP and PE-based NFCs (with a wood filler content of 30–50 wt %), adapted from [4, 13].

Composite type	Tensile strength (MPa)	Tensile Young's modulus (GPa)	Elongation at break (%)
PLA-NFC	37–77	1.2–8.9	1.0–3.1
PHA-NFC	13–26	0.4-5.9	0.5-6.6
PP-NFC (commercial)	18–47	1.8–5.6	1.6–5.4
HDPE-NFC (commercial)	12–39	0.8–4.7	1.4–2.9

2. Materials and Methods

2.1. Materials

2.1.1. Biopolyester and Additives. Four commercially available biodegradable/compostable polyesters with different mechanical and thermal properties were selected: polylactic acid (PLA), polybutylene succinate (PBS), polybutylene adipate-co-terephtalate (PBAT), and polybutylene succinate-co-adipate (PBSA). The biopolyesters used in this study are described in Table 2. Some properties of the selected polymers can be seen in Table 3.

The plastic additive BioAdimide® 500XT, typically used to increase the service life of biobased polyesters and to stabilize melt viscosity during processing, was obtained from Rhein Chemie (Rheinau GmbH, Germany). Joncryl® ADR-4300, a polymeric chain extender designed to reverse the degradation of condensation polymers, was supplied by BASF Corporation (Sturtevant, WI, USA). The dosage for both chain extenders was at the upper limit of the manufacturer's recommendations.

2.1.2. Lignocellulosic Material. The MDF fibres were produced from *Pinus radiata* wood chips at Scion's fibre processing plant (Rotorua, New Zealand). Their average length was 1.44 mm, width 34 μm, and aspect ratio 42, and the fines content was 5.7%. Bleached kraft pulp was obtained from Carter Holt Harvey Pulp & Paper Kinleith Mill (Oji Fibre Solutions, New Zealand). Flax fibre slivers, intended for use in textile fabrics, had an unknown source. Wood flour was obtained from Carter Holt Harvey Pulp & Paper Kinleith Mill (New Zealand) and was sieved to be between 40 and 80 meshes.

TABLE 2: Biopolyesters used in this study.

Name	Acronym	Reference	Grade	Company	Country
Polylactic acid	PLA	5051D (now 3052D)	Injection	NatureWorks	USA
Polybutylene succinate	PBS	Enpol G4560 J	Injection	Ire Chemical	Korea
Polybutylene adipate-co-terephthalate	PBAT	Enpol 8060	Extrusion	Ire Chemical	Korea
Polybutylene succinate-co-adipate	PBSA	Bionolle 3020	Injection	Showa Highpolymer	Japan

TABLE 3: Properties as specified by the manufacturers. Glass transition temperature (T_g), melting point (T_m), Young's modulus (E), stress at maximum load (TS_{max}), and impact strength (IS). NI (not indicated by the manufacturer). Courtesy of the manufacturers.

Polymer	T_g (°C)	T_m (°C)	E (GPa)	TS_{max} (MPa)	IS (kJ/m^2)	MFI (g*10 min^{-1})
PLA 3052	55/60	145/160	3.6 (flex)	57	15.8	25
PBS	−35/−30	112/117	0.78/0.8 (flex)	40/45	38/42 .10^{-3}	25
PBAT	−30/−25	127/132	0.11/0.15 (flex)	44	No break	2
PBSA	−45	94	0.34	34	NI	25

2.2. Methods

2.2.1. MDF Fibre Characterisation. Fibre size and fine content of the original MDF fibres were measured using an FQA-360 fibre analyser. The resolution of this instrument is 14 μm/pixel (OpTest Equipment Inc., Canada).

2.2.2. Extrusion Compounding. The polymers and MDF fibres were dried overnight at 50°C and 105°C, respectively, before extrusion compounding. Polymers were combined with fibres in a twin screw LABtech™ extruder type LTE 26-40 (26 mm corotating screws; $l/d = 40$) (LabTech Engineering Co. Ltd., Praksa, Muang, Samutprakarn, Thailand) with an atmospheric vent used to vent volatiles. The polymer was gravimetrically fed into the main feed. Simultaneously, the fibre was hand fed at the same port to obtain a 30 wt % fibre content. The fibre content was selected due to its wide market application [26]. The compounded materials were extruded at 180°C using a rotational screw speed of 150 rpm. The total feed rate was 2.5 kg h^{-1} for the MDF fibres; all other materials' feed rate was 5 kg h^{-1}. The compound went through a two-strand die and was water cooled before being pelletized into 3 mm long wood plastic pellets.

2.2.3. Injection Moulding. The virgin biopolyester was extruded into plain or wood plastic strands and pelletised. The pellets were dried overnight in an oven at 60°C for the PLA composites and 80°C for the PBS, PBSA, and PBAT. The dried pellets were injection moulded on a BOY 35 machine (BOY Spritzgiessautomaten, Neustadt-Fernthal, Germany) into tensile and flexural specimens (sample shape according to ASTM D 638-03 type I). Injection moulding settings had to be adjusted according to the formulation in order to completely fill the mould cavities without flash and to produce satisfactory test specimens. The PLA, PBS, and PBAT composites were injection moulded at 180°C, and the PBSA samples at 110°C. A mould temperature of 30°C was chosen in order to obtain short cycle times. A cooling time of 20 seconds was used for all polymers except PBAT, where 10 seconds were used due to a longer packing time. Composite densities were measured to ensure that the filler loading was accurate and no voids were present in the composites.

2.2.4. Tensile Properties. Tensile specimens were tested using an Instron 5566 machine (Instron, USA). The tensile properties Young's modulus and maximum strength were obtained from stress-strain curves according to ASTM D638 [27] ($n = 10$).

2.2.5. Impact Strength. Notched Izod impact specimens were tested using a CEAST Resil impactor 6957 (CCSi, USA) according to ASTM 256 [28]. Ten specimens of each composite were tested to obtain the impact strength average value.

2.2.6. Molecular Weight Measurements. Gel permeation chromatography (GPC) was used to determine the molecular weight distribution of the plain and processed biopolyesters. GPC was carried out on a PL-GPC-50 system (Polymer Laboratories, UK, now Varian Inc.) with a Knauer K-301 refractive index detector using two PLgel Mixed-E (300 × 7.5 mm, 3 μm) columns connected in series and protected by a guard column (50 × 7.5 mm) of the same material. The system was run at 30°C with analytical grade chloroform as eluent. Flow rate was maintained at 1 mL*min^{-1}. The system was calibrated using polystyrene standards. Composite samples were stirred with chloroform for four hours at room temperature and filtered to remove the wood fibre (Watermann, ashless 40). The biopolyesters were then recovered with a Rotovapor®. Two technical repeats were taken from each extract and prepared with a chloroform concentration of 10 mg mL^{-1}. Molecular weights are expressed as the weight-average molecular weight (MW) which is defined as

$$\mathrm{MW} = \frac{\sum_{i=1}^{N}(h_i M_i)}{\sum_{i=1}^{N} h_i}, \quad (1)$$

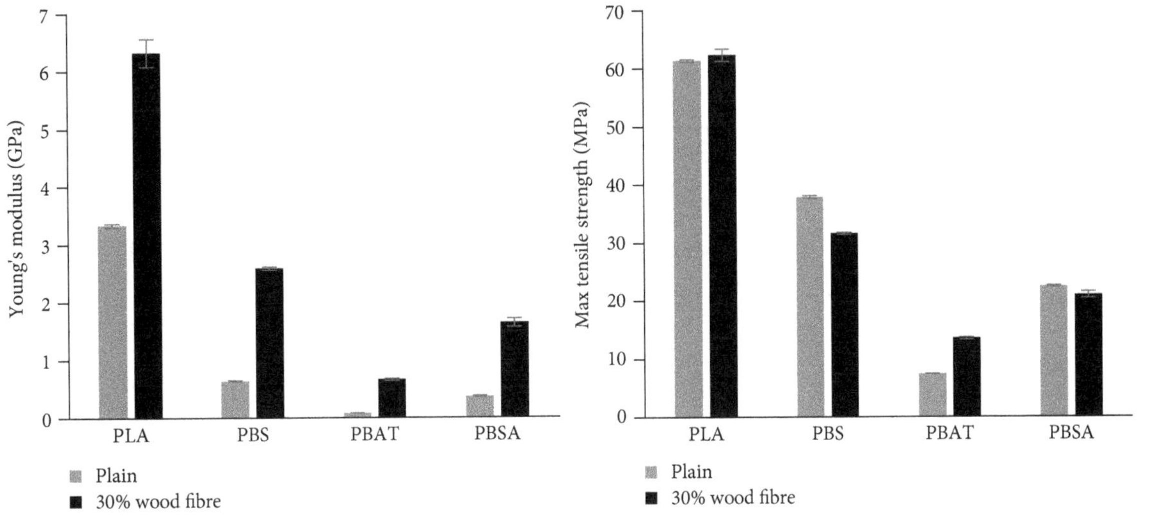

FIGURE 1: Tensile modulus and maximum tensile strength of injection moulded test samples with four biopolymers, plain or filled with 30 wt % MDF fibre (error bars = ±95% confidence interval) ($n = 10$).

where h_i is the height response of the concentration detector and M_i is the molecular weight at ith retention volume increment

2.2.7. Fibre Characterisation. The fibres were extracted from composites by chloroform extraction. One gram of sample was extracted using a Soxhlet apparatus with 200 mL of chloroform added. Samples were first boiled in the chloroform for two hours and then extracted in a Soxhlet for four hours under reflux.

The length of the extracted fibres was measured using a Fiberscan instrument (Andritz-Sprout-Bauer). The length-weighted fibre length (LWL) average was used to compare samples. It puts a greater weight on longer fibres and is given by

$$\mathrm{LWL} = \frac{\sum x^2}{\sum x}, \quad (2)$$

where x is the individual fibre length.

The measurements, including fibre extraction, were performed in triplicate. A rayon standard of known length and distribution (Kajaani, #2) was used to ensure instrument performance.

2.2.8. Rheology. The rheological properties of the materials were assessed in dynamic oscillation mode using a strain controlled rheometer AR2000 (TA instruments). The rheometer was equipped with 25 mm parallel plate geometry and an environmental test chamber to control the temperature. The compounded materials were dried in an oven at 103°C for four hours prior testing. To evaluate the thermal stability of the various compounds, the complex viscosity was monitored as a function of time. The time sweep measurements were conducted at 180°C using a strain of 2%, frequency of 1 Hz, and measurement gap of 1 mm.

2.2.9. Microscopy. A fracture surface was prepared for each treatment, from notched impact testing samples. Samples were sputter coated with chromium using an Emitech K575X coater and imaged using a Jeol JSM 6700F field emission scanning electron microscope (SEM). Images were recorded at 5 kV acceleration voltage and a probe current of 10 μA. The entire fracture surface was assessed, and then a representative area was imaged at 50x and 200x magnifications.

3. Results and Discussion

3.1. Mechanical Properties. The effect of MDF fibre on the tensile properties of four biopolyesters was tested. As expected, the stiffness of the four NFCs was increased by up to sixfold compared to the plain polyesters. However, the strength of the composites was not significantly changed for PLA and even decreased slightly for PBS and PBSA (Figure 1). The strength of the PBAT composite increased, albeit from a very low base level. This result was unexpected as the reinforcing potential of natural fibres in biopolyester has previously been demonstrated, e.g., for TMP fibre in PLA [13] and flax fibre in PLA [29].

The addition of MDF fibres did not improve the impact strength (Figure 2) but caused a moderate (PLA) to significant decrease (PBS, PBSA) of impact strength compared to the plain biopolyesters. Plain PBAT did not break under the testing conditions employed.

When MDF fibres were incorporated within a PP matrix with the addition of a compatibiliser such as MA-PP, a substantial tensile strength increase and impact strength decrease are observed [2, 30]. Therefore, the decrease of both values with the MDF filled biopolyesters was unexpected and prompted us to investigate why MDF fibres were unable to reinforce the biopolyesters.

3.2. Fibre Length. One hypothesis for the lack of reinforcement was a drastic reduction of the MDF fibre aspect ratio.

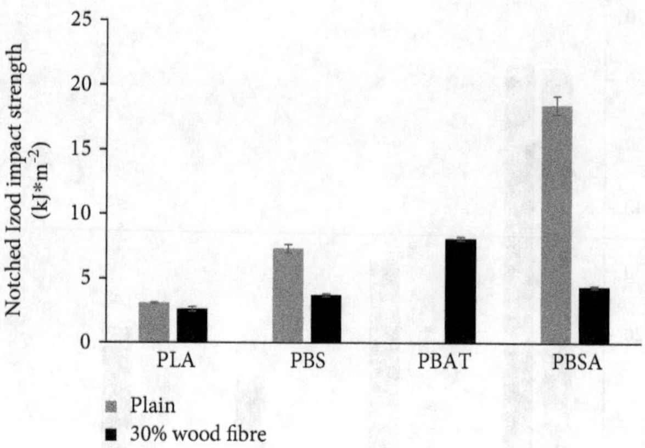

Figure 2: Impact strength of 4 biopolyesters, plain and filled with 30 wt % MDF fibre (error bars = ±95% confidence interval) ($n = 10$).

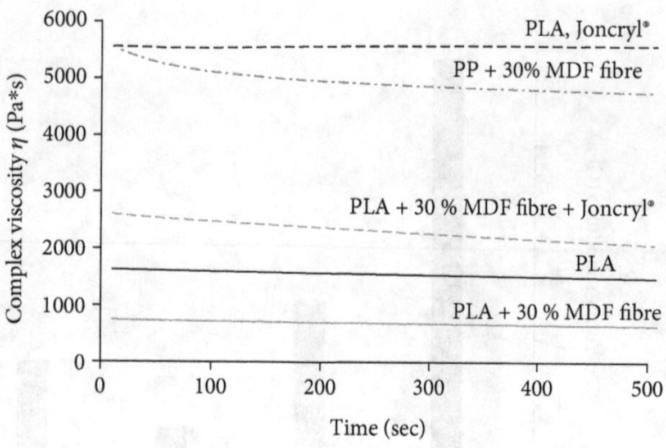

Figure 3: Apparent viscosity of plain PLA and PLA filled with 30 wt % MDF fibre (180°C and 1 Hz). The same materials containing Joncryl®, a chain extender, are also represented. For reference, PP containing 30 wt % MDF fibre is included.

Extrusion compounding and consequent injection moulding typically lead to shortening of natural fibres [13, 31]. Typically, when processed with PP, loose MDF fibre undergoes a fibre length reduction from 1.7–1.5 down to 0.5–0.3 mm [31, 32]. Therefore, the possibility of further MDF fibre length reduction during their processing with biopolyester was tested. It was found that after injection moulding with all biopolyesters, the MDF fibre length was reduced to 0.4 mm (data not shown). The greater decrease in fibre length with PLA compared to PP has been reported by Peltola et al. [13] and attributed to the higher viscosity of the PLA. However, even at a fibre length of 0.4 mm and an aspect ratio of 8.8, some reinforcement potential would be expected from the fibres. For example, Guo et al. [33] added a range of wood fibres of a similar length to PP and showed increases in strength, modulus, and impact. Therefore, the fibre attrition was considered insufficient to explain the reduction of tensile strength generated by the addition of MDF fibre in the biopolyester matrices.

3.3. Rheology. During extrusion and injection moulding, the biopolyesters were difficult to process in the presence of MDF fibres. This was most striking for PLA as it was not possible to pull strands from the extruder. These observations are consistent with a reduction of the polymer viscosities. Therefore, the melt viscosity behaviours of PLA and PLA containing MDF fibre were compared using a rheometer mimicking extrusion conditions (180°C, 1 Hz). The addition of MDF fibre reduced the melt viscosity of the plain PLA by half (Figure 3). This observation was unexpected as the addition of MDF fibres typically increases PP melt viscosity [13]. PLA, containing MDF fibre, is seven times less viscous than its PP reference counterpart. The low melt viscosity of PLA/MDF fibres explains its unstable behaviour during extrusion (Figure 3).

3.4. Molecular Weight. Polymers typically show a correlation between viscosity and molecular weight (MW) as described by a power law [34]: $\eta_0 \propto M^{3.4}$, where η_0 is the melt viscosity and M is the molecular weight.

Rheology observations were compared to molecular weight by assessing molecular weight distributions with gel permeation chromatography (GPC). The extrusion and injection moulding steps did not affect plain PLA and PBSA but led to a 20 and 28% MW reduction for PBS and PBAT, respectively (Figure 4). The presence of MDF fibre during processing negatively affected the MW of all biopolymers with the most drastic effect observed for PLA. After injection moulding, there was a 36% reduction in MW of PLA filled with MDF fibres compared to plain PLA. Overall, it was found that, in accordance with the power law, the magnitude of molecular weight reduction of PLA containing MDF fibres (Figure 4) was consistent with its reduction in viscosity (Figure 3).

Next, it was investigated whether the molecular weight reduction of biopolyesters due to MDF fibres could be observed with other lignocellulosic fibres. PLA was selected as it exhibits the strongest MW decrease during compounding in the presence of MDF fibres. Bleached kraft fibres and flax fibres were the lignocellulosic fibres used for comparison. Bleached *Pinus radiata* kraft fibres were selected as they had the same origin as the MDF fibres but differed by their high cellulose content, having only traces of lignin and hemicelluloses [35]. The flax fibres, a nonwood fibre with very low lignin content [36], represents an agricultural fibre that often directly competes against wood fibre. Both kraft and flax fibres caused a slight MW reduction of PLA during extrusion and injection moulding, but neither of the fibres had an effect as large as that of MDF fibres on PLA (Figure 5).

3.5. Influence of Chain Extender Additives on PLA's Molecular Weight during Compounding. Substantial degradation of the PLA polymer was specifically observed when processed with MDF fibres. Compensation for this detrimental effect on MW and mechanical properties was explored by using chain extender additives to counteract this degradation during extrusion compounding. Two commercial chain extenders were chosen: BioAdimide® which is a product specifically designed to improve melt stability of biobased polyesters and Joncryl®, a BASF product designed to reverse the

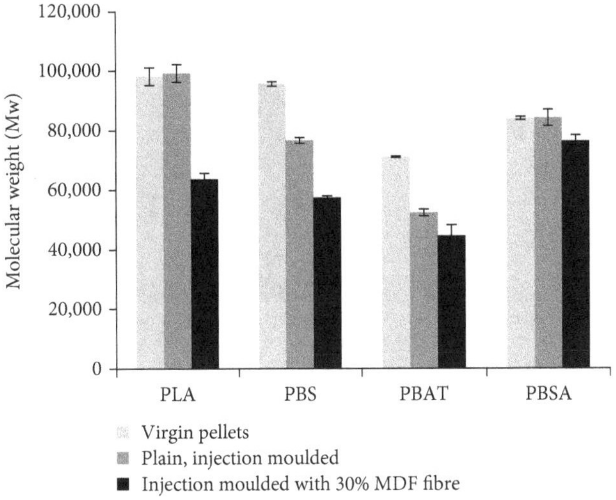

FIGURE 4: Molecular weight reduction of four biopolyesters before and after extrusion with or without 30 wt % MDF fibres (error bars = ±95% confidence interval) ($n = 2$).

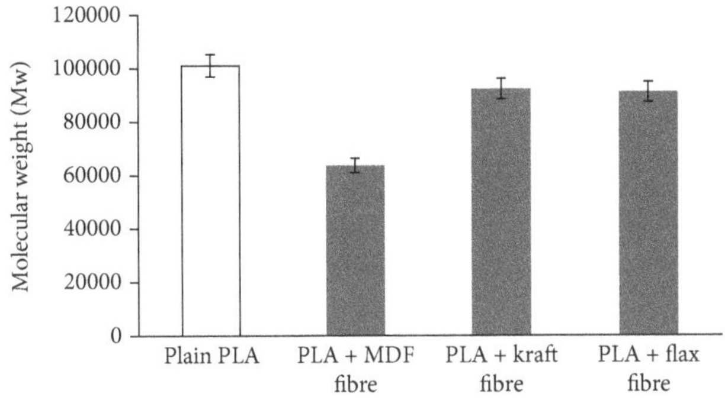

FIGURE 5: Effect of three natural fibre types on PLA MW after extrusion and injection moulding (all fibres at 30 wt %) (error bars = ±95% confidence interval).

degradation of poly-ethylene-terephthalate (PET) and polyester packaging products.

The addition of both chain extenders during extrusion led to an increase of molecular weight beyond the original MW of PLA (Figure 6). The strongest effect was observed with the addition of Joncryl® leading to a 150% increase compared with the plain PLA MW. In the presence of MDF fibres, the addition of chain extenders leads to a doubling of the MW value, with values greater than that of plain PLA. This was reflected by the improved processability of the materials. Melt viscosity was high enough (Figure 3), to allow strands to be pulled from the extruder.

It is unclear why the MDF fibres led the molecular weight decrease of PLA during compounding. When comparing MDF to bleached kraft pulp and flax fibres, the main difference is the presence of lignin. Softwood bleach kraft pulp and MDF fibres contain 0.8 and 31.2% of lignin, respectively [9]. Several studies have shown a positive effect of lignin in PLA blends. Lignin greatly improved the thermal stability

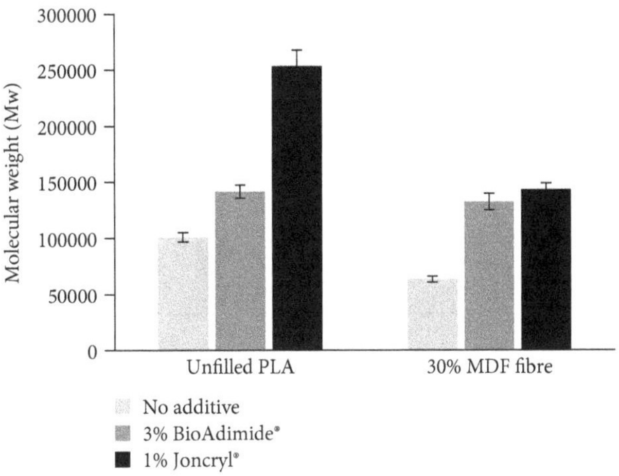

FIGURE 6: Influence of chain extenders on PLA molecular weight (error bars = ±95% confidence interval) ($n = 2$).

of PLA, and acetylated lignins appear to prevent hydrolytic degradation of PLA [37]. Another difference is the relatively high acidity of MDF fibre (pH 4.0) [38] compared to kraft (pH 5.5) and flax (pH 5.5). However, lactic acid pKa is 3.84 [39] and it was reported that PLA was the most stable at pH = 4 [40]. Therefore the pH of the MDF fibre cannot be solely responsible for molecular weight reduction. Finally, extractives are virtually absent from bleached kraft pulp [9], while due to the high temperature used during the softwood MDF manufacture process, the MDF fibre contains high levels of water and DCM extractives, i.e., 7.2 and 0.4 wt %, respectively [41]. A GC-MS study showed that MDF fibre extractives contain a wide range of triglycerides, sterol esters, fatty acids, and resin acids [42]. It can be hypothesised that these extractives were released during extrusion and led to the degradation of the PLA chains. The unique high levels of extractives in MDF would also explain why it has been possible to reinforce PLA with TMP without using additives [13]. TMP is refined at lower temperatures which leads to a substantial reduction in extractives [41].

3.6. Influences of Chain Extender Additives on PLA NFC's Mechanical Properties. As the ability of chain extenders to increase the molecular weight of PLA containing MDF fibres was demonstrated, the next step was to determine if this effect was translated into an improvement of NFC's mechanical properties. Wood flour, a low aspect ratio filler, was used as a comparison to MDF fibre. Young's modulus of the plain PLA was increased by the presence of MDF fibre (Figure 1) and wood flour (Figure 7), but neither of the additives was able to significantly increase this effect further (Figure 7).

Despite leading to the increased molecular weight of PLA, the addition of BioAdimide® did not translate into an improvement of mechanical strength in either NFC (Figure 7). It has previously been found [43] that the molecular weight of amorphous PLA does not always correlate with its tensile strength. The addition of 30 wt % MDF and 1 wt % Joncryl® was the only combination able to increase the tensile strength of PLA (+16%). This effect was not seen with the

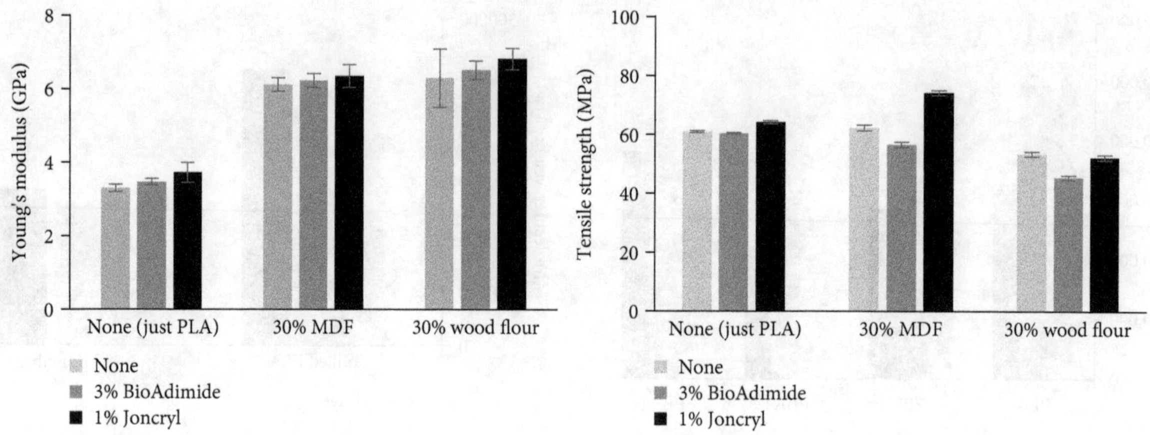

FIGURE 7: Effect of chain extender additives on tensile modulus and maximum tensile strength of injection moulded PLA NFC test samples. 1 wt % Joncryl® and 3 wt % BioAdimide®, on the modulus and strength of a plain PLA. PLA containing 30 wt % of MDF fibre and PLA containing 30 wt % wood flour (error bars = ±95% confidence interval) ($n = 10$).

FIGURE 8: Scanning electron microscopy images of fracture surfaces; effect of chain extender additives on fibre pull-out. (a, b) Plain PLA containing 30 wt % of MDF fibre. (c, d) PLA containing 30 wt % of MDF fibre and 3 wt % BioAdimide®. (e, f) PLA containing 30 wt % of MDF fibre and 1 wt % Joncryl®. Asterisks indicate a fibre pull-out. Scale bar = 100 μm ($n = 1$).

wood flour, as in all the cases, addition of wood flour leads to a decrease in tensile strength. This illustrates that tensile strength increase requires not a simple filler but high aspect ratio fibres.

3.7. Fractography Studies. The fracture surface of injection moulded specimens was observed by scanning electron microscopy (SEM) in order to understand why despite both chain extenders increasing the MW of the MDF fibre PLA composite, only Joncryl® is able to improve tensile properties (Figure 8). A large number of pulled-out fibres can be observed for samples without any additive and in the presence of 3 wt % BioAdimide® (Figures 8(a)–8(d)). On the other hand, the addition of Joncryl® led to a near absence of visible fibre pull-out (Figures 8(e) and 8(f)). This observation was unexpected as Jaszkiewicz et al. [44] observed that the addition of 3 wt % Joncryl® 3229, with man-made cellulose (20 wt %), manufactured by Cordenka, in a PLA matrice led to a massive fibre pull-out. The authors assumed that the maleinated groups of Joncryl® 3229 reacted chemically with the hydroxyl groups of the cellulosic fibres making the fibres smoother and more slippery. In the case of this study, it can be assumed that Joncryl® is improving the interfacial adhesion between the fibre and the matrix which enabled the fibre to reinforce the matrix and led to the increase in tensile strength.

4. Conclusions

In order to obtain a fully biodegradable composite, natural fibres need to be used as reinforcements of the biodegradable plastic matrix. Successful integration of cost competitive MDF fibre with biopolyesters is an attractive prospect.

(i) This study demonstrates that the addition of 30 wt % of MDF fibre with a biopolyester matrix led to unexpected difficulties of compounding due to a reduction in melt viscosity

(ii) This phenomenon can be explained by the fact that MDF fibre reduced the MW of the four biopolyesters tested, with the strongest reduction observed for PLA

(iii) The MW reduction specifically caused by the addition of MDF fibre explains the absence of strength reinforcement of the biopolyester matrix

(iv) The addition of Joncryl®, a chain extender, resulted in a PLA polymer with a higher MW than the starting PLA material

(v) Joncryl® reinforced the interface between PLA and MDF fibres

(vi) This PLA/MDF composite demonstrated an increase in strength

(vii) The MW reduction could be due to the presence of extractives in MDF fibre released during compounding

(viii) This work demonstrates the possibility to obtain biopolyester composites reinforced with MDF fibres

Conflicts of Interest

Scion is the proprietor of a family of registered and licensed patents relating to a reinforcing wood fibre product and process for use in NFCs.

Acknowledgments

The authors would like to acknowledge Dawn Smith, Marc Gaugler, and Gareth Lloyd-Jones for their extensive review of this manuscript. This work was funded by the New Zealand Ministry of Business, Innovation and Employment (MBIE).

References

[1] K. L. Pickering, M. G. A. Efendy, and T. M. Le, "A review of recent developments in natural fibre composites and their mechanical performance," *Composites Part A: Applied Science and Manufacturing*, vol. 83, pp. 98–112, 2016.

[2] A. Fernyhough and M. Markotsis, "Long biofibers and engineered pulps for high performance bioplastics and biocomposites," in *Handbook of Bioplastics and Biocomposites Engineering Applications: Pilla/Handbook*, pp. 555–579, John Wiley & Sons, Inc., 2011.

[3] P. Nygård, B. S. Tanem, T. Karlsen, P. Brachet, and B. Leinsvang, "Extrusion-based wood fibre-PP composites: wood powder and pelletized wood fibres - a comparative study," *Composites Science and Technology*, vol. 68, no. 15-16, pp. 3418–3424, 2008.

[4] C. M. Chan, L. J. Vandi, S. Pratt et al., "Composites of wood and biodegradable thermoplastics: a review," *Polymer Reviews*, vol. 58, no. 3, pp. 444–494, 2017.

[5] M. Sorieul, A. Dickson, S. Hill, and H. Pearson, "Plant fibre: molecular structure and biomechanical properties, of a complex living material, influencing its deconstruction towards a biobased composite," *Materials*, vol. 9, no. 8, p. 618, 2016.

[6] L. A. Donaldson and T. D. Lomax, "Adhesive/fibre interaction in medium density fibreboard," *Wood Science and Technology*, vol. 23, no. 4, pp. 371–380, 1989.

[7] Z. Shao and K. Li, "The effect of fiber surface lignin on interfiber bonding," *Journal of Wood Chemistry and Technology*, vol. 26, no. 3, pp. 231–244, 2007.

[8] A. McDonald, A. Clare, and B. Dawson, "Surface characterisation of Radiata pine high-temperature TMP fibres by X-ray photo-electron spectroscopy," in *Proceeding of the 53rd General APPITA Conference*, Rotorua, New Zealand, 1999.

[9] K. L. Spence, R. A. Venditti, Y. Habibi, O. J. Rojas, and J. J. Pawlak, "The effect of chemical composition on microfibrillar cellulose films from wood pulps: mechanical processing and physical properties," *Bioresource Technology*, vol. 101, no. 15, pp. 5961–5968, 2010.

[10] R. D. Burdon, R. P. Kibblewhite, and M. J. C. Riddell, "Wood density and kraft fibre and pulp properties of four Pinus radiata provenances," *New Zealand Journal of Forestry Science*, vol. 29, no. 2, pp. 214–224, 1999.

[11] A. R. Dickson and D. Sandquist, "Mode of wood fibre breakage during thermoplastic melt processing," *Composites Part A: Applied Science and Manufacturing*, vol. 112, pp. 496–503, 2018.

[12] O. Faruk, A. K. Bledzki, H. P. Fink, and M. Sain, "Biocomposites reinforced with natural fibers: 2000-2010," *Progress in Polymer Science*, vol. 37, no. 11, pp. 1552–1596, 2012.

[13] H. Peltola, E. Pääkkönen, P. Jetsu, and S. Heinemann, "Wood based PLA and PP composites: effect of fibre type and matrix polymer on fibre morphology, dispersion and composite properties," *Composites Part A: Applied Science and Manufacturing*, vol. 61, pp. 13–22, 2014.

[14] K. G. Satyanarayana, G. G. C. Arizaga, and F. Wypych, "Biodegradable composites based on lignocellulosic fibers-an overview," *Progress in Polymer Science*, vol. 34, no. 9, pp. 982–1021, 2009.

[15] T. Gurunathan, S. Mohanty, and S. K. Nayak, "A review of the recent developments in biocomposites based on natural fibres and their application perspectives," *Composites Part A: Applied Science and Manufacturing*, vol. 77, pp. 1–25, 2015.

[16] T. L. do Amaral Montanheiro, F. R. Passador, M. P. de Oliveira, N. Durán, and A. P. Lemes, "Preparation and characterization of maleic anhydride grafted poly (hydroxybutirate-CO-hydroxyvalerate)-PHBV-g-MA," *Materials Research*, vol. 19, no. 1, pp. 229–235, 2016.

[17] D. Plackett, "Maleated polylactide as an interfacial compatibilizer in biocomposites," *Journal of Polymers and the Environment*, vol. 12, no. 3, pp. 131–138, 2004.

[18] L. Quiles-Carrillo, N. Montanes, D. Garcia-Garcia, A. Carbonell-Verdu, R. Balart, and S. Torres-Giner, "Effect of different compatibilizers on injection-molded green composite pieces based on polylactide filled with almond shell flour," *Composites Part B: Engineering*, vol. 147, pp. 76–85, 2018.

[19] M. Kazayawoko, J. J. Balatinecz, R. T. Woodhams, and S. Law, "Effects of wood fiber surface chemistry on the mechanical properties of wood fiber-polypropylene composites," *International Journal of Polymeric Materials and Polymeric Biomaterials*, vol. 37, no. 3-4, pp. 237–261, 2006.

[20] A. Jaszkiewicz, A. Meljon, A. K. Bledzki, and M. Radwanski, "Gaining knowledge on the processability of PLA-based short-fibre compounds - a comprehensive comparison with their PP counterparts," *Composites Part A: Applied Science and Manufacturing*, vol. 83, pp. 140–151, 2016.

[21] L. Zhang, S. Lv, C. Sun, L. Wan, H. Tan, and Y. Zhang, "Effect of MAH-g-PLA on the properties of wood fiber/polylactic acid composites," *Polymer*, vol. 9, no. 11, 2017.

[22] K. M. Bogren, E. K. Gamstedt, R. C. Neagu, M. AÅkerholm, and M. LindstroÖm, "Dynamic-mechanical properties of wood-fiber reinforced polylactide: experimental characterization and micromechanical modeling," *Journal of Thermoplastic Composite Materials*, vol. 19, no. 6, pp. 613–637, 2006.

[23] B. L. Shah, S. E. Selke, M. B. Walters, and P. A. Heiden, "Effects of wood flour and chitosan on mechanical, chemical, and thermal properties of polylactide," *Polymer Composites*, vol. 29, no. 6, pp. 655–663, 2008.

[24] L. Wei and A. McDonald, "A review on grafting of biofibers for biocomposites," *Materials*, vol. 9, no. 4, p. 303, 2016.

[25] B. A. Ahmed Ali, S. M. Sapuan, E. S. Zainudin, and M. Othman, "Implementation of the expert decision system for environmental assessment in composite materials selection for automotive components," *Journal of Cleaner Production*, vol. 107, pp. 557–567, 2015.

[26] A. El-Sabbagh, "Effect of coupling agent on natural fibre in natural fibre/polypropylene composites on mechanical and thermal behaviour," *Composites Part B: Engineering*, vol. 57, pp. 126–135, 2014.

[27] ASTM International, "Standard test method for tensile properties of plastics," ASTM International, Pennsylvania, USA, 2010.

[28] ASTM International, "Standard test methods for determining the Izod pendulum impact resistance of plastics," ASTM International, Pennsylvania, USA, 2010.

[29] K. Oksman, M. Skrifvars, and J. F. Selin, "Natural fibres as reinforcement in polylactic acid (PLA) composites," *Composites Science and Technology*, vol. 63, no. 9, pp. 1317–1324, 2003.

[30] M. D. H. Beg and K. L. Pickering, "Mechanical performance of Kraft fibre reinforced polypropylene composites: influence of fibre length, fibre beating and hygrothermal ageing," *Composites Part A Applied Science and Manufacturing*, vol. 39, no. 11, pp. 1748–1755, 2008.

[31] M. D. H. Beg and K. L. Pickering, "Reprocessing of wood fibre reinforced polypropylene composites. Part I: effects on physical and mechanical properties," *Composites Part A Applied Science and Manufacturing*, vol. 39, no. 7, pp. 1091–1100, 2008.

[32] A. R. Dickson, D. Even, J. M. Warnes, and A. Fernyhough, "The effect of reprocessing on the mechanical properties of polypropylene reinforced with wood pulp, flax or glass fibre," *Composites Part A: Applied Science and Manufacturing*, vol. 61, pp. 258–267, 2014.

[33] Q. Guo, B. Cheng, M. Kortschot et al., "Performance of long Canadian natural fibers as reinforcements in polymers," *Journal of Reinforced Plastics and Composites*, vol. 29, no. 21, pp. 3197–3207, 2010.

[34] P. J. Flory and T. G. Fox, "Treatment of intrinsic viscosities," *Journal of the American Chemical Society*, vol. 73, no. 5, pp. 1904–1908, 1951.

[35] J. Laine, P. Stenius, G. Carlsson, and G. Ström, "The effect of ECF and TCF bleaching on the surface chemical composition of kraft pulp as determined by ESCA," *Nordic Pulp and Paper Research Journal*, vol. 11, no. 3, pp. 201–210, 1996.

[36] A. Lefeuvre, A. Bourmaud, C. Morvan, and C. Baley, "Elementary flax fibre tensile properties: correlation between stress-strain behaviour and fibre composition," *Industrial Crops and Products*, vol. 52, pp. 762–769, 2014.

[37] O. Gordobil, I. Egüés, R. Llano-Ponte, and J. Labidi, "Physico-chemical properties of PLA lignin blends," *Polymer Degradation and Stability*, vol. 108, pp. 330–338, 2014.

[38] C. Xing, S. Y. Zhang, J. Deng, B. Riedl, and A. Cloutier, "Medium-density fiberboard performance as affected by wood fiber acidity, bulk density, and size distribution," *Wood Science and Technology*, vol. 40, no. 8, pp. 637–646, 2006.

[39] A. M. Eyal and R. Canari, "pH dependence of carboxylic and mineral acid extraction by amine-based extractants: effects of pKa, amine basicity, and diluent properties," *Industrial and Engineering Chemistry Research*, vol. 34, no. 5, pp. 1789–1798, 1995.

[40] M. A. Elsawy, K. H. Kim, J. W. Park, and A. Deep, "Hydrolytic degradation of polylactic acid (PLA) and Its composites," *Renewable and Sustainable Energy Reviews*, vol. 79, pp. 1346–1352, 2017.

[41] K. D. Murton, "The effect of preheating and refining conditions on radiata pine MDF fibre quality," in *52nd Appita annual conference*, Melbourne, 1998.

[42] A. G. McDonald, T. R. Stuthridge, A. B. Clare, and M. J. Robinson, "Isolation and analysis of extractives from radiata pine HTMP fibre," in *9th International Symposium of Wood and Pulping Chemistry*, pp. 9–12, Montreal, Canada, June 1997.

[43] G. Perego, G. D. Cella, and C. Bastioli, "Effect of molecular weight and crystallinity on poly(lactic acid) mechanical properties," *Journal of Applied Polymer Science*, vol. 59, no. 1, pp. 37–43, 1996.

[44] A. Jaszkiewicz, A. K. Bledzki, and P. Franciszczak, "Improving the mechanical performance of PLA composites with natural, man-made cellulose and glass fibers- a comparison to PP counterparts," *Polimery*, vol. 58, no. 6, pp. 435–442, 2013.

19

Preparation of Miscible PVA/PEG Blends and Effect of Graphene Concentration on Thermal, Crystallization, Morphological, and Mechanical Properties of PVA/PEG (10wt%) Blend

Fahad H. Falqi,[1] Osamah A. Bin-Dahman,[2] M. Hussain,[3] and Mamdouh A. Al-Harthi[1]

[1]*Department of Chemical Engineering, King Fahd University of Petroleum and Minerals, Dhahran 31261, Saudi Arabia*
[2]*Department of Chemical Engineering, Faculty of Engineering and Petroleum, Hadhramout University, Mukalla, Yemen*
[3]*Department of Materials and Chemical Engineering, Hanyang University, Seoul, Republic of Korea*

Correspondence should be addressed to Mamdouh A. Al-Harthi; mamdouh@kfupm.edu.sa

Academic Editor: Yulin Deng

Water-soluble polymers such as poly(vinyl alcohol) (PVA) and poly(ethylene glycol) (PEG) and their nanocomposites with graphene were prepared by using a solution mixing and casting technique. The effect of different PEG loadings was investigated to determine the optimum blend ratio. The films were characterized using Fourier transform infrared spectroscopy (FTIR), differential scanning calorimetry (DSC), and thermogravimetric analyzer (TGA) methods. Also, the mechanical properties including tensile strength and elongation at break were measured using a universal tensile testing machine. FTIR results confirmed the formation of the H-bond between PEG and PVA. DSC studies revealed that PEG has a significant plasticization effect on PVA as seen by the drop in the glass transition temperature (T_g). The blend with 10 wt% PEG loading was found to be the optimum blend because of good compatibility as shown by FTIR and SEM results and improved thermal properties. PVA/PEG (10%) nanocomposites were prepared using graphene as a nanofiller. It was found that the elongation at break increased by 62% from 147% for the PVA/PEG (10%) blend to 209% for the nanocomposite with graphene loading of 0.2 wt%. The experimental values of tensile strength were compared using the predictive model of Nicolais and Narkis.

1. Introduction

Usage of polymeric thermoplastics has increased remarkably over the past four decades. Applications include grocery, merchandise, trash bags, food containers, caps and closures, plastic fuel tanks, drink cups, home appliances, cooking utensils, children's toys, and construction materials, among others. Water-soluble polymers such as PEG and PVA and their blends have found many applications such as packaging [1], cosmetics [2], and emulsifiers and adhesives [3], among others. PVA and its composites have been extensively studied by our group [1, 4–15]. With some modifications of the existing polymers, products with low cost and light weight may be developed for a wide range of applications with desirable properties such as mechanical, thermal, and optical properties. This makes them suitable as replacements for other traditional engineering materials such as metals. In recent years, biodegradable polymeric materials have been found useful for drug delivery, pharmaceuticals, paints, textiles, and tissue engineering applications [7, 16]. They have many uses, with many new applications involving blends of these polymers with other polymers and other materials [17]. Different physical and chemical treatment methods are available to modify these materials and make them successful in different fields. Blending is a simple and effective method to develop new materials with tailored properties. Recently, nanocomposites with improved properties have been prepared by mixing different polymers through the physical blending method. This is now a promising trend in the science of nanocomposites [18]. Studies on PVA/PEG blends have been an active area for research that has attracted great commercial interest.

Studies have considered the potential to obtain materials with improved properties for various applications. Abdel Tawab et al. [19] studied the influence of the PEG content on PVA and found that the PEG content up to 60% can provide the best compatibility of the PVA/PEG blend. Other authors report PVA/PEG blends with a good film and reported that blends with PEG content above 40% do not form a cast film due to phase separation [19]. However, the main disadvantage of hydrophilic polymers such as PVA and PEG studied here is their weak mechanical properties [20]. The nanomaterials at very low loading content revealed improved properties for nanocomposite systems when incorporated into the polymer matrix [8, 21]. Nanofillers, especially carbon nanomaterials, have led to remarkably improved properties, at very low loading content, when incorporated into the polymer matrix. In particular, polymer composites of carbon nanotubes (CNT) and graphene have attracted a wide range of applications in modern science and technology. These fillers can provide unique properties to the polymer matrices [22]. Graphene has been given special attention due to its unique properties [23]. It is fabricated from natural graphite and is cheaper compared to other fillers such as CNT [24].

PVA can be effectively reinforced by using graphene nanosheets [1]. El Sayed and Morsi [25] prepared PVA/PEG nanocomposites based on a hematite (α-Fe_2O_3) nanofiller and reported improvement in optical and dielectric properties of the prepared nanocomposites. They found that both the refractive index and ac conductivity increased significantly as a function of increasing the filler content [25]. It was also reported that addition of α-Fe_2O_3 nanoparticles improved the thermal stability of PVA [22]. Sengwa et al. [18] used nanoclay and studied the effect of its addition on PVA/polyethylene oxide (PEO) blend. PVA/PEG nanocomposites have also been prepared and studied using montmorillonite (MMT) clay as nanofiller [20]. The hydrophilic characteristic of MMT clay helps their dispersion into water-soluble polymers such as PVA and PEG [20]. The addition of MMT up to 4 wt% improves the mechanical properties of the PVA matrix [26].

To the best of our knowledge and survey of literature, the study of the PVA/PEG nanocomposite based on graphene has not been previously reported. Therefore, in this work, the effect of PEG loadings on PVA/PEG blends is first studied in an attempt to obtain cast films with good physical properties to be used for further investigation. Also, PVA/PEG nanocomposites were prepared using graphene as a nanofiller to investigate its influence on the mechanical properties of the PVA/PEG blend.

2. Experimental

2.1. Materials. Polyethylene glycol (PEG) (powder, molecular $weight = 4000\ g/mol$) and polyvinyl alcohol (PVA) (degree of $hydrolysis = 99\%$, molecular $weight = 27{,}000\ g/mol$) were supplied by Sigma-Aldrich Company, USA. Both homopolymers were used for the preparation of PVA/PEG blends and PVA/PEG nanocomposites as received from the supplier without further purification. Graphene, trade name GRAFEN®-iGP, of a purity of 96–99% and thickness of 50–100 nanometers, was procured from Grafen Chemical Industries, Turkey.

2.2. Preparation of PVA/PEG Blends. The films for pure PVA and PEG/PVA blends were prepared using the solution cast method. After many experimental trials, acceptable films with good transparency were only obtained for pure PVA and PVA/PEG blends with PEG loadings at 5, 10, 15, and 20 wt%, while loss of transparency and occurrence of phase separation were observed for the prepared blends containing PEG above this loading range. Films possessing acceptable transparency (as seen by visual observation) were obtained for the blends shown in Table 1, which have PEG loading (<20 wt%). The obtained films of these blends along with that of pure PVA were characterized and investigated. However, for other blends prepared as part of this experimental work, PEG loadings included 30, 40, 50, 60, and 70 wt%. These films were very brittle due to the higher PEG content which caused a phase separation between the two blended polymers, PVA and PEG. For this reason, these polymer blends were not considered to be suitable for characterization and were discarded.

For the preparation of PEG/PVA blends, different weighted amounts of PEG and PVA were separately dissolved in deionized water based on the formulations shown in Table 1. A fixed amount of PVA (3 g) was used in all PVA/PEG blends. PVA was used in the blends since PVA gives good film with improved flexibility and mechanical stability [20]. Mixtures of the polymers were stirred at 400 rpm and 80°C for 2 hr to ensure complete dissolution/mixing. Then, the whole mixture was degassed in a vacuum oven for 10 min. The solution was then poured into a petri dish and left to dry at room temperature for 6 days. Finally, the obtained film was peeled off from the petri dish and preserved in a desiccator filled with silica gel to avoid moisture uptake [1]. The obtained films for all samples were kept and used for further characterization.

2.3. Preparation of PVA/PEG Nanocomposites. The same procedure mentioned above for the preparation of the films of PVA/PEG blends was followed for the preparation of PVA/PEG nanocomposite films. PVA/PEG nanocomposites were prepared based on the optimum blend ratio of PVA/PEG found in the previous step and using graphene as a nanofiller. Graphene at different loadings (0.1, 0.2, 0.3, 0.5, and 1 wt%) was separately mixed with water, and the aqueous suspension was sonicated for 2 min. Then, the solution was mixed with that of the PVA/PEG blend and stirred for 2 h at 80°C followed by a degassing step and casting similar to the above procedure. The obtained films of PVA/PEG nanocomposites were kept in the same desiccator utilized for PVA/PEG blends and used for further characterization.

2.4. Characterization

2.4.1. Thermal Properties. Differential scanning calorimetry (DSC) and thermogravimetric (TG) studies of the obtained

TABLE 1: The formulations for the prepared PVA/PEG blends.

Sample code	PEG content (wt%)	PVA content (wt%)
Pure PVA	0	100
PVA/PEG (5%)	5	95
PVA/PEG (10%)	10	90
PVA/PEG (15%)	15	85
PVA/PEG (20%)	20	80

films for pure PVA and PVA/PEG blends were carried out to evaluate their thermal and degradation behavior. The thermal properties such as heat of fusion or melting enthalpy (ΔH_{fus} or ΔH_m), melting temperature (T_m), crystallization temperature (T_c), and glass transition temperature (T_g) of all obtained films were evaluated by the DSC method. The DSC thermograms were obtained using DSC-Q1000, Universal V4.2E from TA Instruments. The DSC procedure followed three steps: heating/cooling/heating. In the first step, the films were heated from a temperature of −100 to 250°C at a heating rate of 10°C/min and held at this temperature for 5 min to eliminate the effects from the thermal history. Then, cooling was performed at a cooling rate of 10°C/min followed by a final heating step at the same heating rate of 10°C/min. The DSC experiments were carried out under nitrogen atmosphere with the flow rate of nitrogen gas set at 50 mL/min. Integration of the DSC thermograms over the plot range of 190–240°C allows determination of the heats of fusion [2, 16]. The degree of crystallinity (X_c) was calculated using the values of the measured heats of fusion according to (1) [27]:

$$X_c(\%) = \frac{\Delta H_{\text{fus}}/\varnothing_{\text{PVA}}}{\Delta H^o_{\text{fus}}} \times 100, \qquad (1)$$

where ΔH_{fus} is the heat of fusion of PVA measured from DSC thermograms, $\varnothing_{\text{PVA}}$ is the PVA fraction in the composites, and ΔH^o_{fus} is the melting enthalpy of the totally crystallized PVA that is assumed to be 142 J/g [28].

The thermal behavior and stability were also evaluated for all obtained films using a thermogravimetric analyzer (TGA, Perkin Elmer Pyris 6). All samples were heated from 35 up to 600°C at a heating rate of 10°C/min. From the obtained TGA curves, the weight loss at different temperatures and the thermal degradation temperatures were measured for all samples. The differential thermogravimetric (DTG) thermograms were also developed by taking the first-order derivative of the corresponding TGA curves.

2.4.2. Fourier Transform Infrared (FTIR) Spectroscopy. FTIR (Nicolet 6700 Spectrometer) was used for obtaining the absorption spectra for the obtained films. All spectra were obtained by averaging 32 scans at a resolution of 4 cm^{-1} and recorded in the wave numbers ranging from 4000 to 500 cm^{-1}.

2.4.3. Mechanical Properties. The mechanical properties such as tensile strength and percentage of elongation at break were measured for the obtained films using a Universal Testing Instrument (Instron 3366). The films were cut (dog-bone shape with a gauge length of 25 mm) and tested at a constant cross-head speed of 10 mm/min at room temperature of 25°C. The recorded value for the measured parameters obtained from the stress-strain curves was taken as the average of at least three measurements to obtain reliable data.

The experimental data of the tensile strength were analyzed by comparing to some predicted models to understand the formation of weak structure in the composite [29, 30]. The tensile strength of the composites was predicted using the following empirical equation developed by Nicolais and Narkis [31] as described in (2).

$$\sigma_c = \sigma_m \left(1 - V_f^{2/3}\right), \qquad (2)$$

where σ_c is the TS of the composite, σ_m is the TS of the polymer matrix, and V_f is the volume fraction of the filler. The mass fraction of graphene was converted to volume fraction using the value of graphene density 1.18 g/m^3. The assumptions made for this model include the following: (1) there is no adhesion between the filler particle and polymer matrix with a coefficient value of 1.21 for $V_f^{2/3}$, and (2) the particles are spherical.

2.4.4. Morphological Analysis. The morphology of the fracture surfaces of the films was examined by using a scanning electron microscope (LAYRA3 TESCAN FE-SEM). The samples were cryofractured by immersing the films in liquid nitrogen for 2 min and coated with a thin layer of gold to render the surface conductive prior to taking scanning electron micrograph images at an accelerating voltage of 5 kV.

3. Results and Discussion

3.1. Effect of PEG Loadings on PVA/PEG Blends. While PEG is hydrophilic and gives a clear solution, it is known to be difficult to achieve a good cast film using PEG [19]. On the other hand, PVA is known to form a good film when it is blended with other polymers [7, 8, 10]. It was noticed that at higher PEG loadings (above 20 wt%), it was difficult to get acceptable films that can be utilized for characterization. While blending PEG with PVA in the presence of water using a solution casting technique forms an H-bonded network to achieve miscibility, it was found difficult to obtain cast films due to phase separation [18, 32]. Abd Alla et al. [32] report that the PVA/PEG blend has shown limited compatibility in the range of 0–30 wt% of PEG. It was also reported that PVA/PEG blends containing 60, 70, and 80% PVA form acceptable cast films with acceptable physical properties [19].

The miscibility is an important factor in the study of polymer blends as the final properties of the blends are directly related to the degree of their miscibilities [2]. Generally, miscible polymer blends show one phase while immiscible blends show separated domains. Also, miscible polymeric blends show a decrease in the Gibbs energy of mixing [33]. Phase separation was observed for PVA/PEG blends with PEG content higher than 40%, while good films were obtained at lower PEG content. Transparent films were obtained for PVA/PEG blends containing PEG up to 30%, which was attributed to the miscibility of both PEG and PVA homopolymers. However, the miscibility decreased with increasing PEG content above that limit. At higher PEG loading (above 30%), incompatibility was observed in PVA/PEG blends [32]. A similar finding was reported for poly(lactic acid) (PLA)/PEG blend which showed brittle nature at higher PEG content due to the lack of adhesion between the separate phase [34]. In another study, a phase separation was found in the PVA/starch blend at higher glycerol loading (glycerol was used as a plasticizer) which caused inhomogeneity and crystallinity drop [4]. This phenomenon was also observed in this work for the blends with PEG loading above 20 wt%, as explained earlier.

3.1.1. FTIR Analysis. FTIR spectroscopy is one of the most common techniques which provides useful information regarding the interactions between their functional groups. It is also considered a powerful method for studying the conformational changes in biopolymer systems [35]. The FTIR spectra of pure PVA and its blends at various PEG loadings in the wave number range of 4000–500 cm^{-1} are shown in Figure 1(a). Both polymers have good solubility in water. This is because of the formation of strong H-bonds between the functional group of the polymers and water molecules (hydrophilic property). It is clear from the spectra of all PVA/PEG blends that extensive H-bonding exists in the range 3000–3500 cm^{-1} due to stretching vibration of -OH groups which results from strong H-bonds that form during blending. For pure PVA, the spectra show a broad band centered at 3300 cm^{-1} associated with the stretching vibration of hydroxyl groups (–OH) present in the PVA structure. Similar results are reported by others [2, 32, 36]. Moreover, the FTIR spectra for all blends displayed almost similar absorption bands as pure PVA in the entire wavelength range, which indicates that the main functional groups in PEG and PVA are preserved [37]. However, there was a slight shift observed for peak position and relative intensity of the stretching vibration of the hydroxyl group (–OH) for all blends, which is attributed to the H-bonding formation between PVA and PEG at different PEG contents [38].

The characteristic absorption peaks of PVA/PEG blends that were observed at wavelengths 1330, 1421, 1100, 3300, and 851 cm^{-1} correspond respectively to C–O–C, C_6H_6, C–C, -OH, and C–H [39]. A strong absorption peak was also observed at 2900 cm^{-1} and linked to the stretching mode of the CH_2 group [36]. The absorption band at 1650 cm^{-1} was due to C=O stretching of the ester group formed in the PVA polymer during its preparation process. At higher PEG loadings above 10%, this peak disappeared due to the presence of a high amount of PEG. The stretching vibrations of C–C and C–O–C were observed at 1100 and 1330 cm^{-1} in the spectra, and their intensity values were changed at different PEG loadings. From Figure 1(b), the absorption band of -OH groups observed in the range 3000–3500 cm^{-1} got wider and with higher intensity in the spectra of the nanocomposites indicating occurrence of H-bond interactions between the -OH groups present in the blend and the oxygen groups of graphene [1, 9, 23]. However, there is a decrease in the intensity of the -OH band in the spectra of the nanocomposites with higher content of graphene. This may be attributed to the agglomeration of graphene sheets which causes a weakness in H-bond interactions.

3.1.2. Differential Scanning Calorimetry (DSC) Analysis. DSC is a very useful technique used to investigate the thermal properties of polymeric materials. The DSC thermograms obtained for PVA and its blends at different PEG loadings are shown in Figures 2(a) and 2(b). The thermal properties obtained from these thermograms are summarized in Table 2. PEG loading clearly has influence on the thermal properties of PVA/PEG blends.

Upon addition of PEG to the blends, the melting temperatures for all PVA/PEG blends were not much changed as compared to that of pure PVA, indicating that the T_m of PVA was not affected by PEG. From the heating curves shown in Figure 2(a), the measured melting enthalpy of the pure PVA film was 51.35 J/g. Blending PEG with PVA affected the crystallization behavior of PVA, and the crystallinity changed at different PEG loadings. A drop in crystallinity was observed for almost all blends containing PEG. The decrease in the crystallinity was attributed to the reduction of PVA chain mobility and the formation of H-bonding between PEG and PVA during the blending process, which hindered the crystallization step. Similar findings are reported, where a decrease in crystallinity was found for PVA/PEG blends when PEG of 10 wt% was used in the blends [2].

The endothermic peaks observed in the low-temperature region in the DSC heating curves of all PVA/PEG blends shown in Figure 2(a) were possibly associated with the T_g of the blends [40]. In another study, the melting of PEG was assumed to be responsible for these peaks [37]. As can be seen from Table 2, T_g decreased when adding PEG into PEG/PVA blends. Similar results were reported by others indicating the plasticization effect of PEG on PVA [41]. Li et al. [3] reported that effective plasticization for PVA was observed with PEG content <20 wt% as indicated by DSC results. Little difference in T_g values was observed for PVA/PEG blends at all of the different PEG loadings [3]. The cooling curves shown in Figure 2(b) for the PVA/PEG blend with 5 and 10 wt% PEG show two main crystallization events related to the crystallization temperatures of PEG and PVA. Since the difference between the two temperature values is high, it is reported that the crystallization of PEG had no effect on the crystallization behavior of PVA [3]. It was also noticed that the blends containing 15 and 20 wt% of PEG showed another smaller peak at around 25°C which could be due to the presence of incompatibility in these blends containing high PEG content [32].

3.1.3. Thermogravimetric Analysis (TGA). The integral (TGA) and derivative (DTG) thermograms can provide useful information regarding the thermal stability and the degradation behavior of the polymeric materials [24]. The TGA and DTG thermograms of the obtained films of the pure PVA and PVA/PEG blends at different PEG loadings are shown in Figure 3. The figure clearly shows the dependency of the thermal behavior on PEG loading. It was observed that

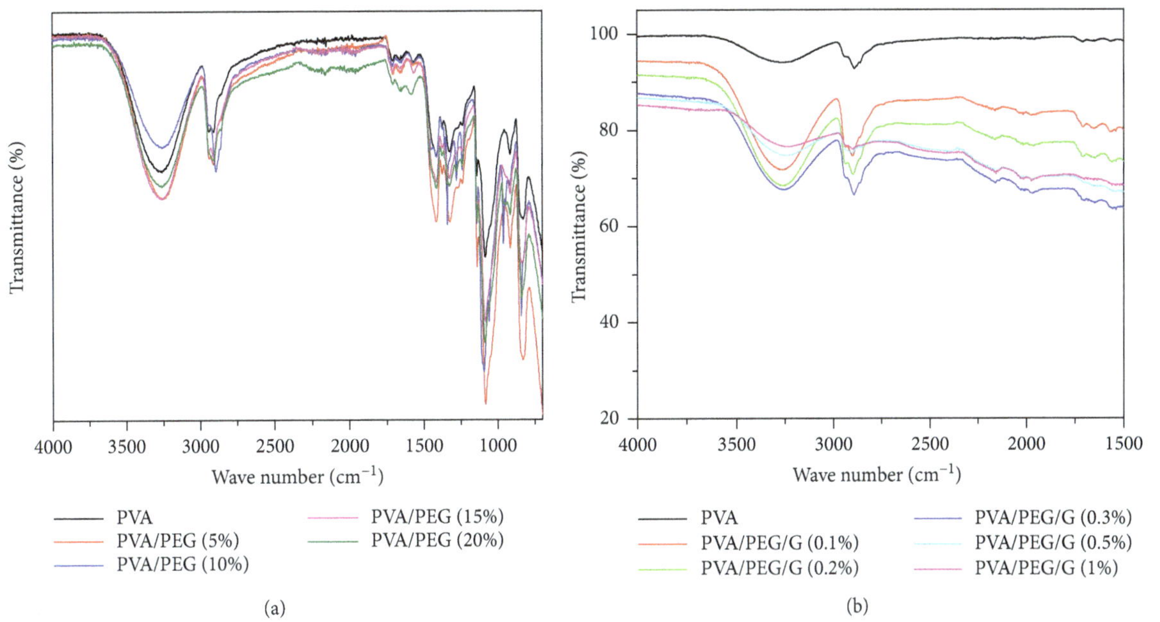

FIGURE 1: FTIR spectra for (a) PVA and its blends at different PEG loadings and (b) PVA/PEG nanocomposites at different graphene loadings.

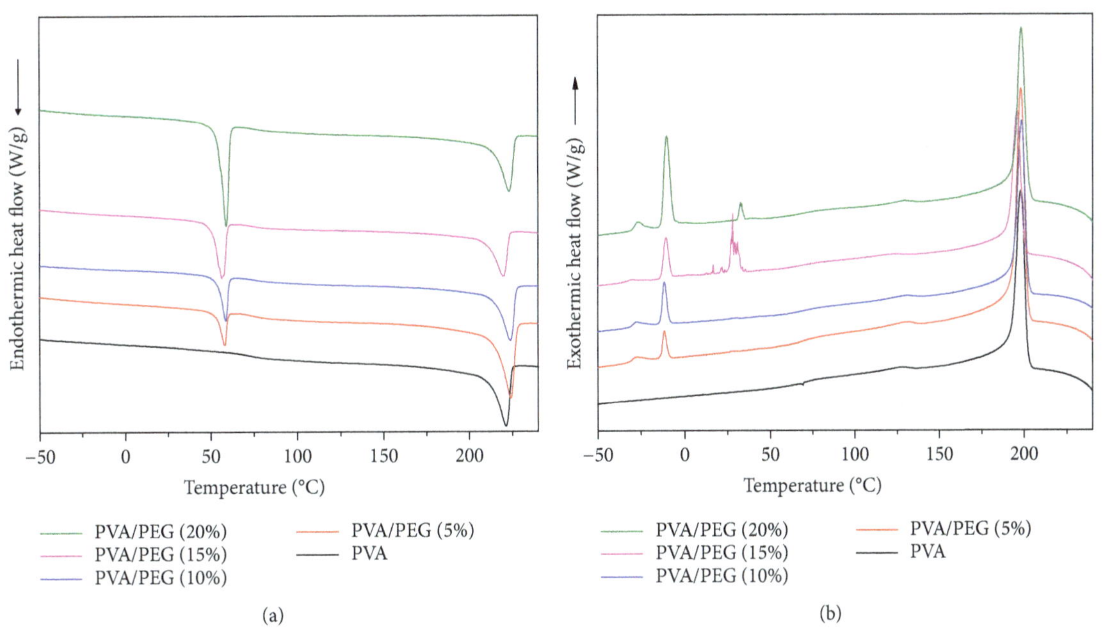

FIGURE 2: DSC thermograms for PVA and its blends at different PEG loadings: (a) heating curves and (b) cooling curves.

most of the blends showed better thermal stability compared to the pure PVA resulting from the shift of the degradation temperatures to higher values as shown in Figure 3.

Various steps can be seen in the TGA curves shown in Figure 3(a) for PVA/PEG blends; for example, the first process occurred up to around 90°C. The first one was related to the loss of physisorbed water [38, 42]. The major degradation of pure PVA was observed in a temperature range of 243–387°C [6]. PEG has higher thermal stability than PVA, and its thermal decomposition begins above 330°C [37, 38]. The products resulting from the thermal degradation of PEG were reported to include aldehydes, ketones, and ethers. On the other hand, the thermal decomposition products from PVA have been found to be water, aldehydes, and ketones up to a temperature of 240°C and above 240°C were alkanes, alkenes, and aromatic or unsaturated hydrocarbons [43, 44].

From the DTG curves shown in Figure 3(b), the maximum weight loss rates and the percent of residue at 550°C for all samples were obtained and are provided in Table 3.

Table 2: DSC data for PVA and PVA/PEG blends.

Sample code	T_m (°C)[a]	ΔH_m (J/g)[b]	T_c (°C)[c]	ΔH_c (J/g)[d]	T_g (°C)[e]	X_c (%)[f]
Pure PVA	211.78	51.35	198.12	84.74	83	36.16
PVA/PEG (5%)	214.43	55.21	197.81	80.88	53.85	38.88
PVA/PEG (10%)	214.79	45.29	198.52	73.94	54.39	31.89
PVA/PEG (15%)	210.15	31.98	195.67	63.16	51.34	22.52
PVA/PEG (20%)	214.21	42.23	198.07	68.22	54.47	29.74

[a]Melting temperature. [b]Melting enthalpy. [c]Crystallization temperature. [d]Enthalpy of crystallization.[e]Glass transition temperature. [f]Degree of crystallinity calculated by using (1).

For pure PVA, two peaks were observed at 280 and 440°C, which is in agreement with other findings [3]. A gradual increase in the peak width was observed with the increase in PEG content which is attributed to the changes in the heterogeneity of the blend composition [5]. It was observed that the weight loss rate decreased with the increase in PEG loading with a maximum value of about 1.7 wt%/°C (reached for the PVA/PEG blend with 5% PEG content). The addition of PEG to the blends formed H-bonds with PVA segments causing the decrease in the weight loss rate [5]. It was also noticed that the blend with 10% PEG showed the lowest residue value compared to other blends. While the other two blends loaded with PEG of 15 and 20% show lower loss rate values, their obtained films tend to be brittle due to phase separation and less compatibility resulting in weak mechanical properties as discussed below.

3.1.4. Mechanical Analysis. Since the blending process is expected to have an influence on the mechanical properties of the blends, a mechanical test was carried out for the pure PVA and its blends. However, it was not possible to perform the mechanical test for the obtained film of the blend containing 20 wt% PEG due to its high brittleness which prevented obtaining the required dog-bone shape of film for testing. However, the obtained films for the pure PVA and remaining blends at PEG loadings of 5, 10, and 15 wt% were successfully tested, and their mechanical properties including the tensile strength and the elongation at break were obtained from their corresponding tensile stress-strain curves as provided in Table 4.

The addition of PEG caused a drop in the tensile strength as compared to that for pure PVA due to the brittle PEG component. This is in agreement with other reported findings [19, 32]. This drop can also be attributed to the plasticizing effect of PEG with the pure PVA matrix [43]. Also, the tensile strength of PVA/PEG blends decreased with the increase in PEG loading. On the contrary, the elongation at break increased with the addition of PEG up to 10 wt% (PEG was used as a plasticizer to PVA). When PEG with moderate molecular weight ($MW = 2000$ and 4000) was used as a plasticizer for PLA/starch blend, the elongation at break was increased for the blend [45].

PVA/PEG blends with 5 and 10 wt% PEG loadings showed better tensile strength and elongation-at-break values compared to that at 15 wt% PEG loading, which can be attributed to the improvement of miscibility [19]. These improved mechanical properties can make the PVA/PEG blend a suitable candidate for industrial applications requiring better bending and strength features [39]. For the blend with 15 wt% PEG, the obtained film was observed to be brittle and showed phase separation due to the high content of PEG. This resulted in a significant drop in its measured mechanical properties including tensile strength and elongation at break.

3.2. Effect of Graphene Loadings on PVA/PEG Nanocomposites. Based on the results obtained and discussed for PVA/PEG blends, it can be concluded that the PEG loading at 10 wt% gives the optimum PVA/PEG blend and this was used for preparation of nanocomposites based on graphene for further studies. For simplicity, the optimum blend and graphene are abbreviated in the next sections as PVA/PEG and G, respectively.

3.2.1. Mechanical Properties. The mechanical properties of the blend and its nanocomposites at different graphene loadings were obtained from the mechanical test curves and are provided in Table 5. It is expected that some of the mechanical properties of the polymer blend can be improved by the addition of graphene as a nanofiller owning to the dispersion of the graphene nanoparticles having a large interfacial area and high aspect ratio into the polymer matrix. The homogenous dispersion of graphene in the polymer matrix is considered a crucial factor to improving the mechanical properties of graphene-based polymer nanocomposites [23]. Table 5 shows the influence of graphene on the mechanical properties of the PVA/PEG blend. Incorporation of graphene especially at low loading levels caused a fall in the tensile strength of the PVA/PEG nanocomposites. It is believed that the graphene nanosheets disperse individually at low graphene loading [1]. Also, it was observed that no significant difference in tensile properties of the PVA/PEG nanocomposites loaded with graphene up to 0.5 wt%. On the contrary, the elongation at break for the PEG/PVA nanocomposites was significantly increased as compared to that of the unfilled polymer blend. For example, the elongation at break increased by 62% from 147% for the blend to 209% for the nanocomposite with graphene loading of 0.2 wt%. Similar findings were reported, where the elongation at break increased by 70% when 0.7 wt% of reduced graphene oxide (RGO) was added to the PVA matrix [46]. However, with further addition of graphene above 0.3 wt%, the elongation at break gradually decreased which could be attributed to the increase in the brittleness of the obtained nanocomposite films at higher content of graphene.

Table 5 shows also the predicated tensile strength for PVA/PEG nanocomposites at different graphene loadings calculated by (2). A good agreement between the experimental and the predicated data was observed up to a graphene loading of 0.5 wt%. However, the increase in graphene loading above this limit caused a deviation between predicated tensile strength and its corresponding experimental data. This indicates the presence of weak interactions between the blend and graphene [47] as reported earlier in FTIR results.

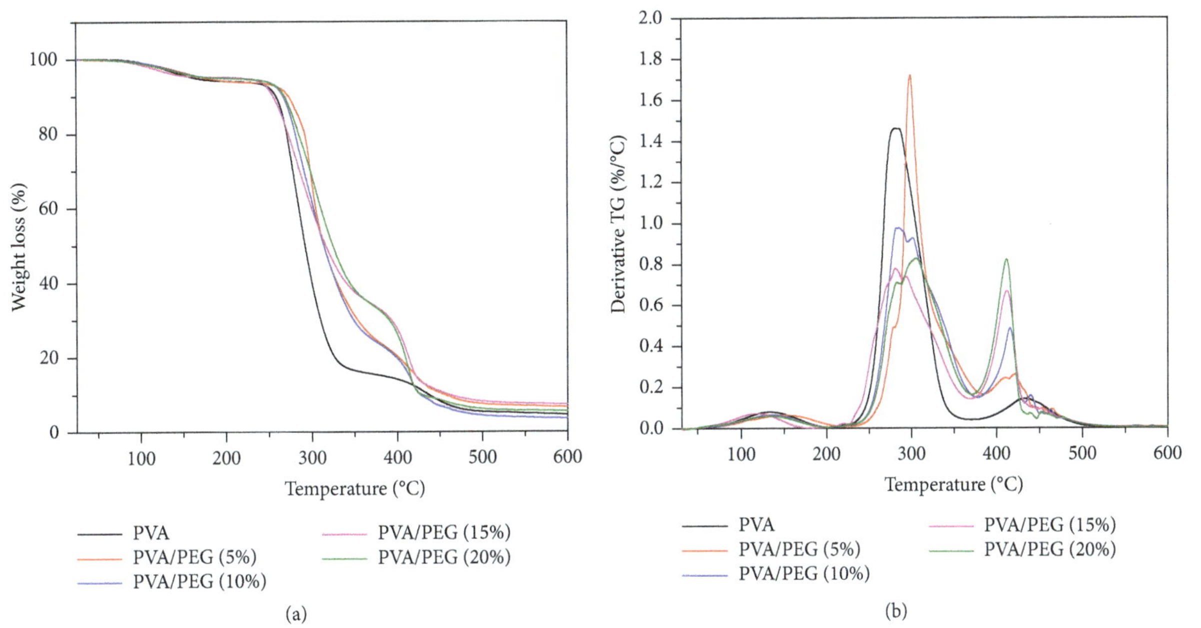

FIGURE 3: Thermograms for PVA and its blends at different PEG loadings: (a) TGA and (b) DTG.

TABLE 3: The maximum weight loss rates of PVA and its blends at different PEG loadings.

Sample code	Maximum weight loss rate (wt%/°C)	Residue at 550°C (%)
Pure PVA	1.46	5
PVA/PEG (5%)	1.7	6.96
PVA/PEG (10%)	0.98	3.88
PVA/PEG (15%)	0.76	7.6
PVA/PEG (20%)	0.82	5.7

TABLE 4: Mechanical properties of PVA and the blends at different PEG loadings.

Sample code	Tensile strength (MPa)	Elongation at break (%)
Pure PVA	45.86	101.99
PVA/PEG (5%)	33.24	172.73
PVA/PEG (10%)	30.67	147.34
PVA/PEG (15%)	27.88	27.89
PVA/PEG (20%)	nd	nd

Note: nd: not determined.

TABLE 5: Mechanical properties of PVA/PEG blend and its nanocomposites at different graphene loadings.

Sample code	Tensile strength (MPa)	Predicated tensile strength (MPa)	Elongation at break (%)
PVA/PEG	30.67	30.67	147.34
PVA/PEG/G (0.1%)	24.68	25.62	190.82
PVA/PEG/G (0.2%)	22.12	22.97	209.8
PVA/PEG/G (0.3%)	23.3	20.58	201.6
PVA/PEG/G (0.5%)	23.20	16.31	179.32
PVA/PEG/G (1%)	28.68	8.15	163.24

3.2.2. Scanning Electron Microscopic (SEM) Studies. The SEM image of the PVA/PEG blend containing no graphene presented in Figure 4(a) shows a homogeneous phase with no occurrence of phase separation indicating the excellent miscibility of PVA and PEG at this blend ratio. There is definite change in surface morphology with the incorporation of graphene into the blend as shown in Figures 4(b) and 4(f), resulting in an increase in the elongation at break for the nanocomposites with graphene loading up to 0.2% compared to the PVA/PEG blend reported previously in mechanical property studies [1]. The obtained films were seen not brittle and can be easily folded indicating higher flexibility and hence higher elongation at break [48]. However, at higher graphene loading, the elongation at break was decreased due to the agglomeration of graphene sheets which was clearly observed at a graphene loading of 1% as shown in Figure 4(f).

4. Conclusions

PVA/PEG blends at different PEG loadings were prepared using a solution casting method. A drop in T_g of the PVA/PEG blends was observed as indicated by DSC studies that revealed PEG has a significant plasticization effect on PVA. Such effect was also noticed in FTIR results which showed that the addition of PEG into the blend formed a hydrogen

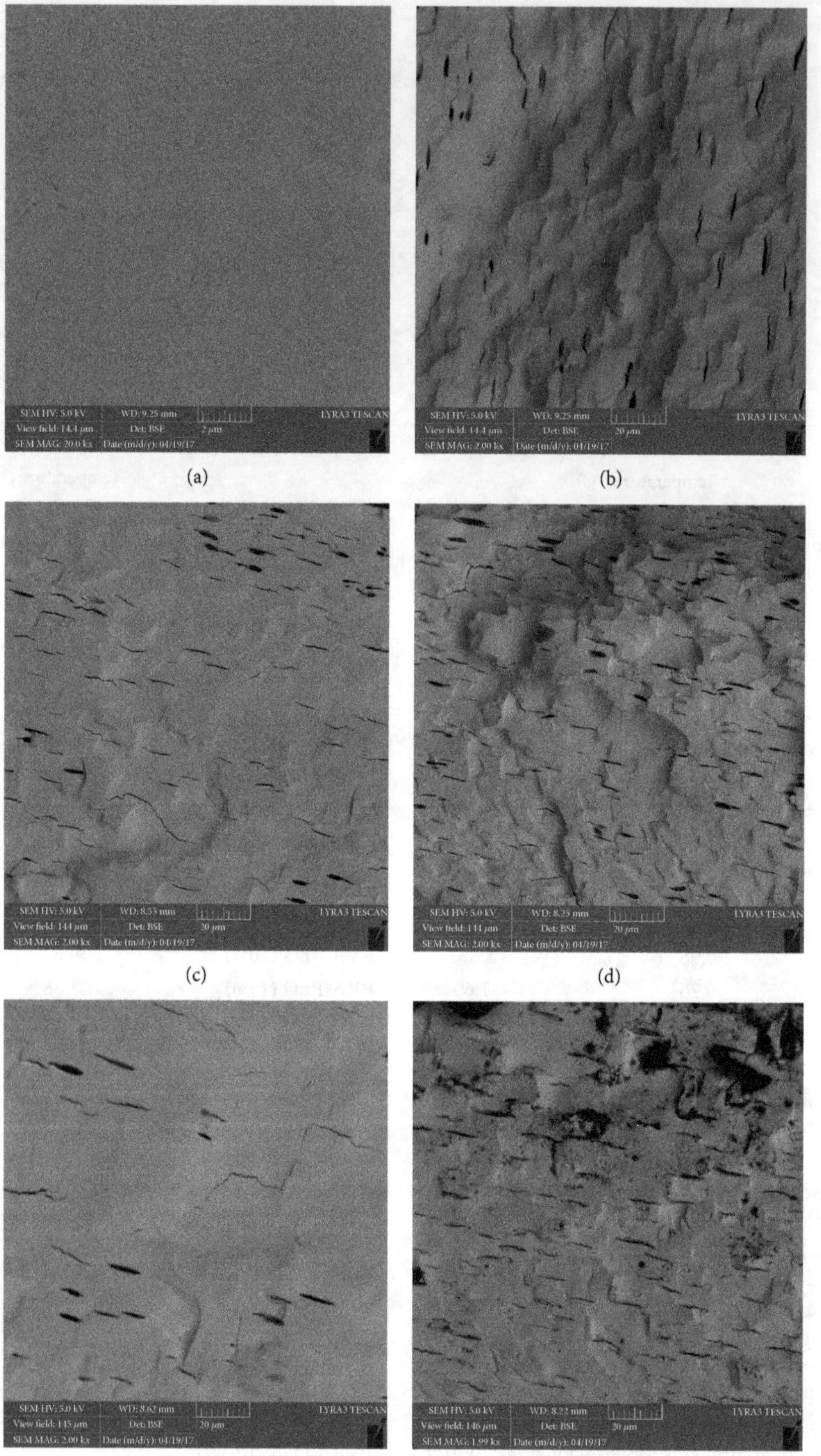

Figure 4: SEM micrographs of PVA/PEG nanocomposites at graphene loadings of (a) 0%, (b) 0.1%, (c) 0.2%, (d) 0.3%, (e) 0.5%, and (f) 1.0%.

bonding between the two polymers. The optimum blend was found at 10 wt% PEG loading, owing to the good compatibility and the improved thermal degradation properties as indicated by SEM and TGA, respectively. The incorporation of 0.2 wt% graphene into the optimum blend increased its elongation at break by 62%. However, not much improvement in the mechanical properties was obtained at higher loading of graphene above 0.3 wt% due to the increase in the brittleness of the obtained nanocomposite films and agglomeration of graphene sheets.

Conflicts of Interest

The authors declare that there is no conflict of interest regarding the publication of this paper.

Acknowledgments

The authors wish to acknowledge the assistance of Deanship of Scientific Research, King Fahd University of Petroleum and Minerals, for their support for providing adequate funds and infrastructure under Project no. FT161010.

References

[1] J. Jose, M. A. Al-Harthi, M. A.-A. AlMa'adeed, J. B. Dakua, and S. K. De, "Effect of graphene loading on thermomechanical properties of poly(vinyl alcohol)/starch blend," *Journal of Applied Polymer Science*, vol. 132, no. 16, 2015.

[2] Z. I. Ali and W. H. Eisa, "Characterization of Electron Beam Irradiated Poly Vinyl Alcohol/Poly Ethylene Glycol Blends," *Journal of Scientific Research*, vol. 6, no. 1, pp. 29–42, 2014.

[3] Y. Li, W. Wu, F. Lin, and A. Xiang, "The interaction between poly(vinyl alcohol) and low-molar-mass poly(ethylene oxide)," *Journal of Applied Polymer Science*, vol. 126, no. 1, pp. 162–168, 2012.

[4] P. A. Sreekumar, M. A. Al-Harthi, and S. K. De, "Effect of glycerol on thermal and mechanical properties of polyvinyl alcohol/starch blends," *Journal of Applied Polymer Science*, vol. 123, no. 1, pp. 135–142, 2012.

[5] P. A. Sreekumar, M. A. Al-Harthi, and S. K. De, "Studies on compatibility of biodegradable starch/polyvinyl alcohol blends," *Polymer Engineering and Science*, vol. 52, no. 10, pp. 2167–2172, 2012.

[6] J. Jose, F. Shehzad, and M. A. Al-Harthi, "Preparation method and physical, mechanical, thermal characterization of poly(vinyl alcohol)/poly(acrylic acid) blends," *Polymer Bulletin*, vol. 71, no. 11, pp. 2787–2802, 2014.

[7] O. A. Bin-Dahman, J. Jose, and M. A. Al-Harithi, "Compatibility of poly(acrylic acid)/starch blends," *Starch - Stärke*, vol. 67, no. 11-12, pp. 1061–1069, 2015.

[8] O. A. Bin-Dahman, J. Jose, and M. A. Al-Harithi, *Starch - Stärke*, vol. 69, no. 7-8, 2017.

[9] J. Jose and M. A. Al-Harthi, "Citric acid crosslinking of poly(-vinyl alcohol)/starch/graphene nanocomposites for superior properties," *Iranian Polymer Journal*, vol. 26, no. 8, pp. 579–587, 2017.

[10] O. Bin-Dahman, M. Rahaman, D. Khastgir, and M. A. Al-Harithi, "Electrical and dielectric properties of poly(vinyl alcohol)/starch/graphene nanocomposites," *The Canadian Journal of Chemical Engineering*, vol. 96, no. 4, pp. 903–911, 2018.

[11] P. A. Sreekumar, M. A. Al-Harthi, and S. K. De, "Reinforcement of starch/polyvinyl alcohol blend using nano-titanium dioxide," *Journal of Composite Materials*, vol. 46, no. 25, pp. 3181–3187, 2012.

[12] S. P. Appu, S. K. De, M. J. Khan, and M. A. Al-Harthi, "Natural weather ageing of starch/polyvinyl alcohol blend: effect of glycerol content," *Journal of Polymer Engineering*, vol. 33, no. 3, pp. 257–263, 2013.

[13] P. A. Sreekumar, M. A. Al-Harthi, M. A. Gondal, and S. K. De, "Heterogeneity of laser-irradiated films of polyvinyl alcohol/starch blends: effect of glycerol content," *Surface and Interface Analysis*, vol. 45, no. 6, pp. 1047–1051, 2013.

[14] J. Jose, S. K. de, M. A. A. AlMa'adeed et al., "Compatibilizing role of carbon nanotubes in poly(vinyl alcohol)/starch blend," *Starch - Stärke*, vol. 67, no. 1-2, pp. 147–153, 2015.

[15] O. A. Bin-Dahman, F. Shehzad, and M. A. al-Harthi, "Influence of graphene on the non-isothermal crystallization kinetics of poly(vinyl alcohol)/starch composite," *Journal of Polymer Research*, vol. 25, no. 1, 2018.

[16] C. K. Chan and I. M. Chu, *Biomaterials*, vol. 23, no. 11, pp. 2353–2358, 2002.

[17] G. Swift, "Water-soluble polymers," *Polymer Degradation and Stability*, vol. 45, no. 2, pp. 215–231, 1994.

[18] R. J. Sengwa, S. Choudhary, and S. Sankhla, "Dielectric properties of montmorillonite clay filled poly(vinyl alcohol)/poly(-ethylene oxide) blend nanocomposites," *Composites Science and Technology*, vol. 70, no. 11, pp. 1621–1627, 2010.

[19] K. Abdel Tawab, M. M. Magida, and S. M. Ibrahim, "Effect of ionizing radiation on the morphological, thermal and mechanical properties of polyvinyl alcohol/polyethylene glycol blends," *Journal of Polymers and the Environment*, vol. 19, no. 2, pp. 440–446, 2011.

[20] R. J. Sengwa and S. Choudhary, "Structural characterization of hydrophilic polymer blends/montmorillonite clay nanocomposites," *Journal of Applied Polymer Science*, vol. 131, no. 16, 2014.

[21] A. M. El Sayed and W. M. Morsi, "Dielectric relaxation and optical properties of polyvinyl chloride/lead monoxide nanocomposites," *Polymer Composites*, vol. 34, no. 12, pp. 2031–2039, 2013.

[22] Z. Guo, D. Zhang, S. Wei et al., "Effects of iron oxide nanoparticles on polyvinyl alcohol: interfacial layer and bulk nanocomposites thin film," *Journal of Nanoparticle Research*, vol. 12, no. 7, pp. 2415–2426, 2010.

[23] X. Yuan, "Enhanced interfacial interaction for effective reinforcement of poly(vinyl alcohol) nanocomposites at low loading of graphene," *Polymer Bulletin*, vol. 67, no. 9, pp. 1785–1797, 2011.

[24] B. W. Chieng, N. A. Ibrahim, W. M. Z. Wan Yunus, and M. Z. Hussein, "Plasticized poly(lactic acid) with low molecular weight poly(ethylene glycol): mechanical, thermal, and morphology properties," *Journal of Applied Polymer Science*, vol. 130, 2013.

[25] A. M. El Sayed and W. M. Morsi, "α-Fe2O3 /(PVA + PEG) nanocomposite films; synthesis, optical, and dielectric characterizations," *Journal of Materials Science*, vol. 49, no. 15, pp. 5378–5387, 2014.

[26] S. G. Abd Alla, H. M. Nizam El-Din, and A. W. M. El-Naggar, "Electron beam synthesis and characterization of poly(vinyl alcohol)/montmorillonite nanocomposites," *Journal of Applied Polymer Science*, vol. 102, no. 2, pp. 1129–1138, 2006.

[27] J. Cheng, J. Zhang, and X. Wang, "Investigation on crystallization behavior and hydrophilicity of poly(vinylidene fluoride)/poly(methyl methacrylate)/poly(vinyl pyrrolidone) ternary blends by solution casting," *Journal of Applied Polymer Science*, vol. 127, no. 5, pp. 3997–4005, 2013.

[28] Y. Nishio, T. Haratani, T. Takahashi, and R. S. J. Manley, "Cellulose/poly(vinyl alcohol) blends: an estimation of ther-

modynamic polymer-polymer interaction by melting-point-depression analysis," *Macromolecules*, vol. 22, no. 5, pp. 2547–2549, 1989.

[29] D. M. Bigg, "Mechanical properties of particulate filled polymers," *Polymer Composites*, vol. 8, no. 2, pp. 115–122, 1987.

[30] K. Ghosh and S. N. Maiti, "Mechanical properties of silver-powder-filled polypropylene composites," *Journal of Applied Polymer Science*, vol. 60, no. 3, pp. 323–331, 1996.

[31] L. Nicolais and M. Narkis, "Stress-strain behavior of styrene-acrylonitrile/glass bead composites in the glassy region," *Polymer Engineering and Science*, vol. 11, no. 3, pp. 194–199, 1971.

[32] S. G. Abd Alla, H. M. Said, M. El-Naggar, and A. W. M. El-Naggar, "Structural properties of γ-irradiated poly(vinyl alcohol)/poly(ethylene glycol) polymer blends," *Journal of Applied Polymer Science*, vol. 94, no. 1, pp. 167–176, 2004.

[33] C. S. Reddy, P. K. Babu, K. Sudhakar et al., "Miscibility studies of hydroxyethyl cellulose and poly (ethylene glycol) polymer blends," *Journal of Polymer Research*, vol. 7, pp. 253–266, 2013.

[34] A. K. Mohaparta, S. Mohanty, and S. K. Nayak, "Effect of PEG on PLA/PEG blend and its nanocomposites: a study of thermo-mechanical and morphological characterization," *Polymer Composites*, vol. 35, no. 2, pp. 283–293, 2014.

[35] B. J. Goodfellow and R. H. Wilson, "A fourier transform IR study of the gelation of amylose and amylopectin," *Biopolymers*, vol. 30, no. 13-14, pp. 1183–1189, 1990.

[36] T. Rajavardhana Rao, I. Omkaram, B. Sumalatha, K. Veera Brahmam, and C. Linga Raju, "Electron paramagnetic resonance and optical absorption studies of manganese ions doped in polyvinyl(alcohol) complexed with polyethylene glycol polymer films," *Ionics*, vol. 18, no. 7, pp. 695–701, 2012.

[37] C. Wang, L. Feng, H. Yang et al., "Graphene oxide stabilized polyethylene glycol for heat storage," *Physical Chemistry Chemical Physics*, vol. 14, no. 38, pp. 13233–13238, 2012.

[38] M. Abu Ghalia and Y. Dahman, "Radiation crosslinking polymerization of poly (vinyl alcohol) and poly (ethylene glycol) with controlled drug release," *Journal of Polymer Research*, vol. 22, no. 11, p. 218, 2015.

[39] S. A. Nouh, A. G. Nagla, M. H. Othman, S. A. Eman, and Z. I. Lotfi, *Journal of Applied Polymer Science*, vol. 124, no. 1, pp. 654–660, 2012.

[40] L. Sperling, *Introduction to Physical Polymer Science*, Wiley, New York, 1986, Chapter 4–6.

[41] P. Sakellariou, A. Hassan, and R. C. Rowe, "Interactions and partitioning of diluents/plasticizers in hydroxypropyl methylcellulose and polyvinyl alcohol homopolymers and blends. Part II: Glycerol," *Colloid & Polymer Science*, vol. 272, no. 1, pp. 48–56, 1994.

[42] V. D. Monopoli, L. R. Pizzio, and M. N. Blanco, "Polyvinyl alcohol–polyethylenglycol blends with tungstophosphoric acid addition: Synthesis and characterization," *Materials Chemistry and Physics*, vol. 108, no. 2-3, pp. 331–336, 2008.

[43] Y. Tsuchiya and K. Sumi, "Thermal decomposition products of poly(vinyl alcohol)," *Journal of Polymer Science Part A-1: Polymer Chemistry*, vol. 7, no. 11, pp. 3151–3158, 1969.

[44] A. Ballistreri, S. Foti, G. Montaudo, and E. Scamporrino, "Evolution of aromatic compounds in the thermal decomposition of vinyl polymers," *Journal of Polymer Science: Polymer Chemistry Edition*, vol. 18, no. 4, pp. 1147–1153, 1980.

[45] Y. Yu, Y. Cheng, J. Ren, E. Cao, X. Fu, and W. Guo, "Plasticizing effect of poly(ethylene glycol)s with different molecular weights in poly(lactic acid)/starch blends," *Journal of Applied Polymer Science*, vol. 132, no. 16, 2015.

[46] S. Lee, H. Jin-Yong, and J. Jang, "The effect of graphene nanofiller on the crystallization behavior and mechanical properties of poly(vinyl alcohol)," *Polymer International*, vol. 62, no. 6, pp. 901–908, 2013.

[47] N. C. Bleach, S. N. Nazhat, and K. E. TannerM. Kellomaki and P. T. Tormala, "Effect of filler content on mechanical and dynamic mechanical properties of particulate biphasic calcium phosphate—polylactide composites," *Biomaterials*, vol. 23, no. 7, pp. 1579–1585, 2002.

[48] K. Deshmukh, M. B. Basheer Ahamed, K. K. Sadasivuni et al., "Graphene oxide reinforced polyvinyl alcohol/polyethylene glycol blend composites as high-performance dielectric material," *Journal of Polymer Research*, vol. 23, no. 8, p. 159, 2016.

20

Synthesis, Characterization, and Aqueous Properties of an Amphiphilic Terpolymer with a Novel Nonionic Surfmer

Qin Yu,[1] Xiaolin Wu,[2] Yanqiu Li,[1] Tenglong Gao,[1] Siliang Liu,[1] Caiqing Hou,[1] and Zirong Zheng[1]

[1]*Department of Chemistry and Chemical Engineering, Northeast Petroleum University, 163318, China*
[2]*Research Institute of Exploration and Exploitation, PetroChina Daqing Oilfield Company, 163318, China*

Correspondence should be addressed to Qin Yu; y187459762379@163.com

Academic Editor: Nobuhiro Kawatsuki

A series of surfactive amphiphilic polymer PAADs were prepared from the copolymerization of sodium acrylate, dodecyl polyoxyethylene acrylate (DPA, a surfmer), and acrylamide under the action of a mixed initiating agent consisting of ammonium persulfate-sodium bisulfite/2,2′-azobis[2-(2-imidazolin-2-yl) propane] dihydrochloride. The aggregative behaviors of PAADs were explored by ^{13}C nuclear magnetic resonance, a viscometer, and a surface tension instrument. It was found that the apparent viscosity and surface activity of PAADs were significantly improved by the increase of average sequence length of hydrophobic micro blocks, strong intermolecular hydrophobic association, or the formation of mixed micelles between hydrophobic micro blocks and micromolecular surface-active agent. The introduction of long-chain alkyls on molecular chains prolonged the average sequence length of hydrophobic micro blocks in molecular chains and enhanced the hydrophobic association between molecular chains and the tight arrangement of molecular chains on water surfaces, thereby increasing the surface activity. Moreover, the anionic monomer sodium acrylate on molecular chains, via electrostatic repulsion, promoted the conversion from intrachain association to intermolecular association and thereby facilitated the formation of dense interfacial films, enhancing the surface activity of water solutions. Then, the anion surfmer sodium dodecylbenzenesulfonate interacted with the hydrophobic micro blocks on the molecular chains to form mixed micelles, which accelerated the interchain association and enhanced the polymer surface activity. The novel polymeric micelle with higher viscosifying ability and surface activity was expected to be a promising oil drive agent for tertiary oil recovery.

1. Introduction

Amphiphilic water-soluble polymer molecules contain incompatible hydrophilic chain segments and hydrophobic chain segments. The higher viscosifying ability [1] and lower molecular surface activity [2] of polymers are mainly induced by the intramolecular association and intermolecular association generated by weak intermolecular forces, such as hydrophobic interaction of hydrophobic groups in aqueous solutions [3]. Especially, the hydrophobic association polymers containing long-chain hydrophobic alkyls are prone to generating intermolecular association and significant surface activity under low concentrations [4]. Thus, amphiphilic water-soluble polymers can be widely used as thermal-resistant and antisalt oil displacement agents in tertiary oil recovery [5–7], watercraft painting polymer dispersing agents [8], drug-controlled release [9–11], and other technical fields [12]. In the past decade, such polymers have received wide attention from both the academia and industry fields.

The properties of hydrophobically associated polymers in water solutions are closely related to their molecular structures, especially the structures, concentrations, and distribution of hydrophobic groups. Generally, a hydrophobic monomer, such as N-alkyl acrylamide (C4-12) [13, 14], N,N-dialkylacrylamide (C 4-12) [15, 16], styrene derivatives [17], or F-containing acrylate [18–20], was used to react with anionic monomer (e.g., acrylamide, sodium acrylate, or 2-methyl-2-acrylamide propyl sulfonic sodium) via ternary copolymerization to form a polymer aqueous solution with

Table 1: Reaction conditions of PAAD.

Sample	DPA (mol%)	NaAA (mol%)	SDS (wt%)	Yield (%)	Intrinsic viscosity (dl/dg)
PAAD-R-1	0	15	3	98.67	6.54
PAAD-R-2	0.1	15	3	97.88	9.36
PAAD-R-3	0.3	15	3	97.24	12.85
PAAD-R-4	0.5	15	3	98.82	18.46
PAAD-R-5	0.7	15	3	98.54	14.12
PAAD-R-6	0.9	15	3	97.17	13.84
PAAD-ID-1	0.5	0	3	98.01	13.02
PAAD-ID-2	0.5	10	3	98.15	13.76
PAAD-ID-3	0.5	15	3	98.82	18.46
PAAD-ID-4	0.5	20	3	98.80	20.32
PAAD-ID-5	0.5	25	3	97.67	19.90
PAAD-S-1	0.5	15	0	98.69	16.16
PAAD-S-2	0.5	15	1	98.50	16.53
PAAD-S-3	0.5	15	2	99.50	16.94
PAAD-S-4	0.5	15	3	98.82	18.46
PAAD-S-5	0.5	15	4	98.70	16.69

high surface activity and viscosifying ability. Hill et al. found that when the concentrations of hydrophobic groups were the same, the hydrophobically associated polymers with hydrophobic micro blocks outperformed irregular hydrophobic association polymers in hydrophobic association [13]. Polymers with hydrophobic micro blocks are conventionally prepared via micelle polymerization. Specifically, researchers alter the ratio of hydrophobic monomers to micromolecular surfactant and adjust the number of hydrophobic monomer molecules in each micelle before copolymerization, so as to control the length of hydrophobic monomer sequences (namely, the length of hydrophobic micro blocks) on the macromolecular chains and thereby to study the effects of macromolecular chain microstructures on hydrophobic association. Though long-chain alkyls or alkyl aryl compounds as hydrophobic components can improve the hydrophobic association performance, they also reduce the water solubility of copolymers. For instance, Xue et al. found that the structures of hydrophobic components largely affected the hydrophobic association substances and their surface activity [21]. Moreover, at the same concentration (1 wt%), the dodecyl acrylamide/AM copolymer was water-insoluble, the viscosity of bihexyl acrylamide/AM copolymer was 30000 mPa s, but that of methyl-hexyl acrylamide/AM copolymer was only 10 mPa s [21]. It was indicated that the hydrophobic association ability of hydrophobic components was enhanced with the prolonging of chain length, but at the cost of polymer insolubility.

This problem was not overcome until the appearance of polymerizable surface-active monomers, which provided a new route for preparation of polymers with hydrophobic micro block structures. Generally, introducing a surfmer monomer could enhance the water solubility of the copolymer and endow the copolymer with polyelectrolyte properties. The commonly used surfactive macromolecular monomers include anionic monomers (e.g., dialkyl maleate cationic [22] and 2-acrylamide long-chain alkyl sulfonic acid [23]), cationic monomers (e.g., (2-methacry-loyloxyethyl) dodecyldimethyl ammonium bromide [24] and (4-(2-(acryloyloxy) ethoxy) benzyl triethyl ammonium bromide) [25]), and F-containing monomers (e.g., fluorocarbon cationic surfmer [26, 27]).

Dodecyl polyoxyethylene acrylate (DPA) is a nonionic polymerizable surfmer monomer, and its water solution even has high dissolving ability and low surface tension under high mineralization condition. In this study, a series of novel amphiphilic water-soluble polymers were synthesized from the copolymerization of DPA, carboxylic acid monomer, and acrylamide. The hydrophobic association performances of the polymers were investigated by a viscometer and a surface tension instrument. Then, we studied how the dose of the surfmer monomer, the dose of the ionic monomer, and the dose of sodium dodecyl sulfonate (SDS) would affect the aggregation in the polymer solutions, as well as the relationship between polymer structures and solution performances.

2. Experimental

2.1. Materials. DPA (chemical grade) was bought from Zouping Mingxing Chemical Co. Ltd. (Shandong, China). Acrylamide (AM), sodium acrylate (NaAA), ethylene diamine tetra-acetic acid (EDTA), urea, ammonium persulfate (APS), sodium bisulfite (NaHSO3), and 2,2′-azobis[2-(2-imidazolin-2-yl)propane] dihydrochloride (AIBA.2HCl) (all analytical grade) were purchased from Tianjin Damao Chemical Reagent Factory (China). Anhydrous ethanol (analytical grade) was bought from Liaoning Quanrui Reagent Co. Ltd. (China).

2.2. Polymerization. The preparation of polymers P(AM/NaAA/DPA) is illustrated in Table 1. Specifically, into a jar containing 100 ml of distilled water, solutions containing different proportions of AM, NaAA, and DPA were added. Then, 2.6% urea and 50 mg/l EDTA were added and mixed well. Then, the initiating agent was added at 25°C, followed by nitrogen ventilation for 30 min. After that, the jar was sealed with rubber and placed in a water bath at 45 ± 1°C, followed by reaction for 12 h. The final conversion rate was above 97%. The specific reactions are illustrated in Figure 1.

2.3. Characterization

2.3.1. Measurement of PAAD Surfactivity. Surface tension (γ) of polymer solutions was measured by a ring method (Shanghai Zhongchen Digital Technology Equipment Co. Ltd., China) at room temperature. In volumetric flasks, a series of polymer solutions at different concentrations were prepared.

2.3.2. Measurement of the Apparent Viscosity of PAAD Polymer. The apparent viscosity was measured by a DV-II viscometer (Brookfield, America) at 45°C and cut speed of 0.6 s^{-1}.

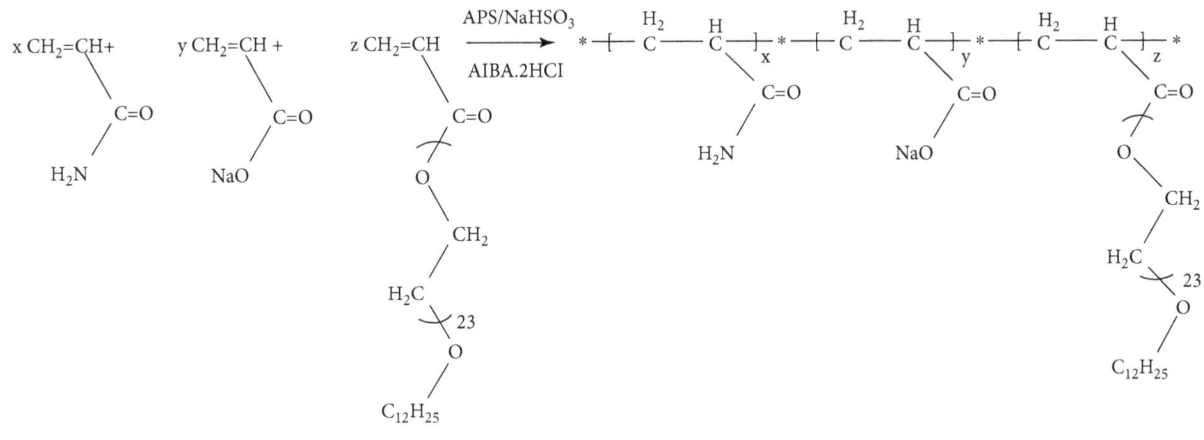

FIGURE 1: Equations of polymer synthesis.

2.3.3. Structural Characterization of PAAD Polymer. ^{13}C NMR spectra in D_2O and solid-state ^{13}C NMR were measured on a Bruker Avance III spectrometer 500 Mhz (Bruker, Swiss) and a Bruker Avance 400 spectrometer 400 Mhz (Bruker, Swiss) at constant temperature (25°C), respectively.

The mass fractions of APS, $NaHSO_3$, and AIBA. 2HCl were 0.36%, 0.36%, and 0.18%, respectively The mass fractions of monomers were all 30%

3. Results and Discussion

3.1. Nuclear Magnetic Characterization. Figure 2 shows the ^{13}C NMR images of monomers and their copolymers. As showed in Figure 2(a), the peaks at 171.95, 136.29 and 126.64, 69.62, and 29.97–17.41 ppm are characteristic of COO, the unsaturated -CH_2=CH-, alkoxy, and saturated alkyl, respectively. As showed in Figure 2(b), the characteristic peak of -CH_2=CH- (at 136.29 and 126.64 ppm) disappeared, but instead, characteristic peaks of COONa (sharp peak at 180 ppm), $CONH_2$ (sharp peak at 180 ppm), and carbon atom on PAM macromolecules at 42.41 ppm appeared. It is confirmed that acrylamide, NPAAD, and DPA have copolymerized.

3.2. Effect of Surfmer DPA Concentration on Intrinsic Viscosity. The intrinsic viscosity of polymers was enhanced by the increasing dose of DPA (Table 1). This was mainly because the hydrophilic and hydrophobic groups of DPA could induce hydrophobic association in solutions, forming dynamic network structures, which enlarged the macromolecular hydrodynamic volume. Thus, the intrinsic viscosity and molecular weight of polymers were increased with the increasing dose of DPA.

3.3. Effect of NaAA Concentration on Intrinsic Viscosity. The molecular weight of polymers significantly increased with the rise in dose of the ion monomer, but declined when the acrylic dose exceeded 20%. This was due to the effect of the electrostatic repulsion on the hydrodynamic volume. When sodium acrylate in water solutions was ionized into -COO-, the electrostatic repulsion between ion groups in the macromolecular chain was gradually intensified and the interchain hydrodynamic volume and molecular weight increased accordingly with the rise of ion group dose. However, the excessive addition of -COO- led to a significant salt effect, and the cations in water solution shielded the static repulsion between anions, which thereby decreased the interchain hydrodynamic volume and intrinsic viscosity.

3.4. Effects of SDS Concentration on Intrinsic Viscosity. The intrinsic viscosity of polymers was first enhanced and then reduced by the increasing dose of SDS (Table 1). This was mainly because SDS could associate with the hydrophobic blocks of molecular chains and thereby accelerate the transition from intramolecular association to intermolecular association, which enlarged the hydrodynamic volume of macromolecular chains. However, excessive addition of SDS would shorten the hydrophobic micro blocks of mixed micelles and thereby weaken the intermolecular association and the macromolecular hydrodynamic volume.

3.5. Effect of DPA Concentration on Apparent Viscosity. The apparent viscosity of PAAD is gradually enhanced with the rise of PAAD concentration (Figure 3). The curves of PAAD-R-3, PAAD-R-4, PAAD-R-5, and PAAD-R-6 all show typical behaviors of hydrophobically associated polymer solutions and all exhibit a turning point of concentration, which is called the critical association concentration (CAC). The viscosity is intensified slowly when the concentration is below CAC, but abruptly when the concentration is above CAC. This phenomenon is related to the intramolecular and intermolecular hydrophobic associations of hydrophobic groups [28]. When the PAAD concentration is low, the hydrophobic groups are dominated by intramolecular association, which leads to molecular chain shrinkage, a decrease of hydrodynamic dimensions, and thereby a decline of apparent viscosity. When the PAAD concentration is above CAC, the intermolecular hydrophobic association is intensified, which is contributive to intermolecular association and formation of reversible physically crosslinked network structures, which result in the enlargement of hydrodynamic dimensions and apparent viscosity.

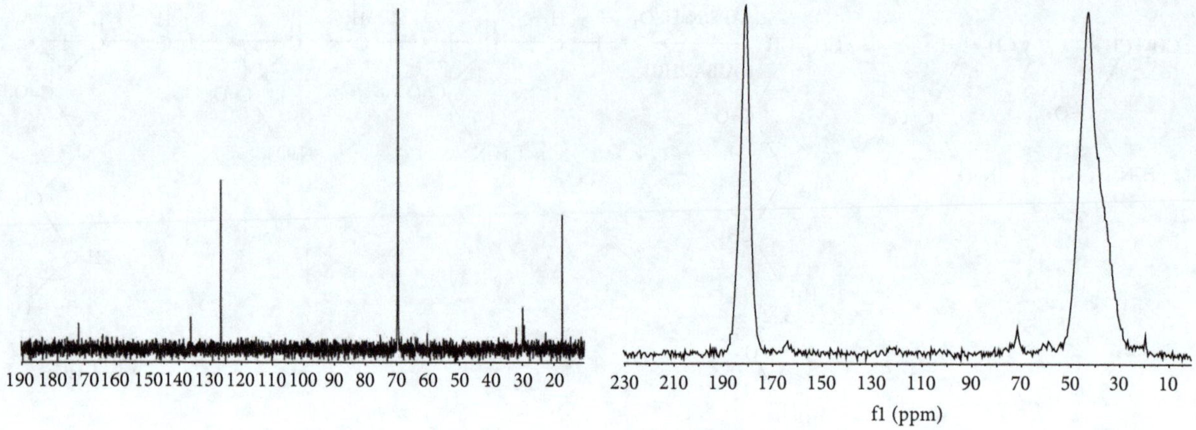

(a) (b)

FIGURE 2: ^{13}C NMR spectra of DPA in D_2O and solid-state ^{13}C NMR of PAAD at 25°C: (a) DPA and (b) PAAD.

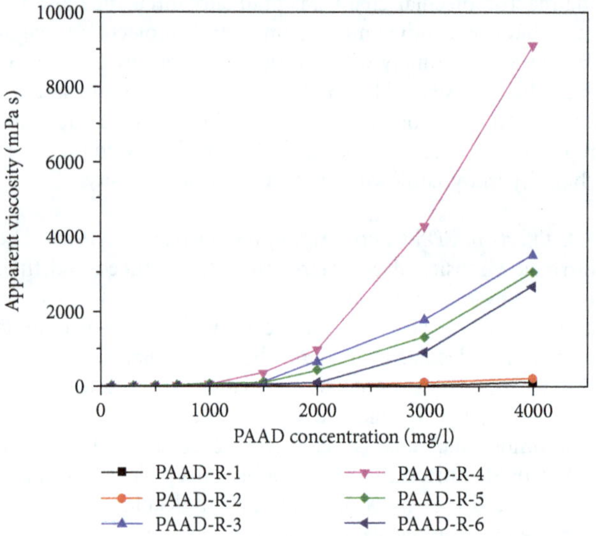

FIGURE 3: Variation of the apparent viscosity with concentration of PAAD at the shear rate of $0.6\,s^{-1}$ in water with the NaCl concentration of 1000 mg/l at 25°C.

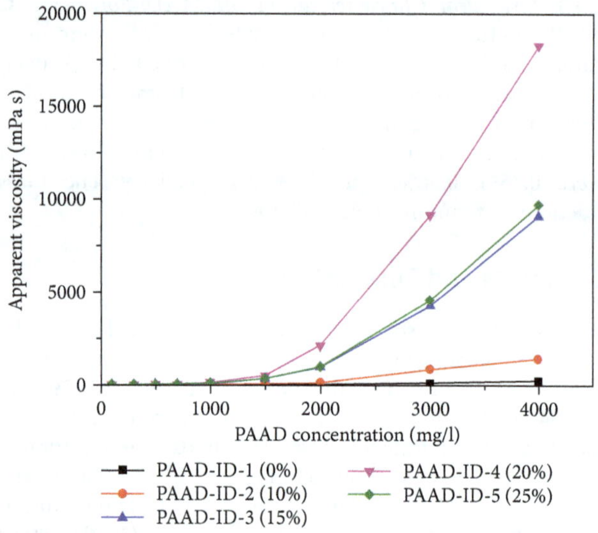

FIGURE 4: Variation of the apparent viscosity with concentration of PAAD at the shear rate of $0.6\,s^{-1}$ in water with the NaCl concentration of 1000 mg/l at 25°C.

However, apparent viscosity of PAAD declined when the DPA dose exceeded 0.5%. This was due to intramolecular association. With the monomer weight ratio, the dose of the surfmer was gradually increased to the number of micelle aggregates before copolymerization, and the length of hydrophobic micro blocks of the corresponding copolymer was prolonged after copolymerization [29]. The behaviors promoted to intermolecular association. However, too long hydrophobic micro blocks of molecular chains easily turned the amphiphilic polymers from intermolecular association to intramolecular association, leading to a reduction of viscosity.

3.6. Effect of NaAA Concentration on Apparent Viscosity. The apparent viscosity of PAAD solutions was first enhanced and then reduced with the rise of NaAA concentration; at low dose of NaAA, no CAC was observed on the curve (Figure 4). When the dose of NaAA exceeded 10%, the apparent viscosity was significantly intensified, which was a typical behavior of hydrophobically associated polymer solution. This was because at low dose of acrylic acid, the intrachain association of polymers was significant, which reduced the apparent viscosity of polymers. With the rise of NaAA dose, the static repulsion between carboxylic acid anions on macromolecular chains was significant, which accelerated the transfer from intramolecular to intermolecular association and thereby enhanced the apparent viscosity of polymer solutions and reduced their CACs [30]. However, the salt effect could shield charge and weaken intermolecular association when the dose of NaAA exceeded 20%, which led to the reducing of apparent viscosity.

3.7. Effects of SDS Concentration on Apparent Viscosity. The apparent viscosity of polymers was first enhanced and then

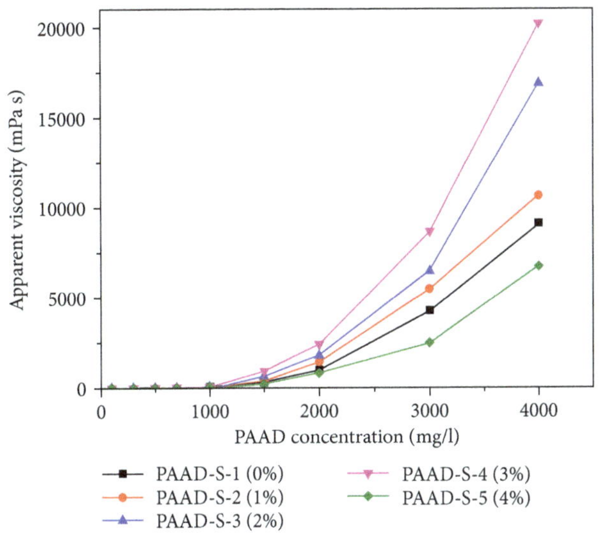

Figure 5: Variation of the apparent viscosity with concentration of PAAD at the shear rate of $0.6\,s^{-1}$ in water with the NaCl concentration of 1000 mg/l at 25°C.

reduced by the increasing surfmer dose (Figure 5). This was related to the interaction between SDS and macromolecules as well as the aggregation shapes. In water solutions, SDS and surfmer associated to form mixed micelles. With the rise of SDS dose, the number of mixed micelles was increased and the intermolecular associated aggregates were enlarged, which led to the forming of tighter space network structures in the system and the large enhancement of apparent viscosity of polymer solutions accordingly [31]. However, when the SDS dose was excessive, the number of hydrophobic side chains contained in one aggregate was less than 2; both intrachain and interchain associations were broken and the system viscosity rapidly declined.

3.8. Effect of Surfmer Concentration on Surface Tension. The surface tension of polymers was gradually reduced and then stabilized with the rise of polymer concentration; thus, a turning point appears on each curve, namely, the critical micelle concentration (CMC) of water solutions. When the polymer concentration was equal to CMC, the surface tension was significantly reduced, minimized, and then stabilized despite further addition of polymers. This was because when the polymer concentration was low, the molecules tended to arrange on the solution surfaces; when the polymer concentration reached a certain level, the newly added polymers, under the action of hydrophobic effect, formed molecular aggregates inside the solution, while the surface polymer molecules were already full. Namely, the surfmer concentration on the solution surfaces was basically stable, so the surface tension was generally constant. The CMCs of PAAD were gradually reduced with the increasing of surfmer (Figure 6).

The surface tension of PAAD on the CMCs was firstly reduced and then increased with the increasing of DPA. The behaviors were related to the length and ratio of hydrophobic blocks. When the dose of DPA was 0.5%, the surface

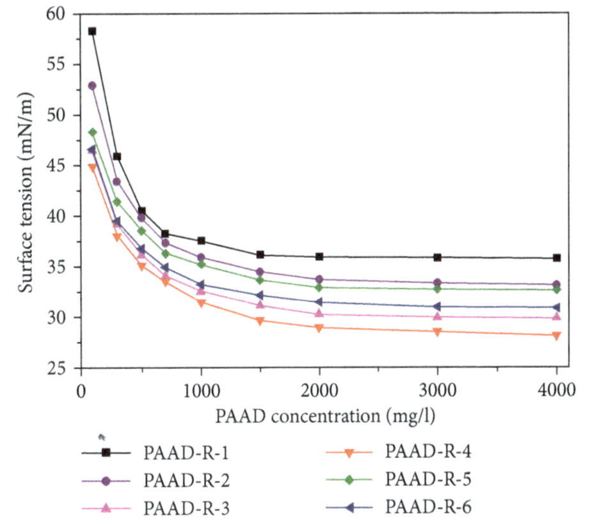

Figure 6: Variation of the surface tension with concentration of PAAD in water with the NaCl concentration of 1000 mg/l at 25°C.

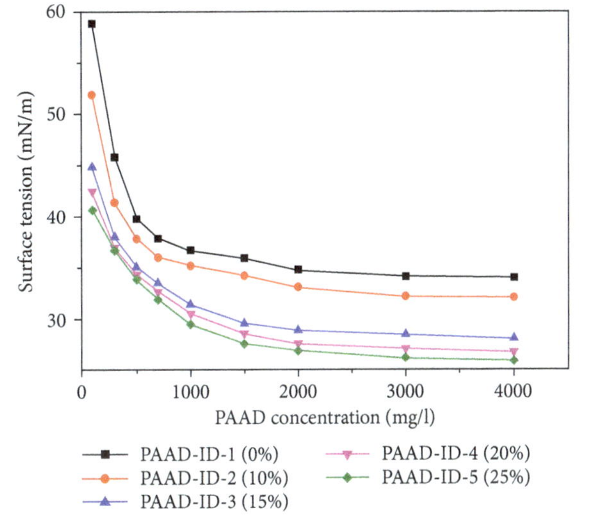

Figure 7: Variation of the surface tension with concentration of PAAD in water with the NaCl concentration of 1000 mg/l at 25°C.

tension of PAAD-R-4 minimized to 28.1 mN/m. It was indicated appropriate dose of surfmer contributed to increasing the length and ratio of hydrophobic blocks and the hydrophobic association ability, tightening the water surface arrangement of polymer molecules and reducing surface tension. However, excessive DPA resulted in the forming of the long hydrophobic blocks, which were prone to intramolecular association to form monomolecular micelles.

3.9. Effect of NaAA Concentration on Surface Tension. The surface tension of polymer solutions first declined with the increase of NaAA dose (Figure 7). The behavior was jointly induced by electrostatic repulsion and hydrophobic interaction. This was because the polymer was gradually transited from intramolecular association to intermolecular association with the rise of NaAA dose, under electrostatic

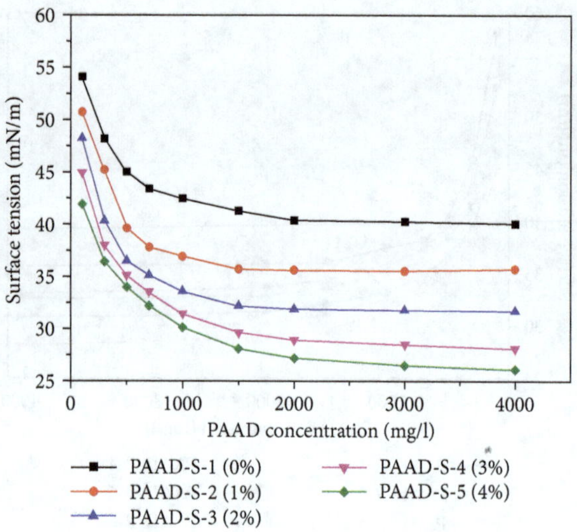

FIGURE 8: Variation of the surface tension with concentration of PAAD in water with the NaCl concentration of 1000 mg/l at 25°C.

repulsion. Thus, intermolecular association between macromolecule leaded to the forming of tight interfacial films at the interface and the reducing of the surface tension in the water solutions. When the dose of PAAD was 25%, the surface tension of PAAD-ID-5 minimized to 25.9 mN/m.

3.10. Effect of SDS Concentration on Surface Tension. The surface tension of PAAD solutions first declined and then stabilized with the increase of PAAD dose (Figure 8). The surface tension first decreased with the rise of SDS dose and minimized to 26.1 mN/m when the dose was 4%. This was related to the mixed micelles formed from the SDS and DPA association. This was because the increasing of mixed micelle aggregation number and the enlarging of DPA capacity in each micelle before aggregation resulted in elongating of the hydrophobic micro blocks on copolymer molecular chains with the increasing of the SDS dose. Thus, the surface tension was reduced [32], accordingly.

4. Conclusions

A series of surfactive amphiphilic polymer PAADs were prepared from acrylic acid, dodecyl polyoxyethylene acrylate (DPA), and acrylamide under the action of a mixed initiating agent consisting of APS-$NaHSO_3$/2,2′-azobis[2-(2-imidazolin-2-yl)propane] dihydrochloride (AIBA.2HCl). The conversion rate was above 97%. The surfmer DPA, through intermolecular association, increased the molecular weights of the polymers and the viscosity and surface activity of polymer solutions. The NaAA repeated units on molecular chains, through static repulsion between carboxylic acid anions, accelerated the transfer from intramolecular to intermolecular association and thereby enhanced the molecular weights, apparent viscosity, and surface activity of polymer solutions. The SDS and DPA repeated units on molecular chains accelerated the hydrophobic blocks from intramolecular to intermolecular association and thereby enhanced the molecular weights, apparent viscosity, and surface activity of polymer solutions.

Conflicts of Interest

The authors declare that they have no conflicts of interest.

Acknowledgments

This work was supported by the major national science and technology projects of the China National Petroleum Corporation (2016E-0206). The fund name is research and industrialization application of amphiphilic polymers, China.

References

[1] A. Maestro, C. González, and J. M. Gutiérrez, "Thickening mechanism of associative polymers," *Macromolecular Symposia*, vol. 187, no. 1, pp. 919–928, 2002.

[2] S. Li, C. Ye, G. Zhao, M. Zhang, and Y. Zhao, "Synthesis and properties of monocleavable amphiphilic comblike copolymers with alternating PEG and PCL grafts," *Journal of Polymer Science Part A: Polymer Chemistry*, vol. 50, no. 15, pp. 3135–3148, 2012.

[3] Z. Li, S. Ma, G. Zhang, D. Wang, and F. Zhou, "Soft/hard-coupled amphiphilic polymer nanospheres for water lubrication," *ACS Applied Materials & Interfaces*, vol. 10, no. 10, pp. 9178–9187, 2018.

[4] M. Cao, C. Cao, P. Zhou et al., "Self-assembly of amphiphilic peptides: effects of the single-chain-to-gemini structural transition and the side chain groups," *Colloids and Surfaces A: Physicochemical and Engineering Aspects*, vol. 469, pp. 263–270, 2015.

[5] M. S. Azad, Y. K. Dalsania, and J. J. Trivedi, "Capillary breakup extensional rheometry of associative and hydrolyzed polyacrylamide polymers for oil recovery applications," *Journal of Applied Polymer Science*, vol. 135, no. 22, article 46253, 2018.

[6] H. X. Chen, H. M. Tang, X. Y. Wu, Y. G. Liu, J. H. Bai, and F. Zhao, "Synthesis, characterization, and property evaluation of a hydrophobically modified polyacrylamide as enhanced oil recovery chemical," *Journal of Dispersion Science and Technology*, vol. 37, no. 4, pp. 486–495, 2016.

[7] Y. Ji, D. Wang, X. Cao, L. Guo, and Y. Zhu, "Both-branch amphiphilic polymer oil displacing system: molecular weight, surfactant interactions and enhanced oil recovery performance," *Colloids and Surfaces A: Physicochemical and Engineering Aspects*, vol. 509, pp. 440–448, 2016.

[8] P. A. Fulmer and J. H. Wynne, "Development of broad-spectrum antimicrobial latex paint surfaces employing active amphiphilic compounds," *ACS Applied Materials & Interfaces*, vol. 3, no. 8, pp. 2878–2884, 2011.

[9] J. Kopecek and P. Kopeckova, "HPMA copolymers: origins, early developments, present, and future," *Advanced Drug Delivery Reviews*, vol. 62, no. 2, pp. 122–149, 2010.

[10] A. Matelová, G. Huerta-Angeles, D. Šmejkalová et al., "Synthesis of novel amphiphilic hyaluronan containing-aromatic fatty

acids for fabrication of polymeric micelles," *Carbohydrate Polymers*, vol. 151, pp. 1175–1183, 2016.

[11] Y. Bai, F. Y. Xie, and W. Tian, "Controlled self-assembly of thermo-responsive amphiphilic H-shaped polymer for adjustable drug release," *Chinese Journal of Polymer Science*, vol. 36, no. 3, pp. 406–416, 2018.

[12] A. Checkervarty, M. Werner, and J. U. Sommer, "Formation and stabilization of pores in bilayer membranes by peptide-like amphiphilic polymers," *Soft Matter*, vol. 14, no. 13, pp. 2526–2534, 2018.

[13] A. Hill, F. Candau, and J. Selb, "Aqueous solution properties of hydrophobically associating copolymers," in *Trends in Colloid and Interface Science V*, vol. 84 of Progress in Colloid & Polymer Science, pp. 61–65, Steinkopff, Heidelberg, Germany, 1991.

[14] L. M. Gouveia, S. Paillet, A. Khoukh, B. Grassl, and A. J. Müller, "The effect of the ionic strength on the rheological behavior of hydrophobically modified polyacrylamide aqueous solutions mixed with sodium dodecyl sulfate (SDS) or cetyltrimethylammonium *p*-toluenesulfonate (CTAT)," *Colloids and Surfaces A: Physicochemical and Engineering Aspects*, vol. 322, no. 1–3, pp. 211–218, 2008.

[15] T. E. Lara-Ceniceros, G. Cadenas-Pliego, C. C. Rivera-Vallejo, R. E. D. de León-Gómez, A. Coronado, and E. J. Jiménez-Regalado, "Synthesis and characterization of thermo-insensitive, water-soluble associative polymers with good thickening properties at low and high temperatures," *Journal of Polymer Research*, vol. 21, no. 7, p. 511, 2014.

[16] Z. Zhu, W. Kang, B. Sarsenbekuly et al., "Preparation and solution performance for the amphiphilic polymers with different hydrophobic groups," *Journal of Applied Polymer Science*, vol. 134, no. 20, 2017.

[17] C. Zhong, W. Wang, and X. Lian, "Associative behavior in aqueous solutions for a salt-thickening tetra-polymer with allyl-capped macro-monomer," *Polymer Science Series A*, vol. 56, no. 4, pp. 422–433, 2014.

[18] X. Yang, J. Liu, P. Li, and C. Liu, "Self-assembly properties of hydrophobically associating perfluorinated polyacrylamide in dilute and semi-dilute solutions," *Journal of Polymer Research*, vol. 22, no. 6, p. 103, 2015.

[19] J. He, P. Ni, and C. Liu, "Synthesis and characterization of amphiphilic fluorinated pentablock copolymers based on pluronic F127," *Journal of Polymer Science Part A: Polymer Chemistry*, vol. 46, no. 9, pp. 3029–3041, 2008.

[20] C. Song, D. Ao, X. Wang, J. Zhang, and Y. Tan, "Synthesis, characterization, and aqueous properties of new amphiphilic copolymer with fluorocarbon groups," *Colloid and Polymer Science*, vol. 291, no. 12, pp. 2815–2823, 2013.

[21] W. Xue, I. W. Hamley, V. Castelletto, and P. D. Olmsted, "Synthesis and characterization of hydrophobically modified polyacrylamides and some observations on rheological properties," *European Polymer Journal*, vol. 40, no. 1, pp. 47–56, 2004.

[22] A. Montoya-Goñi and D. C. Sherrington, "Reactive surfactants in heterophase polymerisation: XXIII. Synthesis and characterisation of novel dialkyl maleate cationic surfmers," *Polymer*, vol. 40, no. 4, pp. 1067–1079, 1999.

[23] L. Jiang, B. Gao, and H. Lei, "Microstructure of macromolecular chains and hydrophobic association properties for copolymers of surface active monomer $NaAMC_{14}S$ and acrylamide," *Acta Polymerica Sinica*, vol. 7, no. 5, pp. 484–490, 2007.

[24] D. Pawlowski and B. Tieke, "Copolymerization of a cationic surfmer with HEMA in three-component lyotropic mesophase: a route to nanostructured polymer gels," in *Mesophases, Polymers, and Particles*, vol. 129 of Progress in Colloid and Polymer Science, pp. 24–31, Springer, Berlin, Heidelberg, 2004.

[25] J. Li, F. P. Wu, and E. J. Wang, "Hydrophobically associating polyacrylamides modified by a novel self-associative cationic monomer," *Chinese Journal of Polymer Science*, vol. 28, no. 2, pp. 137–145, 2010.

[26] Y. J. Che, Q. Gao, M. Fang, Y. Tan, X. Wang, and F. Meng, "Synthesis, characterization and aqueous properties of a new hybrid fluorocarbon cationic surfmer," *Journal of Surfactants and Detergents*, vol. 18, no. 2, pp. 233–244, 2015.

[27] Y. Sun, H. Du, and Y. Lan, "Preparation and properties of fluorinated amphiphilic copolymers via iron-mediated AGET ATRP," *Iranian Polymer Journal*, vol. 24, no. 2, pp. 95–103, 2015.

[28] S. Díaz-Silvestre, E. Saldívar-Guerra, C. Rivera-Vallejo et al., "Synthesis of associative block copolymers electrolytes via RAFT polymerization," *Polymer Bulletin*, vol. 75, no. 3, pp. 891–907, 2018.

[29] F. Li, Y. Luo, P. Hu, and G. Su, "Study on the self-assembly properties of surface active hydrophobically associating polyacrylamide with nonionic surfmer units," *Polymer Bulletin*, vol. 74, no. 7, pp. 2873–2886, 2017.

[30] H. Khakpour and M. Abdollahi, "Rheological properties of acrylamide/butyl acrylate/2-acrylamido-2-methyl-1-propane sulfonic acid co- and terpolymers synthesized by heterogeneous and micellar methods," *Polymer Bulletin*, vol. 74, no. 12, pp. 5145–5161, 2017.

[31] M. S. Alam, G. Ghosh, A. B. Mandal, and Kabir-ud-Din, "Aggregation behavior and interaction of an amphiphilic drug imipramine hydrochloride with cationic surfactant cetyltrimethylammonium bromide: light scattering studies," *Colloids and Surfaces B: Biointerfaces*, vol. 88, no. 2, pp. 779–784, 2011.

[32] N. Azum, M. A. Rub, A. M. Asiri, and W. A. Bawazeer, "Micellar and interfacial properties of amphiphilic drug–non-ionic surfactants mixed systems: surface tension, fluorescence and UV–vis studies," *Colloids and Surfaces A: Physicochemical and Engineering Aspects*, vol. 522, pp. 183–192, 2017.

21

Improving Functional Properties of Cassava Starch-Based Films by Incorporating Xanthan, Gellan, or Pullulan Gums

Mayra Sapper, Pau Talens, and Amparo Chiralt

Instituto Universitario de Ingeniería de Alimentos para el Desarrollo, Universitat Politècnica de València, Camino de Vera s/n, 46022 Valencia, Spain

Correspondence should be addressed to Mayra Sapper; maysap@etsiamn.upv.es

Academic Editor: Markus Schmid

The effect of the partial substitution of cassava starch in edible films for 10 and 20 wt% of gellan, xanthan, or pullulan gums was analysed in films obtained by casting. The tensile properties, barrier capacity to water vapour, and oxygen and water sorption isotherms of the samples were analysed. The blend of starch with gellan gum was effective to reduce the moisture sorption capacity of starch films while reducing water vapour permeability, enhancing the film strength and resistance to break and preserving films against starch retrogradation throughout the storage time. Xanthan gum improved the tensile behaviour of the starch films, but did not reduce their water sorption capacity and water vapour permeability. Pullulan did not notably improve the functional properties of the starch films. Gellan gum at 10 and 20 wt% in the blend could be used to obtain starch films with more adequate properties for food packaging purposes.

1. Introduction

The growth in environmental concerns over nonbiodegradable petrochemical-based plastics has been the driving force behind studies seeking to solve the problems associated with the plastic waste by developing biobased/biodegradable polymers as more environmentally friendly materials, which can be produced from a variety of agricultural commodities and/or of food waste products [1, 2].

Due to its availability, biodegradability, renewability, film-forming ability, and low cost [3], starch from different botanical sources (cassava, corn, wheat, rice, potato, pea, and others) is one of the most promising natural polymers for packaging applications [4, 5]. Starch-based films are transparent [6], odorless, tasteless, and highly impermeable to oxygen [7]. However, they have some limitations such as their highly hydrophilic nature (water sensitivity), which negatively affects tensile and barrier properties as compared with conventional synthetic polymers [8]. Therefore, various approaches, such as blending with additives and/or other biopolymers [9–15] or the starch chemical modification, have been studied to improve these characteristics [16].

Polysaccharides produced by microorganisms, such as pullulan, gellan, or xanthan, have found a wide range of applications in the food, pharmaceutical, and other industries due to their particular functional properties. Some of these applications include their use as emulsifiers, stabilizers, gelling agents, film formers, or thickening agents [17]. Gellan gum is a microbial polysaccharide secreted by the bacterium *Sphingomonas elodea*. Gellan polymer consists of a linear tetrasaccharide repeat unit composed of two molecules of D-glucose, one of D-glucuronic acid and one of L-rhamnose [18, 19]. Kim et al. [13] reported that gellan gum was effective at improving the storage stability and mechanical properties of the starch films.

Xanthan gum is a high molecular weight extracellular polysaccharide produced by the bacterium *Xanthomonas campestris* and is one of the most important commercial microbial hydrocolloids used in the food industry as thickening agent and stabilizer [17]. It is composed by a linear β-1,4-linked D-glucose chain substituted on every second glucose unit by a charged trisaccharide side chain with glucuronic acid residue between two mannose units. The inner mannose residue is normally acetylated at C(6), which is located at the external part of the helical conformation of the polysaccharide. About half of the terminal mannoses are linked to pyruvyl residues. Deprotonation of acetyl and pyruvyl residues at pH > 4.5 increases the negative charge density of the xanthan chain. It has been reported that xanthan influenced the mechanical parameters and moisture adsorption isotherm of cassava starch films [9, 15]. The xanthan gum enhanced film traction resistance but generated a less deformable matrix, which was associated to the gum and starch chain interaction through hydrogen bonds developing a more resistant network but with lower deformability due to the impossibility of polymeric chains to slide for maintaining the ductile behaviour observed in the absence of gum [9]. In accordance with Veiga-Santos et al. [15], the interaction between the xanthan gum and the starch might avoid the occurrence of amylose : amylose interaction, inhibiting the starch retrogradation. The prevention of the amylose : amylose interaction provoked by this gum could also give rise to a more unfolded network with weaker interaction forces, resulting in films with lower elongation at break [15]. Likewise, the development of gum-starch hydrogen bonds might interfere with amylose packing, inhibiting the formation of polymer-water hydrogen bonds in the amorphous areas, thus limiting the water sorption capacity of the blend films [9].

On the other hand, pullulan is a neutral, linear, and water-soluble polysaccharide [20, 21], composed of maltotriose units linked by α-1,6-glycosidic bonds [11]. It is produced by *Aureobasidium pullulans* in starch and sugar cultures. Pure pullulan films are transparent, water-sensitive, and mechanically weak [22]. The hydrophilic nature of pullulan, however, often makes its films sticky when exposed to high relative humidity conditions [13]. Kim et al. [12] found that tapioca starch-pullulan blend films exhibited greater mechanical strength and stability during storage under humid conditions and at the same time the composite films were less water-soluble.

The blending of starch with small amounts of the above polysaccharides could represent a simple method that has the potential of improving the starch film properties while keeping the competitive cost of the material. Although there are several studies which analyse the effect of the partial substitution of starch by different gums on blend films, in terms of the improvement of the starch-based film functionality, few comparative analyses have been reported by using the same film obtaining process [13], in order to select the most appropriate blend to effectively reduce the water sensitivity of the starch-based materials. It is well known that differences in the polymer characteristics, as well as different processing techniques to obtain films, can affect the final properties of the material [23, 24]. Therefore, for an adequate comparison, standardization of the raw materials and methods must be carried out. Likewise, no studies were found into the effect of the gums on the starch retrogradation phenomena, responsible for changes in the film mechanical properties throughout the storage time.

In this sense, the objective of this study was to analyse the effect of the partial starch substitution by three different microbial gums (xanthan, gellan, and pullulan) on the relevant physical properties of the cassava starch-based films (tensile, barrier, and water adsorption capacity), as well as the ageing behaviour of the films through the changes in the film tensile properties during storage, when equilibrated at different relative humidities.

2. Materials and Methods

2.1. Materials. Cassava starch (S) was supplied by Quimidroga S.A. (Barcelona, Spain). The amylose content (*w/w*, %) of cassava starch was 10%, with an amylose : amylopectin ratio of 1 : 9.9 (determined by using an Amylose/Amylopectin Assay Procedure enzymatic kit, Megazyme International Ireland, Bray Business Park, Bray, Co. Wicklow, Ireland). Xanthan gum (X) (high molecular weight, ~10^6 Da) was supplied by EPSA (Valencia, Spain). Negatively charged, low acyl gellan gum (G) KELCOGEL F (M_W $3-5\times10^5$ Da) was purchased from Premium Ingredients (Murcia, Spain), and pullulan (P) (M_W 2×10^5 Da) was provided by Nagase GmbH (Dusseldorf, Germany). Glycerol, used as plasticizer, and P_2O_5, $MgCl_2$, K_2CO_3, $Mg(NO_3)_2$, $CuCl_2$, NaCl, and KCl salts were provided by PanReac Química S.A. (Castellar del Vallés, Barcelona, Spain).

2.2. Preparation of Films. Films were obtained by casting after the preparation of the film-forming solutions (FFSs). Starch (2% *w/w*) was dispersed in distilled water at 95°C for 30 min, while hand-stirring, as described by other authors [23, 25, 26], to induce complete starch gelatinization and cooled down to reach room temperature. This low starch concentration was selected in order to make the blending with the gum solutions easier due to their high viscosity. Glycerol was added to the aqueous gelatinized starch at 0.25 g/g of starch, on the basis of previous studies using glycerol plasticized starch-based films [27, 28]. The starch-glycerol aqueous system was homogenized using a rotor-stator homogenizer (Ultra Turrax D125, Janke and Kunkel, Germany) at 13,500 rpm for 3 min. Aqueous solutions of pullulan (2% *w/w*), xanthan (1% *w/w*), and gellan (2% *w/w*) gums were prepared in distilled water. Gum concentration was selected on the basis of the high viscosity of the respective solutions, especially for xanthan and gellan gums. Pullulan was dissolved in water under magnetic stirring (400 rpm) at room temperature. Xanthan and gellan gums were heated while magnetically stirred (400 rpm) at 90°C for 60 min until the complete dissolution. Glycerol was also added to the gum solutions (0.25 g/g polymer). The solution heating was carried out to accelerate the dissolution kinetics. For the starch-gum blends, the polymer solutions were mixed in the adequate proportion to obtain starch : gum ratios of 90 : 10 and 80 : 20. The FFSs were stirred for at least 60 min and degassed at room temperature by using a vacuum pump (MZ 2C NT, Vacuubrand GMBH + CO KG, Wertheim, Germany). Additionally, the apparent viscosity of the newly prepared FFSs at steady shear rate of 100 s^{-1} was characterised in triplicate at 25°C using a rotational rheometer (HAAKE RheoStress 1, Thermo Electric Corporation, Karlsruhe, Germany) with a sensor system of coaxial cylinders, type Z34DIN Ti. Table 1 shows the apparent viscosity values, where the effect of each gum on the

Table 1: Mean values and standard deviation of the apparent viscosity (η_{100} s^{-1} Pa s) of the film-forming solutions with pure polymers and the different starch : gum ratios.

Sample		Starch : gum ratio	
		90 : 10	80 : 20
S	0.0118 ± 0.0003^{ab}	—	—
X	0.0838 ± 0.0006^{e}	0.0331 ± 0.0001^{c}	0.0619 ± 0.0005^{d}
G	1.62 ± 0.03^{f}	0.0220 ± 0.0001^{bc}	0.0301 ± 0.0002^{c}
P	0.00321 ± 0.00003^{a}	0.00780 ± 0.00002^{ab}	0.0083 ± 0.0001^{ab}

Different superscript letters (a-f) indicate significant differences among formulations ($p < 0.05$).

starch solution viscosity can be observed. Likewise, it shows the much higher viscosity of gellan gum due to its tendency to gelation at the used concentration.

To obtain the films, a mass of the FFSs containing 1.5 g of total solids was poured onto polytetrafluorethylene (PTFE) plates (15 cm diameter) and allowed to dry under natural convection for approximately 48 h at 25°C and 45% relative humidity (RH). Films with 100% starch, gellan gum, and pullulan were also analysed for comparison purposes, whereas it was not possible to obtain pure xanthan gum films, since a homogeneous film could not be obtained.

2.3. Film Conditioning and Storage. Before characterising the films, samples were equilibrated for one week at two different relative humidities (53 and 75%) at 25°C by using oversaturated solutions of $Mg(NO_3)_2$ and NaCl, respectively. Net pullulan films showed deformations during conditioning at high RH, which complicated their handling. One part of the samples was stored under the same conditions for five weeks in order to perform the second series of analyses. Film thickness was measured using a Palmer digital micrometer (Comecta, Barcelona, Spain) to the nearest 0.001 mm. Six random positions in each film sample were considered.

2.4. Characterisation of Films

2.4.1. Water Vapour Permeability. Water vapour permeability (WVP) was determined gravimetrically at 25°C and RH gradients of 53-100% and 75-100%, following the ASTM E96-95 gravimetric method [29], applying the correction proposed by Gennadios et al. [30]. Payne permeability cups, 3.5 cm in diameter (Elcometer Sprl, Hermelle/s Argenteau, Belgium), were filled with 5 mL of distilled water (100% RH). Four circular samples of each formulation (3.5 cm in diameter) were secured in the cups and placed in preequilibrated cabinets containing oversaturated solutions of magnesium nitrate or sodium chloride to generate 53% or 75% RH inside the cabinet, with a fan on top of the cup in order to reduce the resistance to water vapour transport. The cups were weighed every 1.5 h for 24 h by using an analytical balance (ME36S, Sartorius, Germany, 0.0001 g). The slope of the weight loss vs. time curve at stationary state was used to determine the water vapour transmission rate (WVTR) and the water vapour permeability [31].

2.4.2. Oxygen Permeability (OP). The oxygen permeation rate of the films was determined by following the ASTM standard method D3985-05 [32]. Measurements were taken at 53% RH and 25°C in films previously equilibrated in the same conditions, using an OX-TRAN 1/50 system (MOCON, Minneapolis, USA). The transmission values were determined every 20 min until equilibrium was reached. The exposure area of each sample during the tests was 50 cm^2. To obtain the oxygen permeability, the film thickness was considered in every case. At least two replicates per formulation were taken into account.

2.4.3. Tensile Properties. A texture analyser (Stable Micro Systems, TA.XTplus, Haslemere, England) was used to measure the mechanical properties of the films equilibrated at 53 and 75% RH and 25°C, following the ASTM standard method D882 [33]. Film strips (25 mm × 100 mm) were mounted in tensile grips (A/TG model), exposing a length of 5 cm for stretching at 50 mm min^{-1}, until breaking. The elastic modulus (EM), tensile strength at break (TS), and percentage of elongation at break (%E) were determined from the stress–Hencky strain curves, obtained from force–deformation data.

2.4.4. Isothermal Water Sorption Capacity. Triplicate film samples (15-20 mg), accurately weighed, were placed in desiccators at 25°C and equilibrated at different RH (or water activities: a_w = RH/100) by using oversaturated solutions of $MgCl_2$, K_2CO_3, $Mg(NO_3)_2$, $CuCl_2$, NaCl, and KCl (% RH: 32.8, 43.2, 52.9, 67.5, 75.3, and 84.3, respectively). Samples were weighed periodically (0.00001 g precision) for 3 weeks, when equilibrium was reached. Finally, the equilibrium moisture content was determined by drying in a vacuum oven (Vacioterm-T, JP Selecta S.A., Barcelona, Spain) at 60°C for 24 h and placed in desiccators containing P_2O_5 until reaching constant weight.

2.5. Statistical Analysis. A statistical analysis of data was performed through analysis of variance (ANOVA) using Statgraphics Centurion XVI.II (StatPoint Technologies Inc., Warrenton, VA, USA). Fisher's least significant difference (LSD) was used at the 95% confidence level.

3. Results and Discussion

3.1. Barrier Properties. The water vapour permeability (WVP) of polymeric films depends on many factors, including the solubility coefficient, integrity of the film matrix, hydrophobicity, diffusion rate, thickness, polymeric chain mobility, and interactions between the functional groups of polymers and the permeant. Packaging has to be able to prevent moisture and gas exchange between the environment and the product. Then, the water vapour and oxygen barrier capacity of the films is key to determining the efficiency of the material as packaging. The values of the WVP of the studied films stored for 1 week under controlled conditions (53 or 75% RH) are shown in Table 2. As expected, the WVP of the films greatly increased for the 100-75% RH gradient, due to the greater degree of plasticization of the material and the consequent rise in molecular mobility, promoting the mass transfer rate. In both RH conditions, starch-gellan blend films exhibited the lowest WVP values. This suggests that interactions between the negatively charged gellan chains and the linear amylose chains, through hydrogen bonds, are enough effective to reduce the possibilities of water molecules to interact with the mixed matrix, which, in turn, would reduce their solubility and permeation capacity. At 53-100% RH, pure pullulan films had the highest WVP of all the formulations, as also reported by Kanmani and Lim [11]. On the other hand, except for the starch-gellan blends, there were no notable differences between formulations at 75-100% RH. Probably, the high plasticization level of the polymers at these high RH conditions makes differences in the polymer packing less relevant to determine the barrier capacity of the films.

The oxygen permeability (OP) values of films conditioned for 1 week at 53% RH are shown in Table 2. The OP values of net gellan and starch-pullulan blend films were significantly higher than those of the other films. The incorporation of xanthan and gellan gums contributed to a decrease in the oxygen permeability of the films, although all the films had notably low OP values, as previously reported for starch-based films [34]. The reduction in the OP values provoked by the incorporation of both negatively charged chains, gellan, and xanthan could be attributed to the promotion of the hydrogen bond interactions between the gum and amylose, as previously reported by other authors [9, 15], creating a more compact network where the permeation of the gas molecules would be more hindered.

Table 2: Mean values and standard deviation of water vapour (WVP) and oxygen permeability (OP) of the films conditioned at 25°C and 53 and 75% RH.

Sample	WVP × 10^{10} (g/Pa s m) 53% RH	WVP × 10^{10} (g/Pa s m) 75% RH	OP × 10^{13} (cm^3/m s Pa) 53% RH
S	7.1 ± 0.2^{c}	14.0 ± 1.0^{cd}	3.6 ± 0.2^{c}
$S_{90}:X_{10}$	7.1 ± 0.3^{c}	13.0 ± 1.0^{bc}	2.6 ± 0.1^{b}
$S_{80}:X_{20}$	6.2 ± 0.4^{ab}	13.0 ± 1.0^{b}	2.7 ± 0.1^{b}
G	9.0 ± 0.1^{d}	14.5 ± 0.5^{d}	4.0 ± 0.1^{d}
$S_{90}:G_{10}$	5.9 ± 0.2^{a}	11.2 ± 0.3^{a}	2.7 ± 0.1^{b}
$S_{80}:G_{20}$	6.6 ± 0.3^{b}	10.7 ± 0.3^{a}	2.3 ± 0.1^{a}
P	11.8 ± 0.3^{f}	—	2.7 ± 0.1^{b}
$S_{90}:P_{10}$	10.3 ± 0.5^{e}	13.9 ± 0.2^{bcd}	4.1 ± 0.1^{d}
$S_{80}:P_{20}$	10.5 ± 0.2^{e}	14.0 ± 1.0^{bcd}	4.8 ± 0.1^{e}

Different superscript letters (a-f) within the same column indicate significant differences among formulations ($p < 0.05$).

3.2. Tensile Properties. The elastic modulus (EM), related to the stiffness of the material, the tensile strength (TS), and the deformation percentage (%E) at break, of the films equilibrated a different relative humidities were determined for samples conditioned for 1 and 5 weeks in order to analyse the changes associated with the film hydration and ageing. These mechanical parameters, useful for describing the film tensile behaviour, are closely related to its microstructure [35] and therefore reflect the microstructural changes that occurred during storage. Table 3 shows the values of the mechanical parameters obtained for the pure polymer films at the initial and final times for both conditioning conditions (53 and 75% RH). Films obtained from pure pullulan, equilibrated at 75% RH, exhibited deformations along with the moisture gain which did not permit the adequate measurement of their tensile properties. In contrast, pure gellan films were the stiffest (the highest EM) and most resistant to fracture (the highest TS), regardless of the storage conditions. This mechanical behaviour of the gellan films can be attributed to the high packing capacity of the linear polymer chains, also responsible for its high gelling capacity, where interactions and hydrogen bonds can occur throughout a great extent in the chains. The increase in the moisture content of the films when equilibrated with the highest RH reduced the film strength in all cases due to the enhanced plasticising effect of the water molecules, which promoted molecular mobility and reduced the cohesion force between chains. This effect was more intense for starch (EM was reduced by 4.5) than for gellan films (EM reduction by 1.5), but could not be quantified for pullulan films due to their deformation at the highest RH. Film moisturising also promoted the stretchability, which increased more markedly in starch films.

The film ageing behaviour could also be analysed through the changes in the tensile properties over 5 storage weeks. No significant differences in the initial and final values of the tensile parameters were observed for gellan films, whereas the starch samples became stronger and less extensible under both RH conditions. This change in the mechanical behaviour of starch films could be attributed to the retrogradation phenomenon or partial migration of the plasticizer. Different authors [26, 36–38] reported that the starch retrogradation process in starch films after gelatinization is mainly due to the amylose recrystallization in a relatively short time period, depending on the final water content, whereas amylopectin recrystallization is a long-time process. Apart from crystallization, a progressive aggregation of polymer chains throughout time has been observed for starch films [25] which provokes an increase in the matrix compactness and the consequent changes in the tensile response. In contrast, pullulan films become softer, exhibiting a notable decrease in the EM. This could be attributed to a progressive water gain in the films during storage, which progressively plasticize the polymer matrix to a greater extent, giving rise to a greater chain mobility and more extensible films [13].

Figure 1 shows the elastic modulus (EM), tensile strength (TS), and elongation at break (E) for the different starch-gum blend films. In films conditioned for 1 week, adding gums to the starch matrix provoked a strengthening effect reflected in the higher values of the EM compared to the net starch film, except for starch-pullulan blends conditioned at the highest RH. Likewise, although the film's resistance to break (TS) was considerably increased for xanthan and gellan blends, it

Table 3: Mean values and standard deviation of tensile properties (EM: elastic modulus; TS: tensile strength; %E: percentage of elongation) of the pure polymer films conditioned at 25°C and 53 or 75% RH for 1 (initial) or 5 (final) weeks.

			Sample		
			S	G	P
RH – 53%	Initial	EM (MPa)	$160 \pm 50^{a,1}$	$1300 \pm 150^{c,1}$	$500 \pm 60^{b,2}$
		TS (MPa)	$10 \pm 2^{a,2}$	$80 \pm 30^{b,1}$	$9 \pm 1^{a,2}$
		E (%)	$12 \pm 1^{ab,2}$	$10 \pm 6^{a,1}$	$30 \pm 10^{b,1}$
	Final	EM (MPa)	$340 \pm 50^{b,2}$	$1140 \pm 120^{c,1}$	$52 \pm 8^{a,1}$
		TS (MPa)	$6 \pm 2^{a,1}$	$34 \pm 25^{b,1}$	$2 \pm 1^{a,1}$
		E (%)	$6 \pm 2^{a,1}$	$3 \pm 2^{a,1}$	$50 \pm 20^{b,1}$
RH – 75%	Initial	EM (MPa)	$36 \pm 5^{a,1}$	$840 \pm 50^{b,1}$	—
		TS (MPa)	$4 \pm 1^{a,1}$	$80 \pm 13^{b,1}$	—
		E (%)	$34 \pm 2^{b,2}$	$16 \pm 5^{a,1}$	—
	Final	EM (MPa)	$55 \pm 7^{a,2}$	$1000 \pm 150^{b,1}$	—
		TS (MPa)	$4 \pm 1^{a,1}$	$81 \pm 4^{b,1}$	—
		E (%)	$15 \pm 3^{a,1}$	$15 \pm 2^{a,1}$	—

Different superscript letters (a-c) in the same row indicate significant differences among formulations ($p < 0.05$) for each parameter. Different superscript numbers (1, 2) in the same column for each parameter and film sample indicate significant differences between the two storage times ($p < 0.05$).

was reduced in pullulan mixtures, mainly in the films conditioned at the highest RH. However, the film extensibility was generally reduced in all blend films with respect to net starch samples. For the negatively charged polymers (gellan and xanthan) the strengthening effect could be associated with the gum and starch chains interaction through hydrogen bonds, developing a more resistant network but with lower deformability. The restrictions associated to bonded polymeric chains to slide during the stretching compromised the ductile behaviour observed for the gum-free starch films. Pullulan also exerts a reinforcing effect on the starch matrix, but this disappears when films were conditioned at high RH, which could be attributed to the high water sensitivity of pullulan, which becomes highly deformable at high RH [12]. Likewise, the preferential adsorption of water molecules of this gum could limit its interactions with the starch chains at high RH conditions, thus limiting its strengthening effect, as previously observed by other authors [12].

In terms of ageing, all the blend films exhibited a decrease in EM and TS values after five weeks, in contrast with the increase observed in the net starch matrix. This indicates that blending it with these gums inhibited starch retrogradation. The observed decrease in the film strength could be attributed to the rearrangement of the polymer chains during film storage, obtaining a more open matrix with greater molecular mobility, which could be due, in part, to a progressive film hydration till the true equilibrium moisture content was reached. In the films conditioned/stored at the highest RH, the values of EM and TS also decreased during storage for xanthan and pullulan blends, but increased or did not change significantly for gellan blends. At the highest RH, starch retrogradation kinetics become faster due to the highest molecular mobility associated to the water plasticization [36]. In this sense, this phenomenon seems to occur to a certain extent in gellan blends whereas it was limited in the other blend films, where a decrease in the film strength was observed. The hydration kinetics of the films till equilibrium would be affected by the gradient of water activity (a_w) between the initial dried film and the external RH; the lower the driving force (at the lowest RH), the slower the hydration rate, and by the compactness of the film structure which affects the diffusion rate of water molecules. Gellan blends, in particular, did not exhibit EM or TS reduction during storage when conditioned at the highest RH, which could indicate that no changes in the film moisture content and the consequent plasticization level occurred during storage, whereas the starch retrogradation effect was prevalent in these conditions with high molecular mobility.

As concerns extensibility, gum blending provoked a reduction with respect to the net starch films, although this was less noticeable when conditioned at 53% RH. As commented on in the next section, the gums probably limit the water sorption capacity of the films, thus reducing the water plasticising effect. This would be more noticeable at the highest RH (75%), when water plasticization was highly intense in net starch films, as previously commented on. Likewise, the water partition between the polymers, according to their relative water affinity, could also affect the plasticization level of the starch matrix, this affecting its mechanical behaviour.

3.3. Isothermal Water Sorption Capacity of the Films. Table 4 shows the equilibrium moisture content (g/100 g solids) obtained for pure and blend films conditioned under different RH (between 32.8 and 84.3%) at 25°C, in order to compare the water binding capacity of the different blends and pure polymer films. At low RH, pure gellan films exhibited similar water sorption capacity to the starch films, whereas pure pullulan films are slightly less hygroscopic. However, from intermediate RH, the water binding capacity of pure gellan overcame that of starch, whereas pullulan exhibits similar or lower values than does starch. The different water interactions with polar groups of the polymer chains determine the water binding capacity of the matrices depending on the equilibrium relative humidity. As concerns the blend

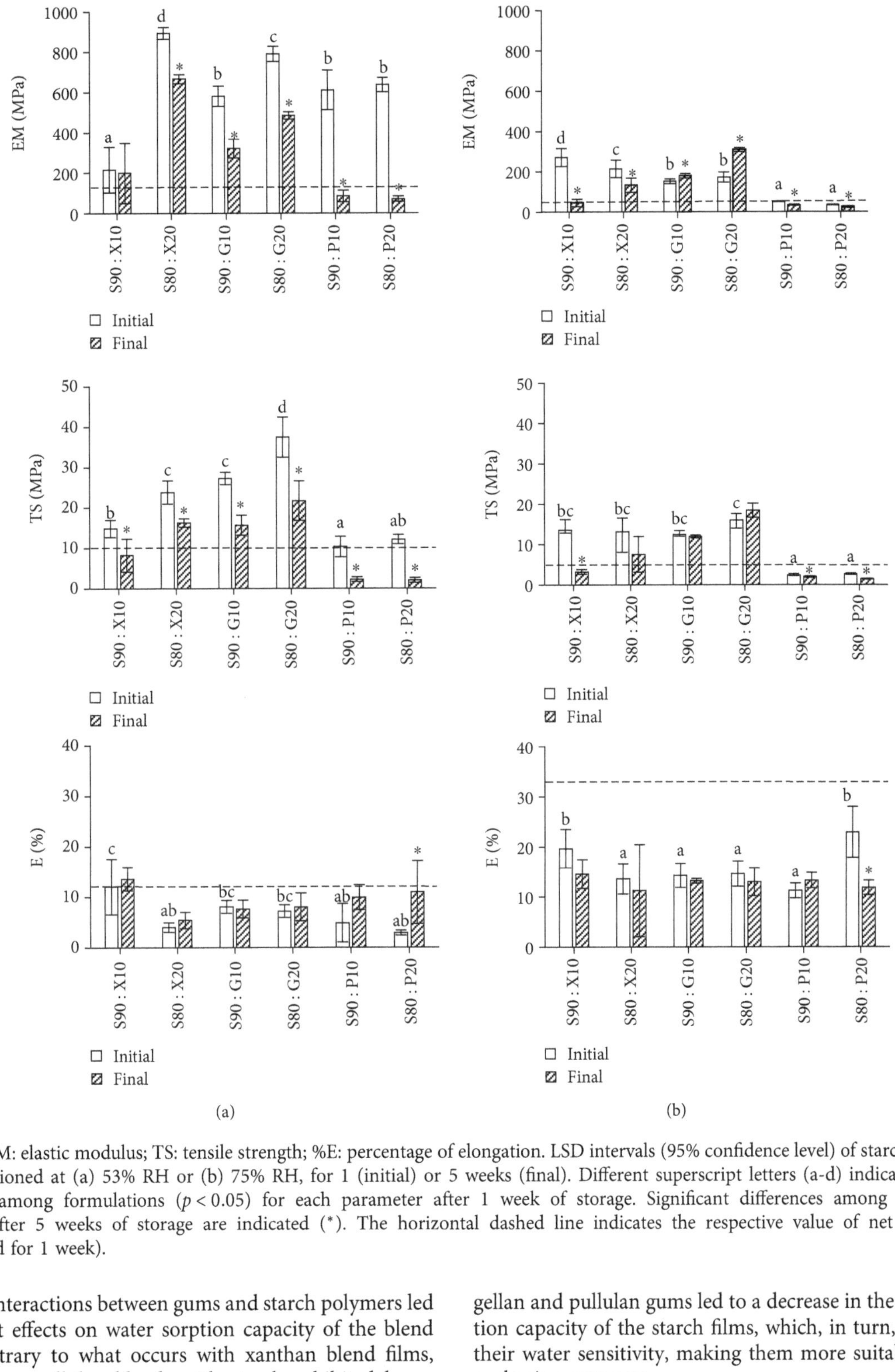

Figure 1: EM: elastic modulus; TS: tensile strength; %E: percentage of elongation. LSD intervals (95% confidence level) of starch-gum blend films conditioned at (a) 53% RH or (b) 75% RH, for 1 (initial) or 5 weeks (final). Different superscript letters (a-d) indicate significant differences among formulations ($p < 0.05$) for each parameter after 1 week of storage. Significant differences among formulations ($p < 0.05$) after 5 weeks of storage are indicated (*). The horizontal dashed line indicates the respective value of net starch films (conditioned for 1 week).

films, the interactions between gums and starch polymers led to different effects on water sorption capacity of the blend films. Contrary to what occurs with xanthan blend films, the gellan or pullulan blends with starch exhibited lower water sorption capacity than did the corresponding films with pure polymers. This could be attributed to the establishment of more efficient hydrogen bonds between the chains of the different polymers, which could reduce the number of active points for water sorption. Then, the incorporation of gellan and pullulan gums led to a decrease in the water sorption capacity of the starch films, which, in turn, will reduce their water sensitivity, making them more suitable for food packaging purposes.

4. Conclusions

Blending starch with gellan gum was effective at reducing the moisture sorption capacity of starch films and also reduced

Table 4: Equilibrium moisture content (g water/100 g solids) for the different film formulations conditioned at different relative humidities (RH) or a_w (RH/100).

Sample	RH (%)					
	32.8	43.2	52.9	67.5	75.3	84.3
S	8.2 ± 0.1^{c}	11.3 ± 0.5^{cd}	12.5 ± 1.3^{c}	21.4 ± 1.0^{de}	26.2 ± 0.3^{c}	39.8 ± 1.5^{d}
$S_{90}:X_{10}$	11.2 ± 1.0^{e}	12.3 ± 0.6^{d}	14.9 ± 1.0^{d}	22.2 ± 0.7^{e}	26.4 ± 0.4^{cd}	37.1 ± 3.6^{cd}
$S_{80}:X_{20}$	10.0 ± 0.3^{d}	11.3 ± 0.9^{cd}	14.8 ± 0.7^{d}	18.2 ± 1.3^{bc}	26.4 ± 0.4^{cd}	35.6 ± 1.4^{bc}
G	8.8 ± 0.4^{c}	10.5 ± 0.1^{c}	11.3 ± 1.5^{bc}	26.3 ± 1.6^{f}	29.9 ± 0.3^{e}	47.2 ± 101^{e}
$S_{90}:G_{10}$	7.2 ± 0.2^{b}	7.7 ± 0.8^{ab}	9.0 ± 0.5^{a}	17.9 ± 1.0^{b}	26.1 ± 0.5^{c}	33.4 ± 1.2^{b}
$S_{80}:G_{20}$	6.3 ± 0.4^{a}	7.9 ± 1.5^{ab}	8.7 ± 1.0^{a}	17.3 ± 1.5^{ab}	27.7 ± 1.6^{d}	35.3 ± 1.8^{bc}
P	6.5 ± 0.3^{ab}	8.5 ± 0.3^{b}	9.7 ± 0.2^{ab}	19.8 ± 0.2^{cd}	22.2 ± 1.0^{b}	39.4 ± 1.2^{d}
$S_{90}:P_{10}$	6.1 ± 0.1^{a}	7.2 ± 0.1^{a}	8.2 ± 0.4^{a}	16.9 ± 0.7^{ab}	19.1 ± 0.7^{a}	29.9 ± 0.4^{a}
$S_{80}:P_{20}$	5.9 ± 0.1^{a}	6.7 ± 0.4^{a}	9.0 ± 1.2^{a}	15.6 ± 0.4^{a}	19.2 ± 0.5^{a}	28.5 ± 1.1^{a}

Different superscript letters (a-f) within the same column indicate significant differences among formulations ($p < 0.05$).

the water vapour and oxygen permeability and increased the film strength and resistance to break, without markedly limiting the film stretchability. Likewise, gellan-starch blends preserve films against the phenomenon of retrogradation. Xanthan gum was effective at increasing the tensile strength of the starch films but did not reduce their water sorption capacity and water vapour permeability. Pullulan was not so effective at enhancing the barrier capacity of starch films or increasing tensile strength. From the obtained results, starch substitution in the films by gellan gum at both 10 and 20 wt% is the one that most improved the barrier and tensile properties of the films and can be used to obtain films that are more useful for food packaging purposes. Particularly, foods such as nuts or snacks with low oxygen pressure requirements to ovoid oxidations, and without free water to avoid film moisturising, may be successfully packaged, allowing to extend their self-life.

Conflicts of Interest

The authors declare that there is no conflict of interest regarding the publication of this paper.

Acknowledgments

This work was supported by the Ministerio de Economía y Competitividad (MINECO) of Spain as a part of projects [AGL2016-76699-R]. Mayra Sapper thanks the Conselleria de Educación, Investigación, Cultura y Deporte de la Comunitat Valenciana for the Santiago Grisolía grant [GRISOLIA/2015/001].

References

[1] B. Ghanbarzadeh, H. Almasi, and A. A. Entezami, “Improving the barrier and mechanical properties of corn starch-based edible films: effect of citric acid and carboxymethyl cellulose,” *Industrial Crops and Products*, vol. 33, no. 1, pp. 229–235, 2011.

[2] B. Hassan, S. A. S. Chatha, A. I. Hussain, K. M. Zia, and N. Akhtar, “Recent advances on polysaccharides, lipids and protein based edible films and coatings: a review,” *International Journal of Biological Macromolecules*, vol. 109, pp. 1095–1107, 2018.

[3] C. H. Chen and L. S. Lai, “Mechanical and water vapor barrier properties of tapioca starch/decolorized hsian-tsao leaf gum films in the presence of plasticizer,” *Food Hydrocolloids*, vol. 22, no. 8, pp. 1584–1595, 2008.

[4] O. V. López, L. A. Castillo, M. A. García, M. A. Villar, and S. E. Barbosa, “Food packaging bags based on thermoplastic corn starch reinforced with talc nanoparticles,” *Food Hydrocolloids*, vol. 43, pp. 18–24, 2015.

[5] C. L. Luchese, J. C. Spada, and I. C. Tessaro, “Starch content affects physicochemical properties of corn and cassava starch-based films,” *Industrial Crops and Products*, vol. 109, no. 5, pp. 619–626, 2017.

[6] S. Mali, M. V. E. Grossmann, M. A. García, M. N. Martino, and N. E. Zaritzky, “Barrier, mechanical and optical properties of plasticized yam starch films,” *Carbohydrate Polymers*, vol. 56, no. 2, pp. 129–135, 2004.

[7] R. Ortega-Toro, S. Collazo-Bigliardi, P. Talens, and A. Chiralt, “Influence of citric acid on the properties and stability of starch-polycaprolactone based films,” *Journal of Applied Polymer Science*, vol. 133, no. 2, pp. 1–16, 2016.

[8] A. Jiménez, M. J. Fabra, P. Talens, and A. Chiralt, “Edible and biodegradable starch films: a review,” *Food and Bioprocess Technology*, vol. 5, no. 6, pp. 2058–2076, 2012.

[9] C. Arismendi, S. Chillo, A. Conte, M. A. del Nobile, S. Flores, and L. N. Gerschenson, “Optimization of physical properties of xanthan gum/tapioca starch edible matrices containing potassium sorbate and evaluation of its antimicrobial effectiveness,” *LWT - Food Science and Technology*, vol. 53, no. 1, pp. 290–296, 2013.

[10] S. K. Flores, D. Costa, F. Yamashita, L. N. Gerschenson, and M. V. Grossmann, “Mixture design for evaluation of potassium sorbate and xanthan gum effect on properties of tapioca starch films obtained by extrusion,” *Materials Science and Engineering*, vol. 30, no. 1, pp. 196–202, 2010.

[11] P. Kanmani and S. T. Lim, “Development and characterization of novel probiotic-residing pullulan/starch edible films,” *Food Chemistry*, vol. 141, no. 2, pp. 1041–1049, 2013.

[12] J. Y. Kim, Y. G. Choi, S. R. Byul Kim, and S. T. Lim, "Humidity stability of tapioca starch-pullulan composite films," *Food Hydrocolloids*, vol. 41, pp. 140–145, 2014.

[13] S. R. B. Kim, Y. G. Choi, J. Y. Kim, and S. T. Lim, "Improvement of water solubility and humidity stability of tapioca starch film by incorporating various gums," *LWT - Food Science and Technology*, vol. 64, no. 1, pp. 475–482, 2015.

[14] R. M. D. Soares, A. M. F. Lima, R. V. B. Oliveira, A. T. N. Pires, and V. Soldi, "Thermal degradation of biodegradable edible films based on xanthan and starches from different sources," *Polymer Degradation and Stability*, vol. 90, no. 3, pp. 449–454, 2005.

[15] P. Veiga-Santos, L. M. Oliveira, M. P. Cereda, A. J. Alves, and A. R. P. Scamparini, "Mechanical properties, hydrophilicity and water activity of starch-gum films: effect of additives and deacetylated xanthan gum," *Food Hydrocolloids*, vol. 19, no. 2, pp. 341–349, 2005.

[16] H. Samsudin and N. M. Hani, "Chapter 8 - Use of Starch in Food Packaging," in *Starch-Based Materials in Food Packaging*, M. A. Villar, S. E. Barbosa, M. A. García, L. A. Castillo, and O. V. López, Eds., pp. 229–256, Academic Press, 2017.

[17] J. D. Linton, S. G. Ash, and L. Huybrechtst, "4 Microbial Polysaccharides," in *Biomaterials*, D. Byrom, Ed., pp. 215–261, Palgrave Macmillan, London, UK, 1991.

[18] R. Chandrasekaran and A. Radha, "Molecular architectures and functional properties of gellan gum and related polysaccharides," *Trends in Food Science and Technology*, vol. 6, no. 5, pp. 143–148, 1995.

[19] A. M. Fialho, L. M. Moreira, A. T. Granja, A. O. Popescu, K. Hoffmann, and I. Sá-Correia, "Occurrence, production, and applications of gellan: current state and perspectives," *Applied Microbiology and Biotechnology*, vol. 79, no. 6, pp. 889–900, 2008.

[20] L. Chen, F. Ren, Z. Zhang, Q. Tong, and M. M. A. Rashed, "Effect of pullulan on the short-term and long-term retrogradation of rice starch," *Carbohydrate Polymers*, vol. 115, pp. 415–421, 2015.

[21] M. Shibata, M. Asahina, N. Teramoto, and R. Yosomiya, "Chemical modification of pullulan by isocyanate compounds," *Polymer*, vol. 42, no. 1, pp. 59–64, 2001.

[22] F. F. Shih, K. W. Daigle, and E. T. Champagne, "Effect of rice wax on water vapour permeability and sorption properties of edible pullulan films," *Food Chemistry*, vol. 127, no. 1, pp. 118–121, 2011.

[23] O. Moreno, R. Díaz, L. Atarés, and A. Chiralt, "Influence of the processing method and antimicrobial agents on properties of starch-gelatin biodegradable films," *Polymer International*, vol. 65, no. 8, pp. 905–914, 2016.

[24] F. M. Fakhouri, D. Costa, F. Yamashita et al., "Comparative study of processing methods for starch/gelatin films," *Carbohydrate Polymers*, vol. 95, no. 2, pp. 681–689, 2013.

[25] A. Jiménez, M. J. Fabra, P. Talens, and A. Chiralt, "Influence of hydroxypropylmethylcellulose addition and homogenization conditions on properties and ageing of corn starch based films," *Carbohydrate Polymers*, vol. 89, no. 2, pp. 676–686, 2012.

[26] A. Cano, A. Jiménez, M. Cháfer, C. Gónzalez, and A. Chiralt, "Effect of amylose:amylopectin ratio and rice bran addition on starch films properties," *Carbohydrate Polymers*, vol. 111, pp. 543–555, 2014.

[27] A. Cano, E. Fortunati, M. Cháfer, J. M. Kenny, A. Chiralt, and C. González-Martínez, "Properties and ageing behaviour of pea starch films as affected by blend with poly(vinyl alcohol)," *Food Hydrocolloids*, vol. 48, pp. 84–93, 2015.

[28] A. Jiménez, M. J. Fabra, P. Talens, and A. Chiralt, "Effect of re-crystallization on tensile, optical and water vapour barrier properties of corn starch films containing fatty acids," *Food Hydrocolloids*, vol. 26, no. 1, pp. 302–310, 2012.

[29] ASTM, "Standard test methods for water vapour transmission of materials. Standard designations: E96-95," in *Annual Book of ASTM Standards*p. 1995.

[30] A. Gennadios, C. L. Weller, and C. H. Gooding, "Measurement errors in water vapor permeability of highly permeable, hydrophilic edible films," *Journal of Food Engineering*, vol. 21, no. 4, pp. 395–409, 1994.

[31] M. Vargas, A. Albors, A. Chiralt, and C. González-Martínez, "Characterization of chitosan-oleic acid composite films," *Food Hydrocolloids*, vol. 23, no. 2, pp. 536–547, 2009.

[32] ASTM, "Standard test method for oxygen gas transmission rate through plastic film and sheeting using a colourimetric sensor. Standard designations: 3985-3995," in *Annual Book of ASTM Standards*p. 2002.

[33] ASTM, "Standard test method for tensile properties of thin plastic sheeting. Standard designations: D882," in *Annual Book of ASTM Standards*p. 2001.

[34] P. Forssell, R. Lahtinen, M. Lahelin, and P. Myllärinen, "Oxygen permeability of amylose and amylopectin films," *Carbohydrate Polymers*, vol. 47, no. 2, pp. 125–129, 2002.

[35] T. H. McHugh and J. M. Krochta, "Sorbitol-vs glycerol-plasticized whey protein edible films: integrated oxygen permeability and tensile property evaluation," *Journal of Agricultural and Food Chemistry*, vol. 42, no. 4, pp. 841–845, 1994.

[36] P. M. Forssell, S. H. D. Hulleman, A. P. J. Myllärinen, G. K. Moates, and R. Parker, "Ageing of rubbery thermoplastic barley and oat starches," *Carbohydrate Polymers*, vol. 39, no. 1, pp. 43–51, 1999.

[37] P. Myllärinen, A. Buleon, R. Lahtinen, and P. Forssell, "The crystallinity of amylose and amylopectin films," *Carbohydrate Polymers*, vol. 48, no. 1, pp. 41–48, 2002.

[38] A. Sa Rindlav-Westling, M. Stading, A.-M. Hermansson, and P. Gatenholm, "Structure, mechanical and barrier properties of amylose and amylopectin films," *Carbohydrate Polymers*, vol. 36, no. 2-3, pp. 217–224, 1998.

Synthesis of an Acrylamide Copolymer Containing Nano-SiO_2 by Ex Situ Cu(0)-Mediated SET-LRP

Ling Wang, Zhendong Li, Pingsen Huang, and Wei Ding

Provincial Key Laboratory of Oil & Gas Chemical Technology, Chemistry and Chemical Engineering College of Northeast Petroleum University, Daqing 163318, China

Correspondence should be addressed to Wei Ding; dingwei40@126.com

Academic Editor: Cornelia Vasile

We report herein the synthesis of a novel star-shaped copolymer containing nano-SiO_2 by single-electron transfer living radical polymerization (SET-LRP) in aqueous solution. The effects of polymerization conditions, such as the total amounts and molar ratios of the monomer, initiator, catalyst, ligand, and modified nano-SiO_2, have been investigated through a series of experiments. The prepared acrylamide copolymers have been characterized by FTIR spectroscopy and ^1H NMR spectrometry. The properties of the copolymers have been assessed by viscometry and rheometry. The results confirmed that the nano-SiO_2 functional monomer was successfully combined in the SET-LRP. The optimum polymerization conditions were established through orthogonal experiments as a ratio of [AM] : [DMAEMA] : [I] : [CuBr] : [Me_6TREN] of 674.4 : 35.5 : 1 : 1 : 2 at a total concentration of [AM] + [DMAEMA] of 2.5 mol/L. The appropriate concentration of the nano-SiO_2 functional monomer (NSFM) was 0.5 wt% with respect to AM + DMAEMA. The rheology of the star-shaped copolymer exhibited a shear-thickening property when the shear rate exceeded a critical value ($100\ s^{-1}$). The AM/DMAEMA/NSFM copolymer displayed a higher viscosity than AM/DMAEMA at the same concentration. It was found that AM/DMAEMA/NSFM exhibited better salt and temperature tolerances.

1. Introduction

Polymer flooding plays an important role in the field of enhanced oil recovery (EOR) [1, 2]. However, the polyacrylamide (PAM) used has poor temperature and salt tolerances with low shear resistances, making it difficult to adapt to increasingly harsh reservoir conditions.

To address the above problems, researchers have recently sought a technique for modifying oil displacement using polyacrylamide, and to this end, two main strategies have been pursued. The first approach involves the preparation of copolymerizing monomers that can inhibit hydrolysis of the amide group or the introduction of a small amount of hydrophobic functional groups through monomers that can coordinate divalent metal ions, as in the copolymerization of acrylamide (AM) with compounds such as *N,N*-dimethylacrylamide, methyl acrylamide, *N*-vinylpyrrolidone (NVP), 2-acrylamido-2-methylpropane sulfonic acid (AMPS), allyl sulfonate, or vinyl sulfonate. The modified polyacrylamide should then display excellent temperature tolerance and salt resistance [3–8]. However, the molecular weights of polymers prepared by various modifications are highly variable, and their performances are fundamentally restricted by their structures, such that it is difficult to obtain satisfactory results. The second approach is to modify the molecular structure through the preparation of dendritic, star-shaped, or hyperbranched polymers, which exhibit excellent shear resistances due to their special reticular structures [9–16]. A special reticular structure can reduce the shear effect on the polymer molecular chain and impart a high-viscosity retention rate. However, the preparation of hyperbranched polymers is relatively complex and incurs high costs.

Cu(0)-mediated living radical polymerization or single-electrontransfer living radical polymerization (Cu(0)-mediated LRP or SET-LRP) is a versatile approach that has long attracted considerable interest. It has been widely used for the synthesis of star-shaped, branched, and hyperbranched polymers in recent years [17–20]. Importantly, it can be efficiently applied to a wide variety of activated monomers (e.g., acrylates, methacrylates, acrylamides, and

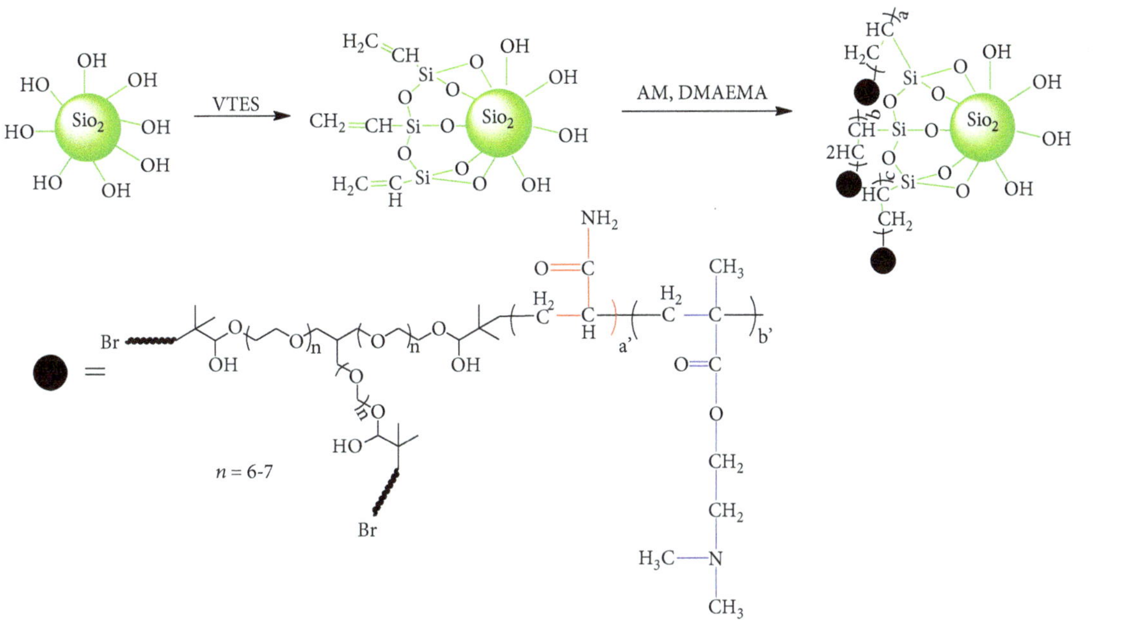

SCHEME 1: Synthesis of NSFM and AM/DMAEMA/NSFM.

methacrylamides), as well as nonactivated monomers, such as vinyl chloride, that are not polymerizable by atom-transfer radical polymerization (ATRP).

Recently, many studies have demonstrated that composite materials containing nano-SiO_2 may exhibit favorable properties, such as superior thermal stability, toughness, and strength, owing to the effects of physical adsorption, hydrogen bonding, Si-O bonds, and C-Si bonds [21–25]. In the literature, the synthesis of silica–polymer composites initiated by a SiO_2-Br macroinitiator via SET-LRP has been reported [26–28]. However, to the best of our knowledge, there have been no papers concerning the application of nano-SiO_2 as a monomer in SET-LRP.

In this work, a nano-SiO_2 functional monomer (NSFM) has been introduced into an AM/DMAEMA copolymer by Cu(0)-mediated LRP at room temperature (25°C) with the aim of obtaining more favorable temperature tolerance, salt tolerance, and shear resistance.

2. Experimental

2.1. Chemicals and Reagents. Acrylamide (AM, >99.0% Sinopharm Chemical Reagent, China) was recrystallized twice from acetone prior to use; 2-(dimethylamino)ethyl methacrylate (DMAEMA, 99%) was purchased from Macklin (Shanghai, China) and distilled under reduced pressure before use. Glycerol ethoxylate ($M_n \approx 1000$ g/mol), α-Bromoisobutyryl bromide (BIBB, 98%), triethylamine (TEA, ≥99%), vinyltriethoxysilane (VTES, ≥98.0%), nano-SiO_2 (10-20 nm), and Tris-(2-dimethylaminoethyl) amine (Me_6TERN) were purchased from Sigma-Aldrich and used as received. Water-soluble star initiator (Gly-Br_3) was synthesized as described in the literature [29] and stored at 4°C prior to use. Copper(I) bromide (CuBr, 98%, Sigma-Aldrich) was purified by stirring in acetic acid overnight and washing with copious amounts of ethanol before drying under vacuum to constant weight. All other chemicals and solvents were purchased from Sigma-Aldrich (China) at the highest purity available and used as received unless stated otherwise. Water was deionized by passing through an ion exchange column and doubly distilled.

2.2. Preparation of the Nano-SiO_2 Functional Monomer (NSFM). The nano-SiO_2 functional monomer was synthesized as described in the literature [22]. Firstly, ethanol (83.6 mL), nano-SiO_2 (1.5 g), distilled water (13.6 mL), and aqueous ammonia (1.3 mL) were placed in a 250 mL round-bottomed flask, and the mixture was dispersed by ultrasonication for 30 min. VTES (2.0 mL) was then added to the stirred solution, and reaction was allowed to proceed for 18 h at 30°C. Thereafter, the produced NSFM was separated by centrifugation and washed with distilled water.

The substitution level (ω) of the silica was determined by sodium hydroxide titration. Ground VTS-SiO_2 powder was placed in a 200 mL beaker, and then absolute ethanol (25 mL) and 20 wt% aqueous NaCl solution (75 mL) were added. In order to disperse the VTS-SiO_2, the beaker was ultrasonicated for 10 min. The dispersion system was then adjusted to pH 4 with 0.1 mol/L HCl. Then, 0.1 mol/L aqueous NaOH solution was used to adjust the dispersion system to pH 9. For comparison, dry SiO_2 (2 g) was subjected to the same treatment.

$$\omega = \frac{L_2 - L_1}{L_2} \times 100\%, \quad (1)$$

where L_1 is the volume of NaOH solution consumed by VTS-SiO_2 (mL) and L_2 the volume of NaOH solution consumed by SiO_2 (mL).

In this study, L_1 and L_2 were determined as 9.3 mL and 10.8 mL, respectively; $\omega = 13.9\%$.

2.3. SET-LRP Synthesis of AM/DMAEMA/NSFM. The requisite amount of CuBr was placed in a Schlenk tube equipped with a magnetic stirrer bar and deoxygenated for 20 min. The requisite amount of Me_6TREN was placed in a sealable vial with 1 mL of H_2O and deoxygenated for 15 min (Vial 1). The contents of Vial 1 were then transferred into the Schlenk tube by means of a degassed syringe, and disproportionation was allowed to proceed for 30 min at room temperature (25°C). In another vial (Vial 2), Gly-Br_3, AM, DMAEMA, 0.5 wt% NSFM, and a certain amount of distilled water (total volume 4 mL) were combined and deoxygenated with nitrogen for 15 min. The contents of Vial 2 were then transferred by means of a degassed syringe into the Schlenk tube to start the polymerization. Transfer of the contents of Vial 2 defined t_0. After the desired polymerization time, the solution was immediately filtered to remove copper residues and insoluble impurities. The solution was extensively dialyzed (MWCO = 5000 g/mol) against water for 3 days to remove residual monomer and other small molecules, and then the product was recovered by freeze drying. AM/DMAEMA copolymer was synthesized according to the same method. All copolymers were prepared in partially hydrolyzed form with the same hydrolysis level. The synthetic routes of NSFM and AM/DMAEMA/NSFM were described in Scheme 1.

2.4. Characterization. Infrared (IR) spectra of AM/DMAEMA and AM/DMAEMA/NSFM were measured from samples in KBr pellets on a Perkin-Elmer 1700 spectrometer. The ^{1}H NMR spectrum of AM/DMAEMA/NSFM was recorded on a Bruker DRX spectrometer operated at 500 MHz from a solution of the copolymer in D_2O.

2.5. Rheological Properties and Viscoelasticity. Measurements of the rheology and viscoelasticity of the polymer products were conducted on a Discovery HR-1 rheometer (USA). The shear rate was varied from 2 s^{-1} to 1300 s^{-1}, and the temperature was 30°C, attained at a heating rate of 1.0°C/min. The test system was a binocular tube. A DG41Ti rotor was used for rheological measurements.

3. Results and Discussion

3.1. IR Spectral Analysis. The structures of AM/DMAEMA and AM/DMAEMA/NSFM were fully supported by their IR spectra, as illustrated in Figure 1. The proposed structure of AM/DMAEMA/NSFM, prepared using an acrylamide, dimethylaminoethyl methacrylate, and nano-SiO_2 functional monomer by SET-LRP was confirmed by strong absorptions at $\nu = 3421\ cm^{-1}$ (-NH stretching vibration), 2929 cm^{-1} (-CH_2 stretching vibration), 1664 cm^{-1} (-C=O stretching vibration), 1452 cm^{-1} (-CH_3 stretching vibration), 1216 cm^{-1} (C-O stretching vibration), 1100 cm^{-1} (Si-O-Si asymmetric stretching vibration) [30, 31], and 640 cm^{-1} (C-Br stretching vibration).

The peak at $\nu = 3421\ cm^{-1}$ in the IR spectrum of AM/DMAEMA/NSFM was broad, partly due to hydroxyl groups on the nano-SiO_2 surface [31]. As expected, the IR spectrum confirmed the presence of all monomers in AM/DMAEMA/NSFM.

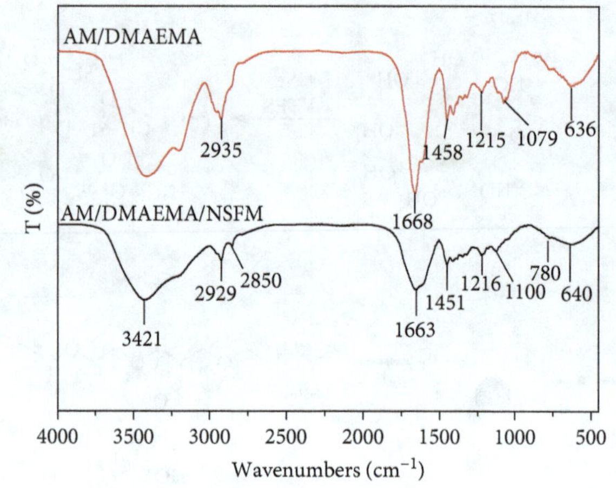

Figure 1: FTIR spectra of AM/DMAEMA and AM/DMAEMA/NSFM copolymers.

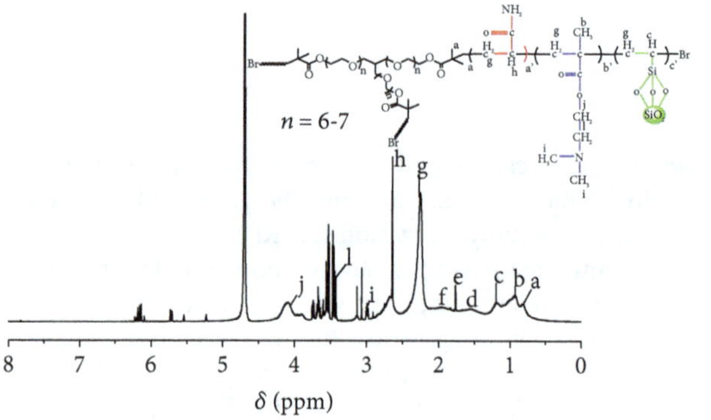

Figure 2: ^{1}H NMR spectrum of AM/DMAEMA/NSFM.

3.2. ^{1}H NMR Analysis. The ^{1}H NMR spectrum of AM/DMAEMA/NSFM is shown in Figure 2. The signals at $\delta = 1.1$–1.3 ppm can be attributed to the protons of type [-CH_2-CH(Si(O-)$_3$)-]. The signals at $\delta = 2.1$–2.5 ppm can be assigned to protons of types [-$\mathbf{CH_2}$-CH(CONH$_2$)-], [-$\mathbf{CH_2}$-CCH$_3$(COO)-], and [-$\mathbf{CH_2}$CH(Si(O-)$_3$)-]. The protons of type [-CH$_2$-**CH**(CONH$_2$)-] give rise to signals at $\delta = 2.5$–2.9 ppm. The signals at $\delta = 2.8$–3.0 ppm can be ascribed to the [-N(CH$_3$)$_2$] protons. Signals in the region $\delta = 6.0$–6.3 ppm can be ascribed to [-OH] on the NSFM. No signal due to -NH$_2$ is seen in the ^{1}H NMR spectrum, because these protons are labile and exchange with the solvent.

The spectrometric analysis confirmed that each of the monomers successfully participated in the SET-LRP.

3.3. SET-LRP of AM, DMAEMA, and NSFM in H_2O

3.3.1. Conditions for Terpolymerization. Orthogonal experiments (L5^4) were arranged to investigate the main effects

TABLE 1: Factors of the orthogonal experiments ($L5^4$).

Levels	1	2	3	4
Total monomer concentration (mol·L^{-1})	1.5	2.5	3.5	4.5
Amount of initiator (g)	0.0075	0.0150	0.0225	0.0300
Amount of catalyst (g)	0.0011	0.0018	0.0025	0.0033
Amount of ligand (μL)	4	7	10	13
The molar ratio of monomers	95 : 5	85 : 15	75 : 25	60 : 40

TABLE 2: Results of the orthogonal experiments ($L5^4$).

No.	Total monomer concentration (mol·L^{-1})	The molar ratio of monomers	Amount of initiator (g)	Amount of catalyst (g)	Amount of ligand (μL)	Viscosity (mPa·s)
1	1.5	95 : 5	0.0075	0.0011	4	6.7
2	1.5	85 : 15	0.0150	0.0018	7	5.6.
3	1.5	75 : 25	0.0225	0.0025	10	6.3
4	1.5	60 : 40	0.0300	0.0033	13	7.8
5	2.5	95 : 5	0.0150	0.0025	13	17.4
6	2.5	85 : 15	0.0075	0.0033	10	10.1
7	2.5	75 : 25	0.0300	0.0011	7	7.8
8	2.5	60 : 40	0.0225	0.0018	4	9.2
9	3.5	95 : 5	0.0225	0.0033	7	3.1
10	3.5	85 : 15	0.0300	0.0025	4	5.5
11	3.5	75 : 25	0.0075	0.0018	13	7.8
12	3.5	60 : 40	0.0150	0.0011	10	5.8
13	4.5	95 : 5	0.0300	0.0018	10	14.6
14	4.5	85 : 15	0.0225	0.0011	13	14.9
15	4.5	75 : 25	0.0150	0.0033	4	5.8
16	4.5	60 : 40	0.0075	0.0025	7	9.2
K1	6.6	10.4	8.4	8.8	6.8	
K2	11.2	9.1	8.7	9.3	6.4	
K3	5.6	7.0	8.4	9.6	9.2	
K4	11.2	8.0	9.0	6.7	11.9	
R	5.6	3.4	0.6	2.9	5.5	

Note: the addition of NSFM is 0.5% of the total mass of monomers; K1: average viscosity of level one; K2: average viscosity of level two; K3: average viscosity of level three; K4: average viscosity of level four; *R*: range.

(see Table 1) of the total monomer concentration; the amounts of initiator, catalyst, and ligand; and the molar ratio of monomers, on the viscosity of the resulting copolymer. Polymer solutions at a concentration of 2 g/L were prepared in deionized water. The viscosities of these solutions were measured by means of a Brookfield DV-II viscometer (USA) at $T = 30°C$ using a no. 63 rotor (rotation speed: 30 rpm). Higher viscosity indicated better performance. From the results of orthogonal experiments shown in Table 2, the optimal polymerization conditions were obtained as follows: total monomer concentration 2.5 mol/L, molar ratio of monomers (AM/DMAEMA), 95 : 5; amount of initiator, 0.0300 g; amount of catalyst, 0.0025 g; amount of ligand 13 μL, namely, $[AM] + [DMAEMA] = 2.5$ mol/L; $[AM] : [DMAEMA] : [I] : [CuBr] : [Me_6TREN] = 674.4 : 35.5 : 1 : 1 : 2$.

The optimally synthesized sample had a viscosity of 19.50 mPa·s under the same test conditions. Comparing the *R* values associated with five factors, it was found that their impacts increased in the following sequence: amount of initiator < amount of catalyst < molar ratio of monomers < amount of ligand < total monomer concentration. We also investigated the effect of the amount of NSFM on the viscosity of the copolymer.

3.3.2. Effect of NSFM Dosage. Based on Figures 1 and 2, it is clear that NSFM successfully participated in the SET-LRP. The effect of the dosage of NSFM on the viscosity of the product was investigated, and the results are presented in Figure 3.

It can be seen from Figure 3 that the viscosity of the copolymer first increased and then decreased with increasing NSFM content. The maximum viscosity of 19.5 mPa·s was seen with an NSFM loading of 0.5 wt%. This may have been due to the terpolymer having the maximum molecular weight at this point.

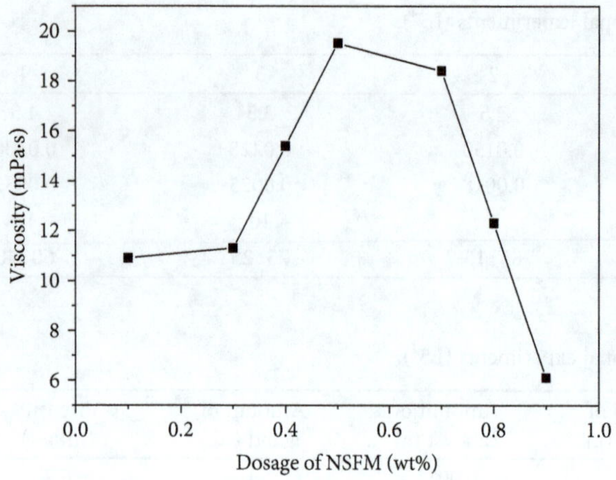

FIGURE 3: Effect of the dosage of NSFM on the performance of AM/DMAEMA/NSFM; [AM] + [DMAEMA] = 2.5 mol/L, [AM] : [DMAEMA] : [I] : [CuBr] : [Me_6TREN] = 674.4 : 35.5 : 1 : 1 : 2.

When the dosage of NSFM was less than 0.5 wt%, more monomers polymerized to form hyperbranched products. With the addition of NSFM, the hydrodynamic volume of the molecules increased, the molecules became interwoven to form more high-strength reticular structures, and as a result, the solution viscosity increased. With the addition of more than 0.5 wt% NSFM, more branched products and shorter molecular chains were formed. As a result, the molecular weight of the polymer decreased, the entanglement between molecules became weak, the hydrodynamic volumes of the molecules decreased, and the solution viscosity decreased.

3.4. Temperature Tolerance. For the samples used to assess temperature tolerance, salt resistance, and shear resistance, the composition of the reagents for producing the AM/DMAEMA copolymer was 95 mol% AM and 5 mol% DMAEMA. The composition of the reagents for the preparation of the AM/DMAEMA/NSFM copolymer was 89 wt% AM, 10.5 wt% DMAEMA, and 0.5 wt% NSFM (the molar ratio of AM to DMAEMA was still 95 : 5).

AM/DMAEMA and AM/DMAEMA/NSFM solutions were prepared in distilled water. The viscosities of the copolymer solutions were measured by means of a Brookfield DV-II+ viscometer at different temperatures. The viscosity versus temperature curves of the AM/DMAEMA and AM/DMAEMA/NSFM solutions are shown in Figure 4. The results show that the introduction of NSFM improved the thermostability of the copolymer. With increasing temperature, the viscosities of the two kinds of polymers gradually decreased to different levels. The AM/DMAEMA/NSFM solution consistently showed higher viscosity compared to that of the AM/DMAEMA solution. However, the viscosity of the AM/DMAEMA/NSFM solution decreased less than that of the AM/DMAEMA solution when the temperature exceeded 50°C. It was also found that AM/DMAEMA/NSFM could achieve up to 47.1% viscosity retention at 90°C, whereas AM/DMAEMA showed just 14.2% viscosity retention at this temperature. This lends support to the view

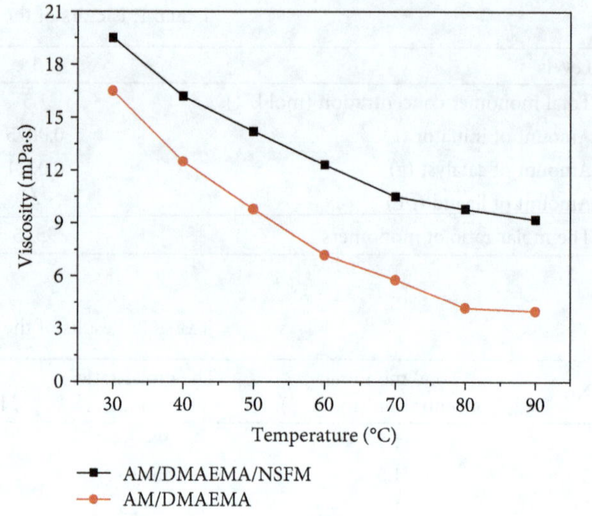

FIGURE 4: Viscosity versus temperature for AM/DMAEMA and AM/DMAEMA/NSFM solutions. The viscosities of copolymer solutions (0.2 wt%) were measured with a Brookfield DV-II viscometer using a no. 63 rotor (rotation speed: 30 rpm).

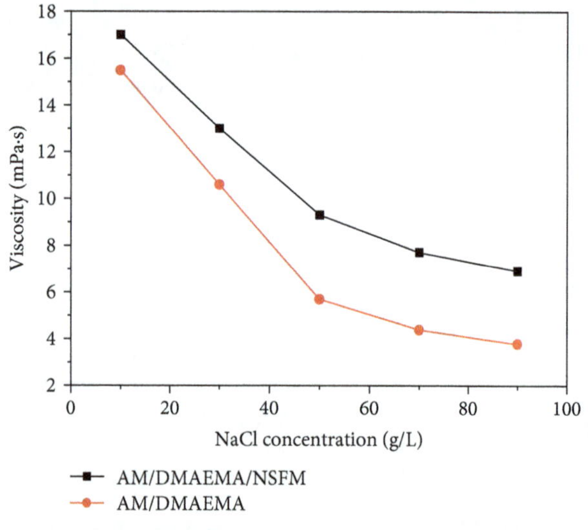

FIGURE 5: Salt tolerances of AM/DMAEMA and AM/DMAEMA/NSFM solutions (0.2 wt%) at 30°C. The viscosities of copolymer solutions (0.2 wt%) were measured with a Brookfield DV-II viscometer using a no. 63 rotor (rotation speed: 30 rpm).

that stable Si-O and C-Si bonds improve the temperature tolerance of AM/DMAEMA/NSFM.

3.5. Salt Tolerance. The influences of electrolyte concentration on the viscosities of AM/DMAEMA and AM/DMAEMA/NSFM solutions are shown in Figure 5. With increasing NaCl concentration, the viscosities of the copolymer solutions gradually decreased. This could be attributed to compression of the diffuse electric double layer of the carboxylate ions by electrolyte ions, thereby decreasing the

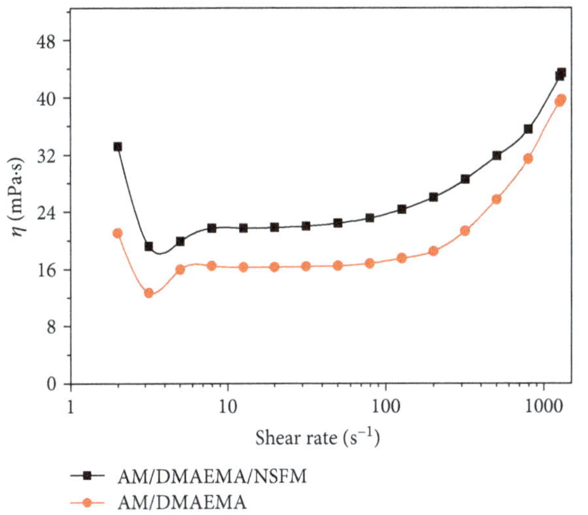

Figure 6: Effect of shear rate on the viscosities of AM/DMAEMA and AM/DMAEMA/NSFM solutions. The copolymer solutions (0.2 wt%) were prepared in distilled water.

electrostatic repulsive force [17]. The polymer chains are in a coiled rather than extended conformation, which leads to a decrease in viscosity. Compared with AM/DMAEMA, AM/DMAEMA/NSFM exhibited better salt tolerance. This could be attributed to the hyperbranched structure of the modified nanostar polymer molecule. The reticular structure formed in the solution can weaken the shielding effect of the small-molecule electrolyte on the charge of the copolymer molecular chain to some extent, reducing the crimp degree of the molecular chains, thereby improving the salt-resisting property of the copolymer.

3.6. Shear Resistance. Viscosity versus shear rate curves for AM/DMAEMA and AM/DMAEMA/NSFM are shown in Figure 6. At lower shear rates, the viscosities of the copolymer solutions decreased slightly. On increasing the shear rate from 2 to 100 s^{-1}, the viscosities of the copolymer solutions remained essentially constant but then increased when the shear rate exceeded 100 s^{-1}.

This may have been due to the hyperbranched structures of the copolymers. The results also indicated that AM/DMAEMA/NSFM had better viscosifying property than AM/DMAEMA, and the viscosity of the former was higher than that of the latter at shear rates less than 100 s^{-1}.

At a certain shear rate, shear stress destroys the intramolecular associative structure of polymer molecules, and the viscosity of a solution decreases. Conversely, due to expansion and extension of the polymer chains, intermolecular association increases the viscosity of the solution. As the net result of these two effects, the solutions showed either a slight decrease or no change in viscosity. At a high shear rate, however, the molecular chains of the polymers were highly stretched in a short time, resulting in intermolecular association. Hence, the viscosity of the polymer solution was increased, and shear thickening occurred.

4. Conclusions

A novel terpolymer containing nano-SiO_2 has been synthesized at ambient temperature by SET-LRP, employing a nano-SiO_2 functional monomer. The product has been characterized by FTIR spectroscopy and ^1H NMR spectrometry. The optimum polymerization conditions were identified as an [AM] : [DMAEMA] : [I] : [CuBr] : [Me_6TREN] ratio of 674.4 : 35.5 : 1 : 1 : 2 in a total concentration of [AM] + [DMAEMA] of 2.5 mol/L, with a concentration of NSFM of 0.5 wt% (with respect to AM + DMAEMA). The rheology of the copolymer displayed shear thickening above a critical shear rate (100 s^{-1}). The AM/DMAEMA/NSFM copolymer showed a higher viscosity retention of 47.1% than AM/DMAEMA (14.2%) at the same concentration. The copolymer containing nano-SiO_2 also exhibited better salt tolerance. AM/DMAEMA/NSFM would thus be expected to enhance oil recovery in high-temperature, high-salinity, and low-permeability reservoirs.

Conflicts of Interest

The authors declare that they have no competing interests.

Acknowledgments

We are grateful for the financial support from the National Nature Science Foundation of China (51474068) which funded the present study. We thank International Science Editing (http://www.internationalscienceediting.com) for editing this manuscript.

References

[1] S. Mahran, A. Attia, and B. Saha, *Proceedings of 11th International Sustainable Energy & Environmental Protection Conference: A Review on Polymer Flooding in Enhanced Oil Recovery under Harsh Conditions*, Paisley, Scotland, 2018.

[2] J. J. Sheng, B. Leonhardt, and N. Azri, "Status of polymer-flooding technology," *Journal of Canadian Petroleum Technology*, vol. 54, no. 2, pp. 116–126, 2015.

[3] Z. Ye, G. Gou, S. Gou, W. Jiang, and T. Liu, "Synthesis and characterization of a water-soluble sulfonates copolymer of acrylamide and N-allylbenzamide as enhanced oil recovery chemical," *Journal of Applied Polymer Science*, vol. 128, no. 3, pp. 2003–2011, 2012.

[4] A. Sabhapondit, A. Borthakur, and I. Haque, "Water soluble acrylamidomethyl propane sulfonate (AMPS) copolymer as an enhanced oil recovery chemical," *Energy and Fuels*, vol. 17, no. 3, pp. 683–688, 2003.

[5] C.-S. Yang, K. Shin, and H. K. Jeong, "Thermal analysis of poly(sodium 4-styrenesulfonate) intercalated graphite oxide composites," *Chemical Physics Letters*, vol. 517, no. 4–6, pp. 196–198, 2011.

[6] X.-J. Liu, W.-C. Jiang, S.-H. Gou, Z. B. Ye, and X. D. Xie, "Synthesis and evaluation of a water-soluble acrylamide binary sulfonates copolymer on MMT crystalline interspace and EOR," *Journal of Applied Polymer Science*, vol. 125, no. 2, pp. 1252–1260, 2012.

[7] X. Liu, W. Jiang, S. Gou, Z. Ye, and C. Luo, "Synthesis and clay stabilization of a water-soluble copolymer based on acrylamide, modular β-cyclodextrin, and AMPS," *Journal of Applied Polymer Science*, vol. 128, no. 5, pp. 3398–3404, 2013.

[8] X. Liu, W. Jiang, S. Gou et al., "Synthesis and evaluation of novel water-soluble copolymers based on acrylamide and modular β-cyclodextrin," *Carbohydrate Polymers*, vol. 96, no. 1, pp. 47–56, 2013.

[9] D. A. Tomalia, H. Baker, J. Dewald et al., "A new class of polymers: starburst-dendritic macromolecules," *Polymer Journal*, vol. 17, no. 1, pp. 117–132, 1985.

[10] D. A. Tomalia, H. Baker, J. Dewald et al., "Dendritic macromolecules: synthesis of starburst dendrimers," *Macromolecules*, vol. 19, no. 9, pp. 2466–2468, 1986.

[11] J. C. Hummelen, J. L. J. Van Dongen, and E. W. Meijer, "Electrospray mass spectrometry of poly(propylene imine) dendrimers—the issue of dendritic purity or polydispersity," *Chemistry*, vol. 3, no. 9, pp. 1489–1493, 1997.

[12] L. J. Twyman, A. E. Beezer, R. Esfand, M. J. Hardy, and J. C. Mitchell, "The synthesis of water soluble dendrimers, and their application as possible drug delivery systems," *Tetrahedron Letters*, vol. 40, no. 9, pp. 1743–1746, 1999.

[13] M. S. Diallo, S. Christie, P. Swaminathan et al., "Dendritic chelating agents. 1. Cu(II) binding to ethylene diamine core poly(amidoamine) dendrimers in aqueous solutions," *Langmuir*, vol. 20, no. 7, pp. 2640–2651, 2004.

[14] L. Xue, U. S. Agarwal, M. Zhang et al., "Synthesis and direct topology visualization of high-molecular-weight star PMMA," *Macromolecules*, vol. 38, no. 6, pp. 2093–2100, 2005.

[15] D. S. Achilleos, T. K. Georgiou, and C. S. Patrickios, "Amphiphilic model conetworks based on cross-linked star copolymers of benzyl methacrylate and 2-(dimethylamino)ethyl methacrylate: synthesis, characterization, and DNA adsorption studies," *Biomacromolecules*, vol. 7, no. 12, pp. 3396–3405, 2006.

[16] X. Wang, Y. Zhang, T. Li, W. Tian, Q. Zhang, and Y. Cheng, "Generation 9 polyamidoamine dendrimer encapsulated platinum nanoparticle mimics catalase size, shape, and catalytic activity," *Langmuir*, vol. 29, no. 17, pp. 5262–5270, 2013.

[17] M. R. Whittaker, C. N. Urbani, and M. J. Monteiro, "Synthesis of linear and 4-arm star block copolymers of poly(methyl acrylate-b-solketal acrylate) by SET-LRP at 25 °C," *Journal of Polymer Science Part A: Polymer Chemistry*, vol. 46, no. 18, pp. 6346–6357, 2008.

[18] W. Zhang, W. Zhang, J. Zhu, Z. Zhang, and X. Zhu, "Controlled synthesis of pH-responsive amphiphilic A2B2miktoarm star block copolymer by combination of SET-LRP and RAFT polymerization," *Journal of Polymer Science Part A: Polymer Chemistry*, vol. 47, no. 24, pp. 6908–6918, 2009.

[19] W. Ding, C. Lv, Y. Sun, H. Luan, T. Yu, and G. Qu, "Synthesis of star polymethyl acrylate by SET-LRP at ambient temperature," *Polymer Bulletin*, vol. 67, no. 8, pp. 1499–1505, 2011.

[20] Q. Zhang, P. Wilson, Z. Li et al., "Aqueous copper mediated living polymerization: exploiting rapid disproportionation of CuBr with Me_6TREN," *Journal of the American Chemical Society*, vol. 135, no. 19, pp. 7355–7363, 2013.

[21] Z. Ye, X. Qin, N. Lai, Q. Peng, X. Li, and C. Li, "Synthesis and performance of an acrylamide copolymer containing nano-SiO2 as enhanced oil recovery chemical," *Journal of Chemistry*, vol. 2013, Article ID 437309, 10 pages, 2013.

[22] X. Wang, X. Zhao, M. Wang, and Z. Shen, "The effects of atomic oxygen on polyimide resin matrix composite containing nanosilicon dioxide," *Nuclear Instruments and Methods in Physics Research B*, vol. 243, no. 2, pp. 320–324, 2006.

[23] V. A. Bershtein, L. M. Egorova, P. N. Yakushev, P. Pissis, P. Sysel, and L. Brozova, "Molecular dynamics in nanostructured polyimide–silica hybrid materials and their thermal stability," *Journal of Polymer Science Part B: Polymer Physics*, vol. 40, no. 10, pp. 1056–1069, 2002.

[24] N. D. Alberola, K. Benzarti, C. Bas, and Y. Bomal, "Interface effects in elastomers reinforced by modified precipitated silica," *Polymer Composites*, vol. 22, no. 2, pp. 312–325, 2001.

[25] A. Voronov, A. Kohut, A. Synytska, and W. Peukert, "Mechanochemical modification of silica with poly(1-vinyl-2-pyrrolidone) by grinding in a stirred media mill," *Journal of Applied Polymer Science*, vol. 104, no. 6, pp. 3708–3714, 2007.

[26] X. Liu, D. Appelhans, T. Zhang, and B. Voit, "Rapid synthesis of dual-responsive hollow capsules with controllable membrane thickness by surface-initiated SET-LRP polymerization," *Macromolecules*, vol. 51, no. 3, pp. 1011–1019, 2018.

[27] T. K. L. Nguyen, X. T. Cao, Y. T. Jeong, H. Heo, Y. S. Gal, and K. T. Lim, "Synthesis and characterization of core-shell type silica-polymer composites by surface-initiated SET-LRP," *Molecular Crystals and Liquid Crystals*, vol. 654, no. 1, pp. 174–180, 2017.

[28] C. Heng, M. Liu, K. Wang et al., "Fabrication of silica nanoparticle based polymer nanocomposites via a combination of mussel inspired chemistry and SET-LRP," *RSC Advances*, vol. 5, no. 111, pp. 91308–91314, 2015.

[29] R. Aksakal, M. Resmini, and C. R. Becer, "Pentablock star shaped polymers in less than 90 minutes via aqueous SET-LRP," *Polymer Chemistry*, vol. 7, no. 1, pp. 171–175, 2016.

[30] G.-H. Hsiue, W.-J. Kuo, Y.-P. Huang, and R.-J. Jeng, "Microstructural and morphological characteristics of PS–SiO2 nanocomposites," *Polymer*, vol. 41, no. 8, pp. 2813–2825, 2000.

[31] B. J. Kim and K. S. Kang, "Fabrication of a crack-free large area photonic crystal with colloidal silica spheres modified with vinyltriethoxysilane," *Crystal Growth & Design*, vol. 12, no. 8, pp. 4039–4042, 2012.

Permissions

All chapters in this book were first published in IJPS, by Hindawi Publishing Corporation; hereby published with permission under the Creative Commons Attribution License or equivalent. Every chapter published in this book has been scrutinized by our experts. Their significance has been extensively debated. The topics covered herein carry significant findings which will fuel the growth of the discipline. They may even be implemented as practical applications or may be referred to as a beginning point for another development.

The contributors of this book come from diverse backgrounds, making this book a truly international effort. This book will bring forth new frontiers with its revolutionizing research information and detailed analysis of the nascent developments around the world.

We would like to thank all the contributing authors for lending their expertise to make the book truly unique. They have played a crucial role in the development of this book. Without their invaluable contributions this book wouldn't have been possible. They have made vital efforts to compile up to date information on the varied aspects of this subject to make this book a valuable addition to the collection of many professionals and students.

This book was conceptualized with the vision of imparting up-to-date information and advanced data in this field. To ensure the same, a matchless editorial board was set up. Every individual on the board went through rigorous rounds of assessment to prove their worth. After which they invested a large part of their time researching and compiling the most relevant data for our readers.

The editorial board has been involved in producing this book since its inception. They have spent rigorous hours researching and exploring the diverse topics which have resulted in the successful publishing of this book. They have passed on their knowledge of decades through this book. To expedite this challenging task, the publisher supported the team at every step. A small team of assistant editors was also appointed to further simplify the editing procedure and attain best results for the readers.

Apart from the editorial board, the designing team has also invested a significant amount of their time in understanding the subject and creating the most relevant covers. They scrutinized every image to scout for the most suitable representation of the subject and create an appropriate cover for the book.

The publishing team has been an ardent support to the editorial, designing and production team. Their endless efforts to recruit the best for this project, has resulted in the accomplishment of this book. They are a veteran in the field of academics and their pool of knowledge is as vast as their experience in printing. Their expertise and guidance has proved useful at every step. Their uncompromising quality standards have made this book an exceptional effort. Their encouragement from time to time has been an inspiration for everyone.

The publisher and the editorial board hope that this book will prove to be a valuable piece of knowledge for researchers, students, practitioners and scholars across the globe.

List of Contributors

Michał Strankowski
Chemical Faculty, Polymer Technology Department, Gdansk University of Technology, Narutowicza 11/12, 80-233 Gdansk, Poland

Chatree Homkhiew, Surasit Rawangwong and Worapong Boonchouytan
Department of Industrial Engineering, Materials Processing Technology Research Unit, Faculty of Engineering, Rajamangala University of Technology Srivijaya, Muang District, Songkhla 90000, Thailand

Wiriya Thongruang
Department of Mechanical Engineering, Faculty of Engineering, Prince of Songkla University, Hat Yai, Songkhla 90112, Thailand

Thanate Ratanawilai
Department of Industrial Engineering, Faculty of Engineering, Prince of Songkla University, Hat Yai, Songkhla 90112, Thailand

Soledad-Cecilia Pech-Cohuo, Gonzalo Canche-Escamilla, Alex Valadez-González and Jorge Uribe-Calderon
Centro de Investigacion Cientifica de Yucatan, Calle 43 No. 130 entre 32 y 34. Col. Chuburna de Hidalgo, CP 97205 Mérida, YUC, Mexico

Victor Vladimir Amilcar Fernández-Escamilla
Departamento de Ciencias Tecnológicas, Centro Universitario de la Ciénega, Universidad de Guadalajara, Av. Universidad No. 1115, 47819 Ocotlán, JAL, Mexico

Thi Minh Phuong Ngo
The University of Danang - University of Technology and Education, Da Nang, Vietnam

Thi Mong Quyen Dang
College of Food Industry, Da Nang, Vietnam

Thi Xo Tran
The University of Danang - University of Science and Technology, Da Nang, Vietnam

Pornchai Rachtanapun
Division of Packaging Technology, Faculty of Agro-Industry, Chiang Mai University, Chiang Mai 50100, Thailand
Center of Excellent in Materials Science and Technology, Chiang Mai University, Chiang Mai 50200, Thailand

Tadashi Abe
College of Industrial Technology, Nihon University, Narashino City, Chiba 275-8575, Japan

Kiyoshi Ito
Nihon University Graduate School of Engineering, Nihon University, Narashino City, Chiba 275-8575, Japan

Noriyuki Kozakai
Construction Materials Division, Sumitomo Osaka Cement Co., Ltd, Chiyoda-ku, Tokyo 102-8465, Japan

Takayoshi Kodama
Technical Sales Division, Kajima Road Co., Ltd, Bunkyo-ku, Tokyo 112-8566, Japan

Federico Vozzi, Ilenia Guerrazzi and Claudio Domenici
Laboratory of Biomimetic Materials and Tissue Engineering, Institute of Clinical Physiology IFC-CNR, Pisa, Italy

Tiziana Nardo, Gianluca Ciardelli and Valeria Chiono
Department of Mechanical and Aerospace Engineering, Polytechnic of Turin, Turin, Italy

Silvia Rocchiccioli and Laura Comelli
Laboratory of Proteomics, Institute of Clinical Physiology IFC-CNR, Pisa, Italy

Antonella Cecchettini
Laboratory of Proteomics, Institute of Clinical Physiology IFC-CNR, Pisa, Italy
Department of Clinical and Experimental Medicine, University of Pisa, Pisa, Italy

Giovanni Vozzi, Carmelo De Maria and Francesca Montemurro
Research Center "E. Piaggio", University of Pisa, Pisa, Italy

Suk-Hun Sur, Pil-Jun Choi, Jae-Wang Ko and Jae-Yeon Lee
Industrial Materials Fusion Technology Center, Korea Institute of Footwear and Leather Technology, Busan 47154, Republic of Korea

Young-Hee Lee and Han-Do Kim
Department of Organic Material Science and Engineering, Pusan National University, Busan 46241, Republic of Korea

Inimfon A. Udoetok and Lee D. Wilson
Department of Chemistry, 110 Science Place, University of Saskatchewan, Saskatoon, SK, Canada S7N 5C9

John V. Headley
Water Science and Technology Directorate, 11 Innovation Boulevard, Environment and Climate Change Canada, Saskatoon, SK, Canada S7N 3H5

Huina Qiu
School of Chemistry and Chemical Engineering, Zhongkai University of Agriculture and Engineering, Guangzhou 510220, China

Zhichuan Shen, Xinhua Zhou and Hongjun Zhou
School of Chemistry and Chemical Engineering, Zhongkai University of Agriculture and Engineering, Guangzhou 510220, China
Key Laboratory of Agricultural Green Fine Chemicals of Guangdong Higher Education Institution, Guangzhou 510220, China

Hua Xu and Huayao Chen
School of Chemistry and Chemical Engineering, Zhongkai University of Agriculture and Engineering, Guangzhou 510220, China
Guangzhou Key Lab for Efficient Use of Agricultural Chemicals, Guangzhou 510220, China

Abdirahman Yussuf, Mohammad Al-Saleh, Salah Al-Enezi and Gils Abraham
Polymeric Products and Customization Program, Petroleum Research Center, Kuwait Institute for Scientific Research, 13109 Safat, Kuwait

M. P. Luda, D. Trevisan and V. Pau
Dipartimento di Chimica, Università di Torino, Via P. Giuria 7 10125, Torino, Italy

N. Li Pira
GML, C.R.F. S.C.p.A, Corso Settembrini 40, 10135 Torino, Italy

María D. Torres, Francisco Chenlo and Ramón Moreira
Department of Chemical Engineering, Universidade de Santiago de Compostela, Rúa Lope Gómez de Marzoa, Santiago de Compostela E-15782, Spain

Jian-Min Wang, Xin-Yuan Sun and Jian-Ming Ouyang
Institute of Biomineralization and Lithiasis Research, Jinan University, Guangzhou 510632, China

Ismail Mohd Saaid
Petroleum Engineering Department, Universiti Teknologi PETRONAS, 32610 Bandar Seri Iskandar, Perak Darul Ridzuan, Malaysia

Zulhelmi Amir
Petroleum Engineering Department, Universiti Teknologi PETRONAS, 32610 Bandar Seri Iskandar, Perak Darul Ridzuan, Malaysia
Department of Chemical Engineering, Faculty of Engineering, University of Malaya, 50603 Kuala Lumpur, Malaysia

Badrul Mohamed Jan
Department of Chemical Engineering, Faculty of Engineering, University of Malaya, 50603 Kuala Lumpur, Malaysia
Center Advanced Frontier Science, Faculty of Science, University of Malaya, 50603 Kuala Lumpur, Malaysia
Center for Energy Science, Department of Mechanical Engineering, University of Malaya, 50603 Kuala Lumpur, Malaysia

Suh Cem Pang, Lee Ken Voon and Suk Fun Chin
Resource Chemistry Programme, Faculty of Resource Science & Technology, Universiti Malaysia Sarawak, 94300 Kota Samarahan, Sarawak, Malaysia

Xin Wang, Pin Chen and Xiaohong Hu
School of Material Engineering, Jinling Institute of Technology, Nanjing 211169, China

Mariano Asteasuain
Planta Piloto de Ingeniería Química (PLAPIQUI), Universidad Nacional del Sur (UNS) - CONICET, 8000 Bahía Blanca, Argentina
Departamento de Ingeniería Química, Universidad Nacional del Sur (UNS), 8000 Bahía Blanca, Argentina

Armin Thumm and Mathias Sorieul
Scion, Rotorua 3046, New Zealand

Damien Even
Cabot Performance Materials, 78 Rue Prévochamps, Pepinster B-4860, Belgium

Pierre-Yves Gini
Safran Nacelles, Route du Pont 8, 76700 Gonfreville l'Orcher, France

Fahad H. Falqi and Mamdouh A. Al-Harthi
Department of Chemical Engineering, King Fahd University of Petroleum and Minerals, Dhahran 31261, Saudi Arabia

Osamah A. Bin-Dahman
Department of Chemical Engineering, Faculty of Engineering and Petroleum, Hadhramout University, Mukalla, Yemen

M. Hussain
Department of Materials and Chemical Engineering, Hanyang University, Seoul, Republic of Korea

Qin Yu, Yanqiu Li, Tenglong Gao, Siliang Liu, Caiqing Hou and Zirong Zheng
Department of Chemistry and Chemical Engineering, Northeast Petroleum University, 163318, China

Xiaolin Wu
Research Institute of Exploration and Exploitation, PetroChina Daqing Oilfield Company, 163318, China

Mayra Sapper, Pau Talens and Amparo Chiralt
Instituto Universitario de Ingeniería de Alimentos para el Desarrollo, Universitat Politècnica de València, Camino de Vera s/n, 46022 Valencia, Spain

Ling Wang, Zhendong Li, Pingsen Huang and Wei Ding
Provincial Key Laboratory of Oil & Gas Chemical Technology, Chemistry and Chemical Engineering College of Northeast Petroleum University, Daqing 163318, China

Index